AF608503

Synthesis Lectures on Engineering, Science, and Technology

The focus of this series is general topics, and applications about, and for, engineers and scientists on a wide array of applications, methods and advances. Most titles cover subjects such as professional development, education, and study skills, as well as basic introductory undergraduate material and other topics appropriate for a broader and less technical audience.

Ömer Civalek • Hayri Metin Numanoğlu
Subrat Kumar Jena

Mechanical Behavior and Vibration of Nano-Scaled Rods, Beams, and Frames

Ömer Civalek
Department of Civil Engineering
Akdeniz University
Antalya, Türkiye

Department of Medical Research
China Medical University
Taichung, Taiwan

Subrat Kumar Jena
Department of Civil, Environmental
Engineering, and Architecture
University of Cagliari
Cagliari, Italy

Hayri Metin Numanoğlu
Department of Civil Engineering
Giresun University
Giresun, Türkiye

ISSN 2690-0300 ISSN 2690-0327 (electronic)
ISBN 978-3-032-12022-9 ISBN 978-3-032-12023-6 (eBook)
https://doi.org/10.1007/978-3-032-12023-6

Synthesis Lectures on Engineering, Science, and Technology

This Springer imprint is published by the registered company Springer Nature Switzerland AG
The registered company address is: Gewerbestrasse 11, 6330 Cham, Switzerland

Preface

Analysis and design of nano- and micro-electro-mechanical systems (NEMS and MEMS) are a quite different issue compared to classical engineering problems. One-dimensional nanostructures, such as nanotubes, nanowires, and nanorods, and planar nanostructures, such as nanosheets, are known as indispensable components of NEMS and MEMS technology. In this field, computer chips, biosensors, medical nanorobots, and microscopes are fundamental examples. The designs of NEMS/MEMS exactly depend on the mechanical behavior of nanostructure. In this context, the responses of such structures to external mechanical forces have been examined with various approaches. First, according to experimental studies, it has been determined that the structure's behavior under external forces cannot be explained by the rules of classical physics. Experimental methods have associated this problem with an interaction between the external force on the atomic-scaled structure and the characteristic internal dimensions of structure. In fact, the basis for this finding dates back to the early 1900s, although it did not attract attention at that time. Later, it was understood that nanotechnological outputs were much superior to their antecedents, research on nanotechnology, and therefore the mechanics of nanostructures increased. Unfortunately, experimental methods have required large operation volume, long processes, and specialized knowledge. For this reason, new approaches have been proposed by scientists by enriching classical continuum theories with the internal dimensions of nanostructure. Among the new approaches, the most important one in the scientific literature is undoubtedly Eringen's nonlocal elasticity theory.

The investigation of mechanical behavior of atomic-scaled structures was first carried out via analytical methods. However, more advanced mechanical analyses have made the use of analytical methods difficult, even impossible. To overcome these difficulties in the investigation of mechanical behavior of nanostructures, formulations based on numerical (approximate) methods have been employed. It is also certain that new nanotechnological outputs will be indispensable in the future. For these reasons, the issues of mechanical analysis and design of nanostructures continue to exist as an innovative research field today. As a result, today's conditions have obliged us to learn all aspects of nanomechanics. Frankly, learning nanomechanics from publications such as scientific articles can be tiring because the introduction of the analysis process with very complex mathematical equations and the lack of application details hinder the ability to conduct effective research. There are some book studies in this field that have fallen within the scientific literature. These sources are of course important guides for researchers working on theoretical nanomechanics. However, in general, books consist manily of theoretical explanations, and the lack of application is very evident. The failure to comprehend the application prevents accurate learning of the mechanical behavior of nanostructures and the solving of new problems that may be encountered in the future. This book aims to provide a systematic resource on the explanations and application solutions of the nonlocal free vibration phenomenon, i.e., a branch of nanomechanics, in order to prevent such problems and support researchers in this field. One-dimensional continuous structures, such as axial nanorod, torsional nanorod, and bending nanorod (nanobeam), are investigated analytically, and numerical applications of nonlocal vibration are presented. However, it is mentioned that the analytical solution is insufficient in most cases, and nonlocal finite element method (NL-FEM) based numerical solutions of related nanostructure models are also introduced. NL-FEM examples are solved for mechanical analyses where the analytical solution cannot be applied. Moreover, trusses and frames, which we know from classical structural mechanics and generally examine with the finite element method, are integrated with nonlocal elasticity. In the chapters of this book, the combination of different kinematic theories of nanostructures with nonlocal theory is investigated. Thus, a more comprehensive understanding of the effect of the nonlocal parameter on atomic-scaled structures is provided.

Antalya, Türkiye — Ömer Civalek
Giresun, Türkiye — Hayri Metin Numanoğlu
Cagliari, Italy — Subrat Kumar Jena

Acknowledgments

The corresponding author, **Dr. Subrat Kumar Jena**, expresses his heartfelt gratitude to his younger brother, **Dr. Rajarama Mohan Jena**, and sister-in-law, **Dr. Sujata Swain**, for their encouragement, affection, and unwavering support. He is deeply indebted to his Ph.D. supervisor, **Prof. S. Chakraverty (NIT Rourkela)**, for his invaluable guidance during the formative stages of his research career. He also gratefully acknowledges the mentorship and inspiration of his postdoctoral mentors and gurus—**Prof. Dineshkumar Harursampath (NMCAD Lab, IISc Bangalore), Prof. S. Pradyumna (IIT Delhi), Prof. Victor A. Eremeyev (University of Cagliari), and Prof. Emanuele Reccia (University of Cagliari).**

We, the three authors, extend our sincere gratitude to the **reviewers** for their insightful feedback and constructive suggestions during the development of this book proposal. We also wish to express our deep appreciation to the **Springer Nature team** for their constant support, cooperation, and assistance, which greatly facilitated the timely publication of this book. Finally, we gratefully acknowledge the **authors and researchers cited in the references** at the end of each chapter, whose contributions have enriched and strengthened this volume immensely.

Competing Interests The authors have no competing interests to declare that are relevant to the content of this manuscript.

Contents

1 Introduction and Basic Concepts

1.1 Nanotechnology Science

1.1.1 Introduction to Nanotechnology

The "nano" word is a Greek and translated as "tiny." The division of meter (m) that is the generally approved unit of length measurement in the world, into billions defines the "nanometer" (nm). Nanotechnology is a discipline that handles studies related to the observation of structures with dimensions between 1 and 100 nm, the discovery and control of their properties. In order to describe the working level of nanotechnology, the following typical examples can be given: Length of 10 adjacent hydrogen atoms is 1 nm [1], diameter of a deoxyribonucleic acid (DNA) molecule is 2.5 nm, thickness of a hair strand is 10^5 nm, width of an ant head 10^6 nm [2]. In macromeaning, the diameter of a soccer ball is 30 cm = 3×10^8 nm [3].

According to the general mentality of nanotechnology, it is aimed to obtain a material produced which are more durable, stronger, lighter, more space-saving, with stronger electrical and thermal properties than the previous ones with today's technology by using engineering, scientific, and medical sciences. Today, it is obvious that the devices and substances that are used by humankind are more functional and smaller. There are two main factors in the focus of researchers on nanotechnological studies:

- The atomic structure can be rearranged as desired,
- When the atomic structure is descended, very different properties of the material emerge.

Öm. Civalek et al., *Mechanical Behavior and Vibration of Nano-Scaled Rods, Beams, and Frames*, Synthesis Lectures on Engineering, Science, and Technology,
https://doi.org/10.1007/978-3-032-12023-6_1

Table 1.1 Comparison of nanofabrication methods [4]

Criterion	Bottom-up approach	Top-down approach
Fundamental philosophy	Passing from unit to whole	Passing from whole to unit
Cost	Cheap	Expensive
Production field	Chemical and biological	Physical
Typical example	Scanning tunneling microscopes, carbon nanotubes, and fullerenes	Computer chips, memories, and storage data
Environmental conditions (heat, temperature, stress, etc.)	Nondominant	Dominant

If a new atom is added to the nanostructure, the physical and chemical properties of structure vary depending on the type of atom, the properties and geometry of structure. On the other hand, it has been demonstrated based on experimental studies that nanotechnological studies cannot be performed through classical physics rules. Nanotechnology has affected various applications of many sectors from medicine to aeronautics, from military to cosmetics, from automotive to chemistry, from construction to biotechnology. DNA modification, nanobioreceptors, cancer-fighting nanorobots, nanoprocessors, genetic information diagnostics, scanning tip microscopy, self-cleaning surface coatings, nanogenerators, nanotransistors, nanocomposites, nano-treated concrete are just a few examples that can be given in this context.

Nanoscale production consists of two classes: Bottom-up and top-down. The bottom-up approach states to manufacture nanoscale structures and systems by combining the basic units of atoms and molecules. This method is also called as molecular nanotechnology. On the other hand, top-down approach considers the fabrication of nanoscale structures through machinery, equipment, and chemicals. In other words, the passing from macroscale to nanoscale is mentioned. Generally, while the bottom-up approach relies on methods such as high energy milling, lithography, gas condensation, heavy plastic deformation, the top-down approach is based on methods such as chemical vapor deposition, physical vapor deposition, plasma processing, sol-gel processing, soft lithography, and self-assembly. In addition to these knowledges, when the outputs of today nanofabrication technology are considered, it can be deduced that the majority of these studies are performed via the top-down approach. A general comparison of both approaches is shared as follows (Table 1.1):

1.1.2 Historical Development

The history of nanoscience is quite a broad issue. However, the studies considered to be important in the development of science are given below in chronological order:

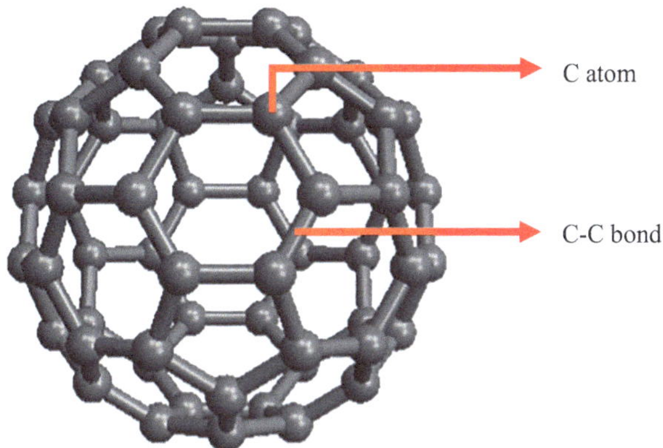

Fig. 1.1 General structure of fullerene molecule [5]

- 1959: R. Feynman carried out his presentation known as *There's Plenty of Room at the Bottom*. In his speech, Feynman said that if materials and devices at the molecular level can be made, they will pave the way for new inventions. Thus, the foundation of nanotechnology was laid.
- 1974: Japanese scientist N. Taniguchi made the definition of nanotechnology in an article that he published. According to this, nanotechnology was defined as the method of separating, combining, and decomposing of substances by an atom or molecule.
- 1981: The first nanotechnology article was published by E. Drexler.
- 1981: An important discovery was made by G.K. Binnig and H. Rohrer for the future of nanotechnological studies. According to this, the scanning tunneling microscope was invented, and it became possible to monitor the atomic movements and reactions thanks to this invention.
- 1985: R.F. Curl, H.W. Kroto, and R.E. Smalley obtained the fullerene molecule that is an allotrope of carbon atom and is named "buckyball" because its shape resembles a soccer ball. Fullerene is 1 nm in size, stronger than steel, and lighter than plastic. The atomic configuration of fullerene is depicted in Fig. 1.1.
- 1986: The atomic force microscope (AFM) was invented by G.K. Binnig, C. Quate, and C. Gerber.
- 1986: E. Drexler published a book describing molecular nanotechnology.
- 1986: The Foresight Institute that is the world's first nanotechnology organization was founded by E Drexler.
- 1987: The quantum property of conductivity was discovered.
- 1988: E. Drexler opened the first nanotechnology course.
- 1989: IBM company employee D. Eigler wrote "IBM" with Xenon atoms on Nickel surface. The height of writing is one 250 thousandth of a millimeter (mm) and its width is one 333 thousandth. The picture of this writing is shown in Fig. 1.2.
- 1991: Perhaps the milestone in the history of nanotechnology was experienced. Japanese scientist S. Iijima discovered the multi-walled carbon nanotube (MWCNT)

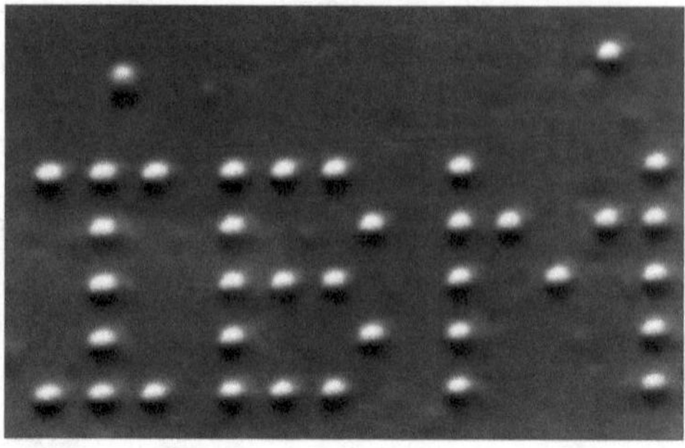

Fig. 1.2 "IBM" typing with Xenon atoms [6]

structure. These structures that consist of multiple cylinders formed by the hexagonal arrays of carbon atoms in space are many times stronger and one sixth lighter than steel.

- 1993: S. Iijima and D.S. Bethune discovered the single-walled carbon nanotubes (SWCNT) with the help of arc-evaporation method.
- 1993: The first nanotechnology laboratory was established at Rice University in the USA.
- 1996: The laser-evaporation technique was discovered at Rice University.
- 1997: N. Seedman made a nano-mechanical device for the first time by utilizing the DNA molecule. In the same year, electric current was measured with the help of nanotubes.
- 1998: C. Dekker and his team invented the carbon nanotube field-effect transistor (CNT-FET) structure.
- 1999: The USA government launched the National Nanotechnology Step that is the first official government program in order to accelerate research and development and commercialization activities in the field of nanotechnology.
- 2001: The EU determined nanotechnology studies as the priority area of their framework program. Additionally, Zinc oxide nanowire (ZnONW) laser was discovered. Moreover, this year, transistors were produced from nanotubes.
- 2005: Some researchers from Rice University have moved a four-wheeled nanocar model [7] for the first time. This car has 3×4 nm measures and is wide of one DNA line. When twenty-thousands of these cars are aligned side by side, a hair thickness is obtained. In addition to these, the atoms of nanocar were brought together to construct molecular axes and shafts. The wheels are moved by turning, not dragging.

1.1.3 Nanotechnology Investments in the Universal Scale

With the understanding of importance of nanoscience, many countries in the world have accelerated their investigations to become prominent in nanotechnology. According to this, the USA declared nanotechnology as a priority area with national nanotechnology declaration that it published in 1999. The USA government allocated 420 million dollars in 2000, 520 million dollars in 2001, and 700 million dollars in 2003 for research and development studies. This amount increased 3.7 billion dollars in 2005. Studies have been carried out in different areas such as molecular electronics, quantum computers, nanoparticles, biosensors, atomic modeling and simulations, and nanorobots [8].

The EU gave support to 80 companies conducting research about the nanotechnology with fourth framework program that it performed between 1994 and 1998. Also, the fund allocated to this field for the fifth framework program between 1998 and 2002 amounted to 45 million Euro. Moreover, nanotechnology was again a priority area and 1.3 billion Euro was allocated in the sixth Framework Program covering the next 4 years. Research have been performed on nanoelectronic systems, carbon nanotubes, biosensors, nanocomposites and, new imaging technologies in the EU [8]. In addition to these, Germany is one of the pioneer countries of Europe in the nanotechnology. 144 million Euro by government and 44 million Euro by industry are spent annually [9].

Finally, when we examine Asian continent, it is seen that Japan takes the lead. Japan is the second country in research and development investments after the USA worldwide. NEC and Sumitomo companies usually come to the fore in Japan. On the other hand, studies in the People's Republic of China are carried out through the Chinese Academy of Sciences. China is working on semiconductor technology and nanoelectronic devices. Additionally, Samsung that is a trademark of South Korea performs its work entirely on micro electromechanical systems (MEMS). Also, it can be expressed that developing countries such as Taiwan, Singapore, Thailand, India, and Vietnam have determined nanotechnology as a priority area [8].

1.2 Nano-Scaled Materials

Nano-scaled materials that are thought to have a significant impact on the development and applications of nanoscience are briefly mentioned in this section.

1.2.1 Nanotubes (NT)

These are defined as one-dimensional structures whose ratio of length-to-diameter can reach millions. According to a descriptive definition, the structures formed by wrapping one or more two-dimensional an atomic plane around a direction predicated on are called as nanotube. If many atomic planes are wrapped, these intertwine to form the nanotube structure. These planes are called as walls. Therefore, if a single plane is wrapped, it refers to single-walled nanotube, and if multiple planes are wrapped, it is called as a multi-walled nanotube. Nanotubes have higher electrical and thermal conductivity, higher stiffness,

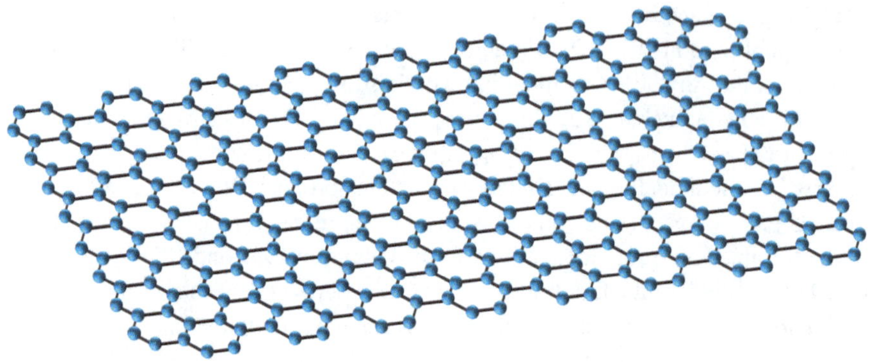

Fig. 1.3 Graphene plane [10]

lower weight, and lower volume than conventional structural and machine materials. Popular nanotubes are introduced as follows.

Carbon nanotubes (CNT): The carbon (C) element is a nonmetal in Group-4A and Period-2 of the periodic table. It participates in the structure of living organisms, inorganic compounds, and some one-dimensional nanomaterials. After the carbon atoms form hexagons in the plane, the structure called graphene, which is formed by these hexagons, is formed by wrapping like a roll in space, which simply defines the carbon nanotube structure. The general configuration of graphene is depicted in Fig. 1.3.

CNT configurations are divided into three as chair, chiral, or zigzag models according to wrapping shape. Different types of nanotubes can be easily identified according to their unit cells. In Fig. 1.4, the hexagonal corners represent carbon atoms, $\mathbf{C_h}$ vector perpendicular to the wrapping direction defines the chord vector of nanotube, and $\mathbf{a_1}$ and $\mathbf{a_2}$ mean unit vectors of hexagonal lattice. So, the beam vector is:

$$\mathbf{C_h} = m\mathbf{a_1} + n\mathbf{a_2} \tag{1.1}$$

where m and n are integers. Also, the angle between $\mathbf{C_h}$ and $\mathbf{a_1}$ vectors is known as the chord angle. The unit vectors of xy-cartesian set are obtained from the unit vectors of hexagonal lattice with the following equations:

$$\mathbf{a}_1 = a_0\left(\frac{\sqrt{3}}{2}\mathbf{e}_1 + \frac{1}{2}\mathbf{e}_2\right), \quad \mathbf{a}_2 = a_0\left(\frac{\sqrt{3}}{2}\mathbf{e}_1 - \frac{1}{2}\mathbf{e}_2\right) \tag{1.2}$$

where $b = 0.142$ nm is the bond length between carbon atoms by means of $a_0 = \sqrt{3}b$ [11]. Substituting the expressions seen in Eq. (1.2) into Eq. (1.1):

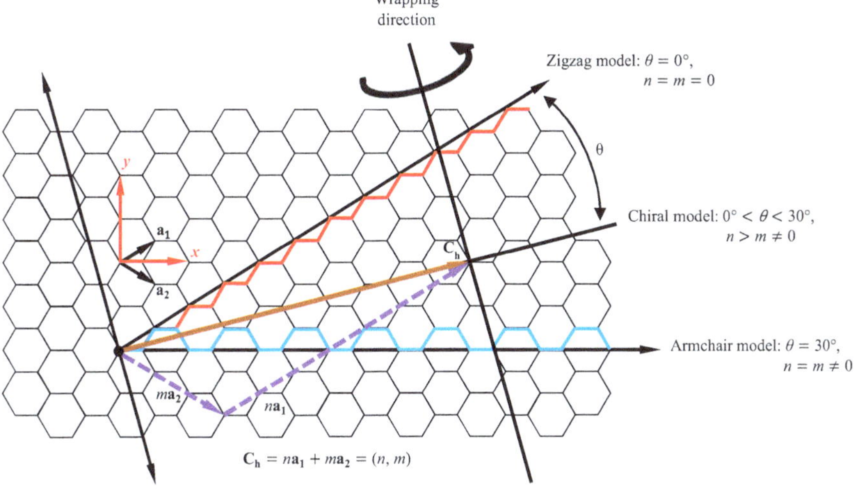

Fig. 1.4 Wrapping of graphene plane [10]

$$\mathbf{C_h} = \frac{\sqrt{3}a_0}{2}(m+n)\mathbf{e_1} + \frac{a_0}{2}(m-n)\mathbf{e_2} \tag{1.3}$$

Hence the circumference of circular section:

$$c = C_h = a_0\sqrt{m^2 + mn + n^2} \tag{1.4}$$

When the graphene sheet is wrapped, the ends of chord vector overlap by means of forming the cylindrical nanotube. The chord vector creates the circumference of circular cross-section of nanotube. Diameter of cross-section of CNT structure is [12]:

$$d = \frac{a_0}{\pi}\sqrt{m^2 + mn + n^2} \tag{1.5}$$

On the other hand, the trigonometric expressions of chirality angle are written as [12]:

$$\sin\theta = \frac{\sqrt{3}n}{2\sqrt{m^2 + mn + n^2}}, \quad \cos\theta = \frac{2m+n}{2\sqrt{m^2 + mn + n^2}}, \quad \tan\theta = \frac{\sqrt{3}n}{2n+m} \tag{1.6}$$

Also, its vector form of **T** unit vector in the wrap direction:

$$\mathbf{T} = \alpha \mathbf{e_1} + \beta \mathbf{e_2} \tag{1.7}$$

Since this vector is perpendicular to the chord vector, theirs dot products should equal to 0:

$$\mathbf{C_h} \cdot \mathbf{T} = 0 \tag{1.8}$$

By using Eqs. (1.3) and (1.7) into Eq. (1.8), the following equations can be reached [12]:

$$\alpha \frac{\sqrt{3} a_0}{2}(m+n) + \beta \frac{a_0}{2}(m-n) = 0 \tag{1.9}$$

Since the lengths of unit vectors in the directions of chord vector and wrapping directions should equal and are 1 unit [12]:

$$\frac{3{a_0}^2}{4c^2}(m+n)^2 + \frac{{a_0}^2}{4c^2}(m-n)^2 = \alpha^2 + \beta^2 \tag{1.10}$$

It is obtained from the common solution of Eqs. (1.9) and (1.10) [12]:

$$\alpha = -\frac{a_0}{2c}(m-n), \ \ \beta = \frac{\sqrt{3} a_0}{2c}(m+n). \tag{1.11}$$

There are different nanotube structures depending on the wrapping configuration. If $n = m$ and chord angle is 30°, the model is chord; $n = 0$, $m = 0$, and chord angle is 0°, the model is zigzag; if chord angle is between 0° and 30°, the model is chiral. Also, wrapping configuration affects the mechanical, physical, and chemical properties. Different wrapping configurations are demonstrated in Fig. 1.5.

We have said that the CNT structures can be single- or multi-walled according to the number of walls. Iijima discovered multi-walled carbon nanotubes (MWCNT) by way of the arc-steam experiments [13]. The discovery of single-walled carbon nanotubes (SWCNT) follows later studies [14, 15]. Single-walled carbon nanotubes consist of a single graphene wrapped. They are usually 0.5–5 nm in diameter and their ends are closed like half the fullerene structure. In the multi-walled carbon nanotube structures formed by intertwining multiple graphene planes, the distance between the layers is usually greater than the distance between two carbon atoms, and their diameter is usually 0.34 nm. The number of walls affects the mechanical and physical properties. Configurations of CNT structures are presented according to the number of walls in Fig. 1.6.

CNTs are used in scanning tip microscopes [17], field-effect transistors (FETs) [18], chemical and biological sensors [19], hydrogen storage [20], thermal conductivity [21], air-water filtration [22], nonvolatile random access memories (NVRAM) [23], and many

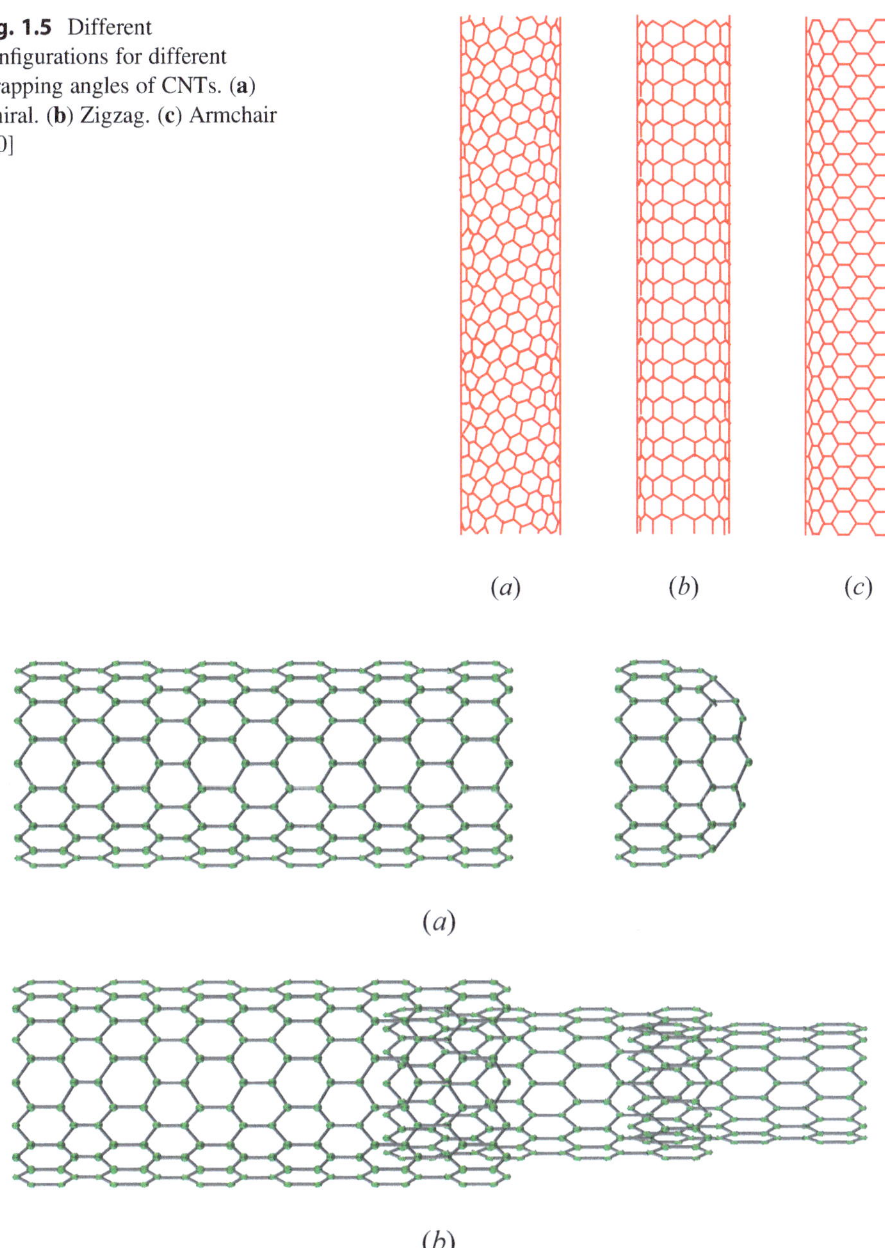

Fig. 1.5 Different configurations for different wrapping angles of CNTs. (**a**) Chiral. (**b**) Zigzag. (**c**) Armchair [10]

Fig. 1.6 CNT configurations according to wall number. (**a**) SWCNT [16]. (**b**) MWCNT [10]

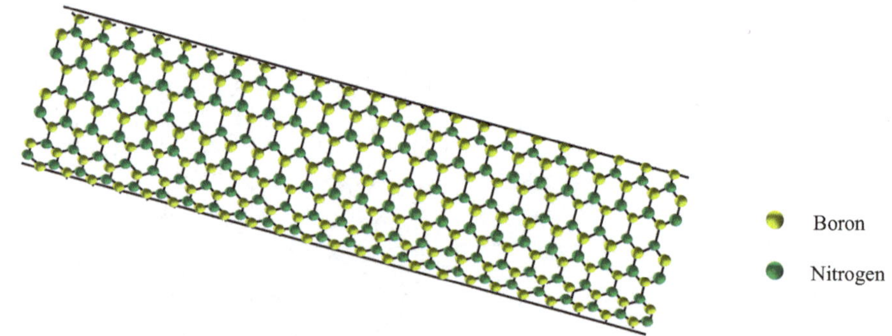

Fig. 1.7 General configuration for BNNT [10]

more. It can also be taken a part in potential electrodes, superconducting cables for space elevators, nanocomposites, solar energy [24].

It can be said that CNTs have very strong mechanical and physical properties. Their weight is 1/6 of the steel. Moreover, those resist temperatures up to 600 °C. Additionally, it has been determined by bending tests that the modulus of elasticity and strength of MWCNT structures are between 1–1.8 TPa and 0.8–150 GPa, respectively [25].

Boron nitride nanotubes (BNNT): Those are one-dimensional nanostructures that occurred after a plane consisting of boron (B) and nitrogen (N) atoms is wrapped in a chiral, zigzag, or chair form as in CNT. Their properties and production methods are like CNT. A general configuration example of BNNT structures is demonstrated in Fig. 1.7.

In a study, the modulus of elasticity of multi-walled BNNTs has been found as 1.22 TPa based on the thermal vibration method [26]. According to experimental studies on purified and precipitated BNNTs, thermal conductivity is determined as 18–46 W/mK [27]. It can be employed in FETs [28], electrochemical insulators [29], biomedicine [30], hydrogen storage [31], surface modification and functionalization [32] due to its very strong mechanical, piezoelectric, electronic, and optical properties.

Silica carbide nanotubes (SiCNT): The one-dimensional nanostructure obtained from silicification of CNTs introduces the silica carbide nanotubes. SiCNT is produced in two techniques. According to this, firstly, the shape memory synthesis, the mixture of silicon oxide and silicon is heated. Silicon is included a solid-gas reaction with a carbon template and thus a SiCNT structure is occurred [33]. On the other hand, in the nanowire core formation method, mixture consisting of silicon carbide and zinc sulfide nanowires is heated with methane gas at high temperature. SiCNT structure is fabricated by way of etching with hydrochloric acid [34].

It is explained that there are two geometries of SiCNT structures. According to the Type-I geometry, each silicon and carbon atom certainly constitute a bond to each other. In Type-II geometry, by constituting a bond each silicon atom in itself, those constitute a bond with carbon atoms. Consequently, the only difference between Type-I and Type-II geometries is that the same atoms constitute a bond together or not. SiCNT structures

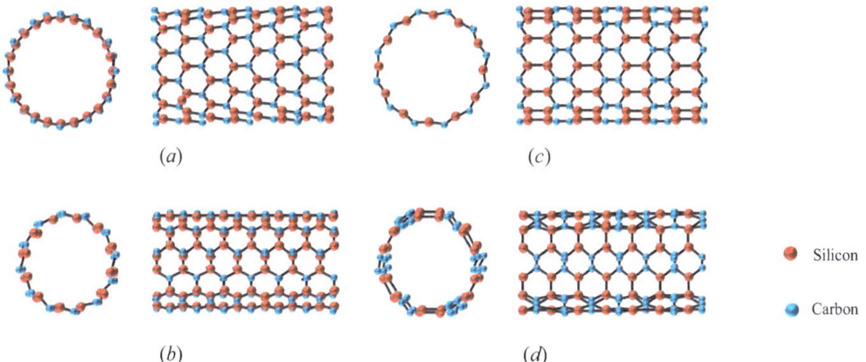

Fig. 1.8 Atomic configurations of SiCNTs. (**a**) Type-I, zigzag. (**b**) Type-I, armchair. (**c**) Type-II, zigzag. (**d**) Type-II, armchair [16]

also exhibit similar chirality to CNTs and the general configurations of two different geometry types of mentioned are given in Fig. 1.8.

Nanowires: Longitudinal nanostructures that have a diameter of 1 nm and have less but have ratio of length-to-diameter that is least 100 are known as nanowires. There are many types of nanowires. Nickel, gold, and platinum are conductor; silicon, inidium, and gallium are semiconductor; silicon dioxide and titanium dioxide are insulator; molybdenum, sulfur, and various molecular formations of iodine atoms exemplify inorganic molecular and also, DNA structure is organic molecular type nanowires. Nanowires are generally used in optoelectronics, biomedical and genetic applications, and solar cells.

Nanorods: One-dimensional nanostructures that are like nanowire but are not as flexible as nanowire, have length between 1 and 100 nm and ratio of length-to-diameter varying between 3–5, belong to the nanorod class. Those are generally produced by synthesis from metal and semiconductor materials. It can be utilized in the imaging technology, electromagnetic applications, cancer treatment, energy generation, and light emitting devices.

Nanorings: This structure consists of nanotubes that become rings. It can be observed in protein and other biomolecules. Additionally, annular nanotubes exhibit high conductivity at low temperature.

1.3 Nano-Electro-Mechanical System (NEMS) Technology

Nano-electro-mechanical systems (NEMS) are a group of structures and devices that perform together electrical and mechanical processes to meet the needs determined by humans and include nano-scaled components that must work in harmony with each other. In other words, NEMS can be defined as the outputs that nanotechnology presents to humanity.

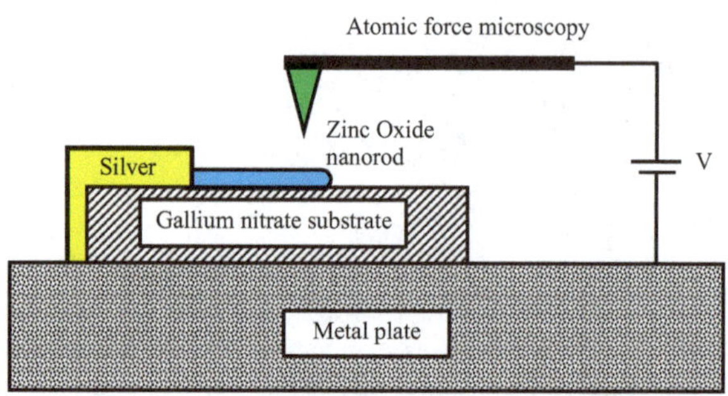

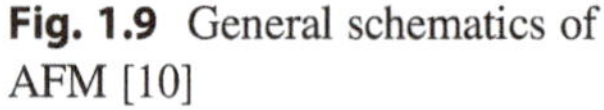
Fig. 1.9 General schematics of AFM [10]

In this section, some examples of NEMS and their usage areas are examined. Here, only a preliminary preparation to the related issue is intended. The mechanical modeling of NEMS is expressed later. The most basic and important example that we can mention about NEMS is the atomic force microscope (AFM). AFM that represents the mechanical structure in the system performs the tasks of imaging, measurement, and atomic arrangement at nanoscale. A general schematic view of AFM is depicted in Fig. 1.9.

Carbon nanotube-based field-effect transistors (CNT-FET) are utilized in logic circuits, biomedical/biotechnology applications, electrical applications, and communications [35]. The CNT-FET structure is typically depicted in Fig. 1.10a. As can be seen, the system labors under electromechanical effects. Silicon nanowire can also be used to represent this mechanical structure [36]. In addition to these, carbon nanotube-based and coaxial cylindrical field-effect transistors can also be mentioned to survey the electrical properties [37]. The CNT-FET models are demonstrated in Fig. 1.10b.

Figure 1.11 displays a light absorption diode (LED) structure [38] consisting of organic and inorganic components. In this nano-scaled system under optoelectronic effects, the zinc oxide nanowire structure embedded in the polymethyl methacrylate medium indicates the mechanical element. Polymethyl methacrylate media provides advantages such as flexibility and adjustable material transport to the system. Examples of piezoelectric composite cantilever nanowire and clamped germanium nanobeams that are other systems common in NEMS are given in Fig. 1.12.

The general schematics of silicon nanowire growth mechanism is demonstrated in Fig. 1.13. Silicon nanowires that generally represent the mechanical element in the system, are firmly embedded in a silver plate at one end and partially firmly embedded in another layer at the other end.

Different from those, NEMS which discrete mechanical elements play a fundamental role are mentioned. According to this, a piezoelectric resonant aluminum nitride plate structure sandwiched between thin aluminum films is used to achieve high operating frequencies in the example presented in Fig. 1.14 [40]. Where platinum that provides ground electrode connection is a discrete system consisting of one-dimensional members. On the other hand, the electrothermal activator V-beam (bend beam) structure seen in Fig. 1.15 is carried out the strain sensing [41]. This structure can be interpreted as a discrete system consisting of two elements.

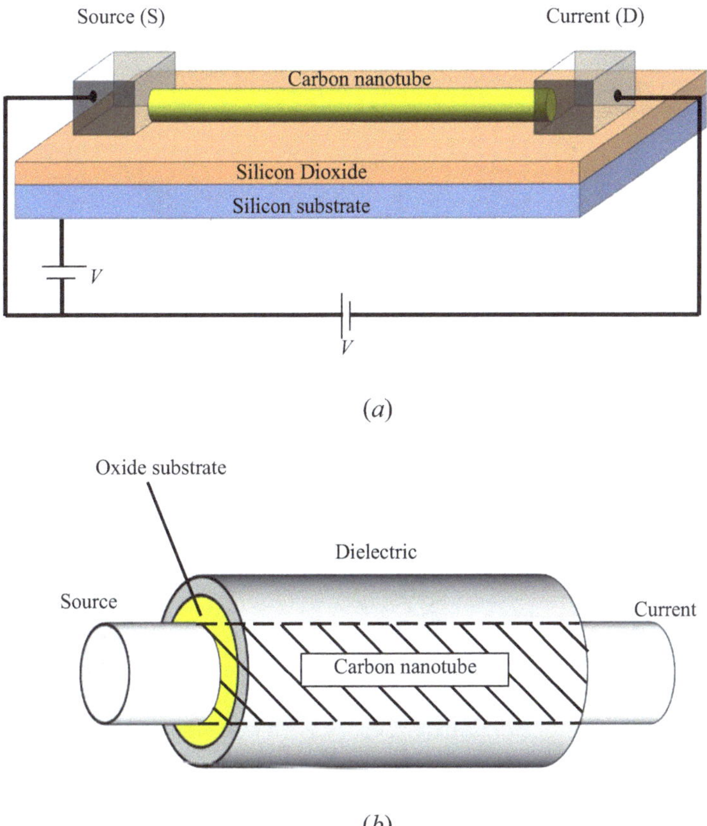

Fig. 1.10 Some CNT-FET schematics. (**a**) CNT resting on Silicon derivatives. (**b**) CNT embedded on Oxide substrate [10]

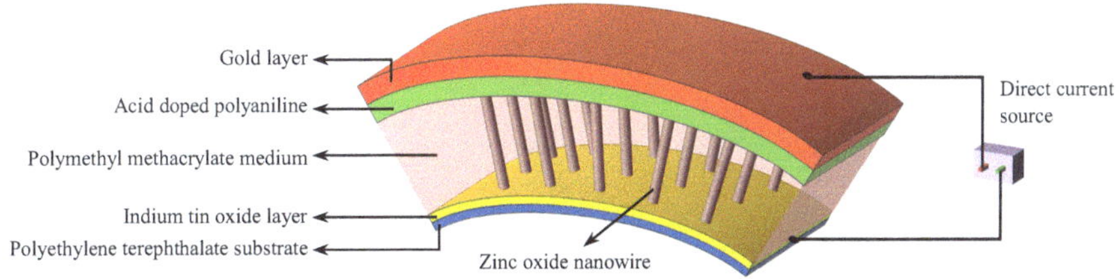

Fig. 1.11 Organic/inorganic light absorption diode system [10]

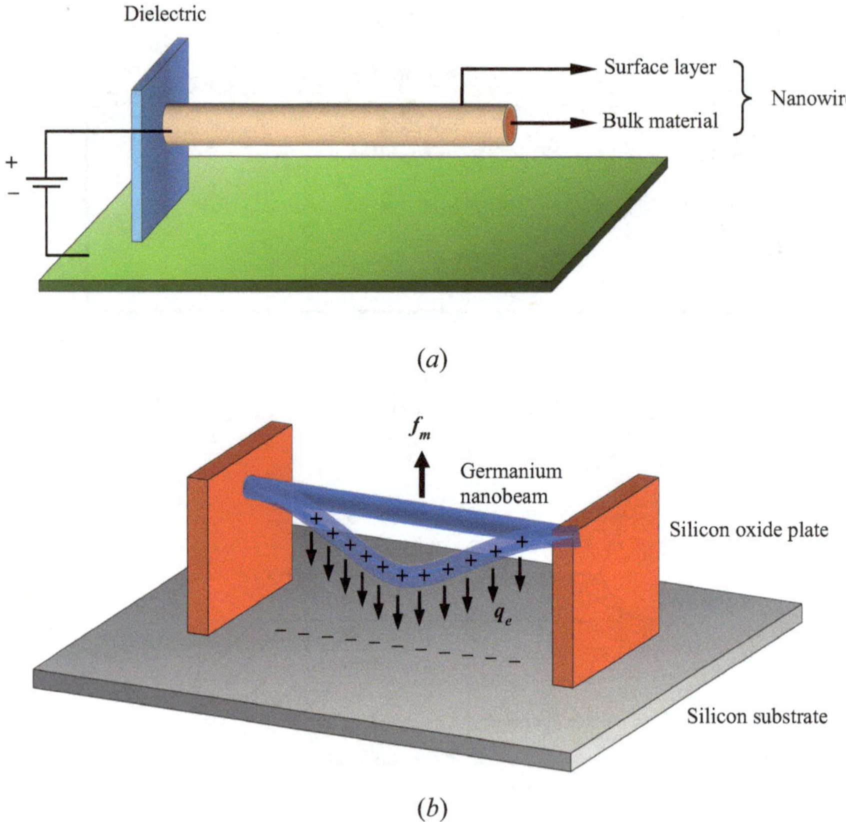

Fig. 1.12 Other common electrical field applications in nano-scaled mechanical systems. (**a**) Composite cantilever nanowire [39]. (**b**) Germanium nanobeam [10]

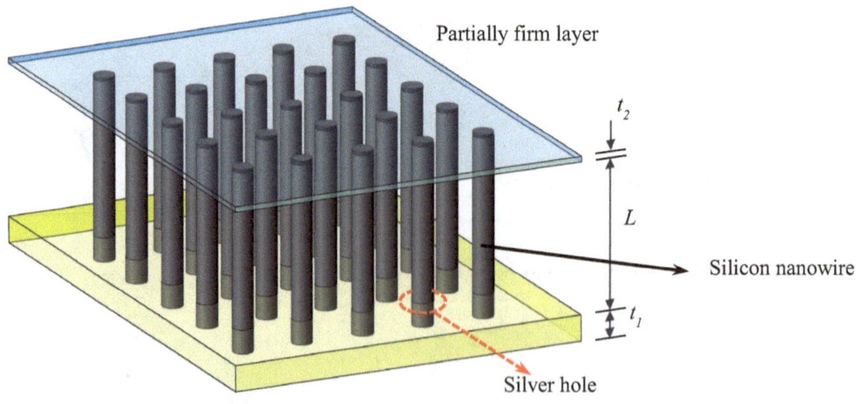

Fig. 1.13 Nanowire growth [39]

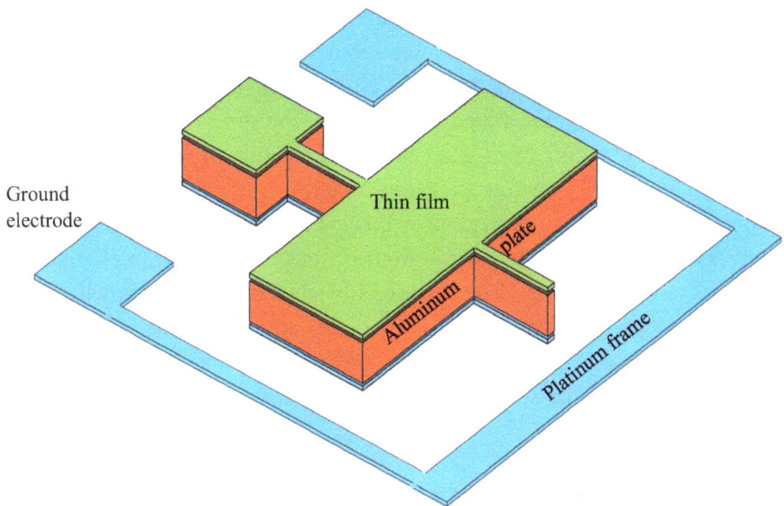

Fig. 1.14 Sandwiched piezoelectrical resonant nanoplate [42]

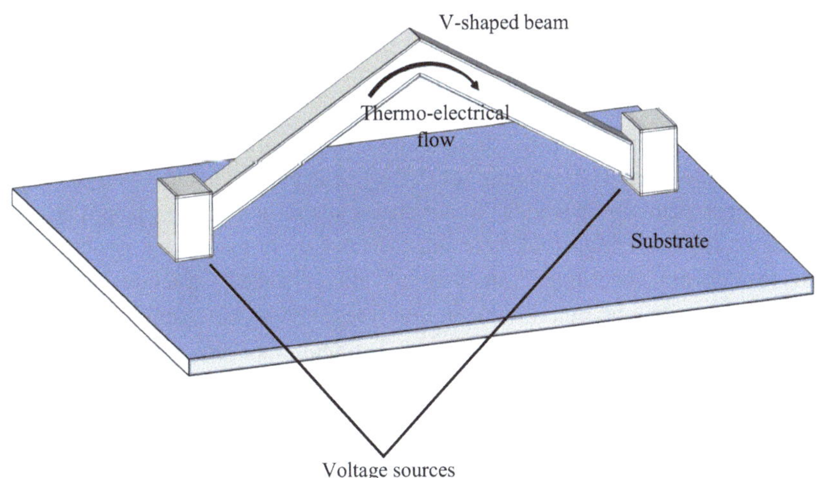

Fig. 1.15 V-shaped beam electrothermal actuator [42]

1.4 Mechanical Analysis of Atomic Systems

1.4.1 Molecular Dynamics Simulation

Examination of any system and its components by using different methods and algorithms in the computer environment specified the simulation. Molecular dynamics simulation (MDS) is a simulation method that provides the numerical understanding of behavior of all atoms of system or structure in respect to the time via Newton's equations of motion. It can be emphasized that computers constitute a semi-experimental environment here. Generally, $\mathbf{F_i} = m_i\mathbf{a_i}$ value that is i. Newtonian force equation of molecule is computed. These expressions are numerically integrated according to the time interval in the simulation. The Newtonian force affecting from other atoms to an atom due to the interaction between atoms, velocity and position parameters can be determined [43, 44].

Computer technology has undergone a great change from the past to the present. However, it is also a fact that there are serious obstacles to obtain an efficient output by modeling atoms on the computer. The need for advanced theoretical knowledge, programming ability, large operation volume of simulations and healthy interpretability of results are the leading of these obstacles. In short, it should be stated that to understand the mechanical analysis of atomic-scaled structures through simulations is some inconvenient.

1.4.2 Higher-Order Continuum Mechanics

In addition to the fact that atomic-scaled structures have different properties compared to macrostructures, it has been revealed by experimental methods that the mechanical behavior of these structures cannot be understood by classical laws. At the beginning of the 1900s, the idea that the atoms in the structure interacted with external factors emerged. According to this idea, internal characteristic dimensions of atomic body have an effect on the mechanical behavior of atoms. The "Théorie Des Corps Déformables" of Cosserat brothers [45] is considered a beginning for size effect in the mechanical analysis of atomic-scaled structures. Although this idea did not attract attention in those years, it has stamped the science of computational solid mechanics, especially in the current period. Since it is difficult to perform mechanical analysis of atomic-scaled structures by means of simulations in the computer environment, many researchers have approached the related problem through analytical methods based on continuum mechanics. Of course, analytical formulations are diverse and have emerged in different periods. Nonlocal elasticity theory [46–49], couple stress elasticity theory [50–52], strain gradient elasticity theory [53, 54], surface energy elasticity theory [55, 56], doublet mechanics elasticity theory [57, 58] are the most basic examples for these formulations.

1.5 Vibration of Mechanical Systems

1.5.1 Fundamental Concepts

The cluster of elements that interact with each other and its environment states a system. This interaction of elements should be understood and expressed most correctly to be realized the purpose of system. Many types of systems can be exemplified such as biological systems, social systems, economic systems, chemical systems, physical systems, mechanical systems, etc. A system transforms these into outputs by processing different inputs. These processes are controlled by the components of system. Since our subject is mechanical systems, we will investigate it. According to these, structures formed by interacting mechanical components with each other denote the mechanical system. The suspension system of car is one of the most typical examples of mechanical system. Suspension consisting of wheel band, air, spring, and shock absorber is a system that connects the wheels to the vehicle and dampens the concussion coming from the wheels. This definition clearly includes the purpose and components of system. Generally, while the spring component of system provides a good roadholding and supports driving comfort, damping eliminates dangerous vibrations arising from concussion. This statement expresses their interaction with each other and with their external environment as well as the tasks of elements. Finally, the inputs are external concussion forces, spring and damping constants. On the other hand, the outputs are the internal reaction force and the natural frequency.

In fact, the suspension system can be described by the simple mass-spring-damper model that is the basis of the science of structural dynamics. Let's depict this system in Fig. 1.16a and ignore the damping effect for now.

The m mass that is allowed to motion only in the horizontal direction and is seen to be at rest in equilibrium position (Fig. 1.16a) is moved away from the system with an external excitation at the initial time ($t = 0$) as $x = x(0)$ (Fig. 1.16b). The elastic strain energy accumulates in the elastic spring during this. If the external excitation is removed, the system starts to move because the strain energy turns into kinetic energy. After the mass

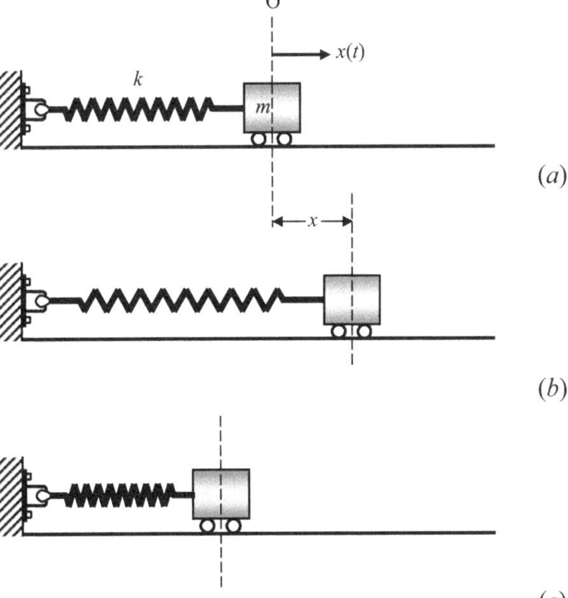

Fig. 1.16 Simple mass-spring model. (**a**) System in equilibrium. (**b**) Displaced system from equilibrium. (**c**) System in repetitive motion with the effect of displacement [10]

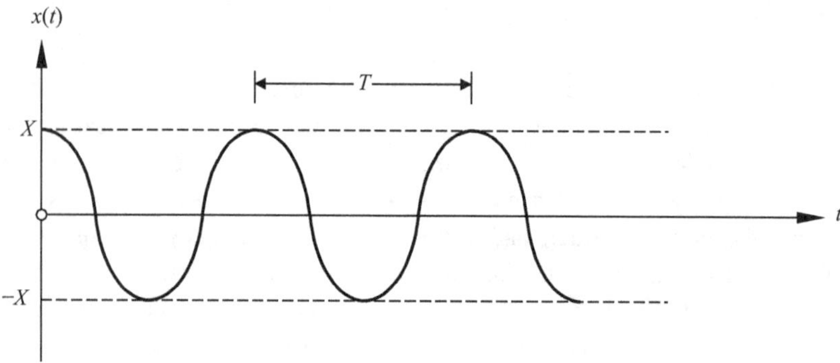

Fig. 1.17 Repetitive (periodic) motion graphics [10]

performs the maximum shortening of spring in Fig. 1.16c, it comes back to the equilibrium point. Because this time the energy arising from the shortening turns into kinetic energy. If there is no friction effect or viscous damping in the system, since the energy transformations will continue without loss, this motion will continue forever. What should be emphasized here is the following: In this motion that will last forever, the variation of distance (position) to the equilibrium changes in a repetitive (cyclic or periodic) manner and this repetitive motion is called vibration. The subject of vibration contains some concepts within itself. Before we talk about these concepts, let explain the repetitive motion that has just been mentioned with the graphic as follows.

The vibrational motion may be nonperiodic or random as well as periodic. We will consider only periodic motion for simplicity. Firstly, Fig. 1.17 shows the change in $x = x(t)$ distance from the equilibrium position in respect to the variation of t time parameter. According to this, the state of coming to any position from the equilibrium, then to the equilibrium and then to the same position again indicates the cycle of motion. The farthest position of a vibrating element to equilibrium is called the amplitude and it is shown by X and $-X$ in the figure. The completion time of a cycle is defined as period. Period is explained as follows by means of the period of motion is T:

$$T = \frac{2\pi}{\omega} \tag{1.12}$$

where ω is an expression known as angular frequency. If the vibration motion is described with a circle which its radius can rotate freely, it defines the angular frequency to the angle that the radius of the circle sweeps per unit time. As can be seen from here,

$$\omega = \frac{\theta}{t} \tag{1.13}$$

where θ is the scanned angle per unit time. In a vibrational motion, the inverse of period according to the multiplication operation is called the frequency. Frequency, with another definition, indicates the number of cycles seen per unit time:

$$f = \frac{1}{T} \tag{1.14}$$

Substituting Eq. (1.12) into Eq. (1.14) yields the relation between angular frequency and frequency:

$$\omega = 2\pi f \tag{1.15}$$

If there is no energy dissipation or loss due to friction and other resistances during the vibration of a system, this system is defined as undamped system. If there is a damping effect, the system is called a damped system. Undamped systems can continue motion indefinitely. When there is no internal or external excitation that may cause vibration in the system, the vibrational motion that occurs in this system denotes the free vibration. On the other hand, if the motion occurs by way of external excitations, this motion defines a forced vibration. Moreover, if all the components (mass, spring, damping, etc.) in a vibrating system can be expressed with linear mathematical models (Newton's equation for mass, Hooke's equation for spring and Coulomb's equation for damping), motion of system reveals a linear vibration. If at least one of them behaves nonlinearly, nonlinear vibration is mentioned.

Another important definition in the vibration phenomenon is the degree of freedom (DOF). According to this, the minimum number of points that must be arranged in system to analyze a motion is called the DOF of system. Let's investigate some examples of this concept. Firstly, in the simple mass-spring-damper model given in Fig. 1.18a, this system is single-DOF since the mass can only move in vertical direction. In Fig. 1.18b, in the single-storey and single-span shear frame model that performs two types of motion namely earthquake (ground motion) and elastic (own motion), the mass and the elastic properties are gathered in the floor beam and the column, respectively. Additionally, damping is constituted from the outside to the structure. In this system, under the earthquake load acting on the extremely rigid floor beam, the beam moves only in the direction of earthquake. Therefore, the system is single-DOF. On the other hand, in the mass-spring-pulley system seen in Fig. 1.18c, the mass moves only in the vertical direction under the external excitation that may be acted to the system. The rotational disc with insignificant thickness attached to the extremely rigid rod in Fig. 1.18d moves only in the direction of rotation. Also, the water tank attached to the rigid tower depicted in Fig. 1.18e moves only in the horizontal direction. Therefore, we can say that these models also constitute examples of single-DOF.

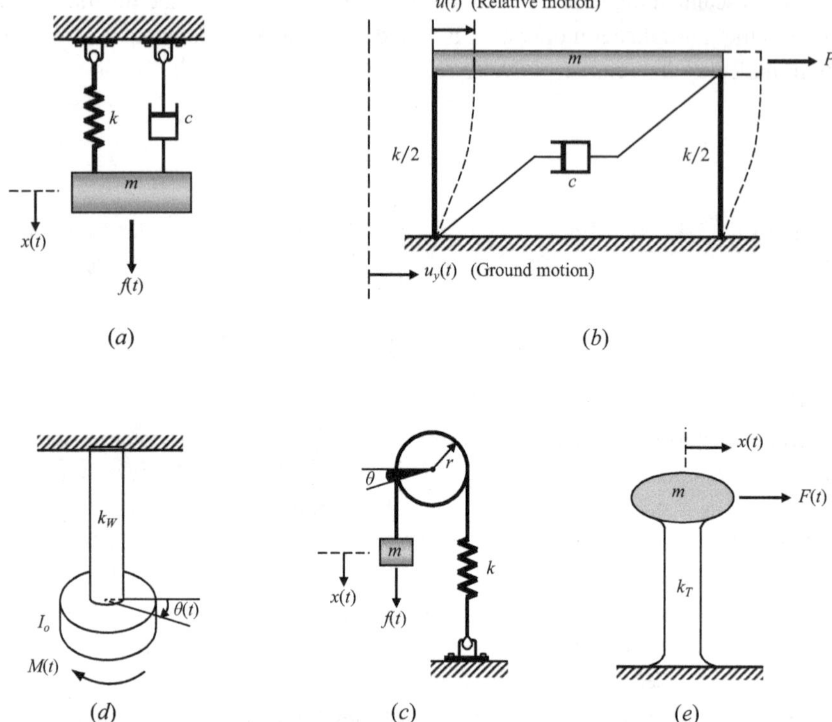

Fig. 1.18 Single degree-of-freedom system examples. (**a**) Simple mass-spring-damping model. (**b**) Simple shear frame. (**c**) Mass-spring-pulley system. (**d**) Rigid rod-disk system. (**e**) Water tank model [10]

In the two-storey and single-span frame demonstrated in Fig. 1.19a, if each storey can move rigidly, slab masses can accumulate at beam-column connections. Considering that the axial elongation in beam elements is generally negligible, lumped masses can only move in the horizontal direction and it can be said that the system has two DOFs. In the example in Fig. 1.19b, the mass of continuous simple beam is agglomerated in two different masses in the middle of span. This structure has also two DOFs. In the system in Fig. 1.19c, this time since the external excitation triggers the horizontal motion of three different masses, the system has three DOFs. Finally, in the floor slab modeled with a rigid slab in Fig. 1.19d, the slab can perform the horizontal and vertical linear motion in the plane. Also, it can also be stated a rotational movement under torsional moment here. Since there are three motion components in total, it can be said that this system has three DOFs.

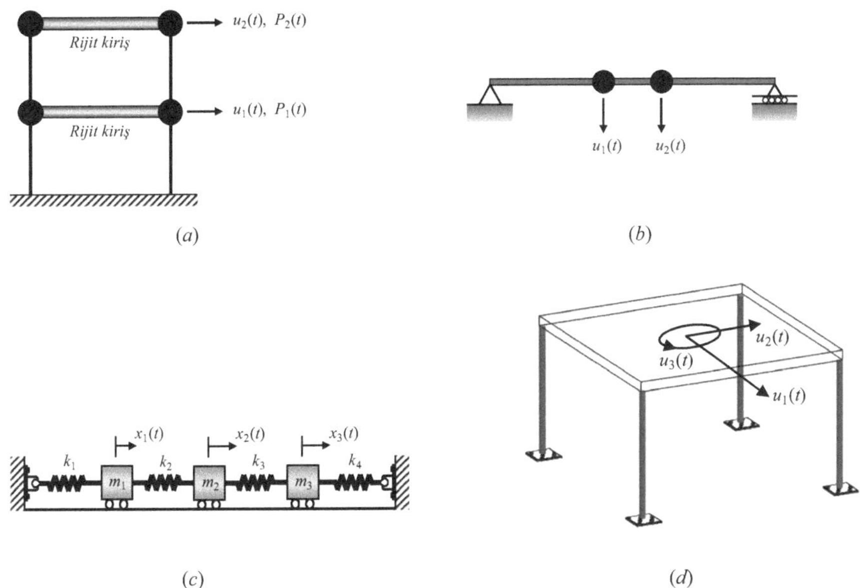

Fig. 1.19 Multi-degree of freedom system examples. (**a**) Two-storey frame. (**b**) Beam with lumped masses. (**c**) Multi-mass-spring system. (**d**) Rigid slab model [10]

1.5.2 Discrete and Continuum Systems

We classified the systems according to the DOF at the end of previous section. Here, we will examine the mechanical systems according to the working principle. Thus, the concepts of discrete system and continuous system will be mentioned. All systems in which we express the DOF with a certain number operate discretely. Although the components interact with each other, mechanical analysis of each can be carried out by oneself. For this, it is sufficient to determine the effects from other components to the component to be analyzed. All of single- and multi-DOF systems that we referred in the previous section are discrete. Also, frame systems, truss systems, or hybrid structures that are their combinations, can be indicated as discrete systems. Truss systems consist of longitudinal members that can only take axial stress and can be deformed axially. On the other hand, frame systems consist of longitudinal members that work under shear and bending effects as well as axial effects and can be deformed in their directions. In addition to these, systems whose members that are different terms of mechanical are combined in are considered as hybrid. The mentioned discrete systems are demonstrated in Fig. 1.20.

It can be expressed that the continuous systems have infinite DOFs. Examples of continuous systems are thin and stretched wires, pipes with variable or constant cross-section, torsion shafts, solid webbed continuous beams/rods, plates, and shells. Such systems can be characterized according to number of directions, geometry, number and

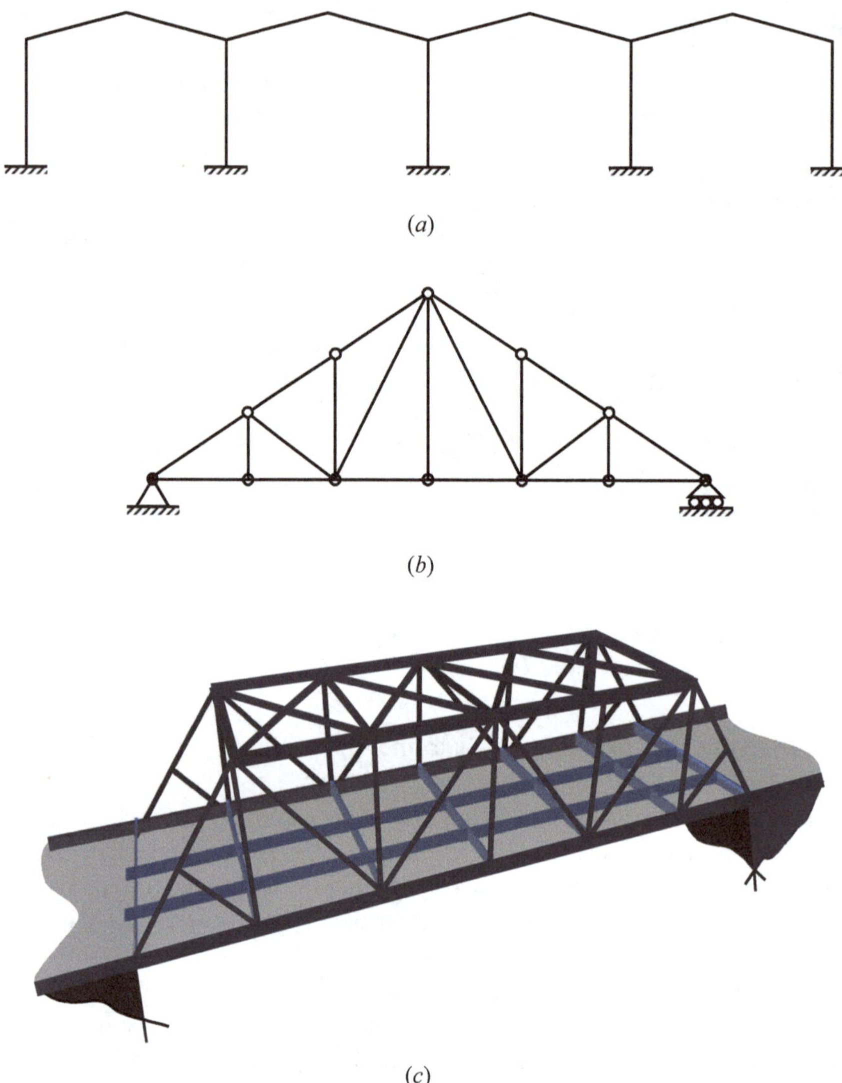

Fig. 1.20 Other discrete system examples. (**a**) Multi-span plane portal frame. (**b**) Plane truss. (**c**) Hybrid system [10]

characteristics of their internal forces and deformations. Examples of continuous systems are drawn in Fig. 1.21.

Analysis of continuous systems is governed through mathematical continuity equations, boundary and compatibility conditions. Let's explain this systematic with a very simple example. The displacement equation of bending members, namely beams, under static loads depends on the solution of a differential equation according to geometric and

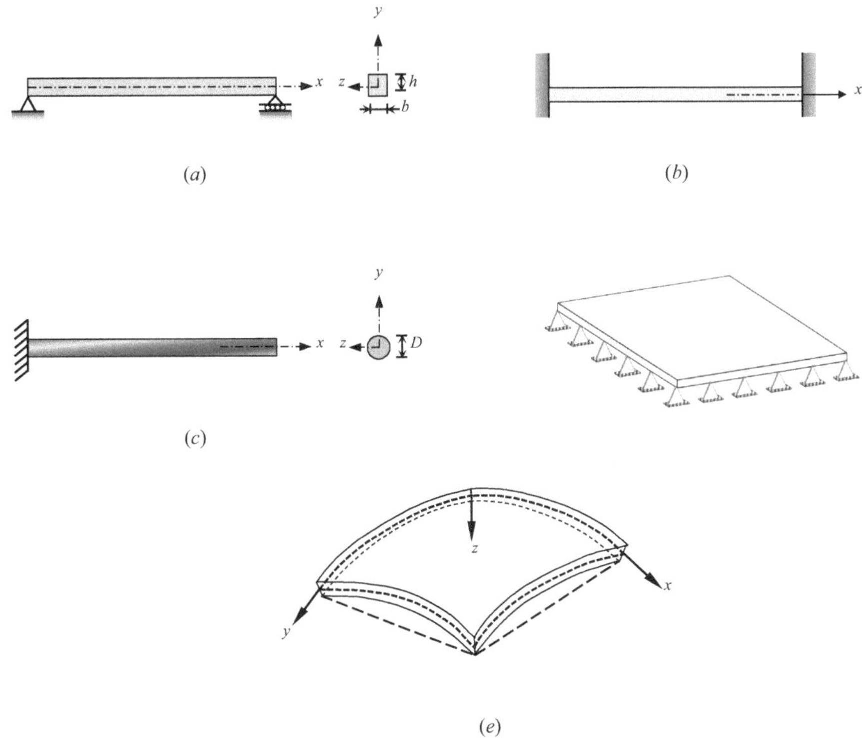

Fig. 1.21 Continuous system examples. (**a**) Simply supported beam. (**b**) Clamped axial rod. (**c**) Cantilever torsional shaft. (**d**) Square plate. (**e**) Curved shell [10]

mechanical boundary conditions. When it is considered that a cantilever beam acts on a uniformly distributed and static load across span, this displacement is continuous throughout the member (Fig. 1.22). In other words, there is no point that is not deflected in the beam (except for fixed end due to the geometric boundary condition). The displacement under such loading at least is expressed as a continuous function.

If the external forces acting on the continuous element are instantaneous or pulsed, a vibration can be mentioned and this vibration is characterized as forced vibration. However, there is a case of vibration that may be seen with natural frequency of structure under the internal equilibrium of structure that is not affected by external forces or immediately lifted as soon as those act, and this vibration defines the free vibration. In the free vibration, a vibrational mode shape occurs and is expected to constitute a continuous function like the elastic deflection under static load given above. For example, the first four mode shapes of free vibration of a cantilever beam are presented in Fig. 1.23.

Based on the characterization of discrete and continuous systems, there are clear distinctions regarding the ability to constitute the mathematical formulations of system. Let's talk about single-DOF systems first. For this, consider the mass and its free body

Fig. 1.22 Continuous displacement of cantilever beam under static uniform distributed load [10]

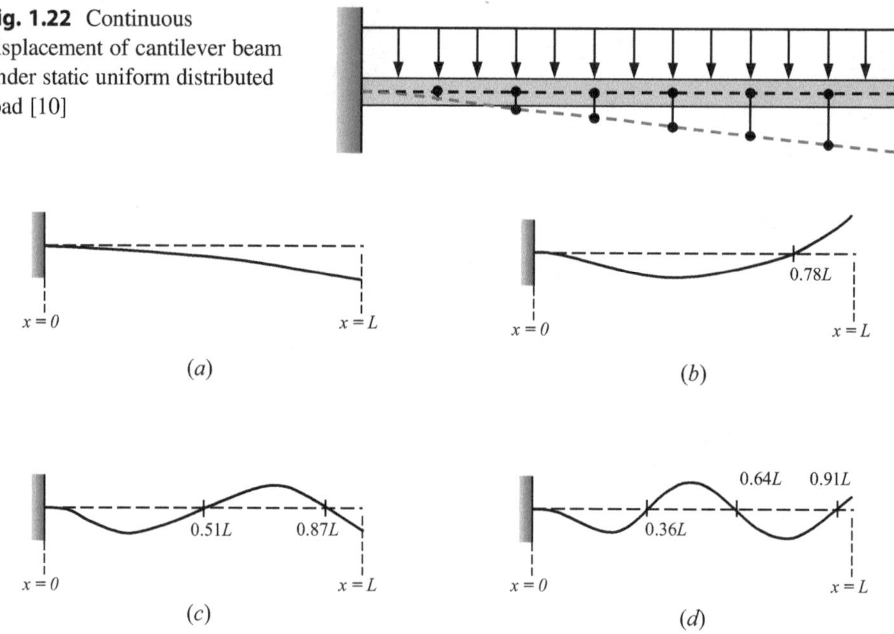

Fig. 1.23 Different mode shapes in free vibration of cantilever beam. (**a**) Fundamental mode shape. (**b**) Second mode shape. (**c**) Third mode shape. (**d**) Fourth mode shape [10]

diagram under the effect of spring with k stiffness constant and damper with c constant shown in Fig. 1.24.

The following internal forces must be defined by means of the displacement of system is $x = x(t)$:

$$F_e(t) = kx, \quad F_s(t) = cv = c\dot{x}, \tag{1.16}$$

where F_e and F_s denote the forces of spring and damper, respectively. $v = \dot{x} = \dot{x}(t) = \mathrm{d}x/\mathrm{d}t$ is the velocity component. The one- and two-point symbolizations specify the first- and second-order derivative of displacement function with respect to time $(\dot{(\cdots)} = \mathrm{d}(\cdots)/\mathrm{d}t$, $\ddot{(\cdots)} = \mathrm{d}^2(\cdots)/\mathrm{d}t^2)$. According to Newton's second law of motion, the following equation is written:

$$\sum F_{net} = F_a = ma = m\ddot{x} \tag{1.17}$$

where m is the mass and $a = \ddot{x}$ is acceleration. As it can be understood from here, the net force is balanced by inertia force. Considering the free body diagram in Fig. 1.25b, Eq. (1.17) can be rewritten as follows:

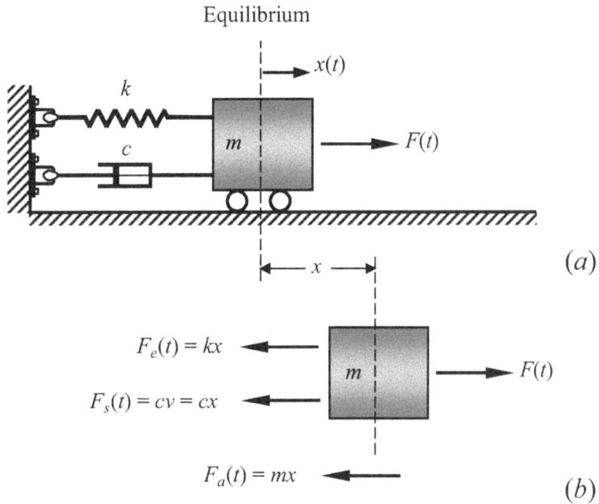

Fig. 1.24 Motion of a single-DOF system. (**a**) Mass exposed to external excitation under the effect of spring and damping. (**b**) Free body diagram [10]

$$F(t) - F_e(t) - F_s(t) = m\ddot{x} \tag{1.18}$$

If the expressions in Eq. (1.16) are substituted into Eq. (1.18), the following equation is obtained:

$$kx(t) + c\dot{x}(t) + m\ddot{x}(t) = F(t) \tag{1.19}$$

This equation is known as the most general equation of motion of vibration. A vibration motion can be characterized according to the following descriptions:

- $c = 0$ and $F = 0$: undamped free vibration,
- $c = 0$ and $F \neq 0$: undamped forced vibration,
- $c \neq 0$ and $F = 0$: damped free vibration,
- $c \neq 0$ and $F \neq 0$: damped forced vibration.

Examination of vibration for single-DOF linear systems can be applied to multi-DOF systems as well. According to this, the multi-DOF spring-damper-mass model and the free body diagram of each mass can be examined (Fig. 1.25).

If Eq. (1.17) is first arranged for the first mass under the assumptions $x_n > x_{n-1} > \cdots > x_3 > x_2 > x_1$, $\dot{x}_{n-1} > \cdots > \dot{x}_3 > \dot{x}_2 > \dot{x}_1$, and $\ddot{x}_n > \ddot{x}_{n-1} > \cdots > \ddot{x}_3 > \ddot{x}_2 > \ddot{x}_1$ for the displacement, velocity, and acceleration components of system:

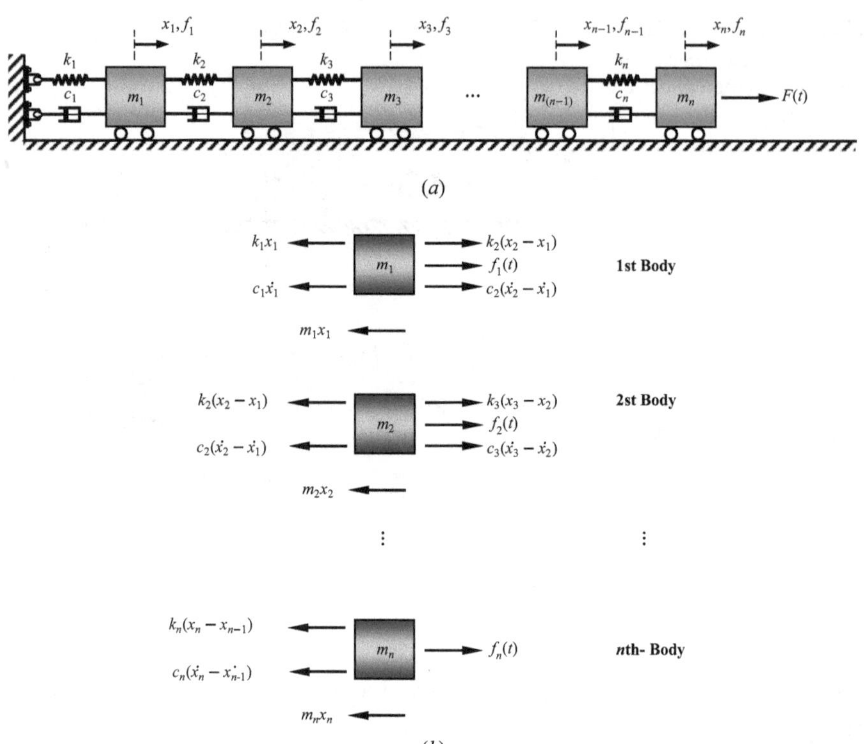

Fig. 1.25 Motion of a multi-DOF system. (**a**) Multi-degree of freedom system model. (**b**) Free body diagram of discrete masses in the system [10]

$$
\begin{aligned}
\sum F_{net,1} &= m_1\ddot{x}_1 - k_1x_1 - c_1\dot{x}_1 + k_2(x_2 - x_1) + c_2(\dot{x}_2 - \dot{x}_1) + f_1 \\
&= m_1\ddot{x}_1(k_1 + k_2)x_1 + (c_1 + c_2)\dot{x}_1 + (-k_1)x_2 + (-c_2)\dot{x}_2 + (m_1)\ddot{x}_1 = f_1
\end{aligned}
\tag{1.20}
$$

Then, equation of motion for the second mass is as follows:

$$
\begin{aligned}
\sum F_{net,2} &= m_2\ddot{x}_2 - k_2(x_2 - x_1) - c_2(\dot{x}_2 - \dot{x}_1) + c_3(\dot{x}_3 - \dot{x}_2) + k_3(x_3 - x_2) + f_2 \\
&= m_2\ddot{x}_2(-k_2)x_1 + (-c_2)\dot{x}_1 + (k_2 + k_3)x_2 + (c_2 + c_3)\dot{x}_2 \\
&+ (m_2)\ddot{x}_2 + (-k_3)x_3 + (-c_3)\dot{x}_3 = f_2
\end{aligned}
\tag{1.21}
$$

Equation (2.19) can be generalized to the free body diagram of nth− mass. Therefore, instead of writing the equilibrium condition of all masses, the equilibrium of nth− mass can be expressed:

$$\sum F_{net,n} = m_n\ddot{x}_n - k_n(x_n - x_{n-1}) - c_n(\dot{x}_n - \dot{x}_{n-1}) + f_n = m_n\ddot{x}_n(-k_n)x_{n-1} + (-c_n)\dot{x}_{n-1} + (k_n)x_n + (c_n)\dot{x}_n + (m_n)\ddot{x}_n = f_n \tag{1.22}$$

Equaitons (1.20)–(1.22) can be indicated in open matrix notation form as follows:

$$[k_{ij}]\{x_i\} + [c_{ij}]\{\dot{x}_i\} + [m_{ij}]\{\ddot{x}_i\} = \{f_i\} \tag{1.23}$$

The following definitions are valid here:

$$[k_{ij}] = \begin{bmatrix} k_1 + k_2 & -k_2 & 0 & \cdots & 0 & 0 \\ -k_2 & k_2 + k_3 & -k_3 & \cdots & 0 & 0 \\ 0 & -k_3 & k_3 + k_4 & \cdots & 0 & 0 \\ \vdots & \vdots & \vdots & \vdots & \vdots & \vdots \\ 0 & 0 & 0 & \cdots & k_{n-1} + k_n & -k_n \\ 0 & 0 & 0 & \cdots & -k_n & k_n \end{bmatrix},$$
$$[c_{ij}] = \begin{bmatrix} c_1 + c_2 & -c_2 & 0 & \cdots & 0 & 0 \\ -c_2 & c_2 + c_3 & -c_3 & \cdots & 0 & 0 \\ 0 & -c_3 & c_3 + c_4 & \cdots & 0 & 0 \\ \vdots & \vdots & \vdots & \vdots & \vdots & \vdots \\ 0 & 0 & 0 & \cdots & c_{n-1} + c_n & -c_n \\ 0 & 0 & 0 & \cdots & -c_n & c_n \end{bmatrix}, \tag{1.24}$$
$$[m_{ij}] = \begin{bmatrix} m_1 & 0 & 0 & \cdots & 0 & 0 \\ 0 & m_2 & 0 & \cdots & 0 & 0 \\ 0 & 0 & m_3 & \cdots & 0 & 0 \\ \vdots & \vdots & \vdots & \vdots & \vdots & \vdots \\ 0 & 0 & 0 & \cdots & m_{n-1} & 0 \\ 0 & 0 & 0 & \cdots & 0 & m_n \end{bmatrix}.$$

where $[k_{ij}]$, $[c_{ij}]$, and $[m_{ij}]$ introduce the stiffness, damping, and mass matrices, respectively. Also, vectorial expressions in Eq. (1.23) are given as follows:

$$\begin{aligned}\{x_i\} &= [x_1 \ x_2 \ x_3 \ \cdots \ x_n \ x_{n-1}]^{\mathrm{T}}, \ \{\dot{x}_i\} = [\dot{x}_1 \ \ \dot{x}_2 \ \ \dot{x}_3 \ \ \cdots \ \ \dot{x}_n \ \ \dot{x}_{n-1}]^{\mathrm{T}} \\ \{\ddot{x}_i\} &= [\ddot{x}_1 \ \ \ddot{x}_2 \ \ \ddot{x}_3 \ \ \cdots \ \ \ddot{x}_n \ \ \ddot{x}_{n-1}]^{\mathrm{T}}, \ \{f_i\} = [f_1 \ \ f_2 \ \ f_3 \ \ \cdots \ \ f_n \ \ f_{n-1}]^{\mathrm{T}},\end{aligned} \tag{1.25}$$

where $\{x_i\}$ is displacement (motion), $\{\dot{x}_i\}$ is velocity, $\{\ddot{x}_i\}$ is acceleration, and $\{\ddot{f}_i\}$ is excitation.

If we look closely, dynamic equilibrium is analyzed depending on the components such as spring and damping beside each discrete mass in the discrete systems. However, the masses of system are not influenced by each other. This fact is different in continuous systems. According to continuum mechanics, the whole body should be considered together for the analysis of behavior of body against mechanical forces. Analyzes are usually performed on a quite small element that can represent the whole body in a mathematically and mechanically. That is, after the accurate description of internal effects and deformations on the infinitesimal element to be extracted from the continuous structure, the mechanical analysis of whole body can be performed with the using of appropriate mathematical operations. These mentioned is exemplified this on the free transverse vibration of elastic rods (beams). According to this, the free body diagram of a straight bending rod and the infinitesimal element extracted are demonstrated in Fig. 1.26.

Bending members may be exposed to vibration under transverse forces. In Fig. 1.26, $w = w(x, t)$ is transverse displacement and $f = f(x, t)$ is transverse external force (excitation). Also, $M = M(x, t)$ means the bending moment and $V = V(x, t)$ denotes the shear force. Firstly, the inertia force of beam is written as follows:

$$\mathrm{d}m \times a = \rho A \mathrm{d}x \frac{\partial^2 w}{\partial t^2} \tag{1.26}$$

where $\mathrm{d}m$ indicates the mass of differential elements, a defines the acceleration, ρ is called as the mass of unit volume, and A is the cross-section area. If we write the equilibrium of transverse forces:

$$-(V + \mathrm{d}V) + f\mathrm{d}x + V = \rho A \mathrm{d}x \frac{\partial^2 w}{\partial t^2} \tag{1.27}$$

On the other hand, the moment equilibrium of internal and external forces according to P point can be expressed as follows:

$$(M + \mathrm{d}M) - (V + \mathrm{d}V) + f(x, t)\mathrm{d}x \frac{\mathrm{d}x}{2} - M = 0 \tag{1.28}$$

Higher-order differential terms removed from equation because these are of quite small order. Consequently, relation between bending moment and shear force is attained:

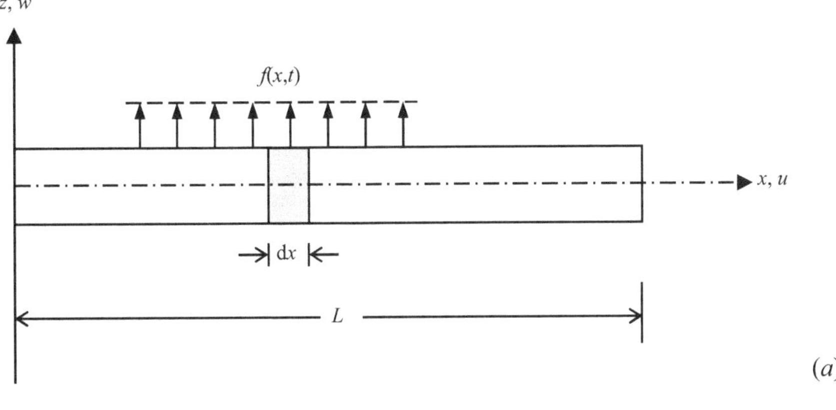

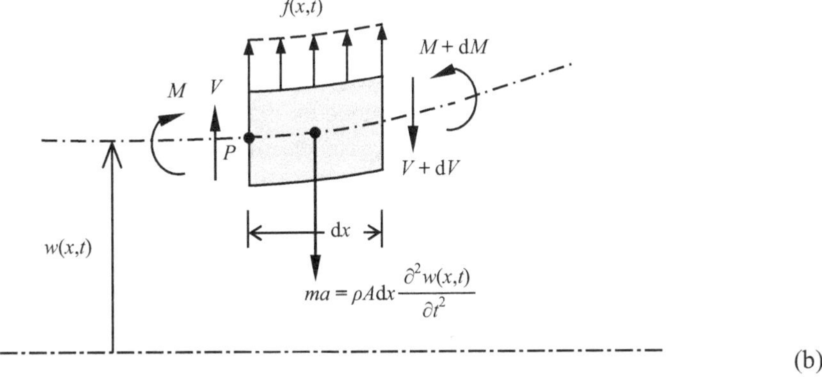

Fig. 1.26 Transverse vibration of a beam. (**a**) Bending rod subjected to transverse external excitation [10]. (**b**) Free body diagram of differential element with dx length extracted from the rod [16]

$$\frac{\mathrm{d}M}{\mathrm{d}x} = V \tag{1.29}$$

The following equation can be given based on Taylor series expansions for differential shear force:

$$\mathrm{d}V = \frac{\partial V}{\partial x}\mathrm{d}x \tag{1.30}$$

If Eq. (1.30) is substituted into Eq. (1.28), the following equation is obtained:

$$-\frac{\partial V}{\partial x} + f = \rho A \frac{\partial^2 w}{\partial t^2} \tag{1.31}$$

We know that an important relationship that emerges according to the Euler-Bernoulli bending assumptions is as follows:

$$M = EI\frac{\partial^2 w}{\partial x^2} \tag{1.32}$$

where E is the modulus of elasticity and I is the moment of inertia. Using Eqs. (1.29) and (1.32) into Eq. (1.31) gives the following expression:

$$EI\frac{\partial^4 w}{\partial x^4} + \rho A\frac{\partial^2 w}{\partial t^2} = f \tag{1.33}$$

This introduces the equation of vibration of continuous beam. If we want to make an analogy with single-DOF systems, we can indicate that here $EI(\partial^4 w/\partial x^4)$ and $\rho A(\partial^2 w/\partial t^2)$ correspond to stiffness and mass inertia, respectively. On the other hand, this equation can be solved by generally known methods such as separation of variables and the fundamentals of vibration motion can be understood for the continuous systems.

In this section, we will talk about the using of modal analysis function that is another solution of vibration. According to this, let specifically consider undamped free vibration for Eq. (1.23):

$$[k_{ij}]\{x_i\} + [m_{ij}]\{\ddot{x}_i\} = 0 \tag{1.34}$$

The displacements of masses representing the nodes in the multi-DOF system are formulated as follows:

$$\{x_i\} = \{X_i\}\sin(\omega_i t + \varphi) \tag{1.35}$$

Here ω_i is the vibration frequency for the ith mode (the mode word means the physical output of vibration). Additionally, φ is the phase angle. It can be stated that Eq. (1.34) is an approach that enables the displacements in a dynamic event to be interpreted according to static principles. The second-order derivative of this harmonic function with respect to time is expressed as follows:

$$\{\ddot{x}_i\} = -\omega_i^2\{X_i\}\sin(\omega_i t + \varphi) \tag{1.36}$$

If Eqs. (1.35) and (1.36) are substituted into Eq. (1.37), the following expression is obtained:

$$([k_{ij}] - \omega_i^2[m_{ij}])\{X_i\} = 0 \tag{1.37}$$

In order to obtain the trivial solution of this equation, the determinant of expression in front of $\{X_i\}$ should be equal to zero:

$$\left|k_{ij} - \omega_i^2 m_{ij}\right| = 0 \tag{1.38}$$

This expression introduces a standard eigenvalue problem. ω_i values that are roots of equation denotes the eigenvalues of each vibration mode, that is, the natural vibrational frequencies. On the other hand, $\{X_i\}$ expression corresponding to each eigenvalue is also defined as the eigenvector, that is, the mode shape.

1.6 Analysis of Mechanical Problems

The vibration of nano- and microscaled structures under some effects is formulated with mathematical equations based on nonclassical physical laws. The equations of motion and mechanical boundary conditions obtained as a result of these equations are in the form of partial differential equation. Generally, three fundamental approaches can be mentioned in the derivation of these equations:

1.6.1 Direct (Equilibrium) Approach

If the equilibrium equations can be constituted correctly thanks to external effects which the structures are exposed and forces arising from the internal effects, the governing equations can be obtained. We can say that it is a quite practical approach for some problems, but using this approach leads to inaccurate results for problems containing much more effects.

1.6.2 Variational Principle

The energy of system is functional since it depends on many functions that can provide the solution. Energy methods search for the extreme value of integral of this functional between definite limits due to the stationarity condition. The variational principle examines the variation of a function within acceptable ranges and provides the required algebraic operations of integral whose extreme value is investigated. According to this, while Hamilton's principle that considers the internal strain energy, kinetic energy, and work energy (the work ability of external force on the structure) of structure is used in the dynamic problems, the minimum total potential energy principle that considers the strain energy and work energy is employed in the static problems.

Since the equilibrium approach may prone to errors of researchers due to the low level of knowledge and interpretation skills in engineering mechanics topics, it can be mentioned that there is a deficiency of this approach. Let's exemplify this with the equation of motion of nonlocal free of nanobeams via the equilibrium approach: The equation of motion that is a partial differential equation can be made an ordinary with the solution of separation of variable procedure that which will be examined presently. Ordinary differential equation forms a basis for derivation of mode shape equation. According to the application of mode equations, for example, we know from geometry that the mode equation of a simply supported nanobeam is equal to zero at the support points and apply this to mode shape. We equate second derivative of mode shape to zero again because we know that these points will not take bending moments. This approach is incorrect systematically, but luckily the solution is correct. Because it is understanding the fourth chapter that support conditions give the researcher this chance accidentally. However, this approach will be incorrect for cantilever beams. Mathematical operation is correct while the mode equation and its derivative are equated to 0 at the free end, but an approach like "Since the free end does not take bending moment and shear force if we equate the second and third derivatives of mode equation to 0, we obtain the nonobvious solution" is incorrect. The reason for this error is that the researcher forgets the bending moment function that is the mechanical boundary condition, although the equation of motion is obtained correctly. The moment function will not be like the classical one and will be influenced by the atomic (nonlocal) parameter. As a result, an important mistake is made because perception, interpretation ability, and level of knowledge are insufficient. It should be emphasized here that the variational principle is a quite systematic approach. It is necessary to express the energy formulations and to use of higher mathematical algebra such as derivative-integral operations. There is almost no chance of mistakes under these skills. As a result of operations, the equation of motion, geometric and mechanical boundary conditions can be attained accurately even if the operation process increases. It is important to use the variational principle to eliminate problems such as "Which forces will the beam or axial rod element take in the equilibrium analysis and what type of deformations will it perform?" that may confuse and cause to make more mistakes especially in higher-order beam and rod problems.

1.6.3 Lagrange's Equations

If the kinetic and potential energies can be constituted through the generalized coordinates of system, the equations of motion and boundary conditions can be determined.

In the second, third, and fourth chapters, various applications of equilibrium approach and variational principle are examined in detail. Especially in continuous system problems, the equations obtained are generally solved by the separation of variables method. It can be said that this method is useful for the solution of partial differential equations that are homogeneous and whose order is not high. The basic logic of method is to make an ordinary and solvable by discretizing the partial equation. According to this, consider template of a homogeneous partial differential equation as follows:

$$\partial_{(n)x}f + \partial_{(n)t}f = 0 \tag{1.39}$$

Here, f represents a differentiable (differentiable) function. ∂ is called as partial derivative operator, n is differentiation amount, x and t are differentiation variables (independent of each other). The solution of this differential equation is given as follows:

$$f(x,t) = D(x)T(t) \tag{1.40}$$

where $f = f(x, t)$ is a multivariate function that must contain and naturally depend on the derivation variables. $D(x)$ and $T(t)$ are discrete functions that depend on x and t expressions that are the discrete differentiation variables, respectively. Physically, while $D(x)$ expresses the mode equation, $T(t)$ indicates the harmonic function. Since the scope of future chapters is modal analysis and frequency calculation, only the mode equation and its ordinary differential equation will be interested in.

Sometimes the application of boundary conditions in the equations obtained through the separation of variables may cause serious complexity, especially in manual calculation. If the exact geometric behavior of structure is known, the following series solution that can be applied to the equation of motion is known as the Fourier Series and is quite useful in analysis of dynamic behaviors of axial rods with simple boundary conditions such as clamped or free, and simply supported beams.

$$W(x,t) = \sum_{n=1}^{\infty} W_n \sin(\alpha x) \sin(\omega t - \varphi) \tag{1.41}$$

where W_n defines the unknown Fourier coefficients, ω is the frequency, t is the time, and φ is the phase angle. Additionally, α is a coefficient that reflects the geometric (mechanically static) behavior of rod. For example, while $\alpha = n\pi/L$ is valid on a fully clamped axial rod, $\alpha = (2n - 1)\pi/L$ is considered for clamped-free axial rod. L is the rod length.

The separation of variables method is an analytical method as can be understood. Of course, this method is insufficient to solve some problems. In the future sections, *the finite element method* will be formulated to eliminate this problem. The finite element method will be based on *the weighted residue method* [60]. According to this:

$$\mathrm{I} = \int_{x_i}^{x_j} hR \, \mathrm{d}x \tag{1.42}$$

where h denotes the weighting function and R indicates the residue. The finite element method labors to equate this weighted residue equal to zero. Also, x_i and x_j mean the weighting boundaries. Weighting functions are transposed appropriate shape functions and equation of motion defines the residue. On the other hand, the general solution equation of finite elements is an eigenvalue problem as follows:

$$\det\left(\sum[K]-\omega^2\sum[M]\right)=0 \tag{1.43}$$

where $[K]$ means the global stiffness matrix consist of all the positive and negative stiffness contribution of structure. $[M]$ explains the global mass matrix containing mass components. In the n-DOF discrete system or continuous system discretized to n-DOF, roots of eigenvalue equation that constitutes a polynomial whose roots are ω_i ($i = 1, 2, 3, \cdots, n$ and $\omega_1 < \omega_2 < \omega_3 < \cdots < \omega_n$) can be computed by using numerical (approximate) methods and computer programs.

1.7 Mechanical Modeling of NEMS Components

The primary principle to obtain the functionality expected from NEMS organizations is correct design. By considering an appropriate mechanical structure representing the components of NEMS and the appropriate theoretical approach of mechanical structure, correct designs of NEMS are ensured. Because the appropriate mechanical model is the one that can resist all mechanical external factors (dynamic excitation, static load, or environmental factors such as thermal, humidity, optical, electrical field, magnetic field, pressure, fluid conveying, etc.).

We mentioned various NEMS examples in Sect. 1.3. Explanations on accurate mechanical modeling of components (nanomaterials) that are contained in these NEMS examples are given here.

The atomic force microscope demonstrated in Fig. 1.10 can form a simple one-dimensional cantilever beam or axial rod model. The microscope component at the end can be thought of as a lumped mass attachment. While the carbon nanotube in Fig. 1.11a represents a one-dimensional continuous mechanical structure, the source and current electrodes can be examples of a fixed support or fully clamped end. Here, silicon dioxide and the substrate can be recommended as the elastic foundation that the structure rests. On the other hand, in the coaxial field-effect transistor in described Fig. 1.11b, the carbon nanotube material depicts the axial or torsional bar and the oxide substrate in which the carbon nanotube is embedded can also be recommended as elastic rigid medium model. Similarly, if we assume that the zinc oxide nanowire in Fig. 1.12 represents the continuous rod, we can state that the polymethyl methacrylate medium constitutes the elastic rigid medium model. It can be added that the examples presented in Fig. 1.13 represent one-dimensional element models. In a nanowire growth mechanism given in Fig. 1.14, the silver substrate can be understood as the fully clamped end. Additionally, since the partially firm layer provides a partial rigidity compared to the fully clamped end, it can be considered as a spring-attached end.

If we want to examine discrete systems, the platin frame effective in electric field formation in Fig. 1.15, resembles the single-storey and single-span frame. On the other hand, the V-shaped beam structure depicted in Fig. 1.16 can be assumed as a frame structure since it can move freely at the joint end. Additionally, voltage sources represent the clamped support.

As a result, firstly it is of great importance to establish the mechanical structure model of components of NEMS organizations. After appropriate mechanical structure model is determined, the mechanical theoretical approach that accurately expresses the behavior of nanostructure should be formulated. The subject of this book is the investigation of importance of nonlocal elasticity theory, which is one of the approaches based on the atomic size effect in the mechanical behavior of nanostructures, and the chapters of book are organized to examine different subfields of this broad subject. The following chapters of book focus on various mechanical analyses of related nanostructure models through the nonlocal elasticity theory. Therefore, it should definitely be noted that it is useful to study the mathematical background of nonlocal elasticity theory.

1.8 Nonlocal Elasticity Theory

According to classical physics theories, equilibrium, compatibility, and energy equations represent the whole of macrostructure. However, the fact that the analyses obtained with mechanical equations based on classical physics at the nano and micro structures do not yield correct results have been proven with experimental results. The reason for this fact is that the characteristic internal dimensions of atomic structure interact with the external forces acting on the structure. In order to remove the faults of classical physics theories in this regard, continuum mechanics formulations have been developed on the condition that to include the size scale parameters of atomic-scaled structures. One of atomic size dependent continuum mechanics formulations is also Eringen's nonlocal elasticity theory. The nonlocal theory mentions that the stresses and strains in one region of the atomic body depend on the stresses and strains in the adjacent regions of point.

In the atomic-scaled structures, geometric irregularities that occur due to the displacement of body under external forces generate additional stresses inside the object and the analysis results via classical theories are not appropriate. As a result of these displacements, stresses have infinity value. This problem is prevented by employing the nonlocal theory [2].

The constitutive equations of nonlocal elasticity theory are introduced by Eringen [47] as follows:

$$\sigma_{ij} + \rho(f_j - \ddot{u}_j) = 0 \tag{1.44}$$

$$\varepsilon_{ij} = \frac{1}{2}\left((\nabla u) + (\nabla u)^{\mathrm{T}}\right) = \frac{1}{2}\left(\frac{\partial u_i}{\partial x_j} + \frac{\partial u_j}{\partial x_i}\right) \tag{1.45}$$

$$\sigma_{ij} = \int_V \alpha(|\mathbf{x} - \mathbf{x}'|, \varphi) C_{ijkl}\varepsilon_{ij} \mathrm{d}V(x') = \int_V \alpha(|\mathbf{x} - \mathbf{x}'|, \varphi) s_{ij} \mathrm{d}V(x') \tag{1.46}$$

$$s_{ij} = \lambda \varepsilon_{rr}\delta_{ij} + \mu \varepsilon_{ij} \tag{1.47}$$

where σ_{ij} and ε_{ij} represent the nonlocal stress and strain tensors, respectively. ρ is the unit volume mass of atomic body, u is the motion component, f_j is the mass force, $\ddot{u}_j$ is the acceleration of the motion, $\alpha(|x - x'|, \varphi)$ is the nonlocal modulus, C_{ijkl} is the fourth-order elasticity tensor, $V(x')$ is the volume. Also, $|x - x'|$ denotes the distance in Euclidean form, δ_{ij} means the Kronecker Delta symbol. λ and μ are Lamé constants and are defined as follows:

$$\lambda = \frac{E\upsilon}{(1+\upsilon)(1-2\upsilon)}, \quad \mu = \frac{E}{2(1+\upsilon)} \tag{1.48}$$

where E is the modulus of elasticity, υ is the Poisson ratio, μ is the shear modulus. Expression in Eq. (1.46) defines the nondimensional atomic parameter:

$$\varphi = \frac{e_0 a}{L} \tag{1.49}$$

where e_0 is an atomic constant that can be determined experimentally, a is the characteristic internal length (such as crack width, distance or bond length between two carbon atoms, lattice length in the lattice model) and L is the external length of atomic body. It was calculated by Eringen [46] for the carbon atom as $e_0 = 0.39$.

Two-dimensional form of nonlocal module is expressed as follows:

$$\alpha(|x|, \varphi) = \frac{1}{2\pi(e_0 a)^2} K_0\left(\frac{\sqrt{\vec{x}\cdot\vec{x}}}{\varphi}\right) \tag{1.50}$$

where K_0 denotes the modified Bessel function and can be given as follows [59]:

$$K_0(r) = \frac{1}{2}\int_{-\infty}^{\infty} \left[\exp\left(-\sqrt{r^2+z^2}\right)/\sqrt{r^2+z^2}\right]\mathrm{d}z \tag{1.51}$$

On the other hand, Eq. (1.50) provides the following equation in most general form:

$$L_0\alpha(|\mathbf{x}-\mathbf{x}'|,\varphi)=\delta(|\mathbf{x}-\mathbf{x}'|) \tag{1.52}$$

where L_0 is the linear differential operator and δ is Dirac function. Eq. (1.52) can be expressed more clearly as follows:

$$L_0\sigma_{ij}=s_{ij} \tag{1.53}$$

The linear differential operator is defined as follows:

$$L_0=1-L^2\lambda^2\nabla^2=1-L^2\lambda^2\frac{\partial^2}{\partial x^2} \tag{1.54}$$

Substituting above expression into Eq. (1.53) yields the following equation:

$$\left(1-L^2\lambda^2\frac{\partial^2}{\partial x^2}\right)\sigma_{ij}=\lambda\varepsilon_{rr}\delta_{ij}+\mu\varepsilon_{ij} \tag{1.55}$$

By using this equation, stress-strain relations of nonlocal elasticity for linear, isotropic, and elastic solids are presented as follows:

$$\sigma_{ii}-(e_0a)^2\frac{\partial^2\sigma_{ii}}{\partial x^2}=s_{ii}=E\varepsilon_{ii} \tag{1.56}$$

$$\tau_{ij}-(e_0a)^2\frac{\partial^2\tau_{ij}}{\partial x^2}=t_{ij}=G\gamma_{ij} \tag{1.57}$$

where σ_{ii} and ε_{ii} explain the nonlocal axial stress and nonlocal axial strain, respectively. τ_{ij} and γ_{ij} denote the nonlocal shear stress and nonlocal shear strain, respectively. Finally, s_{ii} and t_{ij} mean the classical axial stress and classical shear stress, respectively.

References

1. Ş. Erkoç, *Nanoscience and Nanotechnology* (ODTÜ Publishing, Ankara, 2007) (In Turkish)
2. A. Tepe, *A Study of Small Scale Dimensions of Structures in Nonlocal Elasticity*. PhD Thesis, İstanbul Technical University, İstanbul, 2007. (In Turkish)
3. H. Bruus, *Introduction to Nanotechnology* (Technical University of Denmark, Lyngby, 2004)
4. M. Gürses, *Vibration and Bending Analysis of Nano-Sized Sector and Annular Sector Plates Via Nonlocal Elasticity Theory*. PhD Thesis, Akdeniz University, Antalya, 2015. (In Turkish)
5. Anonymous 1, C_n Fullerenes, http://www.nanotube.msu.edu/fullerene/fullerene-isomers.html
6. Anonymous 2, IBM (atoms), https://en.wikipedia.org/wiki/IBM_(atoms)
7. Anonymous 3, Nanocars, https://www.jmtour.com/about/photos_graphics/nanocars/

8. Anonymous 4, Nanotechnology in the World, https://nanoteknoloji.org/dunyada-nano-teknoloji/ (In Turkish)
9. Y. Özer, *Nanoscience and Nanotechnology: Determination of Effective Model from the Perspective of Efficiency/Security of Homeland*. MSc Thesis, Turkish Military Academy, Ankara, 2008. (In Turkish)
10. H.M. Numanoğlu, *Dynamic Analyses of Nano Scaled Continuous and Discrete Structures Based on Nonlocal Finite Element Method*. MSc Thesis, Akdeniz University, Antalya, 2019. (In Turkish)
11. M.S. Dresselhaus, G. Dresselhaus, R. Saito, Physics of carbon nanotubes. Carbon **33**(7), 883–891 (1995)
12. H. Rafii-Tabar, *Computational Physics of Carbon Nanotubes* (Cambridge University Press, Cambridge, 2008)
13. S. Iijima, Helical microtubules of graphitic carbon. Nature **354**, 56–58 (1991)
14. S. Iijima, T. Ichihashi, Single-shell carbon nanotubes of 1-nm diameter. Nature **363**, 603–605 (1993)
15. D.S. Bethune, C.H. Kiang, M.S. de Vries, G. Gorman, R. Savoy, J. Vazquez, R. Beyers, Cobalt-catalysed growth of carbon nanotubes with single-atomic-layer walls. Nature **363**, 605–607 (1993)
16. Ö. Civalek, B. Uzun, M.Ö. Yaylı, B. Akgöz, Size-dependent transverse and longitudinal vibrations of embedded carbon and silica carbide nanotubes by nonlocal finite element method. Eur. Phys. J. Plus **135**(4), 381 (2020)
17. M. Meyyappan, *Carbon Nanotubes: Science and Applications* (CRC Press, Boca Raton, FL, 2005)
18. H.S.P. Wong, D. Akinwande, *Carbon Nanotube and Graphene Device Physics* (Cambridge University Press, New York, 2011)
19. J. Li, Y. Lu, Q. Ye, S. Cinke, J. Han, M. Meyyappan, Carbon nanotube sensors for gas and organic vapor detection. Nano Lett. **3**(7), 929–933 (2003)
20. L. Schlapbach, A. Zuttel, Hydrogen-storage materials for mobile applications. Nature **414**, 353–358 (2001)
21. B. Kumanek, D. Janas, Thermal conductivity of carbon nanotube networks: A review. J. Mater. Sci. **54**(10), 7397–7427 (2019)
22. Anonymous 5, Applications of Carbon Nanotubes, https://www.azonano.com/article.aspx?ArticleID=4842
23. T. Rueckes, K. Kim, E. Joselevich, G.Y. Tseng, C.L. Cheung, M. Lieber, Carbon nanotube-based nonvolatile random access memory for molecular computing. Science **289**(5476), 94–97 (2000)
24. Ç. Işık, *Nonlocal Finite Element Formulation in Modeling Micro and Nano Scaled Mechanical Systems*. PhD Thesis, Akdeniz University, Antalya, 2018. (In Turkish)
25. R.S. Ruoff, D.C. Lorents, Mechanical and thermal properties of carbon nanotubes. Carbon **33**(7), 925–930 (1995)
26. N.G. Chopra, A. Zettl, Measurement of the elastic modulus of a multi-wall boron nitride nanotube. Solid State Commun. **105**(5), 297–300 (1997)
27. L. Hanley, J.L. Whitten, S.L. Anderson, Collision-induced dissociation and ab initio studies of boron cluster ions: Determination of structures and stabilities. J. Phys. Chem. **82**(25), 5803–5812 (1988)
28. M. Radosavljevic, J. Appenzeller, V. Derycke, R. Martel, P. Avouris, Electrical properties and transport in boron nitride nanotubes. Appl. Phys. Lett. **82**(23), 4131 (2003)
29. C.H. Lee, M. Xie, V. Kayastha, J. Wang, Y.K. Yap, Patterned growth of boron nitride nanotubes by catalytic chemical vapor deposition. Chem. Mater. **22**, 1782–1787 (2010)

30. S. Mukopadhyay, R.H. Scheicher, R. Pandey, S.P. Karna, Sensitivity of boron nitride nanotubes toward biomolecules of different polarities. Nanoparticles Nanostruct. **2**, 2442–2447 (2011)
31. S.H. Jhi, Y.K. Kwon, Hydrogen adsorption on boron nitride nanotubes: A path to room-temperature hydrogen storage. Phys. Rev. B **69**, 245407 (2004)
32. J. Kim, S. Lee, D. Seo, Y.S. Seo, Hydrogen adsorption on boron nitride nanotubes: A path to room-temperature hydrogen storage. Mater. Chem. Phys. **137**(1), 182–187 (2004)
33. C. Pham-Huu, N. Keller, G. Ehret, M.J. Ledoux, The first preparation of silicon carbide nanotubes by shape memory synthesis and their catalytic potential. J. Catal. **200**(2), 400–410 (2001)
34. X.H. Sun, C.P. Li, W.K. Wong, N.B. Wong, C.S. Lee, S.T. Lee, B.K. Teo, Formation of silicon carbide nanotubes and nanowires via reaction of silicon (from disproportionation of silicon monoxide) with carbon nanotubes. J. Am. Chem. Soc. **124**(48), 14464–14471 (2002)
35. P. Prakash, K.M. Sundaram, M.A. Bennet, A review on carbon nanotube field effect transistors (CNTFETs) for ultra-low power applications. Renew. Sust. Energ. Rev. **89**, 194–203 (2018)
36. K.I. Chen, B.R. Li, Y.T. Chen, Silicon nanowire field-effect transistor-based biosensors for biomedical diagnosis and cellular recording investigation. Nano Today **6**, 131–154 (2011)
37. A. Singh, M. Khosla, B. Raj, Comparative analysis of carbon nanotube field effect transistors, in *IEEE 4th Global Conference on Consumer Electronics, 27–30 October*, (2015), pp. 552–555
38. Y.Y. Liu, X.Y. Wang, Y. Cao, X.D. Chen, S.F. Xie, X.J. Zheng, H.D. Zeng, A flexible blue light-emitting diode based on ZnO nanowire/polyaniline heterojunctions. J. Nanomater. **2013**, 870254 (2013)
39. H.M. Numanoğlu, Ö. Civalek, On the torsional vibration of nanorods surrounded by elastic matrix via nonlocal FEM. Int. J. Mech. Sci. **161-162**, 105076 (2019)
40. G. Piazza, P.J. Stephanou, A.P. Pisano, Piezoelectric aluminum nitride vibrating contour-mode MEMS resonators. J. Microelectromech. Syst. **15**(6), 1406–1418 (2006)
41. L. Que, J.S. Park, Y.B. Gianchandani, Bent-beam electrothermal actuators–part I: Single beam and cascaded devices. J. Microelectromech. Syst. **10**(2), 247–254 (2001)
42. H.M. Numanoğlu, Ö. Civalek, On the dynamics of small-sized structures. Int. J. Eng. Sci. **145**, 103164 (2019)
43. O. Gürbulak, *The Investigation of Some Thermophysical Properties of n-Heptane+n-Eicosane Mixtures by Molecular Dynamics (MD) Simulation*. MSc Thesis, Ege University, İzmir. (In Turkish)
44. Ö. Civalek, H.M. Numanoğlu, Nonlocal finite element analysis for axial vibration of embedded love–bishop nanorods. Int. J. Mech. Sci. **188**, 105939 (2020)
45. E. Cosserat, F. Cosserat, *Théorie Des Corps Déformables* (Hermann and Sons, Paris, 1909)
46. A.C. Eringen, D.G.B. Edelen, On nonlocal elasticity. Int. J. Eng. Sci. **10**(3), 233–248 (1972)
47. A.C. Eringen, Linear theory of nonlocal elasticity and dispersion of plane waves. Int. J. Eng. Sci. **10**(5), 425–435 (1972)
48. A.C. Eringen, Edge dislocation in nonlocal elasticity. Int. J. Eng. Sci. **15**(3), 177–183 (1977)
49. A.C. Eringen, On differential equations of non local elasticity and solutions of screw dislocation and surface waves. J. Appl. Phys. **54**, 4703 (1983)
50. R.A. Toupin, Elastic materials with couple-stresses. Arch. Ration. Mech. Anal. **11**(1), 385–414 (1962)
51. R.D. Mindlin, H.F. Tiersten, Effects of couple-stresses in linear elasticity. Arch. Ration. Mech. Anal. **11**(1), 415–448 (1962)
52. W.T. Koiter, Couple stresses in the theory of elasticity, I & II. Philos. Trans. R. Soc. Lond. B. **67**, 17–44 (1964)
53. N.A. Fleck, J.W. Hutchinson, A phenomenological theory for strain gradient effects in plasticity. J. Mech. Phys. Solids **41**(12), 1825–1857 (1993)

54. D.C.C. Lam, F. Yang, A.C.M. Chong, J. Wang, P. Tong, Experiments and theory in strain gradient elasticity. J. Mech. Phys. Solids **51**(8), 1477–1508 (2003)
55. M.E. Gurtin, A.I. Murdoch, A continuum theory of elastic material surfaces. Arch. Ration. Mech. Anal. **57**(4), 291–323 (1975)
56. M.E. Gurtin, A.I. Murdoch, Surface stress in solids. Int. J. Solids Struct. **14**(6), 431–440 (1978)
57. V.T. Granik, J.W. Ferrari, Microstructural mechanics of granular media. Mech. Mater. **15**(4), 301–322 (1993)
58. M. Ferrari, V.T. Granik, A. Imam, J.C. Nadeau, *Advances in Doublet Mechanics* (Springer-Verlag, Berlin, 1997)
59. Y.W. Kwon, H. Bang, *The Finite Element Method Using MATLAB* (CRC Press, Boca Raton, FL, 2000)
60. R. Haberman, *Applied Partial Differential Equations: With Fourier Series and Boundary Value Problems* (Pearson, London, 2013)

Axial Vibration of Nonlocal Nanorods

2

2.1 Axial Elements

Axial (or called as "longitudinal") rods are structural members characterized by mechanical effects in the longitudinal direction. In the macrosense, it is the type of element that is generally encountered in planar and space truss elements and hauling cables. The deformation of axial rods should also be in the axial direction of rod. In many cases, while only axial stress and strains are sufficient to analyze the rod, shear stresses may also become meaningful sometimes in mechanical analysis.

In the nanosense, mechanical modeing of materials such as nanotubes, nanowires, and nanorods that can play a role in NEMS organizations as continuous axial rods is important for accurate design. Formations such as nanocantilever [1], nanobridge [2], nanomotor-mass [3], coaxial nanotube [4], atomic force microscope [5], nanotube growth [6] mostly require the using of continuous axial rod model in mechanical analysis.

It can be stated that there are different approaches to investigate the mechanics of axial rods. The most primitive rod formulation is the without shear (simple) rod theory. Additionally, Rayleigh [7] and Love-Bishop [8–10] axial rod theories focus on the effects of shear deformation. On the other hand, three-mode and two-mode Mindlin-Herrmann [11, 12] axial rod theories can also be mentioned.

In this chapter, free vibration analysis of axial nanorods is presented by using the nonlocal elasticity theory. Accordingly, nonlocal equations of motion of axial nanorods embedded in an axial elastic medium are derived and their solutions are carried out according to simple and Love-Bishop rod theories. The atomic size-dependent mechanical behavior of axial nanorods is investigated through several numerical examples.

Öm. Civalek et al., *Mechanical Behavior and Vibration of Nano-Scaled Rods, Beams, and Frames*, Synthesis Lectures on Engineering, Science, and Technology,
https://doi.org/10.1007/978-3-032-12023 6_2

2.2 Recent Contributions

Dynamic analysis of structural models such as axial nanorods of nano-scaled and one-dimensional materials has been extensively investigated through continuum theories based on atomic size effect. In these studies, it can be stated that the nonlocal elasticity theory is more prominent than other continuum theories. The beginning of studies on the nonlocal dynamics of axial nanorods goes back to the investigation of nonlocal free vibration formulations of two different boundary condition models such as clamped-clamped and clamped-free of these structures [13]. According to the studies on the nonlocal frequency equations, the fact that the atomic parameter originates an additional effect on the mass of body in general has exhibited that nonlocality is an effect reducing the natural frequencies. On the other hand, the frequencies of free axial vibration of carbon nanotubes embedded in a medium with elastic resistance against axial motion have also been presented [14]. Additionally, nonlocal free vibration analyses of axial nanorods with tip attachments such as elastic spring and mass have been performed [15]. Moreover, free vibration of axial rod model of carbon nanotube material with the buckyball mass attached at the tip has been given [16].

Nonlocal free vibration of axial nanorods based on Rayleigh theory has been examined with elastic medium [17] and without elastic medium [18]. In addition to this, free [19] and forced [20] vibration behaviors of nanorods with a single crack have been investigated. Moreover, it should be noted that nonlocal finite element solutions have been developed for the axial vibration behavior of simple axial nanorods [21, 22]. Free vibration of axial nanorods with a single crack is also focused on with finite elements [23]. Different from those, the effect of transverse magnetic field based on Maxwell's equation on the free vibration of embedded nanorods has been investigated [24]. Vibration analysis of tapered axial nanorods has also been performed [25, 26]. The natural frequencies of multi-segment axial nanorod embedded in the discrete elastic medium have been computed via the dynamic stiffness matrix method [27]. Nonlocal vibration analysis of elastically deformable axial rod whose position is variable along the span has been studied [28]. On the other hand, dynamics of tapered nanorods with nonlinear spring attachment at the tip and made of axially functionally graded material has also been given [29].

It should also be noted that the effects of lateral deformation (Poisson's effect) and lateral inertia on nonlocal vibration analyses of nanorods have been investigated. A finite difference formulation for nonlocal vibration of multiple axial rods based on Bishop's theory has been introduced [30]. Additionally, radial wave propagation in axial rods has been expressed for Love and Bishop rod theories [31]. On the other hand, a finite element formulation has been developed for the free vibration of nanorods embedded in the axial medium according to the nonlocal Love-Bishop rod theory [32]. Moreover, the nonlocal free vibration of axial nanorods with deformable boundary condition has been formulated by the Stokes' transform [33]. Finally, the nonlocal vibration of axial functionally graded nanorods has been investigated with the Bishop's theory [34, 35].

2.3 Nonlocal Vibration of Simple Rod

Simple (or elementary) axial rods are characterized by only axial deformation since the lateral strain effects are neglected. According to this, the rod shrinks or elongates only in the axial direction (longitudinal) under external influence. No deformation occurs in the directions of cross-section (transverse and lateral). This behavior formulation is more useful for axial rods that are not too short and not too high.

In this subsection, the nonlocal equation of motion of simple nanorods buried in elastic medium will first be derived according to the equilibrium approach and then the algebra of variation. Let begin with the equilibrium approach first.

2.3.1 Equation of Motion: Equilibrium Approach

Let consider a rod that is straight, homogeneous, with constant cross-section, and embedded in a fully axial elastic medium. The mentioned mechanical element model and its free body diagram are given in Fig. 2.1.

The axial equilibrium of rod can be specified as follows:

$$\sum F_{net} = \rho A\mathrm{d}x \frac{\partial^2 u}{\partial t^2} \tag{2.1}$$

where $u = u(x, t)$ denotes the axial displacement of rod. ρ and A represent the mass of unit volume and area of cross-section, respectively. t is the time component. In the axial members, the internal force is symbolized by force $N = N(x, t)$. For equilibrium analysis, the axial equilibrium of differential element with length dx in Fig. 2.1b can be written. According to this:

$$-N + (N + \mathrm{d}N) - k_M u\mathrm{d}x + q\mathrm{d}x = \rho A\mathrm{d}x \frac{\partial^2 w}{\partial t^2} \tag{2.2}$$

where $q = q(x, t)$ is the axial dynamic excitation. Also, k_M is called as the stiffness of axial elastic medium. By arranging the equation:

$$\frac{\partial N}{\partial x} = \rho A \frac{\partial^2 u}{\partial t^2} + k_M u - q \tag{2.3}$$

The derivation of nonlocal equation of motion is possible by integrating the constitutive equation of nonlocal elasticity via the section. That is:

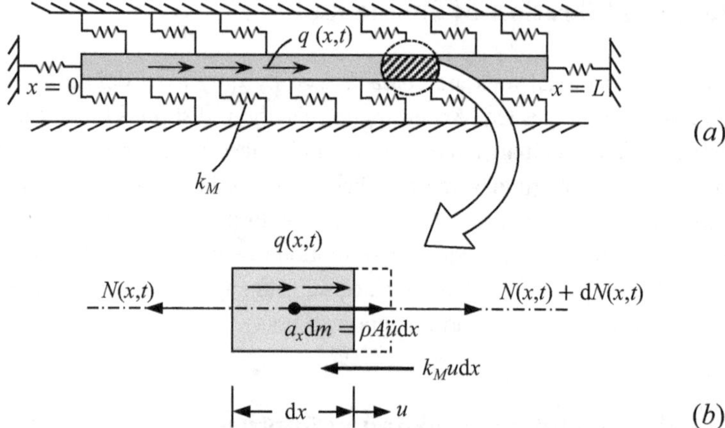

Fig. 2.1 Longitudinal vibration of a nanorod embedded on elastic medium. (**a**) Schematic representation of rod. (**b**) Free body diagram [36]

$$\left(1-(e_0a)^2\frac{\partial^2}{\partial x^2}\right)\int_A \sigma_{ii}\mathrm{d}A=\int_A s_{ii}\mathrm{d}A \tag{2.4}$$

where following expressions are written:

$$\int_A \sigma_{xx}\mathrm{d}A=N_{nl}=N, \qquad \int_A s_{xx}\mathrm{d}A=N_c=EA\frac{\partial u}{\partial x} \tag{2.5}$$

where E defines the modulus of elasticity. Firstly, after substituting Eq. (2.5) into Eq. (2.4), let's derive the expression and use Eq. (2.3). Thus, we can reach the following expression:

$$\begin{aligned}&\left[EA+(e_0a)^2k_M\right]\frac{\partial^2u}{\partial x^2}-\rho A\frac{\partial^2u}{\partial t^2}-k_Mu+(e_0a)^2\rho A\frac{\partial^4u}{\partial x^2\partial t^2}\\&+q-(e_0a)^2\frac{\partial^2q}{\partial x^2}=0\end{aligned} \tag{2.6}$$

This expression specifies the nonlocal equation of motion. However, it will be seen later that a mechanical boundary condition is also required to determine the size effect mechanical behavior of nanorods. The mechanical boundary condition of simple axial nanorods is the internal effect of axial force. According to this, after Eqs. (2.3) and (2.5) are replaced into Eq. (2.4), the axial force is obtained as follows:

$$N=\left[EA+(e_0a)^2k_M\right]\frac{\partial u}{\partial x}+(e_0a)^2\rho A\frac{\partial^3u}{\partial x\partial t^2}-(e_0a)^2\frac{\partial^2q}{\partial x^2} \tag{2.7}$$

2.3.2 Equation of Motion: Hamilton's Principle

Energy methods investigate the extreme value of integral of a functional between certain bounds due to the stationarity condition. Variational algebra examines the variation of a function within admissible ranges and provides the necessary algebraic operations of the integral whose extreme value is investigated.

In dynamical problems according to variational algebra, Hamilton's Principle that considers the strain energy of structure, its kinetic energy, and the work ability on the structure of external force, is used.

The first application of Hamilton's principle is given in this section. According to this, the first variation of potential energy of nonlocal nanorod in the (t_1, t_2) time interval should be equal to 0.

$$\delta\Pi = \delta \int_{t_1}^{t_2} [U - (T + W)]\mathrm{d}t = 0 \tag{2.8}$$

where U, T, and W define the strain energy, kinetic energy, and work energy, respectively. In order to establish energy expressions, the kinematic relations of nanorod, namely displacements, stresses, and deformations, should be known. The displacement components of a simple axial nanorod are as follows:

$$u(x,t) = u, \quad v(x,t) = 0, \quad w(x,t) = w \tag{2.9}$$

where u is the axial displacement, v is the lateral displacement, and w is the transverse displacement. Since Poisson's effects are neglected in the rod, lateral and transverse displacements do not occur during axial displacement. On the other hand, the strain expressions of rod are written as follows:

$$\begin{aligned}
&\varepsilon_{xx} = \frac{1}{2}\left(\frac{\partial u}{\partial x} + \frac{\partial u}{\partial x}\right) = \frac{\partial u}{\partial x}, \quad \varepsilon_{yy} = \frac{1}{2}\left(\frac{\partial v}{\partial y} + \frac{\partial v}{\partial y}\right) = 0, \quad \varepsilon_{zz} = \frac{1}{2}\left(\frac{\partial w}{\partial z} + \frac{\partial w}{\partial z}\right) = 0, \\
&\gamma_{xy} = \frac{\partial u}{\partial y} + \frac{\partial v}{\partial x} = 0, \quad \gamma_{xz} = \frac{\partial u}{\partial z} + \frac{\partial w}{\partial x} = 0, \quad \gamma_{yz} = \frac{\partial w}{\partial y} + \frac{\partial v}{\partial z} = 0
\end{aligned} \tag{2.10}$$

where ε_{xx}, which is the only nonzero strain component, is the axial strain. ε_{yy} and ε_{zz} are lateral and transverse strains, respectively. Also, ε_{xy}, ε_{xz}, and ε_{yz} denote the shear strains. On the other hand, stress components can be specified as follows:

$$\sigma_{xx}=E\varepsilon_{xx}=E\frac{\partial u}{\partial x},\quad \sigma_{yy}=\sigma_{zz}=\sigma_{xy}=\sigma_{xz}=\sigma_{yz}=0, \tag{2.11}$$

where σ_{xx} defines the axial stress. On the other hand, σ_{yy} is the lateral stress and σ_{zz} is the transverse stress. σ_{xy}, σ_{xz}, and σ_{yz} represent the shear stresses.

Energy expressions are constituted as follows:

$$\begin{aligned} U&=\frac{1}{2}\int_V\left(\sigma_{xx}\varepsilon_{xx}+\sigma_{yy}\varepsilon_{yy}+\sigma_{zz}\varepsilon_{zz}+2\sigma_{xy}\gamma_{xy}+2\sigma_{xz}\gamma_{xz}+2\sigma_{yz}\gamma_{yz}\right)\mathrm{d}V\\ &+\frac{1}{2}\int_0^L k_M u^2\mathrm{d}x+\frac{1}{2}\sum_{i=1}^{n}k_{a,i}[u(x_i,t)]^2\\ &=\frac{1}{2}\int_0^L N\frac{\partial u}{\partial x}\mathrm{d}x+\frac{1}{2}\int_0^L k_M u^2\mathrm{d}x+\frac{1}{2}k_a[u(0,t)]^2+\frac{1}{2}k_a[u(L,t)]^2 \end{aligned} \tag{2.12}$$

$$\begin{aligned} T&=\frac{1}{2}\int_V\rho\left[\left(\frac{\partial u}{\partial t}\right)^2+\left(\frac{\partial v}{\partial t}\right)^2+\left(\frac{\partial w}{\partial t}\right)^2\right]\mathrm{d}V+\frac{1}{2}\sum_{i=1}^{n}m_{a,i}\left(\frac{\partial u(x_i,t)}{\partial t}\right)^2\\ &=\frac{1}{2}\int_0^L\rho A\left(\frac{\partial u}{\partial t}\right)^2\mathrm{d}x+\frac{1}{2}m_a\left(\frac{\partial u(0,t)}{\partial t}\right)^2+\frac{1}{2}m_a\left(\frac{\partial u(L,t)}{\partial t}\right)^2 \end{aligned} \tag{2.13}$$

$$W=\int_0^L qu\mathrm{d}x \tag{2.14}$$

In Eqs. (2.13) and (2.14), $k_{a,\,i}$ defines the stiffness of generalized tip spring attachment and $m_{a,\,i}$ defines the mass of generalized tip attachment. In this section, since the attachments are assumed to be at the ends of nanorod, the equations are arranged as seen.

The first variations of energy expressions are given as follows:

$$\begin{aligned} \int_{t_1}^{t_2}\delta U\mathrm{d}t&=\int_{t_1}^{t_2}\int_0^L N\frac{\partial u}{\partial x}\mathrm{d}x\mathrm{d}t+\int_{t_1}^{t_2}\int_0^L k_M u\delta u\mathrm{d}x\mathrm{d}t+\int_{t_1}^{t_2}k_a u(0,t)\delta u(0,t)\mathrm{d}t\\ &+\int_{t_1}^{t_2}k_a u(L,t)\delta u(L,t)\mathrm{d}t \end{aligned} \tag{2.15}$$

$$\int_{t_1}^{t_2} \delta T \mathrm{d}t = \int_{t_1}^{t_2}\int_0^L \rho A \frac{\partial u}{\partial t}\delta\frac{\partial u}{\partial t}\mathrm{d}x\mathrm{d}t + \int_{t_1}^{t_2} m_a \frac{\partial u(0,t)}{\partial t}\delta\frac{\partial u(0,t)}{\partial t}\mathrm{d}t + \int_{t_1}^{t_2} m_a \frac{\partial u(L,t)}{\partial t}\delta\frac{\partial u(L,t)}{\partial t}\mathrm{d}t \tag{2.16}$$

$$\int_{t_1}^{t_2} \delta W \mathrm{d}t = \int_{t_1}^{t_2}\int_0^L q\delta u \mathrm{d}x\mathrm{d}t \tag{2.17}$$

Variation expressions here should be edited as follows:

$$\int_0^L N\delta\frac{\partial u}{\partial x}\mathrm{d}x = N\delta u|_0^L - \int_0^L \frac{\partial N}{\partial x}\delta u\mathrm{d}x,\quad \int_{t_1}^{t_2} \rho A\frac{\partial u}{\partial t}\delta\frac{\partial u}{\partial t}\mathrm{d}t = \rho A\frac{\partial u}{\partial t}\delta u\Big|_{t_1}^{t_2} - \int_{t_1}^{t_2}\rho A\frac{\partial^2 u}{\partial t^2}\delta u\mathrm{d}t,$$
$$\int_{t_1}^{t_2} m_a\frac{\partial u(0,t)}{\partial t}\delta\frac{\partial u(0,t)}{\partial t}\mathrm{d}t = m_a\frac{\partial u(0,t)}{\partial t}\delta u(0,t)\Big|_{t_1}^{t_2} - \int_{t_1}^{t_2} m_a\frac{\partial^2 u(0,t)}{\partial t^2}\delta u(0,t)\mathrm{d}t,$$
$$\int_{t_1}^{t_2} m_a\frac{\partial u(L,t)}{\partial t}\delta\frac{\partial u(L,t)}{\partial t}\mathrm{d}t = m_a\frac{\partial u(L,t)}{\partial t}\delta u(L,t)\Big|_{t_1}^{t_2} - \int_{t_1}^{t_2} m_a\frac{\partial^2 u(L,t)}{\partial t^2}\delta u(L,t)\mathrm{d}t. \tag{2.18}$$

Using Eqs. (2.15)–(2.18) into Eq. (2.8) yields the following equation:

$$\int_{t_1}^{t_2}\int_0^L\left(-\frac{\partial N}{\partial x} + k_M u + \rho A\frac{\partial^2 w}{\partial t^2} - q\right)\delta u\mathrm{d}x\mathrm{d}t + \int_{t_1}^{t_2} N\delta u|_0^L\mathrm{d}t + \int_{t_1}^{t_2}\left(k_a u(0,t) + m_a\frac{\partial^2 u}{\partial t^2}(0,t)\right)\delta u(0,t)\mathrm{d}t + \int_{t_1}^{t_2}\left(k_a u(L,t) + m_a\frac{\partial^2 u}{\partial t^2}(L,t)\right)\delta u(L,t)\mathrm{d}t$$
$$+\int_0^L\left(-\rho A\frac{\partial u}{\partial t}\right)\delta u\Big|_{t_1}^{t_2}\mathrm{d}x + \left(-m_a\frac{\partial u}{\partial t}(0,t)\right)\delta u(0,t)\Big|_{t_1}^{t_2} + \left(-m_a\frac{\partial u}{\partial t}(L,t)\right)\delta u(L,t)\Big|_{t_1}^{t_2} = 0 \tag{2.19}$$

where all parts of equation should be equal to zero. Therefore, first, the following expression is written for the $\delta u \neq 0$ limit in uppermost double integral expression:

$$\frac{\partial N}{\partial x} = k_M u + \rho A \frac{\partial^2 w}{\partial t^2} - q \tag{2.20}$$

It should be noted that this equation is the same as Eq. (2.3). To remember, if Eqs. (2.4) and (2.5) are used here, Eq. (2.6) that expresses the equation of motion could be obtained. On the other hand, by neglecting the attachments, the axial force is as in Eq. (2.7). The internal effects of attached nanorods will be examined in the next subsections.

2.3.3 Solution of Equation of Motion

Firstly, in Eq. (2.6) that is the equation of motion of nonlocal vibration of embedded axial nanorods, $q = 0$ for free vibration is known. The resulting equation is a homogeneous partial differential equation. The following separation of variable expression is used to solve such a differential equation:

$$u(x,t) = U(x)T(t) = U(x)\sin(\omega t - \theta) \tag{2.21}$$

where $u = u(x, t)$ is the motion of axial vibration, $U(x)$ is the mode equation and $T(t)$ is a harmonic function. θ is the phase angle and ω is the natural frequency. This section focuses on the frequency calculations for nonlocal free vibration. After Eq. (2.21) is replaced into Eq. (2.6), the $\sin(\omega t - \theta)$ harmonic expression is simplified and the following expression is attained:

$$\frac{\mathrm{d}^2 U(x)}{\mathrm{d}x^2} + \eta^2 U(x) = 0 \tag{2.22}$$

In which,

$$\eta^2 = \frac{-k_M + \omega^2 \rho A}{EA + (e_0 a)^2 k_M - \omega^2 (e_0 a)^2 \rho A} \tag{2.23}$$

Equation (2.22) is an ordinary, homogeneous, and constant coefficient differential equation. In this case, the solution is as follows:

$$U(x) = Ce^{kx} \tag{2.24}$$

By substituting Eq. (2.24) into Eq. (2.22), the characteristic polynomial of Eq. (2.22) is obtained as follows:

$$k^2 + \eta^2 = 0 \tag{2.25}$$

where the roots of equation can be written as follows:

$$k_1 = \mathbf{i}\eta, \quad k_2 = -\mathbf{i}\eta, \tag{2.26}$$

So, the solution transforms into the following form:

$$U(x) = C_1 e^{\mathbf{i}\eta x} + C_2 e^{-\mathbf{i}\eta x} \tag{2.27}$$

where C_1 and C_2 are the constant numbers. The following transformations can be used to arrange Eq. (2.27):

$$e^{\mathbf{i}\theta} = \cos\theta + \mathbf{i}\sin\theta \tag{2.28}$$

$$e^{-\mathbf{i}\theta} = \cos\theta - \mathbf{i}\sin\theta \tag{2.29}$$

With the using of these expressions into Eq. (2.27), the mode shape takes its final form as follows:

$$U(x) = C_1 \cos\eta x + C_2 \sin\eta x \tag{2.30}$$

2.3.4 Boundary Conditions of Nanorods

The nonlocal free vibration behavior of axial nanorods is formulated with the differential equation given in Eq. (2.6). Free vibration behavior can be explained with the help of Eq. (2.30), which is the solution of this differential equation. Unknown constants or an explicit solution of solution of Eq. (2.30) should be determined. The solution depends on the boundary conditions of the nanorod. In this subsection, general supporting such as free and clamped and dynamic ends such as spring and mass are examined. Next, the nonlocal free vibration behavior of nanorods without Poisson's effect is investigated with several examples given.

1. **Clamped end (C):** Clamped ends do not allow axial displacement. Therefore, the boundary condition is specified as follows:

$$U = 0 \tag{2.31}$$

2. **Free end (F):** If there is no axial dynamic external excitation at the free ends, internal effect of axial force does not occur. Then, considering Eq. (2.7),

$$N = \left[EA + (e_0a)^2 k_M\right] \frac{\partial u}{\partial x} + (e_0a)^2 \rho A \frac{\partial^3 u}{\partial x \partial t^2} = 0 \tag{2.32}$$

If Eq. (2.21) is employed here, for the modal part,

$$B \frac{dU}{dx} = 0 \tag{2.33}$$

Where,

$$B = EA + (e_0a)^2 k_M - \omega^2 (e_0a)^2 \rho A \tag{2.34}$$

Since B cannot be equal to zero

$$U' = 0 \tag{2.35}$$

3. **Attached ends (MA and SA):** End attachment affects the axial force boundary condition of nanorod. End attachments can be considered as an axial elastic spring or mass effect. For example, suppose the nanorod has both mass and spring attachments. According to this, the following expression should be written in Eq. (2.30) whose sum can be 0:

$$\int_{t_1}^{t_2} N\delta u|_0^L dt + \int_{t_1}^{t_2} \left(k_a u(0,t) + m_a \frac{\partial^2 u}{\partial t^2}(0,t)\right) \delta u(0,t) dt + \int_{t_1}^{t_2} (k_a u(L,t) = 0$$
$$+ m_a \frac{\partial^2 u}{\partial t^2}(L,t)\Big) \delta u(L,t) dt = 0 \tag{2.36}$$

Firstly, the following expression can be formed for the right end of the nanorod, namely the $\delta u(L,t) \neq 0$ boundary

$$N(L,t) = -k_a u(L,t) - m_a \frac{\partial^2 u}{\partial t^2}(L,t) \tag{2.37}$$

where the dynamic boundary condition is written with the help of Eqs. (2.21) and (2.33), as follows:

$$\left[EA + (e_0a)^2 k_M - \omega^2 (e_0a)^2 \rho A\right] U'(L) = \left(-k_a + \omega^2 m_a\right) U(L) \tag{2.38}$$

Also, the following equation is given for the left end of nanorod, namely the $\delta u(0, t) \neq 0$ boundary

Fig. 2.2 Clamped nanorod embedded in the elastic medium

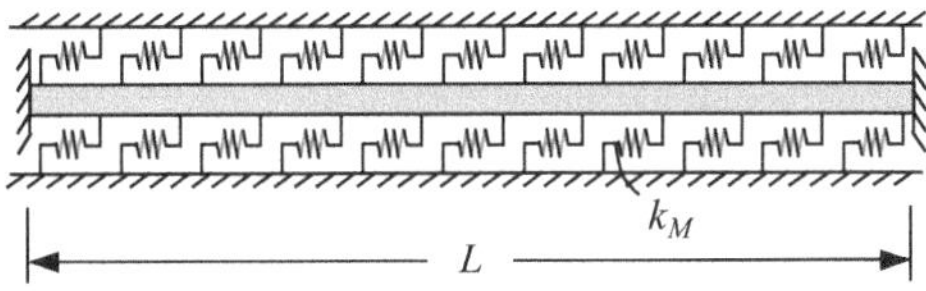

$$N(0,t) = k_a u(0,t) + m_a \frac{\partial^2 u}{\partial t^2}(0,t) \tag{2.39}$$

Thus, the following expression is attained:

$$\left[EA + (e_0 a)^2 k_M - \omega^2 (e_0 a)^2 \rho A\right] U'(0) = \left(k_a - \omega^2 m_a\right) U(0) \tag{2.40}$$

It is impossible to obtain the nonlocal frequency equation directly in attached nanorods, but the roots of frequency equation can be calculated by programming on the computer.

Example 2.1 Determine the frequency equation of nonlocal free vibration of clamped nanorod embedded in the elastic medium shown in Fig. 2.2.

Solution. Boundary conditions are given as follows:

$$U(0) = 0, \quad U(L) = 0 \tag{2.E2.1.1}$$

By using the first boundary condition in Eq. (2.30), the following expression is obtained:

$$C_1 \cos 0 + C_2 \sin 0 = 0 \tag{2.E2.1.2}$$

Thus, one of the solution constants is calculated as follows:

$$C_1 = 0 \tag{2.E2.1.3}$$

If the second boundary condition in Eq. (2.E2.1.1) is applied to the remain of mode equation:

$$C_2 \sin \eta L = 0 \tag{2.E2.1.4}$$

where C_2 cannot be equal to 0. In that case,

$$\sin \eta L = 0 \tag{2.E2.1.5}$$

The periodic solution of this trigonometric equation is presented as:

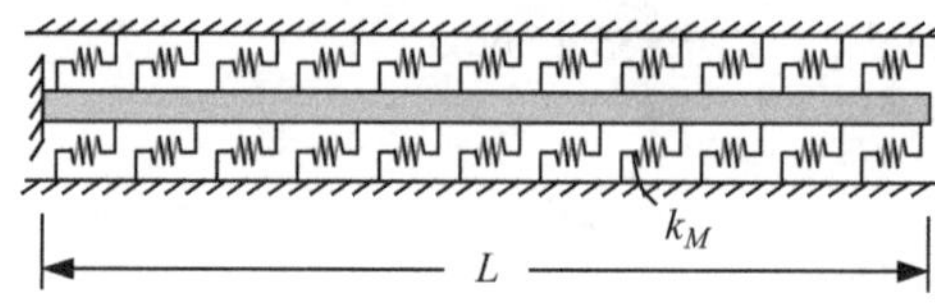

Fig. 2.3 Clamped-free nanorod embedded in the elastic medium

$$\eta L = n\pi \tag{2.E2.1.6}$$

where n is the number of modes. If Eq. (2.23) is used into the square of Eq. (2.E2.1.7), frequency equation is obtained:

$$\omega = \sqrt{\frac{\left[EA + (e_0 a)^2 k_M\right]\left(\frac{n\pi}{L}\right)^2 + k_M}{\rho A\left[1 + (e_0 a)^2\left(\frac{n\pi}{L}\right)^2\right]}} \tag{2.E2.1.7}$$

Example 2.2 Determine the frequency equation of nonlocal free vibration of clamped-free nanorod embedded in the elastic medium shown in Fig. 2.3.

Solution. The boundary conditions for this nanorod are introduced as follows:

$$U(0) = 0, \quad U'(L) = 0 \tag{2.E2.2.1}$$

The result of first boundary condition is known from Eq. (2.E2.1.3). Applying the other boundary condition to the remain of mode equation gives the following expression:

$$C_2 \eta \cos \eta L = 0 \tag{2.E2.2.2}$$

In Example 2.1, it is understood that C_1 is not equal to zero. In addition to this, η cannot be equal to zero. Especially if the elastic environment is neglected, since η is equal to zero, the natural frequency is also equal to zero. The following equation can be written as the last case:

$$\cos \eta L = 0 \tag{2.E2.2.3}$$

The periodic solution of equation is as follows:

$$\eta L = n\pi \tag{2.E2.2.4}$$

Similar to the solution of Example 2.1, the frequency equation is obtained as follows:

$$\omega = \sqrt{\frac{\left[EA + (e_0a)^2 k_M\right]\left(\frac{(2n-1)\pi}{L}\right)^2 + k_M}{\rho A\left[1 + (e_0a)^2\left(\frac{(2n-1)\pi}{L}\right)^2\right]}} \tag{2.E2.2.5}$$

Several numerical results about natural frequency calculation are presented in Example 2.6.

The nonlocal free vibrational frequencies of clamped or clamped-free nanorods can be obtained by a series expansion as well as analytical solution. It is sufficient to use the following series expansions that satisfy the boundary conditions into Eq. (2.6):

$$\begin{aligned} &\text{Clamped-Free}: u(x,t) = \sum_{n=1}^{\infty} u_n \sin\left(\frac{(2n-1)\pi x}{2L}\right)\sin(\omega t - \theta) \\ &\text{Clamped-Clamped}: u(x,t) = \sum_{n=1}^{\infty} u_n \sin\left(\frac{n\pi x}{L}\right)\sin(\omega t - \theta) \end{aligned} \tag{2.41}$$

For example, let's substitute the series expansion for the clamped nanorod into Eq. (2.6):

$$\begin{aligned} &-\left[EA + (e_0a)^2 k_M\right]\left(\frac{n\pi}{L}\right)^2 \sum_{n=1}^{\infty} u_n \sin\left(\frac{n\pi x}{L}\right)\sin(\omega t - \theta) \\ &+\omega^2 \rho A \sum_{n=1}^{\infty} u_n \sin\left(\frac{n\pi x}{L}\right)\sin(\omega t - \theta) - k_M \sum_{n=1}^{\infty} u_n \sin\left(\frac{n\pi x}{L}\right)\sin(\omega t - \theta) \\ &-\omega^2 (e_0a)^2 \rho A\left(\frac{n\pi}{L}\right)^2 \sum_{n=1}^{\infty} u_n \sin\left(\frac{n\pi x}{L}\right)\sin(\omega t - \theta) = 0 \end{aligned} \tag{2.42}$$

where Eq. (2.E2.2.5) is reached with the simplification of series expressions.

Example 2.3 Determine the frequency equation of nonlocal free vibration of clamped-mass attached nanorod embedded in the elastic medium shown in Fig. 2.4.

Solution. The boundary conditions of nanorod are given below:

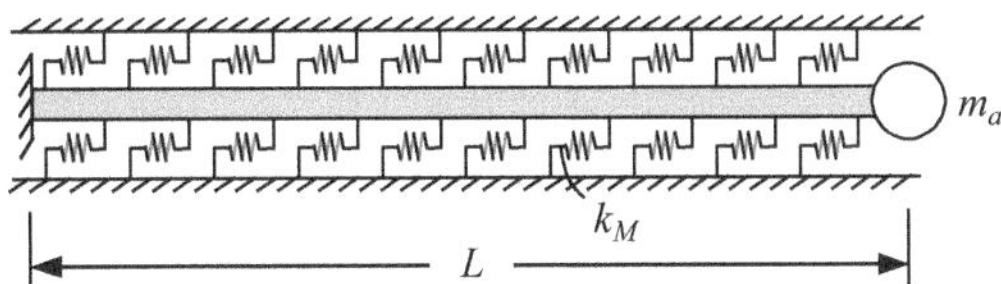

Fig. 2.4 Clamped-mass attached nanorod embedded in the elastic medium

$$U(0) = 0, \quad RU'(L) = \omega^2 m_a U(L) \tag{2.E2.3.1}$$

According to the result of first boundary condition, $C_1 = 0$. Let apply the second boundary condition:

$$R\eta C_2 \cos \eta L = \omega^2 m_a C_2 \sin \eta L \tag{2.E2.3.2}$$

This expression takes the following form after some mathematical operations:

$$R\eta \cot \eta L = \omega^2 m_a \tag{2.E2.3.3}$$

Since the manual solution of this equation is difficult, the frequencies can be calculated by symbolic programming on the computer.

As a special case, the elastic environment parameter is assumed as zero. Then Eqs. (2.33) and (2.23) transform the following forms, respectively:

$$R = EA - \omega^2 (e_0 a)^2 \rho A \tag{2.E2.3.4}$$

$$\eta^2 = \frac{\omega^2 \rho A}{EA - \omega^2 (e_0 a)^2 \rho A} = \frac{\omega^2 \rho A}{R} \tag{2.E2.3.5}$$

Let substitute these expressions into Eq. (2.E2.3.2) and perform some mathematical operations:

$$\cot \eta L = \eta L \beta_m \tag{2.E2.3.6}$$

where the following definition is valid:

$$\beta_m = \frac{m_a}{\rho AL} \tag{2.E2.3.7}$$

Example 2.4 Determine the frequency equation of nonlocal free vibration of free-spring attached nanorod embedded in the elastic medium shown in Fig. 2.5.

Solution. The boundary conditions of nanorod are presented below:

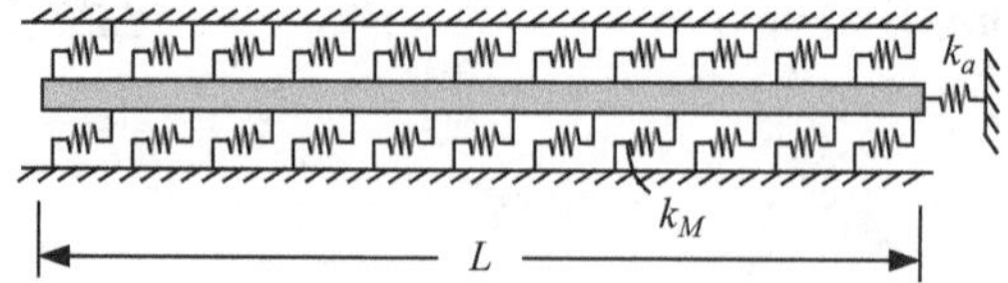

Fig. 2.5 Free-spring attached nanorod embedded in the elastic medium

$$U'(0) = 0, \quad RU'(L) = -k_a U(L) \tag{2.E2.4.1}$$

From the first boundary condition, $C_2 = 0$ is resulted. Let's apply the second boundary condition:

$$-R\eta C_1 \sin \eta L = -k_a C_1 \cos \eta L \tag{2.E2.4.2}$$

If this expression is edited, the frequency equation is obtained as follows:

$$R\eta \tan \eta L = k_a \tag{2.E2.4.3}$$

This is expressed as follows after some mathematical operations:

$$\eta L \tan \eta L = \left(1 + (e_0 a)^2 \eta^2\right) \beta_k \tag{2.E2.4.4}$$

Similar to Example 2.3, programming on the computer should be carried out because the solution of Eq. (2.E2.4.4) via manual solution is difficult. The following definition is valid:

$$\beta_k = \frac{k_a L}{EA} \tag{2.E2.4.5}$$

2.4 Nonlocal Vibration of Love-Bishop Rods

In the Love-Bishop axial rod theory, the rod does not only axial motion but also perform lateral motion due to axial motion (Poisson's effect). The shear inertia from lateral motion and higher-order mass inertia are also considered. Since the Poisson's effect becomes important in cases where the beam height is high and the beam length is low, the determination of nonlocal dynamic behavior in systems modeled as atomic scale and axial rod becomes more meaningful by using the Love-Bishop theory.

Analogies of behavior theories can be made between axial rod theories and bending rods (beams). According to this, since the simple theory in the axial rods does not consider the shear effects, it corresponds to the Euler-Bernoulli theory in the beams. Love-Bishop rod theory is like the Timoshenko beam theory due to the shear effects and mass inertias.

In this subsection, considering the Love-Bishop axial rod theory, the equation of motion of free vibration of nonlocal axial nanorods is derived and solved based on the algebra of variation. Applications of this solution are shown on some examples.

2.4.1 Derivation of Equation of Motion

Only Hamilton's principle will be used to derive the equation of motion. First, let's consider the nanorod model embedded in an elastic medium in Fig. 2.6. Let's start by writing the displacement components of the nanorod:

$$u(x,t)=u, \quad v(x,t)=-\upsilon y\frac{\partial u}{\partial x}, \quad w(x,t)=-\upsilon z\frac{\partial u}{\partial x} \tag{2.43}$$

where u, v, and, w denote axial, lateral, and transverse displacements, respectively. υ is Poisson's ratio. The deformation expressions of rod are as follows:

$$\begin{aligned}
&\varepsilon_{xx}=\frac{1}{2}\left(\frac{\partial u}{\partial x}+\frac{\partial u}{\partial x}\right)=\frac{\partial u}{\partial x},\ \varepsilon_{yy}=\frac{1}{2}\left(\frac{\partial v}{\partial y}+\frac{\partial v}{\partial y}\right)=-\upsilon\frac{\partial u}{\partial x},\ \varepsilon_{zz}=\frac{1}{2}\left(\frac{\partial w}{\partial z}+\frac{\partial w}{\partial z}\right)=-\upsilon\frac{\partial u}{\partial x}\\
&\varepsilon_{xy}=\frac{1}{2}\left(\frac{\partial u}{\partial y}+\frac{\partial v}{\partial x}\right)=-\frac{1}{2}\upsilon y\frac{\partial^2 u}{\partial x^2},\ \varepsilon_{xz}=\frac{1}{2}\left(\frac{\partial u}{\partial z}+\frac{\partial w}{\partial x}\right)=-\frac{1}{2}\upsilon z\frac{\partial^2 u}{\partial x^2},\ \varepsilon_{yz}=\frac{1}{2}\left(\frac{\partial w}{\partial y}+\frac{\partial v}{\partial z}\right)=0
\end{aligned} \tag{2.44}$$

where ε_{xx}, ε_{yy}, and ε_{zz} are axial, lateral, and transverse strain components, respectively. In addition to this, ε_{xy}, ε_{xz}, and ε_{yz} denote shear deformations. On the other hand, classical stress components can be written as follows:

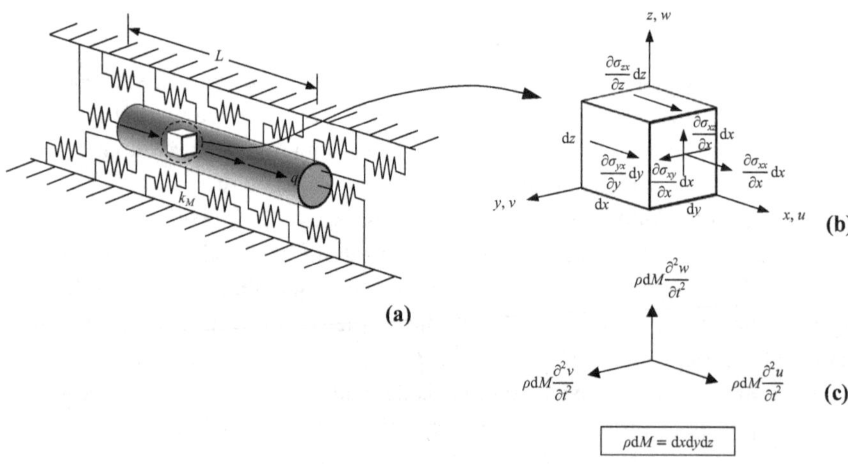

Fig. 2.6 Longitudinal vibration of Love-Bishop nanorod embedded on elastic medium. (**a**) Schematic representation of rod. (**b**) Net internal stresses on a differential prismatic element. (**c**) Newton force [32]

$$s_{xx} = E\varepsilon_{xx} = E\frac{\partial u}{\partial x}, \; s_{xy} = 2G\varepsilon_{xy} = -G\upsilon y\frac{\partial^2 u}{\partial x^2}, \; s_{xz} = 2G\varepsilon_{xz} = -G\upsilon z\frac{\partial^2 u}{\partial x^2}, \; s_{yy} = s_{zz} = s_{yz} = 0, \tag{2.45}$$

where s_{xx} denotes longitudinal stress and s_{xy}, s_{xz} specifies shear stresses from nonzero stress components.

Equation of dynamic equilibrium should be known in this rod formulation before the energy equations are constituted. Since only the axial stress plays a role in a simple rod, the longitudinal stress can be obtained directly by using Eq. (2.3) and thus the vibration analysis can be performed more simply. However, since there are stresses that work to shear the section in the axial direction in the Love-Bishop rod, to establish the equations of dynamic equilibrium is important. According to this [30, 32]:

$$\frac{\partial \sigma_{xx}}{\partial x} + \frac{\partial \sigma_{xy}}{\partial y} + \frac{\partial \sigma_{xz}}{\partial z} = \rho\frac{\partial^2 u}{\partial t^2} - \rho\tilde{q}_x + \frac{k_M}{A}u \tag{2.46}$$

$$\frac{\partial \sigma_{xy}}{\partial x} = \rho\frac{\partial^2 v}{\partial t^2} - \rho\tilde{q}_y \tag{2.47}$$

$$\frac{\partial \sigma_{xz}}{\partial x} = \rho\frac{\partial^2 w}{\partial t^2} - \rho\tilde{q}_z \tag{2.48}$$

Here, k_M is the stiffness of elastic medium and since axial vibration is investigated, it is assumed that there is an elastic rigid medium against longitudinal displacement only. On the other hand, q_x, q_y, and q_z define the mass forces:

$$\tilde{q}_x = \frac{q_x}{\rho A} = \frac{q}{\rho A}, \quad \tilde{q}_y = \frac{q_y}{\rho A}, \quad \tilde{q}_z = \frac{q_z}{\rho A} \tag{2.49}$$

For the x-axis only, since $q_y = q_z = 0$ in the axial vibration analysis, $\tilde{q}_y = \tilde{q}_z = 0$ becomes.

On the other hand, energy expressions are formulated as follows:

$$\begin{aligned} U = {} & \frac{1}{2}\int_V \left(\sigma_{xx}\varepsilon_{xx} + \sigma_{yy}\varepsilon_{yy} + \sigma_{zz}\varepsilon_{zz} + 2\sigma_{xy}\varepsilon_{xy} + 2\sigma_{xz}\varepsilon_{xz} + 2\sigma_{yz}\varepsilon_{yz}\right)\mathrm{d}V \\ & + \frac{1}{2}\int_0^L k_M u^2\mathrm{d}x + \frac{1}{2}\sum_{i=1}^{n} k_{a,i}[u(x_i,t)]^2 \\ = {} & \frac{1}{2}\int_0^L \left(N\frac{\partial u}{\partial x} - P\frac{\partial^2 u}{\partial x^2}\right)\mathrm{d}x + \frac{1}{2}\int_0^L k_M u^2\mathrm{d}x + \frac{1}{2}k_a[u(0,t)]^2 + \frac{1}{2}k_a[u(L,t)]^2 \end{aligned} \tag{2.50}$$

$$
\begin{aligned}
T &= \frac{1}{2}\int_V \rho\left[\left(\frac{\partial u}{\partial t}\right)^2 + \left(\frac{\partial v}{\partial t}\right)^2 + \left(\frac{\partial w}{\partial t}\right)^2\right]\mathrm{d}V + \frac{1}{2}\sum_{i=1}^{n} m_{a,i}\left(\frac{\partial u(x_i,t)}{\partial t}\right)^2 \\
&= \frac{1}{2}\int_0^L \left(\rho A\left(\frac{\partial u}{\partial t}\right)^2 + \rho I_P \upsilon^2\left(\frac{\partial^2 u}{\partial x \partial t}\right)^2\right)\mathrm{d}x + \frac{1}{2} m_a\left(\frac{\partial u(0,t)}{\partial t}\right)^2 \\
&\quad + \frac{1}{2} m_a\left(\frac{\partial u(L,t)}{\partial t}\right)^2
\end{aligned}
\tag{2.51}
$$

$$
W = \int_0^L qu\mathrm{d}x \tag{2.52}
$$

where P seen in the internal strain energy is defined as follows:

$$
P = \upsilon \int_A (\sigma_{xy} y + \sigma_{xz} z)\mathrm{d}A \tag{2.53}
$$

Here I_P is the polar moment of inertia and is given as follows:

$$
I_P = \int_A (y^2 + z^2)\mathrm{d}A \tag{2.54}
$$

The kinetic energy is rearranged by using this expression into Eq. (2.51).

Considering Eq. (2.8), the first variations of energy equations can be obtained as follows:

$$
\begin{aligned}
\int_{t_1}^{t_2} \delta U \mathrm{d}t &= \int_{t_1}^{t_2}\int_0^L \left(N\delta\frac{\partial u}{\partial x} - P\delta\frac{\partial^2 u}{\partial x^2}\right)\mathrm{d}x\mathrm{d}t + \int_{t_1}^{t_2}\int_0^L k_M u\delta u \mathrm{d}x\mathrm{d}t + \int_{t_1}^{t_2} k_a u(0,t)\delta u(0,t)\mathrm{d}t \\
&\quad + \int_{t_1}^{t_2} k_a u(L,t)\delta u(L,t)\mathrm{d}t
\end{aligned}
\tag{2.55}
$$

$$\int_{t_1}^{t_2} \delta T \mathrm{d}t = \int_{t_1}^{t_2}\int_0^L \left(\rho A \frac{\partial u}{\partial t}\delta\frac{\partial u}{\partial t} + \rho I_P \upsilon^2 \frac{\partial^2 u}{\partial x \partial t}\delta\frac{\partial^2 u}{\partial x \partial t}\right)\mathrm{d}x\mathrm{d}t + \int_{t_1}^{t_2} m_a \frac{\partial u(0,t)}{\partial t}\delta\frac{\partial u(0,t)}{\partial t}\mathrm{d}t + \int_{t_1}^{t_2} m_a \frac{\partial u(L,t)}{\partial t}\delta\frac{\partial u(L,t)}{\partial t}\mathrm{d}t \tag{2.56}$$

$$\int_{t_1}^{t_2} \delta W \mathrm{d}t = \int_{t_1}^{t_2}\int_0^L q\delta u \mathrm{d}x\mathrm{d}t \tag{2.57}$$

In addition to the variation calculations in Eq. (2.18), the variation expressions seen here for the first time are organized as follows:

$$\int_0^L P\delta\frac{\partial^2 u}{\partial x^2}\mathrm{d}x = P\delta\frac{\partial u}{\partial x}\bigg|_0^L - \frac{\partial P}{\partial x}\delta u\bigg|_0^L + \int_0^L \frac{\partial^2 P}{\partial x^2}\delta u \mathrm{d}x,$$
$$\int_{t_1}^{t_2}\int_0^L \rho I_P \upsilon^2 \frac{\partial^2 u}{\partial x \partial t}\delta\frac{\partial^2 u}{\partial x \partial t}\mathrm{d}t = \int_0^L \rho I_P \upsilon^2 \frac{\partial^2 u}{\partial x \partial t}\delta\frac{\partial u}{\partial t}\bigg|_{t_1}^{t_2}\mathrm{d}x - \int_{t_1}^{t_2} \rho I_P \upsilon^2 \frac{\partial^3 u}{\partial x \partial t^2}\delta u\bigg|_0^L \mathrm{d}t + \int_{t_1}^{t_2}\int_0^L \rho I_P \upsilon^2 \frac{\partial^4 u}{\partial x^2 \partial t^2}\delta u \mathrm{d}x\mathrm{d}t \tag{2.58}$$

where since there are derivative and integration operations according to both position and time, the term containing the $\rho I_P \upsilon^2$ factor is given with double integral. Since there is no such situation in the previous subsection, mechanical expressions are written with a single integral. Eqs. (2.18) and (2.54) are substituted into Eq. (2.8):

$$
\begin{aligned}
&\int_{t_1}^{t_2}\int_{0}^{L}\left(-\frac{\partial N}{\partial x}-\frac{\partial^2 P}{\partial x^2}+k_M u+\rho A\frac{\partial^2 u}{\partial t^2}-\rho I_P \upsilon^2\frac{\partial^4 u}{\partial x^2\partial t^2}-q\right)\delta u \mathrm{d}x\mathrm{d}t\\
&+\int_{t_1}^{t_2}\left(N+\frac{\partial P}{\partial x}+\rho I_P \upsilon^2\frac{\partial^3 u}{\partial x\partial t^2}\right)\delta u\Big|_0^L \mathrm{d}t+\int_{t_1}^{t_2}(-P)\delta\frac{\partial u}{\partial x}\Big|_0^L \mathrm{d}t\\
&+\int_{t_1}^{t_2}\left(k_a u(0,t)+m_a\frac{\partial^2 u}{\partial t^2}(0,t)\right)\delta u(0,t)\mathrm{d}t+\int_{t_1}^{t_2}\left(k_a u(L,t)+m_a\frac{\partial^2 u}{\partial t^2}(L,t)\right)\delta u(L,t)\mathrm{d}t\\
&+\int_{0}^{L}\left(-\rho A\frac{\partial u}{\partial t}\right)\delta u\Big|_{t_1}^{t_2}\mathrm{d}x+\int_{0}^{L}\left(-\rho I_P \upsilon^2\frac{\partial^2 u}{\partial x\partial t}\right)\delta\frac{\partial u}{\partial t}\Big|_{t_1}^{t_2}\mathrm{d}x+\left(-m_a\frac{\partial u}{\partial t}(0,t)\right)\delta u(0,t)\Big|_{t_1}^{t_2}\\
&+\left(-m_a\frac{\partial u}{\partial t}(L,t)\right)\delta u(L,t)\Big|_{t_1}^{t_2}=0
\end{aligned}
\tag{2.59}
$$

where all parts of this equation are equal to zero. The following expression is written the for $\delta u \neq 0$ limit in the first double integral expression:

$$
\frac{\partial N}{\partial x}+\frac{\partial^2 P}{\partial x^2}-k_M u-\rho A\frac{\partial^2 u}{\partial t^2}+\rho I_P \upsilon^2\frac{\partial^4 u}{\partial x^2\partial t^2}+q=0 \tag{2.60}
$$

This expression is the equation of axial equilibrium and constitutes the basis of equation of motion. The constitutive equation of nonlocal elasticity is used to obtain the equation of motion. According to this, considering Eq. (2.4), following expressions are given as follows:

$$
\left(1-(e_0 a)^2\frac{\partial^2}{\partial x^2}\right)\sigma_{xx}=E\frac{\partial u}{\partial x} \tag{2.61}
$$

$$
\left(1-(e_0 a)^2\frac{\partial^2}{\partial x^2}\right)\sigma_{xy}=-G\upsilon y\frac{\partial^2 u}{\partial x^2} \tag{2.62}
$$

$$
\left(1-(e_0 a)^2\frac{\partial^2}{\partial x^2}\right)\sigma_{xz}=-G\upsilon z\frac{\partial^2 u}{\partial x^2} \tag{2.63}
$$

if the expressions obtained by substituting Eq. (2.45) into Eqs. (2.46)–(2.48) are used into Eqs. (2.61)–(2.63), the following nonlocal stresses are obtained [30, 32]:

$$\sigma_{xx} = (e_0a)^2(1+2\upsilon)\rho\frac{\partial^3 u}{\partial x\partial t^2} - (e_0a)^2\rho\frac{\partial \tilde{q}_x}{\partial x} + (e_0a)^2\frac{k_M}{A}\frac{\partial u}{\partial x} + E\frac{\partial u}{\partial x} \tag{2.64}$$

$$\sigma_{xy} = -(e_0a)^2\rho\upsilon y\frac{\partial^4 u}{\partial x^2\partial t^2} - G\upsilon y\frac{\partial^2 u}{\partial x^2} \tag{2.65}$$

$$\sigma_{xz} = -(e_0a)^2\rho\upsilon z\frac{\partial^4 u}{\partial x^2\partial t^2} - G\upsilon z\frac{\partial^2 u}{\partial x^2} \tag{2.66}$$

Similar to Eq. (2.5), if Eq. (2.53) is substituted into result of some operations that are performed by using Eqs. (2.65) and (2.66):

$$P = -(e_0a)^2\upsilon^2\rho I_P\frac{\partial^4 u}{\partial x^2\partial t^2} - \upsilon^2 GI_P\frac{\partial^2 u}{\partial x^2} \tag{2.67}$$

This expression expresses the shear force in the longitudinal direction. On the other hand, if Eq. (2.64) is integrated via area, the axial normal force is reached:

$$N = (e_0a)^2(1+2\upsilon)\rho A\frac{\partial^3 u}{\partial x\partial t^2} + (e_0a)^2 k_M\frac{\partial u}{\partial x} + EA\frac{\partial u}{\partial x} - (e_0a)^2\frac{\partial q}{\partial x} \tag{2.68}$$

If Eqs. (2.67) and (2.68) are substituted into Eq. (2.60), the equation of motion can be obtained as follows:

$$\begin{aligned} &\upsilon^2 GI_P\frac{\partial^4 u}{\partial x^4} - \left[EA + (e_0a)^2 k_M\right]\frac{\partial^2 u}{\partial x^2} + \rho A\frac{\partial^2 u}{\partial t^2} + k_M u + (e_0a)^2\upsilon^2\rho I_P\frac{\partial^6 u}{\partial x^4\partial t^2} \\ &- \left[\rho I_P\upsilon^2 + (e_0a)^2(1+2\upsilon)\rho A\right]\frac{\partial^4 u}{\partial x^2\partial t^2} + (e_0a)^2\frac{\partial^2 q}{\partial x^2} - q = 0 \end{aligned} \tag{2.69}$$

2.4.2 Solution of Equation of Motion

As is the case with simple axial rod, external excitation is considered as $q = 0$ for the solution of free vibration. According to this, the following homogeneous and ordinary differential equation is obtained by using Eq. (2.21) into Eq. (2.69):

$$A_1\frac{\mathrm{d}^4 U}{\mathrm{d}x^4} + A_2\frac{\mathrm{d}^2 U}{\mathrm{d}x^2} + A_3 U = 0 \tag{2.70}$$

where the following definition is valid:

$$A_1 = \upsilon^2 GI_P - (e_0 a)^2 \upsilon^2 \rho I_P \omega^2, \;\; A_2 = -EA - (e_0 a)^2 k_M + \omega^2 \rho I_P \upsilon^2 + \omega^2 (e_0 a)^2 (1 + 2\upsilon)\rho A, \;\; A_3 = -\omega^2 \rho A + k_M \tag{2.71}$$

It should be noted that the mechanical boundary conditions will be obtained in the next subsection. Eq. (2.70) is a homogeneous differential equation with constant coefficients. Thus, the solution can be determined by the mathematical expression given in Eq. (2.24). But in this rod type, mathematical operations are a bit more. According to this, substituting Eq. (2.24) into Eq. (2.70) yields the following equations:

$$A_1 k^4 + A_2 k^2 + A_3 = 0 \tag{2.72}$$

where a mathematical transformation is used as follows:

$$k^2 = K \tag{2.73}$$

If Eq. (2.73) is substituted into Eq. (2.72):

$$A_1 K^2 + A_2 K + A_3 = 0 \tag{2.74}$$

In the solution od this equation, the discriminant $\Delta = {A_2}^2 - 4A_1A_3$ should be positive definite. Assuming the external factor $k_M = 0$ and nonlocal parameter $e_0 a = 0$, $A_1 > 0$, $A_2 < 0$, and $A_3 < 0$ cases occur due to Poisson's ratio and small vibration frequencies. Thus, we understand that Eq. (2.74) has a negative and a positive roots. The roots of equation are written as follows:

$$K_1 = \frac{-A_2 + \sqrt{{A_2}^2 - 4A_1A_3}}{2A_1}, \;\; K_2 = \frac{-A_2 - \sqrt{{A_2}^2 - 4A_1A_3}}{2A_1}, \tag{2.75}$$

where $K_1 > 0$ and $K_2 < 0$. Considering Eq. (2.73), the following four different roots can be defined:

$$k_{11} = \sqrt{K_1}, \;\; k_{12} = -\sqrt{K_1}, \;\; k_{21} = \mathbf{i}\sqrt{-K_2}, \;\; k_{22} = -\mathbf{i}\sqrt{-K_2}, \tag{2.76}$$

The resulting solution can be attained as follows:

$$U(x) = C_1 e^{\sqrt{K_1}x} + C_2 e^{-\sqrt{K_1}x} + C_3 e^{\mathbf{i}\sqrt{-K_2}x} + C_4 e^{-\mathbf{i}\sqrt{-K_2}x} \tag{2.77}$$

Here $C_i\,(i = 1, 2, 3, 4)$ are unknowns. This equation is rearranged with the help of the Euler transforms shown below:

$$e^{\mathbf{i}\theta} = \cos\theta + \mathbf{i}\sin\theta \tag{2.78}$$

$$e^{-\mathbf{i}\theta} = \cos\theta - \mathbf{i}\sin\theta \tag{2.79}$$

$$e^{\theta} = \cosh\theta + \sinh\theta \tag{2.80}$$

$$e^{-\theta} = \cosh\theta - \sinh\theta \tag{2.81}$$

If Eqs. (2.78)–(2.81) are substituted into Eq. (2.77), the mode equation is obtained in the following form:

$$U(x) = C_1 \cos\left(\sqrt{-K_2}x\right) + C_2 \sin\left(\sqrt{-K_2}x\right) + C_3 \cosh\left(\sqrt{K_1}x\right) + C_4 \sinh\left(\sqrt{K_1}x\right) \tag{2.82}$$

2.4.3 Boundary Conditions of Nanorods

In the simple rod theory, clamped and free ends that are the generally known boundary conditions are easy to express and apply because the geometric and mechanical boundary conditions have a very simple structure and the number of equations is very few. However, the expression of common boundary conditions in Love-Bishop rods is a bit more complicated. It should be noted that it is not sufficient to use only a "clamped" or "free" expression for boundary conditions. In this subsection, first four general boundary conditions and then additional boundary conditions are investigated.

1. **Clamped end (C):** There is not only longitudinal displacement, but also lateral deformation:

$$U = 0, \quad U' = 0 \tag{2.83}$$

2. **Free end (F):** There is no internal force at the free ends. Therefore, the total axial normal force and shear forces are equal to zero. First, the following equation can be given for $\delta u \neq 0$ in Eq. (2.59) for the total normal force:

$$N_x = N + \frac{\partial P}{\partial x} + \rho I_P \upsilon^2 \frac{\partial^3 u}{\partial x \partial t^2} = 0 \tag{2.84}$$

Here, N_x is the total axial force. Substituting Eqs. (2.67) and (2.68) into Eq. (2.84):

$$
\begin{aligned}
&-\upsilon^2 GI_P\frac{\partial^3 u}{\partial x^3}+(e_0a)^2k_M\frac{\partial u}{\partial x}+EA\frac{\partial u}{\partial x}-(e_0a)^2\upsilon^2\rho I_P\frac{\partial^5 u}{\partial x^3\partial t^2}\\
&+(e_0a)^2(1+2\upsilon)\rho A\frac{\partial^3 u}{\partial x\partial t^2}+\rho I_P\upsilon^2\frac{\partial^3 u}{\partial x\partial t^2}=0
\end{aligned}
\tag{2.85}
$$

With the use of Eq. (2.21):

$$
A_1\frac{\mathrm{d}^3 U}{\mathrm{d}x^3}+A_2\frac{\mathrm{d}U}{\mathrm{d}x}=0 \tag{2.86}
$$

On the other hand, the shear force is specified for $\delta u' \neq 0$ in Eq. (2.59) as follows:

$$
P=0 \tag{2.87}
$$

By using Eq. (2.21) into Eq. (2.67), the following expression can be obtained:

$$
A_1\frac{\mathrm{d}^2 U}{\mathrm{d}x^2}=0 \quad \text{or} \quad \frac{\mathrm{d}^2 U}{\mathrm{d}x^2}=0 \tag{2.88}
$$

3. **Soft clamped end (S):** There is no axial deformation and shear force in this supporting type:

$$
U=0, \quad U''=0 \tag{2.89}
$$

4. **Gripped clamped end (G)**: There is no axial force and lateral deformation in this supporting type:

$$
A_1\frac{\mathrm{d}^3 U}{\mathrm{d}x^3}+A_2\frac{\mathrm{d}U}{\mathrm{d}x}=0, \quad U'=0 \tag{2.90}
$$

where the following boundary condition can be reached as a result:

$$
A_1U'''=0 \quad \text{or} \quad U'''=0 \tag{2.91}
$$

5. **Longitudinal attached ends (MA and SA):** It is assumed that the tip attachments simply affect the total axial normal force. In Eq. (2.59), the following expression whose sum can be zero can be written:

$$\int_{t_1}^{t_2}\left(N+\frac{\partial P}{\partial x}+\rho I_P \upsilon^2\frac{\partial^3 u}{\partial x \partial t^2}\right)\delta u\bigg|_0^L \mathrm{d}t+\int_{t_1}^{t_2}\left(k_a u(0,t)+m_a\frac{\partial^2 u}{\partial t^2}(0,t)\right)\delta u(0,t)\mathrm{d}t$$
$$+\int_{t_1}^{t_2}\left(k_a u(L,t)+m_a\frac{\partial^2 u}{\partial t^2}(L,t)\right)\delta u(L,t)\mathrm{d}t=0 \tag{2.92}$$

Let's consider the right end of nanorod,

$$N(L,t)+\frac{\partial P}{\partial x}(L,t)+\rho I_P \upsilon^2\frac{\partial^3 u}{\partial x \partial t^2}(L,t)+k_a u(L,t)+m_a\frac{\partial^2 u}{\partial t^2}(L,t)=0 \tag{2.93}$$

If Eq. (2.86) is used here:

$$A_1 U'''(L)+A_2 U'(L)=\left(-k_a+\omega^2 m_a\right)U(L) \tag{2.94}$$

On the other hand, the following expression can be constituted for the left end of nanorod:

$$-N(0,t)-\frac{\partial P}{\partial x}(0,t)-\rho I_P \upsilon^2\frac{\partial^3 u}{\partial x \partial t^2}(0,t)+k_a u(0,t)+m_a\frac{\partial^2 u}{\partial t^2}(0,t)=0 \tag{2.95}$$

So the boundary condition is reached as:

$$A_1 U'''(0)+A_2 U'(0)=\left(k_a-\omega^2 m_a\right)U(0) \tag{2.96}$$

As a result, the mechanical boundary conditions for the different attached ends of nonlocal Love-Bishop nanorods are outlined below:

$$\begin{array}{l}\text{Mass attachment at } x=0: A_1 U'''+B_2 U'=\omega^2 m_a U,\\ \text{Elastic spring attachment at } x=0: A_1 U'''+B_2 U'=-k_a U,\\ \text{Mass attachment at } x=L: A_1 U'''+B_2 U'=-\omega^2 m_a U,\\ \text{Elastic spring attachment at } x=L: A_1 U'''+B_2 U'=k_a U\end{array} \tag{2.97}$$

In addition to these, the other boundary condition of attached end is introduced as in Eq. (2.88).

Example 2.5 Determine the frequency equation of nonlocal free vibration of clamped nanorod embedded in the elastic medium shown in Fig. 2.2.

Solution. The boundary conditions of rod are as follows:

$$U(0)=0, \quad U'(0)=0, \quad U(L)=0, \quad U'(L)=0 \tag{2.E2.5.1}$$

When these expressions are applied sequentially to Eq. (2.82), the following equations are obtained:

$$C_1 + C_3 = 0 \tag{2.E2.5.2}$$

$$C_2\sqrt{-K_2} + C_4\sqrt{K_1} = 0 \tag{2.E2.5.3}$$

$$\begin{aligned} &C_1 \cos\left(\sqrt{-K_2}L\right) + C_2 \sin\left(\sqrt{-K_2}L\right) + C_3 \cosh\left(\sqrt{K_1}L\right) \\ &\quad + C_4 \sinh\left(\sqrt{K_1}L\right) = 0 \end{aligned} \tag{2.E2.5.4}$$

$$\begin{aligned} &-C_1\sqrt{-K_2}\sin\left(\sqrt{-K_2}L\right) + C_2\sqrt{-K_2}\cos\left(\sqrt{-K_2}L\right) + C_3\sqrt{K_1}\sinh\left(\sqrt{K_1}L\right) \\ &+C_4\sqrt{K_1}\cosh\left(\sqrt{K_1}L\right) = 0 \end{aligned} \tag{2.E2.5.5}$$

This set of equations is presented as $[a]\{C\}$:

$$\begin{bmatrix} a_{11} & a_{12} & a_{13} & a_{14} \\ a_{21} & a_{22} & a_{23} & a_{24} \\ a_{31} & a_{32} & a_{33} & a_{34} \\ a_{41} & a_{42} & a_{43} & a_{44} \end{bmatrix} \begin{Bmatrix} C_1 \\ C_2 \\ C_3 \\ C_4 \end{Bmatrix} = 0 \tag{2.E2.5.6}$$

In which, $[a]$ is the coefficients matrix and $\{C\}$ is the vector of mode equation unknowns. Inputs of coefficient matrix are computed as follows:

$$\begin{aligned} &a_{11}=1, a_{12}=0, a_{13}=1, a_{14}=0, a_{21}=0, a_{22}=\sqrt{-K_2}, a_{23}=0, a_{24}=\sqrt{K_1}, a_{31}=\cos\left(\sqrt{-K_2}L\right), \\ &a_{32}=\sin\left(\sqrt{-K_2}L\right), a_{33}=\cosh\left(\sqrt{K_1}L\right), \\ &a_{34}=\sinh\left(\sqrt{K_1}L\right), a_{41}=-\sqrt{-K_2}\sin\left(\sqrt{-K_2}L\right), a_{42}=\sqrt{-K_2}\cos\left(\sqrt{-K_2}L\right) \\ &a_{43}=\sqrt{K_1}\sinh\left(\sqrt{K_1}L\right), a_{44}=\sqrt{K_1}\cosh\left(\sqrt{K_1}L\right) \end{aligned} \tag{2.E2.5.7}$$

According to the linear equation system given by Eq. (2.E2.5.6), the determinant of $[a]$ matrix gives the frequency equation of clamped nanorod. Because manual solution of frequency is difficult, the solution can be performed by programming on the computer.

Example 2.6 Determine the frequency equation of nonlocal free vibration of clamped-free nanorod embedded in the elastic medium shown in Fig. 2.3.

Solution. The boundary conditions of this rod are written below:

$$U(0)=0, U'(0)=0, A_1U'''(L)+A_2U'(L)=0, U''(L)=0 \tag{2.E2.6.1}$$

where Eqs. (2.E2.5.2) and (2.E2.5.3) are valid due to the clamped end. Applying the third and fourth boundary conditions into Eq. (2.82) give the following expressions, respectively:

$$C_1\left(A_1\sqrt{-K_2}^3-A_2\sqrt{-K_2}\right)\sin\left(\sqrt{-K_2}L\right)+C_2\left(-A_1\sqrt{-K_2}^3+A_2\sqrt{-K_2}\right)\cos\left(\sqrt{-K_2}L\right)+ \\ C_3\left(A_1\sqrt{K_1}^3+A_2\sqrt{K_1}\right)\sinh\left(\sqrt{K_1}L\right)+C_4\left(A_1\sqrt{K_1}^3+A_2\sqrt{K_1}\right)\cosh\left(\sqrt{K_1}L\right)=0 \tag{2.E2.6.2}$$

$$-C_1K_2\cos\left(\sqrt{-K_2}L\right)-C_2K_2\sin\left(\sqrt{-K_2}L\right)+C_3K_1\cosh\left(\sqrt{K_1}L\right) \\ +C_4K_1\sinh\left(\sqrt{K_1}L\right)=0 \tag{2.E2.6.3}$$

In the set of equations written by Eqs. (2.E2.5.2), (2.E2.5.3), (2.E2.6.2), and (2.E2.6.3) similar to Eq. (2.E2.5.6), the inputs of coefficients matrix can be specified as follows:

$$a_{11}=1, a_{12}=0, a_{13}=1, a_{14}=0, a_{21}=0, a_{22}=\sqrt{-K_2}, a_{23}=0, a_{24}=\sqrt{K_1}, \\ a_{31}=\left(A_1\sqrt{-K_2}^3-A_2\sqrt{-K_2}\right)\sin\left(\sqrt{-K_2}L\right), a_{32}=\left(-A_1\sqrt{-K_2}^3+A_2\sqrt{-K_2}\right)\cos\left(\sqrt{-K_2}L\right), \\ a_{33}=\left(A_1\sqrt{K_1}^3+A_2\sqrt{K_1}\right)\sinh\left(\sqrt{K_1}L\right), a_{34}=\left(A_1\sqrt{K_1}^3+A_2\sqrt{K_1}\right)\cosh\left(\sqrt{K_1}L\right) \\ a_{41}=-K_2\cos\left(\sqrt{-K_2}L\right), a_{42}=-K_2\sin\left(\sqrt{-K_2}L\right), a_{43}=K_1\cosh\left(\sqrt{K_1}L\right), a_{44}=K_1\sinh\left(\sqrt{K_1}L\right) \tag{2.E2.6.4}$$

Example 2.7 Determine the frequency equation of nonlocal free vibration of soft clamped nanorods shown in Fig. 2.7.

Solution. Considering Eq. (2.89), the boundary conditions of this rod are expressed as

$$U(0)=0, U''(0)=0, U(L)=0, U''(L)=0 \tag{2.E2.7.1}$$

Applying the first and second expressions to Eq. (2.82) yields the following results, respectively:

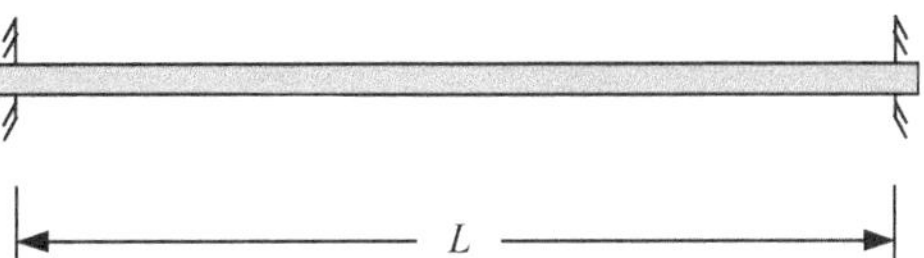

Fig. 2.7 Soft clamped nanorod

$$C_1 \cos 0 + C_2 \sin 0 + C_3 \cos h0 + C_4 \sin h0 = 0 \tag{2.E2.7.2}$$

$$C_1 K_2 \cos 0 - C_2 K_2 \sin 0 + C_3 K_1 \cosh 0 + C_4 K_1 \sinh 0 = 0 \tag{2.E2.7.3}$$

The following expression is obtained from the common solution of Eqs. (2.E2.7.2) and (2.E2.7.3):

$$C_1 = C_3 = 0 \tag{2.E2.7.4}$$

On the other hand, by substituting the third and fourth boundary conditions into Eq. (2.82), the following expressions are obtained, respectively:

$$C_2 \sin\left(\sqrt{-K_2}L\right) + C_4 \sinh\left(\sqrt{K_1}L\right) = 0 \tag{2.E2.7.5}$$

$$-C_2 K_2 \sin\left(\sqrt{-K_2}L\right) + C_4 K_1 \sinh\left(\sqrt{K_1}L\right) = 0 \tag{2.E2.7.6}$$

These equations can be written in matrix form as follows:

$$\left[\sin\left(\sqrt{-K_2}L\right) \sinh\left(\sqrt{K_1}L\right) - K_2 \sin\left(\sqrt{-K_2}L\right) K_1 \sinh\left(\sqrt{K_2}L\right)\right] \begin{Bmatrix} C_2 \\ C_4 \end{Bmatrix} = 0 \tag{2.E2.7.7}$$

where the determinant of coefficient matrix is calculated as follows:

$$(K_1 + K_2) \sin\left(\sqrt{-K_2}L\right) \sinh\left(\sqrt{K_1}L\right) = 0 \tag{2.E2.7.8}$$

This expression defines the frequency equation. An analysis should be made here: Considering Eqs. (2.71) and (2.75) for $K_1 + K_2 = 0$, $A_2 = 0$ is obtained, but a periodic solution cannot be reached. On the other hand, in case of $\sinh\left(\sqrt{K_1}L\right) = 0$ equation, we get $K_1 = 0$. This inference is invalid because a positive and a negative root are defined in Eq. (2.75). In the solution of the equation, the following last alternative should be investigated:

$$\sin\left(\sqrt{-K_2}L\right) = 0 \tag{2.E2.7.9}$$

where the solution of equation can be presented as:

$$-K_2 = \left(\frac{n\pi}{L}\right)^2 \tag{2.E2.7.10}$$

where the integer defined as $n = 1, 2, 3\ldots$ denotes the mode number of vibration. Substituting Eq. (2.E2.7.10) into Eq. (2.75):

$$\frac{A_2 + \sqrt{A_2{}^2 - 4A_1A_3}}{2A_1} = \left(\frac{n\pi}{L}\right)^2 \tag{2.E2.7.11}$$

As a result of a few mathematical operations, the following equation is attained:

$$A_1\left(\frac{n\pi}{L}\right)^4 - A_2\left(\frac{n\pi}{L}\right)^2 + A_3 = 0 \tag{2.E2.7.12}$$

If Eq. (2.71) is substituted into Eq. (2.E2.7.12), the natural frequency equation occurs as follows:

$$\omega^2 = \frac{GI_P\upsilon^2\left(\frac{n\pi}{L}\right)^4 + \left[EA + (e_0a)^2k_M\right]\left(\frac{n\pi}{L}\right)^2 + k_M}{(e_0a)^2\rho I_P\upsilon^2\left(\frac{n\pi}{L}\right)^4 + \left[(e_0a)^2\rho A(1+2\upsilon) + \rho I_P\upsilon^2\right]\left(\frac{n\pi}{L}\right)^2 + \rho A} \tag{2.E2.7.13}$$

The frequency equation can also be obtained with the help of a series expansion that provides boundary conditions. According to this, the following series expansion can be used:

$$u(x,t) = \sum_{n=1}^{\infty} u_n \sin\left(\frac{n\pi x}{L}\right)\sin(\omega t - \theta) \tag{2.E2.7.14}$$

Here U_n is an unknown coefficient. Using Eq. (2.E2.7.14) into Eq. (2.69) gives the following expression:

$$\begin{aligned}
&\left[\upsilon^2 GI_P - \omega^2(e_0a)^2\upsilon^2\rho I_P\right]\left(\frac{n\pi}{L}\right)^4 \sum_{n=1}^{\infty} u_n \sin\left(\frac{n\pi x}{L}\right)\sin(\omega t - \theta) \\
&+\left[EA + (e_0a)^2k_M - \omega^2\rho I_P\upsilon^2 - \omega^2(e_0a)^2(1+2\upsilon)\rho A\right]\left(\frac{n\pi}{L}\right)^2 \sum_{n=1}^{\infty} u_n \\
&\times \sin\left(\frac{n\pi x}{L}\right)\sin(\omega t - \theta) + \left(-\omega^2\rho A + k_M\right)\sum_{n=1}^{\infty} u_n \sin\left(\frac{n\pi x}{L}\right)\sin(\omega t - \theta) = 0
\end{aligned} \tag{2.E2.7.15}$$

After simplifying the series expansions in the above equation, Eq. (2.E2.7.12) is reached. As a result, a faster result can be obtained by using series expansion compared

Table 2.1 Comparisons of nondimensional frequencies of nanorods

Mode	SRT			LBRT		
number	$\alpha = 0$	$\alpha = 0.1$	$\alpha = 0.2$	$\alpha = 0$	$\alpha = 0.1$	$\alpha = 0.2$
1	1.57080	1.55177	1.49858	1.57078	1.54471	1.47363
2	4.71239	4.26279	3.42933	4.71205	4.12256	3.15845
3	7.85398	6.17668	4.21782	7.85240	5.77150	3.74163
4	10.99557	7.39805	4.55152	10.99123	6.72926	3.96836
5	14.13717	8.16401	4.71386	14.12794	7.28947	4.07407

to the separation of variables in the nonlocal free vibration analysis of this nanorod, but unfortunately the boundary condition of each nanorod does not allow for a series solution.

The kinematics of soft clamped Love-Bishop nanorods are the same as the clamped simple nanorods. Since when Eq. (2.E2.7.12) is considered, the differences between Eqs. (2.E2.7.13) and (2.E2.7.12) are due to shear effects and lateral inertia effects.

According to two different nonlocal rod models (SRT: Simple rod theory and LBRT: Love-Bishop rod theory), first five modes free vibration frequencies of clamped-free nanorods are calculated in Table 2.1 by ignoring the elastic medium ($k_M = 0$). Computations are presented as nondimensional form:

$$\overline{\omega} = \omega_i L \sqrt{\frac{\rho}{E}} \tag{2.E2.7.16}$$

where i is the mode number, ω_i is the natural frequency and $\overline{\omega}_i$ is called as nondimensional frequency. Additionally, α indicates the nondimensional nonlocal parameter:

$$\alpha = \frac{e_0 a}{L} \tag{2.E2.7.17}$$

It is noted that the following parameters are valid in the computing of results in Table 2.1: Nanobeam length: $L = 20$ nm, diameter of circular cross-section: $d = 1$ nm, modulus of elasticity: $E = 1$ TPa, Poisson's ratio: $\upsilon = 0.19$, shear modulus: $G = E/2(1 + \upsilon) = 0.42$ TPa, and mass of unit volume: $\rho = 2300$ kg/m^3.

According to the results in Table 2.1, firstly, it is understood that the presence of nonlocal parameter reduces the nondimensional classical frequencies of nanorods ($\alpha = 0$). In addition to this, the Love-Bishop nanorods have lower nondimensional frequencies compared to simple nanorods. It is also observed that the nonlocal parameter is more effective at higher modes. Also, the decrease rate of frequencies increases as the nonlocal parameter increases. Moreover, the effect of nonlocal parameter is more pronounced in Love-Bishop nanorods. All inferences reveal that the nonlocal parameter is an

Fig. 2.8 Soft clamped-gripped clamped nanorod

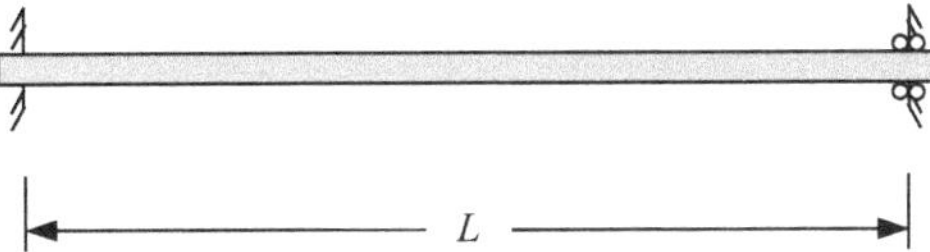

indispensable factor on the dynamics of nanostructures, moreover, the Poisson effect at the nanoscale may gain importance.

Example 2.8 Determine the frequency equation of nonlocal free vibration of soft clamped-gripped clamped nanorod shown in Fig. 2.8.

Solution. According to Eqs. (2.89)–(2.91), the boundary conditions of this rod should be as follows:

$$U(0)=0,\quad U''(0)=0,\quad U'(L)=0,\quad U''(L)=0 \tag{2.E2.8.1}$$

To apply the first two boundary conditions to the equation of motion yields Eq. (2.E2.7.4). In addition to this, the applications of third and fourth boundary conditions are presented as follows:

$$C_2\sqrt{-K_2}\cos\left(\sqrt{-K_2}L\right)+C_4\sqrt{K_1}\cosh\left(\sqrt{K_1}L\right)=0 \tag{2.E2.8.2}$$

$$-C_2\sqrt{-K_2}^3\cos\left(\sqrt{-K_2}L\right)+C_4\sqrt{K_1}^3\cosh\left(\sqrt{K_1}L\right)=0 \tag{2.E2.8.3}$$

These equations can be written in matrix form as follows:

$$\begin{bmatrix} \sqrt{-K_2}\cos\left(\sqrt{-K_2}L\right) & \sqrt{K_1}\cosh\left(\sqrt{K_1}L\right) \\ -\sqrt{-K_2}^3\cos\left(\sqrt{-K_2}L\right) & \sqrt{K_1}^3\cosh\left(\sqrt{K_1}L\right) \end{bmatrix}\begin{Bmatrix} C_2 \\ C_4 \end{Bmatrix}=0 \tag{2.E2.8.4}$$

In the above set of equations, the determinant of coefficient matrix can be written as:

$$(K_1+K_2)\sqrt{-K_1K_2}\cos\left(\sqrt{-K_2}L\right)\cosh\left(\sqrt{K_1}L\right)=0 \tag{2.E2.8.5}$$

Here, when considering Example 2.7, it was stated that $K_1 + K_2 = 0$ option is invalid. Second, the $\sqrt{K_1K_2}=0$ option can be examined. This statement becomes invalid when Eqs. (2.71) and (2.75) are taken into account because vibration frequency cannot be obtained in the absence of elastic medium. Also, even if there is an elastic medium, a periodic frequency calculation cannot be mentioned. Therefore, other options should be investigated. Since the value of cosine hyperbola function can never equal zero, it is not possible to substantiate the equation $\cosh\left(\sqrt{K_1}L\right)=0$. Finally, the following option can be examined:

$$\cos\left(\sqrt{-K_2}L\right)=0 \tag{2.E2.8.6}$$

The periodic solution of this equation can be obtained as

$$-K_2=\left(\frac{(2n-1)\pi}{2L}\right)^2 \tag{2.E2.8.7}$$

Similar to the mathematical operation in Eq. (2.7.11), the following result is explained:

$$A_1\left(\frac{(2n-1)\pi}{2L}\right)^4-A_2\left(\frac{(2n-1)\pi}{2L}\right)^2+A_3=0 \tag{2.E2.8.9}$$

So, the frequency equation is as follows:

$$\omega^2=\frac{GI_P\upsilon^2\left(\frac{(2n-1)\pi}{2L}\right)^4+\left[EA+(e_0a)^2k_M\right]\left(\frac{(2n-1)\pi}{2L}\right)^2+k_M}{(e_0a)^2\rho I_P\upsilon^2\left(\frac{(2n-1)\pi}{2L}\right)^4+\left[(e_0a)^2\rho A(1+2\upsilon)+\rho I_P\upsilon^2\right]\left(\frac{(2n-1)\pi}{2L}\right)^2+\rho A} \tag{2.E2.8.10}$$

On the other hand, frequency solution is also possible via the following series expansion by considering the kinematics of this type of nanorod:

$$u(x,t)=\sum_{n=1}^{\infty}u_n\sin\left(\frac{(2n-1)\pi x}{2L}\right)\sin(\omega t-\theta) \tag{2.E2.8.11}$$

It should be noted here that the series expansion provides the boundary conditions. By substituting this expression into Eq. (2.69), Eq. (2.E2.8.9) is reached.

As in Example 2.7, it should be certainly emphasized that the kinematics of this rod type is the same as cantilever nanorod (Example 2.1) with nonlocal simple axial rod behavior.

Example 2.9 Determine the frequency equation of nonlocal free vibration of clamped-spring attached nanorod shown in Fig. 2.9.

Solution. Based on Eqs. (2.83) and (2.97), the boundary conditions for this rod type can be introduced as follows:

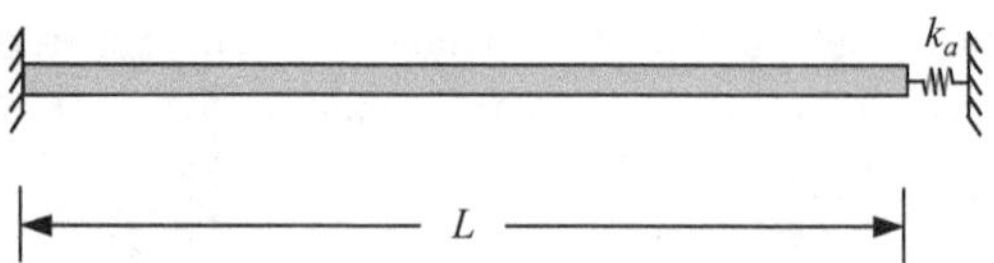

Fig. 2.9 Clamped-spring attached nanorod

$$U(0)=0, U'(0)=0, U''(L)=0, A_1U'''(L)+B_2U'(L)=k_aU(L) \tag{2.E2.9.1}$$

As a result of applying the first and second boundary conditions to Eq. (2.82), Eq. (2.E2.5.2) and (2.E2.5.3) are obtained. Applying the third and fourth boundary conditions to Eq. (2.82), the following expressions are obtained, respectively:

$$\begin{gathered}-C_1K_2\cos\left(\sqrt{-K_2}L\right)-C_2K_2\sin\left(\sqrt{-K_2}L\right)+C_3K_1\cosh\left(\sqrt{K_1}L\right)\\+C_4K_1\sinh\left(\sqrt{K_1}L\right)=0\end{gathered} \tag{2.E2.9.2}$$

$$\begin{gathered}C_1\left(\left(A_1\sqrt{-K_2}^3-B_2\sqrt{-K_2}\right)\sin\left(\sqrt{-K_2}L\right)-k_a\cos\left(\sqrt{-K_2}L\right)\right)\\+C_2\left(\left(-A_1\sqrt{-K_2}^3+B_2\sqrt{-K_2}\right)\cos\left(\sqrt{-K_2}L\right)-k_a\sin\left(\sqrt{-K_2}L\right)\right)\\+C_3\left(\left(A_1\sqrt{K_1}^3+B_2\sqrt{K_1}\right)\sinh\left(\sqrt{K_1}L\right)-k_a\cosh\left(\sqrt{K_1}L\right)\right)\\+C_4\left(\left(A_1\sqrt{K_1}^3+B_2\sqrt{K_1}\right)\cosh\left(\sqrt{K_1}L\right)-k_a\sinh\left(\sqrt{K_1}L\right)\right)=0\end{gathered} \tag{2.E2.9.3}$$

Inputs of coefficients matrix appear as follows for the set of equations in Eq. (2.E2.5.6) from Eqs. (2.E2.5.2), (2.E2.5.3), (2.E2.9.2), and (2.E2.9.3):

$$\begin{aligned}&a_{11}=1, a_{12}=0, a_{13}=1, a_{14}=0, a_{21}=0, a_{22}=\sqrt{-K_2}, a_{23}=0, a_{24}=\sqrt{K_1},\\&a_{31}=-K_2\cos\left(\sqrt{-K_2}L\right), a_{32}=-K_2\sin\left(\sqrt{-K_2}L\right), a_{33}=K_1\cosh\left(\sqrt{K_1}L\right)\\&a_{34}=K_1\sinh\left(\sqrt{K_1}L\right), a_{41}=\left(A_1\sqrt{-K_2}^3-B_2\sqrt{-K_2}\right)\sin\left(\sqrt{-K_2}L\right)-k_a\cos\left(\sqrt{-K_2}L\right)\\&a_{42}=\left(-A_1\sqrt{-K_2}^3+B_2\sqrt{-K_2}\right)\cos\left(\sqrt{-K_2}L\right)-k_a\sin\left(\sqrt{-K_2}L\right)\\&a_{43}=\left(A_1\sqrt{K_1}^3+B_2\sqrt{K_1}\right)\sinh\left(\sqrt{K_1}L\right)-k_a\cosh\left(\sqrt{K_1}L\right)\\&a_{44}=\left(A_1\sqrt{K_1}^3+B_2\sqrt{K_1}\right)\cosh\left(\sqrt{K_1}L\right)-k_a\sinh\left(\sqrt{K_1}L\right)\end{aligned} \tag{2.E2.9.4}$$

Example 2.10 Determine the frequency equation of nonlocal free vibration of free-mass attached nanorod in Fig. 2.10.

Solution. The boundary conditions of rod are written below:

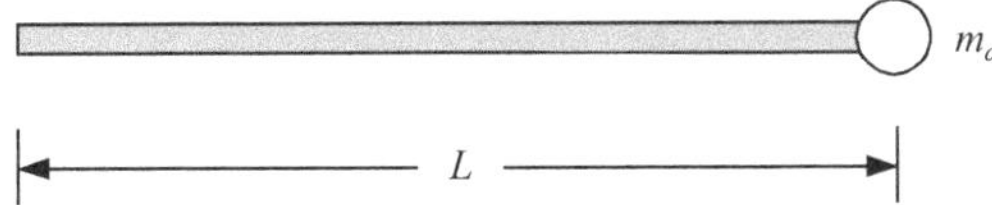

Fig. 2.10 Free-mass attached nanorod

$$A_1U'''(0) + A_2U'(0) = 0, U''(0) = 0, U''(L) = 0, A_1U'''(L) + B_2U'(L) = -\omega^2 m_a U(L) \tag{2.E2.10.1}$$

where the applications of boundary conditions by utilizing Eq. (2.82) are expressed as follows:

$$C_2(-A_1K_2 + A_2) + C_4(A_1K_1 + A_2) = 0 \tag{2.E2.10.2}$$

$$-C_1K_2 + C_3K_1 = 0 \tag{2.E2.10.3}$$

$$\begin{aligned} &-C_1K_2\cos\left(\sqrt{-K_2}L\right) - C_2K_2\sin\left(\sqrt{-K_2}L\right) + C_3K_1\cosh\left(\sqrt{K_1}L\right) \\ &+C_4K_1\sinh\left(\sqrt{K_1}L\right) = 0 \end{aligned} \tag{2.E2.10.4}$$

$$\begin{aligned} &C_1\left(\left(A_1\sqrt{-K_2}^3 - B_2\sqrt{-K_2}\right)\sin\left(\sqrt{-K_2}L\right) + \omega^2 m_a\cos\left(\sqrt{-K_2}L\right)\right) \\ &+C_2\left(\left(-A_1\sqrt{-K_2}^3 + B_2\sqrt{-K_2}\right)\cos\left(\sqrt{-K_2}L\right) + \omega^2 m_a\sin\left(\sqrt{-K_2}L\right)\right) \\ &+C_3\left(\left(A_1\sqrt{K_1}^3 + B_2\sqrt{K_1}\right)\sinh\left(\sqrt{K_1}L\right) + \omega^2 m_a\cosh\left(\sqrt{K_1}L\right)\right) \\ &+C_4\left(\left(A_1\sqrt{K_1}^3 + B_2\sqrt{K_1}\right)\cosh\left(\sqrt{K_1}L\right) + \omega^2 m_a\sinh\left(\sqrt{K_1}L\right)\right) = 0 \end{aligned} \tag{2.E2.10.5}$$

By using the above equations, the inputs of coefficients matrix in Eq. (2.E2.5.6) are as follows for this rod type:

$$\begin{aligned} &a_{11} = 0, a_{12} = -A_1K_2 + A_2, a_{13} = 0, a_{14} = A_1K_1 + A_2, a_{21} = -K_2, a_{22} = 0, a_{23} = K_1, a_{24} = 0, \\ &a_{31} = -K_2\cos\left(\sqrt{-K_2}L\right), a_{32} = -K_2\sin\left(\sqrt{-K_2}L\right), a_{33} = K_1\cosh\left(\sqrt{K_1}L\right), \\ &a_{34} = K_1\sinh\left(\sqrt{K_1}L\right), a_{41} = \left(A_1\sqrt{-K_2}^3 - B_2\sqrt{-K_2}\right)\sin\left(\sqrt{-K_2}L\right) + \omega^2 m_a\cos\left(\sqrt{-K_2}L\right), \\ &a_{42} = \left(-A_1\sqrt{-K_2}^3 + B_2\sqrt{-K_2}\right)\cos\left(\sqrt{-K_2}L\right) + \omega^2 m_a\sin\left(\sqrt{-K_2}L\right), \\ &a_{43} = \left(A_1\sqrt{K_1}^3 + B_2\sqrt{K_1}\right)\sinh\left(\sqrt{K_1}L\right) + \omega^2 m_a\cosh\left(\sqrt{K_1}L\right), \\ &a_{44} = \left(A_1\sqrt{K_1}^3 + B_2\sqrt{K_1}\right)\cosh\left(\sqrt{K_1}L\right) + \omega^2 m_a\sinh\left(\sqrt{K_1}L\right). \end{aligned} \tag{2.E2.10.6}$$

Problems

2.1. Determine the frequency equation of nonlocal free vibration of axial nanorods given in Fig. 2.11 according to nonlocal SRT.

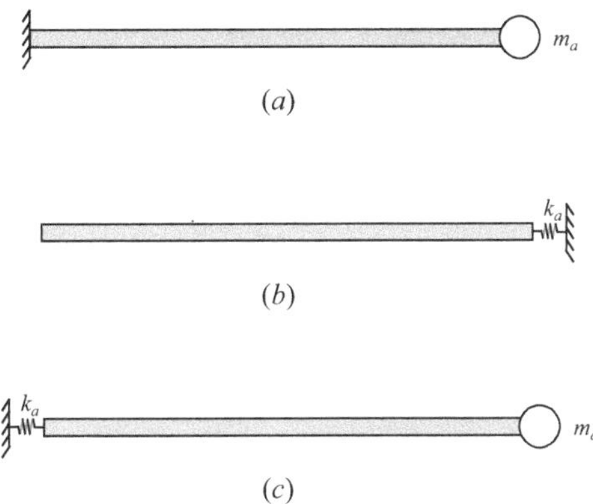

Fig. 2.11 Axial nanorods with different boundary conditions. (**a**) Clamped supported-mass attached (**b**) Free-spring attached (**c**) Spring attached-mass attached

2.2. Compute the first five mode nondimensional frequencies of axial nanorods in Problem 2.1 according to nonlocal SRT under parameters as follows. Discuss the effects of nonlocal parameter and boundary conditions on differences between results.

Nanobeam length: $L = 20$ nm, diameter of circular cross-section: $d = 1$ nm, modulus of elasticity: $E = 1$ TPa, Poisson's ratio: $\upsilon = 0.19$, mass of unit volume: $\rho = 2300$ kg/m^3, stiffness ratio for axial spring attachment $\beta_k = 10$, mass ratio for axial spring attachment $\beta_m = 10$.

Compute according to three different nondimensional nonlocal parameters: $\alpha = 0$, 0.1, 0.2.

Compute the nondimensional frequencies as: $\overline{\omega} = \omega L^2 \sqrt{\rho/E}$.

2.3. Determine the frequency equation of nonlocal free vibration of axial nanorods given in Problem 2.1 by considering also an axial elastic medium according to nonlocal LBRT.

2.4. Determine the frequency equation of nonlocal free vibration of axial nanorods given in Fig. 2.12 according to nonlocal LBRT.

2.5. Compute the first five mode nondimensional frequencies of clamped-clamped axial nanorods according to nonlocal SRT and LBRT under the following parameters. Discuss the effects of nonlocal parameter and rod theory on differences between results.

Nanobeam length: $L = 20$ nm, diameter of circular cross-section: $d = 1$ nm, modulus of elasticity: $E = 1$ TPa, Poisson's ratio: $\upsilon = 0.19$, mass of unit volume: $\rho = 2300$ kg/m^3.

Compute according to three different nondimensional nonlocal parameters: $\alpha = 0$, 0.1, 0.2.

Compute the nondimensional frequencies as: $\overline{\omega} = \omega L^2 \sqrt{\rho/E}$.

2.6. Examine the Problem 2.5 by considering also an axial elastic medium according to two different medium stiffnesses: $k_M = 10$ kN/m^2, 20 kN/m^2. Discuss the effect of nonlocal parameter and medium stiffness on differences between results.

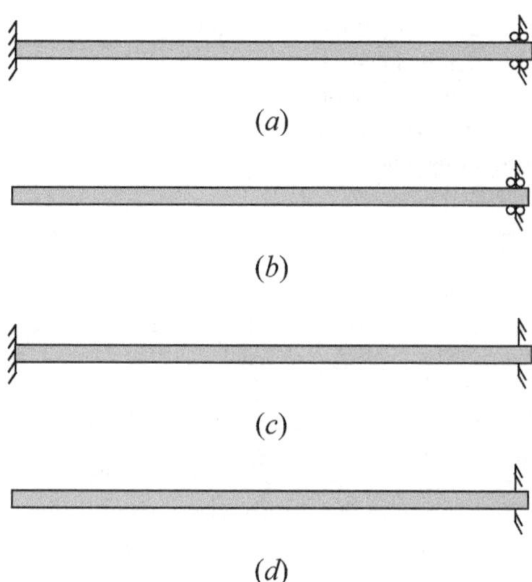

Fig. 2.12 Axial nanorods with different boundary conditions. (**a**) Clamped-Gripped Clamped (**b**) Free-Gripped Clamped (**c**) Clamped-Soft Clamped (**d**) Free-Soft Clamped

2.7. Examine the Problem 2.5 by considering also an axial elastic medium according to three different diameters of circular cross-section: $d = 1.2$ nm, 1.5 nm, 2 nm. Discuss the effect of nonlocal parameter and medium stiffness on differences between results.

References

1. B.N. Johnson, R. Muthasaran, Biosensing using dynamic-mode cantilever sensors: A review. Biosens. Bioelectron. **32**(1), 1–18 (2012)
2. J.W. Kang, J.H. Lee, H.J. Lee, O.K. Kwon, H.J. Hwang, Electromechanical modeling and simulations of nanobridge memory device. Phys. E **28**(3), 273–280 (2005)
3. J. Ahn, Z. Xu, J. Bang, P. Ju, X. Gao, T. Li, Ultrasensitive torque detection with an optically levitated nanorotor. Nat. Nanotechnol. **15**, 89–93 (2020)
4. D.L. John, L.C. Castro, J. Clifford, D.L. Pulfrey, Electrostatics of coaxial Schottky-barrier nanotube field-effect transistors, electrostatics of coaxial Schottky-barrier nanotube field-effect transistors. IEEE Trans. Nanotechnol. **2**(3), 175–180 (2003)
5. F.J. Giessibl, Advances in atomic force microscopy. Rev. Mod. Phys. **75**, 949–983 (2003)
6. N. Kilinc, O. Cakmak, A. Kosemen, E. Ermek, S. Ozturk, Y. Yerli, Z.Z. Ozturk, H. Urey, Fabrication of 1D ZnO nanostructures on MEMS cantilever for VOC sensor application. Sens. Actuators B Chem. **202**, 357–364 (2014)
7. J.W.S. Rayleigh, *The Theory of Sound* (Dover, New York, 1945)
8. A.E.H. Love, *A Treatise on the Mathematical Theory of Elasticity* (Dover, New York, 1944)
9. R.E.D. Bishop, Longitudinal waves in beams. Aeronaut. Q. **3**(2), 280–293 (1952)
10. D.K. Rao, J.S. Rao, Free and forced vibrations of rods according to Bishop's theory. J. Acoust. Soc. Am. **56**(4), 1792–1800 (1974)

11. M. Krawczuk, J. Grabowska, M. Palacz, Longitudinal wave propagation. Part I—Comparison of rod theories. J. Sound Vib. **295**(3–5), 461–478 (2006)
12. C. Mei, Comparison of the four rod theories of longitudinally vibrating rods. J. Vib. Control. **21**(8), 1639–1656 (2015)
13. M. Aydogdu, Axial vibration of the nanorods with the nonlocal continuum rod model. Phys. E **41**(5), 861–864 (2009)
14. M. Aydogdu, Axial vibration analysis of nanorods (carbon nanotubes) embedded in an elastic medium using nonlocal elasticity. Mech. Res. Commun. **43**, 34–40 (2012)
15. H.M. Numanoğlu, B. Akgöz, Ö. Civalek, On dynamic analysis of nanorods. Int. J. Eng. Sci. **130**, 33–50 (2018)
16. T. Murmu, S. Adhikari, Nonlocal vibration of carbon nanotubes with attached buckyballs at tip. Mech. Res. Commun. **38**(1), 62–67 (2011)
17. I. Ecsedi, A. Baksa, Free axial vibration of nanorods with elastic medium interaction based on nonlocal elasticity and Rayleigh model. Mech. Res. Commun. **86**, 1–4 (2017)
18. M.Ö. Yayli, Axial vibration analysis of a Rayleigh nanorod with deformable boundaries. Microsyst. Technol. **26**, 2661–2671 (2020)
19. Ş.D. Akbaş, Axially forced vibration analysis of cracked a nanorod. J. Comput. Appl. Mech. **50**(1), 63–68 (2019)
20. J.C. Hsu, H.L. Lee, W.J. Chang, Longitudinal vibration of cracked nanobeams using nonlocal elasticity theory. Curr. Appl. Phys. **11**(6), 1384–1388 (2011)
21. S. Adhikari, T. Murmu, M.A. McCarthy, Dynamic finite element analysis of axially vibrating nonlocal rods. Finite Elem. Anal. Des. **63**, 42–50 (2013)
22. Ç. Demir, Ö. Civalek, Torsional and longitudinal frequency and wave response of microtubules based on the nonlocal continuum and nonlocal discrete models. Appl. Math. Model. **37**(22), 9355–9367 (2013)
23. H.M. Numanoğlu, Ö. Civalek, Novel size-dependent finite element formulation for modal analysis of cracked nanorods. Mater. Today Commun. **31**, 103545 (2022)
24. T. Murmu, S. Adhikari, M.A. McCarthy, Axial vibration of embedded nanorods under transverse magnetic field effects via nonlocal elastic continuum theory. J. Comput. Theor. Nanosci. **11**(5), 1230–1236 (2014)
25. S.Q. Guo, S.P. Yang, Axial vibration analysis of nanocones based on nonlocal elasticity theory. Acta Mech. Sinica **28**, 801–807 (2012)
26. M. Danesh, A. Farajpour, M. Mohammadi, Axial vibration analysis of a tapered nanorod based on nonlocal elasticity theory and differential quadrature method. Mech. Res. Commun. **39**(1), 23–27 (2012)
27. M.S. Taima, T. El-Sayed, S.H. Farghaly, Longitudinal vibration analysis of a stepped nonlocal rod embedded in several elastic media. J. Vib. Eng. Technol. **10**, 1399–1412 (2022)
28. M. Aydoğdu, I. Elishakoff, On the vibration of nanorods restrained by a linear spring in-span. Mech. Res. Commun. **57**, 90–96 (2014)
29. A. Bahrami, A. Zargaripoor, H. Shiri, N. Khosravi, Size-dependent free vibration of axially functionally graded tapered nanorods having nonlinear spring constraint with a tip nanoparticle. J. Vib. Control. **25**(21–22), 2769–2783 (2019)
30. D.Z. Karličić, S. Ayed, E. Flaieh, Nonlocal axial vibration of the multiple Bishop nanorod system. Math. Mech. Solids **24**(6), 1668–1691 (2019)
31. X.F. Li, Z.B. Shen, K.Y. Lee, Axial wave propagation and vibration of nonlocal nanorods with radial deformation and inertia. J. Appl. Math. Mech. **97**(5), 602–616 (2017)
32. Ö. Civalek, H.M. Numanoğlu, Nonlocal finite element analysis for axial vibration of embedded love–bishop nanorods. Int. J. Mech. Sci. **188**, 105939 (2020)

33. B. Uzun, U. Kafkas, M.Ö. Yaylı, Axial dynamic analysis of a Bishop nanorod with arbitrary boundary conditions. ZAMM **100**(12), e202000039 (2020)
34. M. Arda, Axial dynamics of functionally graded Rayleigh-Bishop nanorods. Microsyst. Technol. **27**, 269–282 (2021)
35. R. Nazemnezhad, K. Kamali, Free axial vibration analysis of axially functionally graded thick nanorods using nonlocal Bishop's theory. Steel Compos. Struct. **28**(6), 749–758 (2018)
36. Ö. Civalek, B. Uzun, M.Ö. Yaylı, B. Akgöz, Size-dependent transverse and longitudinal vibrations of embedded carbon and silica carbide nanotubes by nonlocal finite element method. Eur. Phys. J. Plus **135**(4), 381 (2020)

Torsional Vibration of Nonlocal Shafts 3

3.1 Torsional Elements

Torsion rods are defined by the angular rotational motion of sections perpendicular to the rod axis. The rods can also perform torsional vibration if it is subjected to time-dependent torque on its axis as well as the rods can vibrate axially. If we examine the macrolevel, for example, transmission shafts and axles provide the main movement of macrostructure in the motion mechanism of automobiles and wind turbines. Torsion is also observed in electric generators. In addition to these, the fact that the centers of mass and stiffness in the floor plans of building structures are different from each other may constitute torsional motion during earthquake response which is a structural dynamic investigation.

The mechanics of torsion is studied in the simplest sense through torsion of circular section [1]. However, the solution of torsion is not simple enough to be reduce to a circular section. Warping effects in the torsion of noncircular cross-sections are also examined and analysis is a little more complicated. If we descend from the macro scale to the nanoscale, the cross-sections of one-dimensional nanorods may not be just circular. In the nanotechnological studies, the presences of elliptical [2], triangular [3], and quadrilateral [4] cross-section geometries have been determined in one-dimensional and longitudinal nanorods. When different cross-section geometries and nanomaterials such as nanotubes, nanowires, and nanostrips in various NEMS organizations are considered, it is necessary to investigate the mechanical behavior of these structures under rotational forces.

According to the elementary torsion theory, while only the presence of shear stresses is sufficient for the analysis of circular nanostructures, higher-order mechanical effects due to warping and lateral inertia in noncircular sections are also included in the analysis. Approaches such as Saint-Venant [5] and Timoshenko-Gere [6, 7] are used for torsional vibration of noncircular rods. The warping function, which varies according to the section

Öm. Civalek et al., *Mechanical Behavior and Vibration of Nano-Scaled Rods, Beams, and Frames*, Synthesis Lectures on Engineering, Science, and Technology,
https://doi.org/10.1007/978-3-032-12023-6_3

geometry especially, is a quite different effect than the dynamics of axial nanorods with shear effect in the previous section (Love-Bishop nanorods) and the dynamics of shear deformable nanobeams in the next section (Timoshenko nanobeams). Therefore, it is understood that nonlocal torsion should be considered separately for the mechanical analysis of nanostructures.

This chapter focuses on the nonlocal free torsional vibrations of nanorods. Assuming the torsion rod is in a fully elastic medium, nonlocal equations of motion are obtained under the related mechanical theories of circular rods and elliptical rods. These equations are solved and applications of solution on nanorods with different boundary conditions are given. A few numerical examples are presented and the torsional behavior of nanorods is examined based on the atomic size effect.

3.2 Recent Contributions

Nonlocal torsional dynamics also has an important place in research on the size-dependent mechanics of atomic scaled structures. Researchers have studied the nonlocal torsional free vibration behavior of a one-dimensional and continuous rod (or shaft) model of nanostructures according to different effects. In this regard, the analytical investigation of free vibration of carbon nanotubes embedded in elastic media can be mentioned [8]. In addition to this, there are studies that formulate the free vibration of nanorods by way of semi-analytically different approaches [9, 10]. The free vibration of mass attached rod model of carbon nanotube has been investigated [11]. Mass attached and cracked nanorods has also been studied [12]. Moreover, nonlocal torsional vibration has been formulated for the effects of crack [13] and elastic medium [14] for the nanorods with deformable boundary conditions. Furthermore, deformable nanorods made of functionally graded materials are another subject in this field [15].

A finite element procedure for torsional vibration based on two different forms of nonlocal elasticity has been expressed [16]. The torsional dynamic behavior of embedded nanorods with deformable boundary conditions has also been given via finite elements [17].

On the other hand, it is noted that the nonlocal free torsional vibration of noncircular nanorods has also been a subject studied in the scientific literature. A nonlocal torsional vibration analysis has been presented for elliptical [18] and triangular [19] cross-section of nanorods with general and deformable boundary conditions. Additionally, the atomic size dependent torsional dynamics of nanorods have been researched for rectangular section [20].

3.3 Nonlocal Vibration of Circular Shaft

Elementary torsion theory of nanorods is valid for circular nanorods. In this theory, known as the Bernoulli-Navier approximation, circular sections before torsion still remain circular after torsion. However, before torsion, all fibers parallel to the rod axis transform into an elliptical spring after torsion. Also, the torsional angle of circular section is defined as a linear function of the rod axis.

The subject of this section is the derivation of equation of motion of nanorods embedded in a fully elastic medium according to the nonlocal elasticity theory and presentation of a few applications related to this. The equation of motion will be obtained via first equilibrium approach and then algebra of variation.

3.3.1 Equation of Motion: Equilibrium Approach

The free body diagram of nanorod and its differential element embedded in a fully elastic medium are depicted in Fig. 3.1. First, the rotational stability of differential element is given below:

$$\sum M_{net} = \rho I_p \mathrm{d}x \frac{\partial^2 \theta}{\partial t^2} \tag{3.1}$$

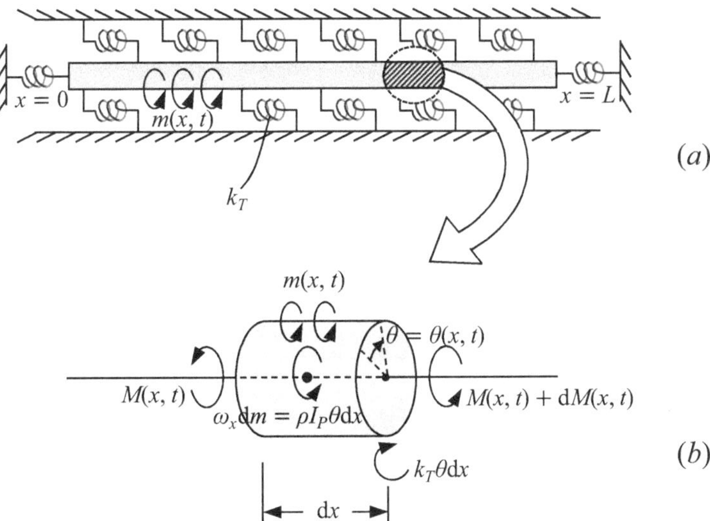

Fig. 3.1 Torsional vibration of a nanorod embedded on elastic medium. (**a**) Schematic representation of rod (**b**) Free body diagram

where $t = t(x, t)$ explains the torsional rotation of rod. ρ and I_p are unit volume mass and polar moment of inertia of rod. The internal force in torsional vibration is written as $M = M(x, t)$.

Considering Fig. 3.1b, the rotational equilibrium for equilibrium analysis can be established as follows:

$$-M + (M + \mathrm{d}M) - k_T\theta \mathrm{d}x + m\mathrm{d}x = \rho I_p \mathrm{d}x \frac{\partial^2 \theta}{\partial t^2} \tag{3.2}$$

where $m = m(x, t)$ denotes the external torsional excitation. In addition, k_T is the stiffness of elastic medium. Editing of equation gives the following expression:

$$\frac{\partial M}{\partial x} = \rho I_p \frac{\partial^2 \theta}{\partial t^2} + k_T\theta - m \tag{3.3}$$

The equation of motion is obtained by integrating the constitutive equation of nonlocal elasticity through the area as in the previous section. Accordingly, firstly the following expression is written:

$$\left(1 - (e_0 a)^2 \frac{\partial^2}{\partial x^2}\right)\sigma_{ij} = s_{ij} \tag{3.4}$$

where s_{ij} denotes classical shear stresses. σ_{ij} is called as nonlocal shear stress components. It should be noted that the indices of stress components are different from each other since the torsional motion is caused by shear stresses. This equation presents the nonlocal constitutive relations for shear stresses that are effective in torsion as follows:

$$\left(1 - (e_0 a)^2 \frac{\partial^2}{\partial x^2}\right)\sigma_{xy} = s_{xy} \tag{3.5}$$

$$\left(1 - (e_0 a)^2 \frac{\partial^2}{\partial x^2}\right)\sigma_{xz} = s_{xz} \tag{3.6}$$

where the following relations need to be known:

$$M_{nl} = M = \int_A (\sigma_{xz} y - \sigma_{xy} z)\mathrm{d}A, \quad M_c = \int_A (s_{xz} y - s_{xy} z)\mathrm{d}A, \quad s_{xy} = -Gz\frac{\partial \theta}{\partial x}, \quad s_{xz} = Gy\frac{\partial \theta}{\partial x} \tag{3.7}$$

where G is the shear modulus and is calculated as follows for isotropic elastic solids:

$$G = \frac{E}{2(1+\upsilon)} \tag{3.8}$$

where υ indicates Poisson's ratio.

Equations (3.5) and (3.6) are multiplied by $-z$ and y, respectively, and integrated according to the area. Then, the following expression can be written by summing the results:

$$\left(1-(e_0a)^2\frac{\partial^2}{\partial x^2}\right)\int_A \left(\sigma_{xz}y-\sigma_{xy}z\right)\mathrm{d}A = \int_A \left(s_{xz}y-s_{xy}z\right)\mathrm{d}A \tag{3.9}$$

If the first derivative of equation obtained after using of Eq. (3.7) into Eq. (3.9) is taken and Eq. (3.3) is used, the equation of motion of nonlocal torsional vibration can be obtained as follows:

$$\left[GI_P+(e_0a)^2k_T\right]\frac{\partial^2\theta}{\partial x^2}-\rho I_P\frac{\partial^2\theta}{\partial t^2}-k_T\theta+(e_0a)^2\rho I_P\frac{\partial^4\theta}{\partial x^2\partial t^2}+m-(e_0a)^2\frac{\partial^2 m}{\partial x^2}=0 \tag{3.10}$$

In the previous section, it was stated that the mechanical boundary condition is necessary to determine the atomic size-dependent mechanical behavior of axial rods. Thus, the bending moment can be reached after Eq. (3.7) is substituted into Eq. (3.9):

$$M=\left[GI_P+(e_0a)^2k_T\right]\frac{\partial\theta}{\partial x}+(e_0a)^2\rho I_P\frac{\partial^3\theta}{\partial x\partial t^2}-(e_0a)^2\frac{\partial^2 m}{\partial x^2}=0 \tag{3.11}$$

3.3.2 Equation of Motion: Hamilton's Principle

According to Hamilton's Principle, the first variation of potential energy of torsional nanorod should equal 0 for (t_1, t_2) time interval:

$$\delta\Pi = \int_{t_1}^{t_2} [U-(T+W)]\mathrm{d}t = 0 \tag{3.12}$$

where U is the internal strain energy, T is the kinetic energy, and W is the work of external forces. The energy expressions for are constituted from the kinematic relationships of torsional nanorod. For this, firstly the displacement components of nanorod are written [17]:

$$u(x,t)=0,\quad v(x,t)=-z\theta(x,t),\quad w(x,t)=y\theta(x,t) \tag{3.13}$$

where u, v, and w express the axial, lateral, and transverse displacements of structure, respectively. Since only the cross-sectional rotation will be seen in the circular nanorod, axial displacement does not occur. The strain expressions are given according to the displacement components as follows:

$$\begin{aligned}
&\varepsilon_{xx}=\frac{1}{2}\left(\frac{\partial u}{\partial x}+\frac{\partial u}{\partial x}\right)=0, \varepsilon_{yy}=\frac{1}{2}\left(\frac{\partial v}{\partial y}+\frac{\partial v}{\partial y}\right)=0, \varepsilon_{zz}=\frac{1}{2}\left(\frac{\partial w}{\partial z}+\frac{\partial w}{\partial z}\right)=0\\
&\gamma_{xy}=\frac{\partial u}{\partial y}+\frac{\partial u}{\partial x}=-z\frac{\partial \theta}{\partial x}, \gamma_{xz}=\frac{\partial u}{\partial z}+\frac{\partial w}{\partial x}=y\frac{\partial \theta}{\partial x}, \gamma_{yz}=\frac{\partial v}{\partial z}+\frac{\partial w}{\partial y}=0
\end{aligned} \tag{3.14}$$

where ε_{xx}, ε_{yy}, and ε_{zz} are the longitudinal, lateral, and transverse strains, respectively. Other expressions describe shear strains. Also, nonlocal stress components can be defined as:

$$\sigma_{xx}=\sigma_{yy}=\sigma_{zz}=\sigma_{yz}=0,\quad \sigma_{xy}=G\gamma_{xy}=-Gz\frac{\partial \theta}{\partial x},\quad \sigma_{xz}=G\gamma_{xz}=Gy\frac{\partial \theta}{\partial x} \tag{3.15}$$

Using these expressions, the energy equations are written as follows:

$$\begin{aligned}
U&=\frac{1}{2}\int_V\left(\sigma_{xx}\varepsilon_{xx}+\sigma_{yy}\varepsilon_{yy}+\sigma_{zz}\varepsilon_{zz}+2\sigma_{xy}\gamma_{xy}+2\sigma_{xz}\gamma_{xz}+2\sigma_{yz}\gamma_{yz}\right)\mathrm{d}V\\
&\quad+\frac{1}{2}\int_0^L k_T\theta^2\mathrm{d}x+\frac{1}{2}\sum_{i=1}^{n}k_{a,i}[\theta(x_i,t)]^2\\
&=\frac{1}{2}\int_0^L M\frac{\partial \theta}{\partial x}\mathrm{d}x+\frac{1}{2}\int_0^L k_T\theta^2\mathrm{d}x+\frac{1}{2}k_a[\theta(0,t)]^2+\frac{1}{2}k_a[\theta(L,t)]^2
\end{aligned} \tag{3.16}$$

$$\begin{aligned}
T&=\frac{1}{2}\int_V\rho\left[\left(\frac{\partial u}{\partial t}\right)^2+\left(\frac{\partial v}{\partial t}\right)^2+\left(\frac{\partial w}{\partial t}\right)^2\right]\mathrm{d}V+\frac{1}{2}\sum_{i=1}^{n}I_{a,i}\left(\frac{\partial \theta(x_i,t)}{\partial t}\right)^2\\
&=\frac{1}{2}\int_0^L \rho I_P\left(\frac{\partial \theta}{\partial t}\right)^2\mathrm{d}x+\frac{1}{2}I_a\left(\frac{\partial \theta(0,t)}{\partial t}\right)^2+\frac{1}{2}I_a\left(\frac{\partial \theta(L,t)}{\partial t}\right)^2
\end{aligned} \tag{3.17}$$

$$W=\int_0^L m\theta\mathrm{d}x \tag{3.18}$$

where $k_{a,\,i}$ defines the stiffness of generalized point torsional spring attachment (rotational spring) and $I_{a,\,i}$ specifies the rotational inertia of generalized torsional mass attachment (rotational disk). In the research in this section, it is assumed that the attachments exist at the ends of nanorod.

The first variations of energy equations given by Eqs. (3.16)–(3.18) are obtained below:

$$\int_{t_1}^{t_2} \delta U \mathrm{d}t = \int_{t_1}^{t_2}\int_0^L M\frac{\partial \theta}{\partial x}\delta\frac{\partial \theta}{\partial x}\mathrm{d}x\mathrm{d}t + \int_{t_1}^{t_2}\int_0^L k_T\theta\delta\theta\mathrm{d}x\mathrm{d}t + \int_{t_1}^{t_2} k_a\theta(0,t)\delta\theta(0,t)\mathrm{d}t + \int_{t_1}^{t_2} k_a\theta(L,t)\delta\theta(L,t)\mathrm{d}t \tag{3.19}$$

$$\int_{t_1}^{t_2} \delta T \mathrm{d}t = \int_{t_1}^{t_2}\int_0^L \rho I_P\frac{\partial \theta}{\partial t}\delta\frac{\partial \theta}{\partial t}\mathrm{d}x\mathrm{d}t + \int_{t_1}^{t_2} I_a\frac{\partial \theta(0,t)}{\partial t}\delta\frac{\partial \theta(0,t)}{\partial t}\mathrm{d}t + \int_{t_1}^{t_2} I_a\frac{\partial \theta(L,t)}{\partial t}\delta\frac{\partial \theta(L,t)}{\partial t}\mathrm{d}t \tag{3.20}$$

$$\int_{t_1}^{t_2} \delta W \mathrm{d}t = \int_{t_1}^{t_2}\int_0^L m\delta\theta\mathrm{d}x\mathrm{d}t \tag{3.21}$$

The partial integration results of some variation expressions in these equations can be presented as follows:

$$\int_0^L M\delta\frac{\partial \theta}{\partial x}\mathrm{d}x = M\delta\theta|_0^L - \int_0^L \frac{\partial M}{\partial x}\delta\theta\mathrm{d}x,\quad \int_{t_1}^{t_2} \rho I_P\frac{\partial \theta}{\partial t}\delta\frac{\partial \theta}{\partial t}\mathrm{d}t = \rho I_P\frac{\partial \theta}{\partial t}\delta\theta\Big|_{t_1}^{t_2} - \int_{t_1}^{t_2}\rho I_P\frac{\partial^2 \theta}{\partial t^2}\delta\theta\mathrm{d}t$$
$$\int_{t_1}^{t_2} I_a\frac{\partial \theta(0,t)}{\partial t}\delta\frac{\partial \theta(0,t)}{\partial t}\mathrm{d}t = I_a\frac{\partial \theta(0,t)}{\partial t}\delta\theta(0,t)\Big|_{t_1}^{t_2} - \int_{t_1}^{t_2} I_a\frac{\partial^2 \theta(0,t)}{\partial t^2}\delta\theta(0,t)\mathrm{d}t$$
$$\int_{t_1}^{t_2} I_a\frac{\partial \theta(L,t)}{\partial t}\delta\frac{\partial \theta(L,t)}{\partial t}\mathrm{d}t = I_a\frac{\partial \theta(L,t)}{\partial t}\delta\theta(L,t)\Big|_{t_1}^{t_2} - \int_{t_1}^{t_2} I_a\frac{\partial^2 \theta(L,t)}{\partial t^2}\delta\theta(L,t)\mathrm{d}t \tag{3.22}$$

Using the equations in Eqs. (3.19)–(3.22) into Eq. (3.13), the following equation is obtained:

$$\int_{t_1}^{t_2}\int_0^L\left(-\frac{\partial M}{\partial x}+k_T\theta+\rho I_P\frac{\partial^2\theta}{\partial t^2}-m\right)\delta\theta \mathrm{d}x\mathrm{d}t+\int_{t_1}^{t_2}M\delta\theta|_0^L\mathrm{d}t+\int_{t_1}^{t_2}\left(k_a\theta(0,t)\right.$$
$$\left.+I_a\frac{\partial^2\theta}{\partial t^2}(0,t)\right)\delta\theta(0,t)\mathrm{d}t+\int_{t_1}^{t_2}\left(k_a\theta(L,t)+m_a\frac{\partial^2\theta}{\partial t^2}(L,t)\right)\delta\theta(L,t)\mathrm{d}t$$
$$+\int_0^L\left(-\rho I_P\frac{\partial\theta}{\partial t}\right)\delta\theta\Big|_{t_1}^{t_2}\mathrm{d}x+\left(-I_a\frac{\partial\theta}{\partial t}(0,t)\right)\delta\theta(0,t)\Big|_{t_1}^{t_2}+\left(-I_a\frac{\partial\theta}{\partial t}(L,t)\right)\delta\theta(L,t)\Big|_{t_1}^{t_2}=0 \tag{3.23}$$

All parts of this equation should be equal to 0. According to this, the following equation is written for the $\delta\theta \neq 0$ limit in the double integral expression:

$$\frac{\partial M}{\partial x}=\rho I_p\frac{\partial^2\theta}{\partial t^2}+k_T\theta-m \tag{3.24}$$

This equation is the same as Eq. (3.3). Using Eqs. (3.4)–(3.7), the nonlocal equation of motion in Eq. (3.10) is reached. Also, the mechanical boundary condition neglecting the attachments is given in Eq. (3.11) and the attached nonlocal torsion nanorods will be investigated in the next sections.

3.3.3 Solution of Equation of Motion

In the solution of equation of motion, the dynamic external excitation is expressed as $m=0$ for the free torsional vibration. In this case, Eq. (3.10) becomes a homogeneous differential equation. We know from the previous section that the solution of homogeneous and ordinary differential equations is provided by the method of separation of variables. According to this:

$$\theta(x,t)=\varphi(x)T(t)=\varphi(x)\sin(\omega t-\alpha) \tag{3.25}$$

where θ is the time-dependent rotation of rod. φ denotes for the static rotation function, that is, the mode shape. T denotes the harmonic function. In addition to these, ω is the natural frequency, t is the time, and α is the phase angle. Using Eq. (3.25) into Eq. (3.10) gives the following equation:

$$\frac{d^2\varphi(x)}{dx^2} + \eta^2\varphi(x) = 0 \tag{3.26}$$

The following definition is valid here:

$$\eta^2 = \frac{-k_T + \omega^2\rho I_P}{GI_P + (e_0a)^2k_T - \omega^2(e_0a)^2\rho I_P} \tag{3.27}$$

The solution for Eq. (3.26) is determined by the following equation:

$$\varphi(x) = Ce^{mx} \tag{3.28}$$

If this expression is substituted into Eq. (3.26), the following equation is obtained:

$$m^2 + \eta^2 = 0 \tag{3.29}$$

The roots of equation are written as follows:

$$m_1 = \mathbf{i}\eta, \quad m_2 = -\mathbf{i}\eta \tag{3.30}$$

Thus, the solution of Eq. (3.26) is indicated as follows:

$$\varphi(x) = C_1e^{\mathbf{i}\eta x} + C_2e^{-\mathbf{i}\eta x} \tag{3.31}$$

where C_1 and C_2 are constant coefficients determined with the help of boundary conditions. A further rearrangement of this equation can be performed with the Euler transformations given in Eqs. (2.27) and (2.28). Thus, the mode shape equation of nonlocal torsional vibration is determined as follows:

$$\varphi(x) = C_1 \cos \eta x + C_2 \sin \eta x \tag{3.32}$$

3.3.4 Boundary Conditions of Nanoshafts

To determinate the nonlocal free torsional vibration behavior of circular nanorods depends on Eq. (3.32). For this, the boundary conditions of the rod should be known. The unknown coefficients in the equation can be calculated by way of boundary conditions. Like simple axial nanorods, the boundary conditions of circular torsional nanorods can be composed of general known conditions such as clamped and free, or they can be defined end attachments such as a rotational spring or a rotational mass. In this section, several boundary conditions for nonlocal torsional vibration of nanorods will be investigated and their free vibration behavior will be examined through some examples.

1. **Clamped end (C):** Clamped ends do not allow rotational motion:

$$\varphi = 0 \tag{3.33}$$

2. **Free end (F):** If the external dynamic torsional force does not act on the free end, there is no internal effect at this point. So, using Eq. (3.11), the nonlocal torsional internal effect at the free end can be given as:

$$M = \left[GI_P + (e_0a)^2 k_T\right] \frac{\partial \theta}{\partial x} + (e_0a)^2 \rho I_P \frac{\partial^3 \theta}{\partial x \partial t^2} = 0 \tag{3.34}$$

where using Eq. (3.25), the following expression is written for the mode equation:

$$B \frac{d\varphi}{dx} = 0 \tag{3.35}$$

where

$$B = GI_P + (e_0a)^2 k_T - \omega^2 (e_0a)^2 \rho I_P \tag{3.36}$$

The value of B cannot be equal to zero. Thus, the boundary condition for the free end is expressed as:

$$\varphi' = 0 \tag{3.37}$$

3. **Attached ends (MA and SA):** The tip attachment affects the nonlocal torsional moment. Similar to axial nanorods, the tip attachment in a torsion nanorod can be modeled as a rotational spring or rotational mass. Now, considering both effects together, let write the following expression which should equal zero in Eq. (3.23)

$$\begin{aligned} &\int_{t_1}^{t_2} M\delta\theta|_0^L dt + \int_{t_1}^{t_2} \left(k_a\theta(0,t) + I_a \frac{\partial^2\theta}{\partial t^2}(0,t)\right)\delta\theta(0,t)dt + \int_{t_1}^{t_2} \left(k_a\theta(L,t)\right. \\ &\left. + m_a \frac{\partial^2\theta}{\partial t^2}(L,t)\right)\delta\theta(L,t)dt = 0 \end{aligned} \tag{3.38}$$

For the right end of nanorod, that is $\delta\theta(L, t) \neq 0$ limit, the following equation can be reached:

$$M(L,t) = -k_a\theta(L,t) - I_a\frac{\partial^2\theta}{\partial t^2}(L,t) \tag{3.39}$$

Using Eqs. (3.25) and (3.35) into this equation, the dynamic boundary condition for the attached right end is written as follows:

$$\left[GI_P + (e_0a)^2k_T - \omega^2(e_0a)^2\rho I_P\right]\varphi'(L) = (-k_a + \omega^2 I_a)\varphi(L) \tag{3.40}$$

According to the left end of nanorod, that is $\delta\theta(0,t) \neq 0$ boundary, the following equation is obtained

$$M(L,t) = k_a\theta(L,t) + I_a\frac{\partial^2\theta}{\partial t^2}(L,t) \tag{3.41}$$

The result is the following equation:

$$\left[GI_P + (e_0a)^2k_T - \omega^2(e_0a)^2\rho I_P\right]\varphi'(L) = (k_a - \omega^2 I_a)\varphi(L) \tag{3.42}$$

It should be certainly expressed that in nonlocal torsion nanorods as well as in axial nanorods, it is impossible to directly derive the nonlocal frequency equation in the case of tip attachment. Of course, the roots of frequency equation that defines the natural frequency values can be calculated by programming on the computer.

Example 3.1 Determine the frequency equation of nonlocal free torsional vibration of clamped circular nanorod embedded in the elastic medium shown in Fig. 3.2.

Solution. The boundary conditions of structure are introduced as follows:

$$\varphi(0) = 0, \quad \varphi(L) = 0 \tag{3.E3.1.1}$$

With the using of first boundary condition into Eq. (3.33):

$$C_1 = 0 \tag{3.E3.1.2}$$

Substituting second boundary condition into the remainder of mode equation yields below:

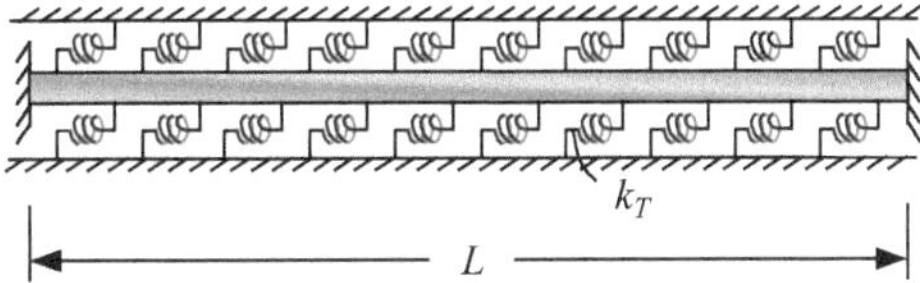

Fig. 3.2 Clamped circular nanorod embedded in the elastic medium

$$C_2 \sin \eta L = 0 \tag{3.E3.1.3}$$

where C_2 cannot be equal to 0. Thus, the following expression can be written:

$$\sin \eta L = 0 \tag{3.E3.1.4}$$

The periodic solution of this equation is expressed as:

$$\eta L = n\pi \tag{3.E3.1.5}$$

where n is the mode number of torsional vibration. Using of Eq. (3.28) into Eq. (3.E3.1.5) gives the following result:

$$\omega = \sqrt{\frac{\left[GI_P + (e_0 a)^2 k_T\right]\left(\frac{n\pi}{L}\right)^2 + k_T}{\rho I_P\left[1 + (e_0 a)^2\left(\frac{n\pi}{L}\right)^2\right]}} \tag{3.E3.1.6}$$

Several numerical calculations about natural frequency are presented in Example 3.5.

Example 3.2 Determine the frequency equation of nonlocal free torsional vibration of clamped-free circular nanorod embedded in the elastic medium shown in Fig. 3.3.

Solution. The boundary conditions of this nanoshaft can be clarified as follows:

$$\varphi(0) = 0, \quad \varphi'(L) = 0 \tag{3.E3.2.1}$$

The result of first boundary condition is as in Eq. (3.E3.1.2). With the substituting of other boundary condition into Eq. (3.33), following equation is occurred:

$$C_2 \eta \cos \eta L = 0 \tag{3.E3.2.2}$$

Since C_1 and η are not equal zero, following equation can be reached:

$$\cos \eta L = 0 \tag{3.E3.2.3}$$

It can be stated here that the periodic solution is as follows:

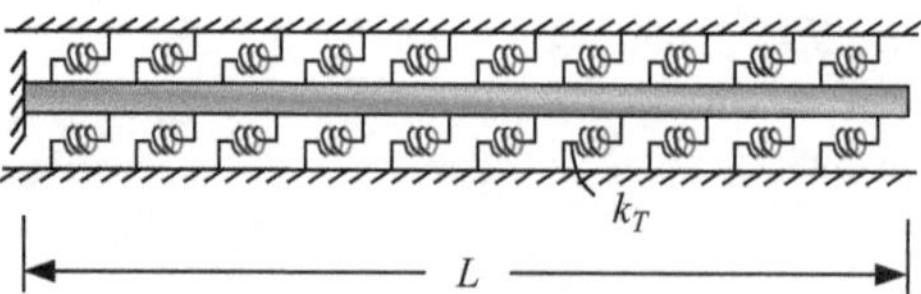

Fig. 3.3 Clamped-free circular nanorod embedded in the elastic medium

$$\eta L = \frac{2n-1}{2}\pi \tag{3.E3.2.4}$$

As in the solution of Example 3.1, the nonlocal frequency equation for this nanoshaft is attained as follows:

$$\omega = \sqrt{\frac{\left[GI_P + (e_0a)^2k_T\right]\left(\frac{(2n-1)\pi}{2L}\right)^2 + k_T}{\rho I_P\left[1 + (e_0a)^2\left(\frac{(2n-1)\pi}{2L}\right)^2\right]}} \tag{3.E3.2.5}$$

It can be stated that there is a series expansion in addition to the analytical solution for the nonlocal free vibration frequencies of clamped or clamped-free torsional nanorods. According to this, the following series expansions providing the geometric boundary conditions should be employed into Eq. (3.10):

$$\begin{aligned}
&\text{Clamped-Free}: \theta(x,t) = \sum_{n=1}^{\infty}\theta_n \sin\left(\frac{(2n-1)\pi x}{2L}\right)\sin(\omega t - \varphi) \\
&\text{Clamped-Clamped}: \theta(x,t) = \sum_{n=1}^{\infty}\varphi_n \sin\left(\frac{n\pi x}{L}\right)\sin(\omega t - \varphi)
\end{aligned} \tag{3.43}$$

where let substitute the series expansion for the clamped nanorod into Eq. (3.10):

$$\begin{aligned}
&-\left[GI_P + (e_0a)^2k_T\right]\left(\frac{(2n-1)\pi}{2L}\right)^2\sum_{n=1}^{\infty}\theta_n \sin\left(\frac{(2n-1)\pi x}{2L}\right)\sin(\omega t - \theta) \\
&+\omega^2\rho I_P\sum_{n=1}^{\infty}\theta_n \sin\left(\frac{(2n-1)\pi x}{2L}\right)\sin(\omega t - \theta) - k_T\sum_{n=1}^{\infty}\theta_n \sin\left(\frac{(2n-1)\pi x}{2L}\right)\sin(\omega t - \theta) \\
&-\omega^2(e_0a)^2\rho I_P\left(\frac{(2n-1)\pi}{2L}\right)^2\sum_{n=1}^{\infty}\theta_n \sin\left(\frac{(2n-1)\pi x}{2L}\right)\sin(\omega t - \theta) = 0
\end{aligned} \tag{3.44}$$

where if we simplify the series expressions, Eq. (3.E3.2.5) is obtained.

Example 3.3 Determine the frequency equation of nonlocal free torsional vibration of free torsional mass attached circular nanorod embedded in the elastic medium shown in Fig. 3.4.

Solution. The boundary conditions of nanorod are defined below based on Eq. (3.40):

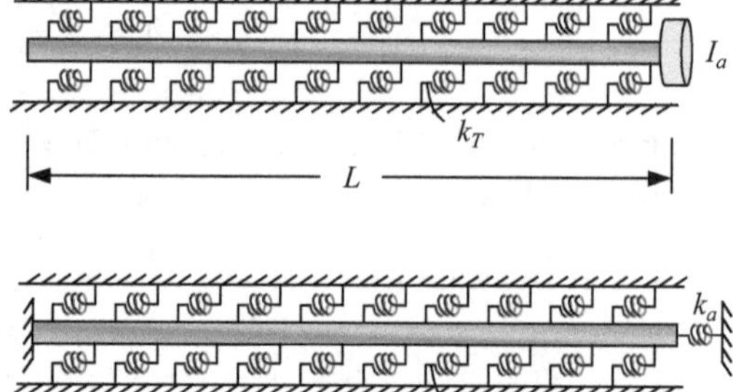

Fig. 3.4 Free torsional mass attached circular nanorod embedded in the elastic medium

Fig. 3.5 Clamped-torsional spring attached circular nanorod embedded in the elastic medium

$$\varphi(0) = 0, \quad B\varphi'(L) = \omega^2 I_a \varphi(L) \tag{3.E3.3.1}$$

The result of first boundary condition is $C_2 = 0$. By applying the second boundary condition:

$$-B\eta C_1 \sin \eta L = \omega^2 I_a C_1 \cos \eta L \tag{3.E3.3.2}$$

After some mathematical operations, this equation is expressed as:

$$\eta B \tan \eta L = -\omega^2 I_a \tag{3.E3.3.3}$$

An analytical solution of this equation is not possible. Instead, the roots of equation can be calculated by programming the equation with symbolic programming on the computer.

Example 3.4 Determine the frequency equation of nonlocal free torsional vibration of clamped-torsional spring attached circular nanorod embedded in the elastic medium shown in Fig. 3.5.

Solution. The boundary conditions of this nanoshaft are specified as follows:

$$\varphi(0) = 0, \quad B\varphi'(L) = -k_a \varphi(L) \tag{3.E3.4.1}$$

When the first boundary condition is applied into the mode equation, it is calculated as $C_1 = 0$. Let's apply the second boundary condition:

$$B\eta C_2 \cos \eta L = -k_a C_2 \sin \eta L \tag{3.E3.4.2}$$

If this expression is rearranged a little more, the frequency equation occurred as follows:

$$B\eta \cot \eta L = -k_a \tag{3.E3.4.3}$$

This expression becomes the following form after some mathematical operations:

$$\eta L \cot \eta L = -\left(1 + (e_0 a)^2 \eta^2\right) \beta_k \tag{3.E3.4.4}$$

Of course, the analytical solution of frequency equation cannot be reached in this nanorod type either. Again, the roots of equation, that is, nonlocal natural frequencies can be calculated by programming on the computer. Also, the following definition is valid in above nonlocal frequency equation:

$$\beta_k = \frac{k_a L}{EA} \tag{3.E3.4.5}$$

3.4 Nonlocal Vibration of Noncircular Shafts

According to the behavior of noncircular rod under torsion, the rod section does not remain straight like a circular section, instead a warping is observed. In Fig. 3.6, in addition to an elliptical torsion rod (Fig. 3.6a), the torsional motions of elliptical (Fig. 3.6b) and rectangular (Fig. 3.6c) rods are depicted. As it is seen, the fibers perpendicular to the axis of rods are not planar.

Different mechanical approaches such as Saint-Venant and Timoshenko-Gere have been developed for torsional vibration of noncircular rods. While the Saint-Venant theory does not consider axial inertia, the Timoshenko-Gere approach examines torsional vibration by considering this mechanical phenomenon. In parallel with the axial vibration investigations that are the subject of previous chapter, analogies between mechanical approaches can be established as follows:

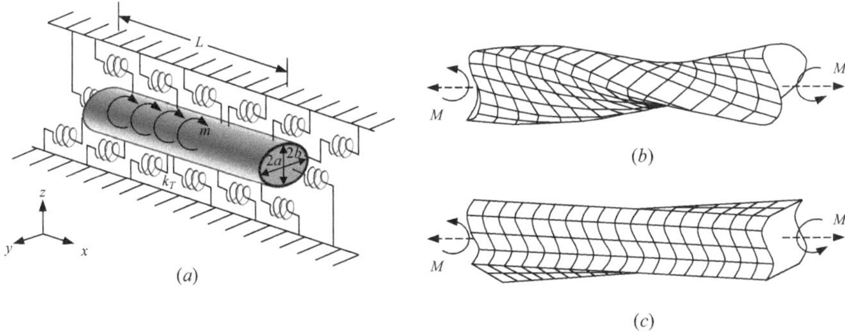

Fig. 3.6 Torsion of nanorods with noncircular cross-section: (**a**) Schematic view of elliptical nanorod, (**b**) Torsion of elliptical nanorod, (**c**) Torsion of rectangular nanorod

- Circular torsional nanorod—Simple (or *without shear*) axial nanorod,
- Saint-Venant torsional nanorod—Rayleigh axial nanorod,
- Timoshenko-Gere torsional nanorod—Love-Bishop axial nanorod.

The fact that the existence of one-dimensional and longitudinal elements with elliptical or rectangular cross-sections in nano-electro-mechanical systems technology research points to the determination of mechanical behavior of such nanostructures. Therefore, this section concerns the nonlocal torsional vibration of noncircular nanorods. Only elliptical cross-section and the Saint-Venant approach will be examined for fundamental investigations of nonlocal torsion with warping effect. Because, for example, in the torsion of rectangular rods, the analyses may become a little more complicated due to the warping function and the addition of lateral inertia.

3.4.1 Derivation of Equation of Motion

The nonlocal free vibration equation of related nanorod model is derived from Hamilton's principle. Torsional nanorod with elliptical cross-section and embedded in elastic medium is given in Fig. 3.6a. Here $2a$ and $2b$ define the section diameters of elliptical section parallel to the $y-$ and $z-$ axes, respectively.

First the displacement components of rod are expressed below:

$$u(x,t) = \psi(y,z)\frac{\partial \theta(x,t)}{\partial x}, v(x,t) = -z\theta(x,t), w(x,t) = y\theta(x,t) \tag{3.45}$$

where ψ defines the warping function seen for the first time compared to the circular bar. The warping function is formulated for an elliptical section [18]:

$$\psi(y,z) = yz\left(\frac{b^2 - a^2}{b^2 + a^2}\right) \tag{3.46}$$

Considering the displacement field, the strain1 components can be written as:

$$\begin{gathered}\varepsilon_{xx} = \frac{1}{2}\left(\frac{\partial u}{\partial x} + \frac{\partial u}{\partial x}\right) = 0, \varepsilon_{yy} = \frac{1}{2}\left(\frac{\partial v}{\partial y} + \frac{\partial v}{\partial y}\right) = 0, \varepsilon_{zz} = \frac{1}{2}\left(\frac{\partial w}{\partial z} + \frac{\partial w}{\partial z}\right) = 0 \\ \gamma_{xy} = \frac{\partial u}{\partial y} + \frac{\partial u}{\partial x} = \left(\frac{\partial \psi}{\partial y} - z\right)\frac{\partial \theta}{\partial x}, \ \gamma_{xz} = \frac{\partial u}{\partial z} + \frac{\partial w}{\partial x} = \left(\frac{\partial \psi}{\partial z} + y\right)\frac{\partial \theta}{\partial x} \ \gamma_{yz} = \frac{\partial v}{\partial z} + \frac{\partial w}{\partial y} = 0\end{gathered} \tag{3.47}$$

Because the first derivative of rotation is assumed as zero, longitudinal strain (ε_{xx}) is equal to zero. Thus, stress components are attained as follows:

$$\sigma_{xx}=\sigma_{yy}=\sigma_{zz}=\sigma_{yz}=0, \sigma_{xy}=G\gamma_{xy}=G\left(\frac{\partial\psi}{\partial y}-z\right)\frac{\partial\theta}{\partial x}, \sigma_{xz}=G\gamma_{xz}=G\left(\frac{\partial\psi}{\partial z}+y\right)\frac{\partial\theta}{\partial x} \tag{3.48}$$

Using Eqs. (3.47) and (3.48), the energy equations can be written as follows:

$$\begin{aligned} U &= \frac{1}{2}\int_V \left(\sigma_{xx}\varepsilon_{xx}+\sigma_{yy}\varepsilon_{yy}+\sigma_{zz}\varepsilon_{zz}+2\sigma_{xy}\gamma_{xy}+2\sigma_{xz}\gamma_{xz}+2\sigma_{yz}\gamma_{yz}\right)\mathrm{d}V \\ &+\frac{1}{2}\int_0^L k_T\theta^2\mathrm{d}x+\frac{1}{2}\sum_{i=1}^{n}k_{a,i}[\theta(x_i,t)]^2 \\ &= \frac{1}{2}\int_0^L M\frac{\partial\theta}{\partial x}\mathrm{d}x+\frac{1}{2}\int_0^L k_T\theta^2\mathrm{d}x+\frac{1}{2}k_a[\theta(0,t)]^2+\frac{1}{2}k_a[\theta(L,t)]^2 \end{aligned} \tag{3.49}$$

$$\begin{aligned} T &= \frac{1}{2}\int_V \rho\left[\left(\frac{\partial u}{\partial t}\right)^2+\left(\frac{\partial v}{\partial t}\right)^2+\left(\frac{\partial w}{\partial t}\right)^2\right]\mathrm{d}V+\frac{1}{2}\sum_{i=1}^{n}I_{a,i}\left(\frac{\partial\theta(x_i,t)}{\partial t}\right)^2 \\ &= \frac{1}{2}\int_0^L \rho I_P\left(\frac{\partial\theta}{\partial t}\right)^2\mathrm{d}x+\frac{1}{2}\int_0^L \rho I_\psi\left(\frac{\partial^2\theta}{\partial x\partial t}\right)^2\mathrm{d}x+\frac{1}{2}I_a\left(\frac{\partial\theta(0,t)}{\partial t}\right)^2 \\ &+\frac{1}{2}I_a\left(\frac{\partial\theta(L,t)}{\partial t}\right)^2 \end{aligned} \tag{3.50}$$

$$W=\int_0^L m\theta\mathrm{d}x \tag{3.51}$$

where I_ψ is known as warping moment of inertia and defined as follows:

$$I_\psi=\int_A \psi^2\mathrm{d}A \tag{3.52}$$

According to the variations of energy equations, first the related operation for the internal strain is given in Eq. (3.19). On the other hand, the first variation of kinetic energy is obtained as follows:

$$\int_{t_1}^{t_2}\delta T\mathrm{d}t=\int_{t_1}^{t_2}\int_{0}^{L}\rho I_P\frac{\partial\theta}{\partial t}\delta\frac{\partial\theta}{\partial t}\mathrm{d}x\mathrm{d}t+\int_{t_1}^{t_2}\int_{0}^{L}\rho I_\psi\frac{\partial^2\theta}{\partial x\partial t}\delta\frac{\partial^2\theta}{\partial x\partial t}\mathrm{d}x\mathrm{d}t$$
$$+\int_{t_1}^{t_2}I_a\frac{\partial\theta(0,t)}{\partial t}\delta\frac{\partial\theta(0,t)}{\partial t}\mathrm{d}t+\int_{t_1}^{t_2}I_a\frac{\partial\theta(L,t)}{\partial t}\delta\frac{\partial\theta(L,t)}{\partial t}\mathrm{d}t \tag{3.53}$$

Finally, the variation of work energy is written in Eq. (3.21). In addition to those given in Eq. (3.22), the first variation of double integral expression containing ρI_ψ factor in the above equation is expressed as:

$$\int_{t_1}^{t_2}\int_{0}^{L}\rho I_\psi\frac{\partial^2\theta}{\partial x\partial t}\delta\frac{\partial^2\theta}{\partial x\partial t}\mathrm{d}x\mathrm{d}t=\int_{0}^{L}\rho I_\psi\frac{\partial^2\theta}{\partial x\partial t}\delta\frac{\partial\theta}{\partial t}\bigg|_{t_1}^{t_2}\mathrm{d}x-\int_{t_1}^{t_2}\rho I_\psi\frac{\partial^3\theta}{\partial x\partial t^2}\delta\theta\bigg|_{0}^{L}\mathrm{d}t$$
$$+\int_{t_1}^{t_2}\int_{0}^{L}\rho I_\psi\frac{\partial^4\theta}{\partial x^2\partial t^2}\delta\theta\mathrm{d}x\mathrm{d}t \tag{3.54}$$

Using Eq. (3.19), (3.21), (3.22), and (3.53) into Eq. (3.12), the following equation can be attained:

$$\int_{t_1}^{t_2}\int_{0}^{L}\left(-\frac{\partial M}{\partial x}+k_T\theta+\rho I_P\frac{\partial^2\theta}{\partial t^2}-\rho I_\psi\frac{\partial^4\theta}{\partial x^2\partial t^2}-m\right)\delta\theta\mathrm{d}x\mathrm{d}t+\int_{t_1}^{t_2}\left(M+\rho I_\psi\frac{\partial^3\theta}{\partial x\partial t^2}\right)\delta\theta\bigg|_{0}^{L}\mathrm{d}t$$
$$+\int_{t_1}^{t_2}\left(k_a\theta(0,t)+I_a\frac{\partial^2\theta}{\partial t^2}(0,t)\right)\delta\theta(0,t)\mathrm{d}t+\int_{t_1}^{t_2}\left(k_a\theta(L,t)+m_a\frac{\partial^2\theta}{\partial t^2}(L,t)\right)\delta\theta(L,t)\mathrm{d}t$$
$$+\int_{0}^{L}\left(-\rho I_P\frac{\partial\theta}{\partial t}\right)\delta\theta\bigg|_{t_1}^{t_2}\mathrm{d}x+\int_{0}^{L}\left(-\rho I_\psi\frac{\partial^2\theta}{\partial x\partial t}\right)\delta\frac{\partial\theta}{\partial t}\bigg|_{t_1}^{t_2}\mathrm{d}x+\left(-I_a\frac{\partial\theta}{\partial t}(0,t)\right)\delta\theta(0,t)\bigg|_{t_1}^{t_2}$$
$$+\left(-I_a\frac{\partial\theta}{\partial t}(L,t)\right)\delta\theta(L,t)\bigg|_{t_1}^{t_2}=0 \tag{3.55}$$

where all summable expressions are equal to zero. Accordingly, first the double integral expression for the limit $\delta\theta' \neq 0$ presents the following expression, in other words, rotational equilibrium:

$$-\frac{\partial M}{\partial x}+k_T\theta+\rho I_P\frac{\partial^2\theta}{\partial t^2}-\rho I_\psi\frac{\partial^4\theta}{\partial x^2\partial t^2}-m=0 \tag{3.56}$$

where the effect of warping according to Eq. (3.24) is as seen above. From this equation, the nonlocal equation of motion is obtained. According to this, the following definitions are given:

$$M_{nl}=M=\int_A\left(\sigma_{xz}y-\sigma_{xy}z+\sigma_{xz}\frac{\partial\psi}{\partial y}+\sigma_{xy}\frac{\partial\psi}{\partial y}\right)\mathrm{d}A, M_c=\int_A\left(s_{xz}y-s_{xy}z+s_{xz}\frac{\partial\psi}{\partial z}+s_{xy}\frac{\partial\psi}{\partial y}\right)\mathrm{d}A,$$
$$s_{xy}=G\left(\frac{\partial\psi}{\partial y}-z\right)\frac{\partial\theta}{\partial x}, s_{xz}=G\left(\frac{\partial\psi}{\partial z}+y\right)\frac{\partial\theta}{\partial x} \tag{3.57}$$

Equations (3.5) and (3.6) are integrated through area and integrated by multiplying with $\left(\frac{\partial\psi}{\partial y}-z\right)$ and $\left(\frac{\partial\psi}{\partial z}+y\right)$, respectively, the following expression can be obtained:

$$\begin{gathered}\left(1-(e_0a)^2\frac{\partial^2}{\partial x^2}\right)\int_A\left(\sigma_{xz}y-\sigma_{xy}z+\sigma_{xz}\frac{\partial\psi}{\partial y}+\sigma_{xy}\frac{\partial\psi}{\partial y}\right)\mathrm{d}A=\\ \int_A\left(s_{xz}y-s_{xy}z+s_{xz}\frac{\partial\psi}{\partial z}+s_{xy}\frac{\partial\psi}{\partial y}\right)\mathrm{d}A\end{gathered} \tag{3.58}$$

Let use here the nonlocal bending moment and classical stress components expressed in Eq. (3.58):

$$\left(1-(e_0a)^2\frac{\partial^2}{\partial x^2}\right)M=\int_A G\frac{\partial\theta}{\partial x}\left(\left(\frac{\partial\psi}{\partial z}+y\right)^2+\left(\frac{\partial\psi}{\partial y}-z\right)^2\right)\mathrm{d}A \tag{3.59}$$

where a definition can be made as follows [18]:

$$I_\varphi=\int_A\left(\left(\frac{\partial\psi}{\partial z}+y\right)^2+\left(\frac{\partial\psi}{\partial y}-z\right)^2\right)\mathrm{d}A \tag{3.60}$$

After the derivative of Eq. (3.59) is computed, the equation of motion is reached by utilizing Eqs. (3.55) and (3.60):

$$\left[GI_{\varphi}+(e_0a)^2k_T\right]\frac{\partial^2\theta}{\partial x^2}-\rho I_P\frac{\partial^2\theta}{\partial t^2}-k_T\theta-(e_0a)^2\rho I_{\psi}\frac{\partial^6\theta}{\partial x^4\partial t^2}$$
$$+\left[\rho I_{\psi}+(e_0a)^2\rho I_P\right]\frac{\partial^4\theta}{\partial x^2\partial t^2}-(e_0a)^2\frac{\partial^2 m}{\partial x^2}+m=0 \tag{3.61}$$

3.4.2 Solution of Equation of Motion

The excitation is $m = 0$ in Eq. (3.61) for nonlocal free torsional vibration. The following homogeneous and ordinary differential equation results from the use of Eq. (3.25):

$$A_1\frac{d^4\varphi}{dx^4}+A_2\frac{d^2\varphi}{dx^2}+A_3\varphi=0 \tag{3.62}$$

Where

$$A_1=\omega^2(e_0a)^2\rho I_{\psi}, A_2=GI_{\varphi}+(e_0a)^2k_T-\omega^2\rho I_{\psi}-\omega^2(e_0a)^2\rho I_P, A_3=-k_T+\omega^2\rho I_P \tag{3.63}$$

Equation (3.6) is quite similar to the differential equation obtained in Sect. 2.4.2 of previous section. According to this, an equation like Eq. (2.74) is obtained. An analysis can be made as follows for the roots of equations given in Eq. (2.75): Discriminant $\Delta = {A_2}^2 - 4A_1A_3$ should have a positive value. Since $A_1 > 0, A_2 > 0$, and $A_3 < 0$ truths are determined for small vibrational frequencies, both the sum of roots and the product of roots is also negative in Eq. (2.74). Therefore, it is concluded that there are a positive and a negative root. Here, the roots of characteristic equation of Eq. (3.62) can be formulated as in Eq. (2.75). Because of this inference, the signs of roots of characteristic equation are detected as $K_1 > 0$ and $K_2 < 0$. In this case, like the roots given in Eq. (2.76), the mode shape solution is obtained as follows:

$$\varphi(x)=C_1e^{\sqrt{K_1}x}+C_2e^{-\sqrt{K_1}x}+C_3e^{i\sqrt{-K_2}x}+C_4e^{-i\sqrt{-K_2}x} \tag{3.64}$$

By using Euler equations given by Eqs. (2.78)–(2.81), the following expression is reached:

$$\varphi(x)=C_1\cos\left(\sqrt{-K_2}x\right)+C_2\sin\left(\sqrt{-K_2}x\right)+C_3\cosh\left(\sqrt{K_1}x\right)+C_4\sinh\left(\sqrt{K_1}x\right) \tag{3.65}$$

3.4.3 Boundary Conditions of Nanoshafts

It is understood that four boundary conditions are required to determine the frequency equation based on the mode shape equation in Eq. (3.65). Unfortunately, the geometric and mechanical boundaries seen in Eq. (3.54) introduce only two boundary conditions for torsional vibration of noncircular nanorods according to the nonlocal Saint-Venant theory. Therefore, boundary conditions do not allow natural frequency analysis. When the classical theory is examined in the analysis ($e_0a = 0$), since $A_1 = 0$ is determined in Eq. (3.62), the related differential equation is reduced to the second degree and the solution in Eq. (3.65) have two unknown constants. Thus, there is no problem with the classical vibration solution of noncircular rods. It is remarkable that this problem that is not occurred in classical theory is experienced in nonlocal theory.

In this section, the boundary conditions of noncircular nanorods are briefly mentioned, but since the boundary conditions are insufficient in practice, applications involving only nanorods with generally known end conditions are examined with series expansions.

1. **Clamped end (C):** The clamped end does not allow angular rotation:

$$\varphi = 0 \tag{3.66}$$

2. **Free end (F):** If there is no dynamic torsional force at the free end, internal effect of torsional moment does not become. So, for $\delta\theta' \neq 0$ in Eq. (3.53):

$$M_t = M + \rho I_\psi \frac{\partial^3 \theta}{\partial x \partial t^2} = 0 \tag{3.67}$$

where if Eq. (3.57) is substituted into Eq. (3.61), the following expression is obtained:

$$M = GI_\varphi \frac{\partial \theta}{\partial x} + (e_0a)^2 k_T \frac{\partial \theta}{\partial x} + (e_0a)^2 \rho I_P \frac{\partial^3 \theta}{\partial x \partial t^2} - (e_0a)^2 \rho I_\psi \frac{\partial^5 \theta}{\partial x^3 \partial t^2} \tag{3.68}$$

where using Eq. (3.25) yields the following equation:

$$M_t = B_1 \frac{d^3 \varphi}{dx^3} + B_2 \frac{d\varphi}{dx} \tag{3.69}$$

where

$$B_1 = \omega^2 \rho I_\psi, \quad B_2 = GI_\varphi + (e_0a)^2 k_T - \omega^2 (e_0a)^2 \rho I_P + (e_0a)^2 \omega^2 \rho I_\psi \tag{3.70}$$

3. **Attached ends (MA and SA):** Similar to the nonlocal torsion of circular nanorods,

considering boundary conditions with tip attachment and Eq. (3.56) that the following expression is equal to zero:

$$\int_{t_1}^{t_2} \left(M + \rho I_\psi \frac{\partial^3 \theta}{\partial x \partial t^2} \right) \delta\theta \bigg|_0^L \mathrm{d}t + \int_{t_1}^{t_2} \left(k_a \theta(0,t) + I_a \frac{\partial^2 \theta}{\partial t^2}(0,t) \right) \delta\theta(0,t) \mathrm{d}t$$
$$+ \int_{t_1}^{t_2} \left(k_a \theta(L,t) + m_a \frac{\partial^2 \theta}{\partial t^2}(L,t) \right) \delta\theta(L,t) \mathrm{d}t = 0 \tag{3.71}$$

The following equation is valid at the right end of nanorod:

$$B_1 \frac{\mathrm{d}^3 \varphi}{\mathrm{d}x^3}(L,t) + B_2 \frac{\mathrm{d}\varphi}{\mathrm{d}x}(L,t) = (-k_a + \omega^2 I_a)\varphi(L,t) \tag{3.72}$$

On the other hand, mechanical boundary condition for the left end of nanorod is expressed as follows:

$$B_1 \frac{\mathrm{d}^3 \varphi}{\mathrm{d}x^3}(0,t) + B_2 \frac{\mathrm{d}\varphi}{\mathrm{d}x}(0,t) = (k_a - \omega^2 I_a)\varphi(0,t) \tag{3.73}$$

In the attached end types, numerical solution methods should be referred to because of the absence of a series expansion used in the solution of equation of motion as well as lack of boundary conditions. Therefore, to derive the frequency equation of attached noncircular nanorods is not the subject of this chapter and it will be discussed in the following sections.

Example 3.5 Determine the frequency equation of nonlocal free torsional vibration of clamped-clamped and clamped-free circular nanorod embedded in the elastic medium shown in Figs. 3.2 and 3.3, respectively.

Solution. The boundary condition for these two rod types can be written as follows:

$$\begin{aligned} &\text{Clamped-Free}: \varphi(0) = 0, B_1 \varphi'''(L) + B_2 \varphi'(L) = 0 \\ &\text{Clamped-Clamped}: \varphi(0) = 0, \varphi(L) = 0 \end{aligned} \tag{3.E3.5.1}$$

Since Eq. (3.64) has four unknown coefficients, above two boundary conditions are insufficient to determine the solution. Therefore, the solution of Eq. (3.60) is searched through the series expansions defined in Eq. (3.43). Where after the series expansions are simplified, the following equations are reached as a result of mathematical operations:

Clamped-Free:

$$\omega^2 = \frac{\left[GI_\varphi + (e_0a)^2k_T\right]\left(\frac{(2n-1)\pi}{2L}\right)^2 + k_T}{(e_0a)^2\rho I_\psi\left(\frac{(2n-1)\pi}{2L}\right)^4 + \left[\rho I_\psi + (e_0a)^2\rho I_P\right]\left(\frac{(2n-1)\pi}{2L}\right)^2 + \rho I_P}$$

Clamped-Clamped:

$$\omega^2 = \frac{\left[GI_\varphi + (e_0a)^2k_T\right]\left(\frac{n\pi}{L}\right)^2 + k_T}{(e_0a)^2\rho I_\psi\left(\frac{n\pi}{L}\right)^4 + \left[\rho I_\psi + (e_0a)^2\rho I_P\right]\left(\frac{n\pi}{L}\right)^2 + \rho I_P} \tag{3.E3.5.2}$$

where the different moments of inertia from the circular nanorod are calculated as follows:

$$I_\varphi = \int_A \left[y^2(S_\psi + 1)^2 + z^2(S_\psi - 1)^2\right]\mathrm{d}A = \left(S_\psi{}^2 + 1\right)I_P + 2S_\psi\left(I_z - I_y\right) \tag{3.E3.5.3}$$

$$I_\psi = S_\psi{}^2 \int_A y^2z^2\mathrm{d}A \tag{3.E3.5.4}$$

where the following expressions can be defined for Eq. (3.E3.5.3):

$$S_\psi = \frac{b^2 - a^2}{b^2 + a^2}, I_y = \int_A z^2\mathrm{d}A = \frac{1}{4}\pi ab^3, I_z = \int_A y^2\mathrm{d}A = \frac{1}{4}\pi a^3b, I_P = I_z + I_y = \frac{1}{4}\pi ab\left(a^2 + b^2\right) \tag{3.E3.5.5}$$

Considering the elliptical section depicted in Fig. 3.6, since $a = b$ for the circular section, it should be noted that $S_\psi = 0$, $I_y = I_z$, $I_\varphi = I_P$, and $I_\psi = 0$. Also, for Eq. (3.E3.5.4), the following definitions can be written by considering the polar coordinates of the ellipse:

$$y = r\cos\alpha, \quad z = r\sin\alpha, \quad r = \frac{ab}{\sqrt{a^2\sin^2\alpha + b^2\cos^2\alpha}}, \quad \mathrm{d}A = r\mathrm{d}\alpha \tag{3.E3.5.6}$$

Here, y and z define the Cartesian coordinates on the $y-$ and $z-$ axes of a point on the ellipse, respectively. r is the length of line segment (radius function) drawn between the point on the ellipse and the origin. Also, α is called as the positive angle of radius function. dA is the area of the differential element describing the infinitesimal ellipse sector. By substituting all these expressions into Eq. (3.E3.5.4), I_ψ can be calculated:

$$I_\psi = \int_A \psi^2 dA = \frac{S_\psi{}^2}{6} \int_0^{2\pi} \left(\frac{ab}{\sqrt{(b\cos\theta)^2 + (a\sin\theta)^2}} \right)^6 \sin^2\theta \cos^2\theta d\theta \qquad (3.E3.5.7)$$

On the other hand, first five modes nonlocal torsional frequencies of clamped nanorods are calculated by considering two different nonlocal torsion theories in Table 3.1. The calculations neglect the elastic medium ($k_T = 0$). Nondimensional frequencies are formulated as follows:

$$\overline{\omega} = \omega_i L \sqrt{\frac{\rho}{G}} \qquad (3.E3.5.8)$$

where i is the mod number and ω_i is the natural frequency. Additionally, nondimensional nonlocal parameter used in calculations is defined as follows:

$$\alpha = \frac{e_0 a}{L} \qquad (3.E3.5.9)$$

The results in Table 3.1 are computed according to the following parameters: Nanorod length, $L = 20$ nm, cross-section radii parallel to $y-$ and $z-$ axes, $a = 1$ nm and $b = 0.5$ nm, modulus of elasticity, $E = 1$ TPa, Poisson's ratio, $\upsilon = 0.19$, mass of unit volume, $\rho = 2300$ kg/m^3.

The results listed in Table 3.1 present the comparison of nonlocal torsional frequencies of an elliptical section nanorod according to the warping effect. The case where the warping effect is neglected indicates the elementary torsion theory and its natural frequency calculation is given in Eq. (3.E3.1.6). When the results are examined, first it is understood that the warping effect is a factor that reduces the nonlocal frequencies. In addition to this, the increasing nonlocal parameter more reduces the frequencies. It is reached that the frequency reduction effect of nonlocal parameter is almost the same in both cases. These inferences suggest that consideration of nonlocal warping will be a correct approach to investigate the torsional dynamics of noncircular nanorods.

Table 3.1 Comparisons of nondimensional torsional frequencies of clamped nanobeams

Mode	Without warping effect			With warping effect		
number	$\alpha = 0$	$\alpha = 0.1$	$\alpha = 0.2$	$\alpha = 0$	$\alpha = 0.1$	$\alpha = 0.2$
1	3.1416	2.9972	2.6601	2.5129	2.3974	2.1278
2	6.2832	5.3202	3.9124	5.0236	4.2536	3.1281
3	9.4248	6.8587	4.4169	7.5298	5.4796	3.5288
4	12.5664	7.8248	4.6458	10.0294	6.2451	3.7078
5	15.7080	8.4356	4.7645	12.5201	6.7237	3.7975

Problems

3.1. Determine the frequency equations of nonlocal torsional free vibration of circular nanorods without the elastic medium given in Fig. 3.7 according to nonlocal EBBT.

3.2. Compute the first five modes nondimensional torsional frequencies of nanorods in Problem 6.1 under parameters as follows. Discuss the effects of nonlocal parameter and boundary conditions on differences between results.

Nanorod length: $L = 20$ nm, diameter of circular cross-section: $d = 1$ nm, modulus of elasticity: $E = 1$ TPa, mass of unit volume: $\rho = 2300$ kg/m^3, Poisson's ratio: $\upsilon = 0.19$, stiffness and mass ratios for torsional spring and torsional mass attachments: $\beta_k = 1$ and $\beta_m = 1$.

Compute according to three different nondimensional nonlocal parameters: $\alpha = 0$, 0.1, 0.2.

Compute the nondimensional frequencies as: $\overline{\omega} = \omega L\sqrt{\rho/G}$.

3.3. Determine the frequency equations of nonlocal torsional free vibration of nanorods in Problem 6.1 by considering torsional elastic medium.

3.4. Compute the first five modes nondimensional frequencies of nonlocal torsional free vibration of clamped-free nanorods under parameters as follows. Discuss the effect of nonlocal parameter on differences between results.

Nanorod length: $L = 20$ nm, diameter of circular cross-section: $d = 1$ nm, modulus of elasticity: $E = 1$ TPa, mass of unit volume: $\rho = 2300$ kg/m^3, Poisson's ratio: $\upsilon = 0.19$.

Compute according to three different nondimensional nonlocal parameters: $\alpha = 0$, 0.1, 0.2.

Compute the nondimensional frequencies as: $\overline{\omega} = \omega L\sqrt{\rho/G}$.

3.5. Examine the Problem 3.4 by considering also torsional elastic medium according to two different nondimensional medium stiffnesses: $K_M = 1$ and 20. Discuss the effect of elastic medium on differences between results.

Compute the elastic medium stiffness: $K_M = k_M L^2/GI_P$.

Compute according to two different nondimensional nonlocal parameters: $\alpha = 0$ and 0.2.

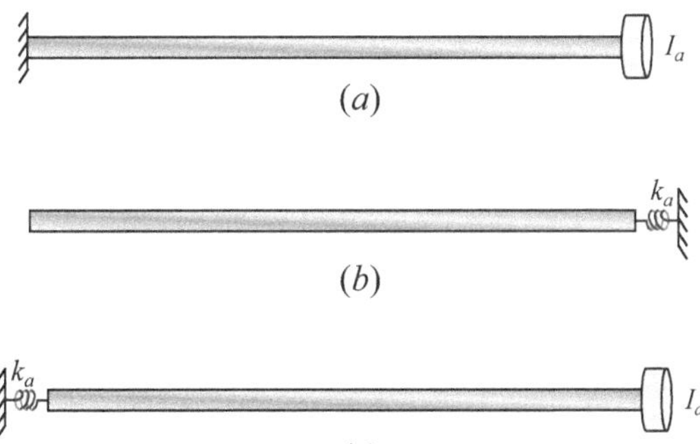

Fig. 3.7 Circular torsional nanorods with different boundary conditions. (**a**) Clamped-torsional mass attached (**b**) Free torsional spring attached (**c**) Torsional spring attached-torsional mass attached

3.6. Compute the first five modes nondimensional frequencies of nonlocal torsional free vibration of elliptical clamped-free nanorods under parameters as follows. Discuss the effect of nonlocal parameter on differences between results.

Nanorod length: $L = 20$ nm, vertical and horizontal radii of elliptical cross-section: $a = 1$ nm, $b = 0.5$ nm, modulus of elasticity: $E = 1$ TPa, mass of unit volume: $\rho = 2300$ kg/m^3, Poisson's ratio: $\upsilon = 0.19$.

Compute according to three different nondimensional nonlocal parameters: $\alpha = 0$, 0.1, 0.2.

Compute the nondimensional frequencies as: $\overline{\omega} = \omega L\sqrt{\rho/G}$.

3.7. Examine the Problem 3.6 by considering also torsional elastic medium according to two different nondimensional medium stiffnesses: $K_M = 1$ and 20. Discuss the effect of elastic medium on differences between results.

Compute the elastic medium stiffness: $K_M = k_M L^2/GI_P$

Compute according to two different nondimensional nonlocal parameters: $\alpha = 0$ and 0.2.

3.8. Compute the first five modes nondimensional frequencies of nonlocal torsional free vibration of rectangular clamped-free and clamped-clamped nanorods under parameters as follows. Discuss the effect of nonlocal parameter and boundary conditions on differences between results (Fig. 3.8).

Nanorod length: $L = 20$ nm, height and thickness of rectangular cross-section: $h = 2$ nm, $b = 1$ nm, modulus of elasticity: $E = 1$ TPa, mass of unit volume: $\rho = 2300$ kg/m^3, Poisson's ratio: $\upsilon = 0.19$.

Compute according to three different nondimensional nonlocal parameters: $\alpha = 0$, 0.1, 0.2.

Compute the nondimensional frequencies as: $\overline{\omega} = \omega L\sqrt{\rho/G}$.

Note that: Polar moment of inertia of elliptical nanorod: $I_P = bh(b^2 + h^2)/12$.

Warping function of rectangular cross-section is given as follows [20]:

$$\psi(y,z) = yz - \frac{8b^2}{\pi^3}\sum_{i=0}^{\infty}\frac{(-1)^i}{(2i+1)^3}\frac{\sinh(k_i z)}{\cosh(k_i h/2)}\sin(k_i y) \tag{3.P3.8.1}$$

Where:

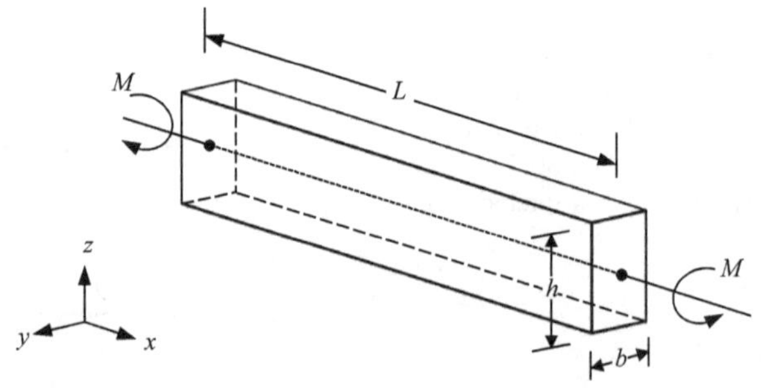

Fig. 3.8 Schematic representation of rectangular torsional nanorod

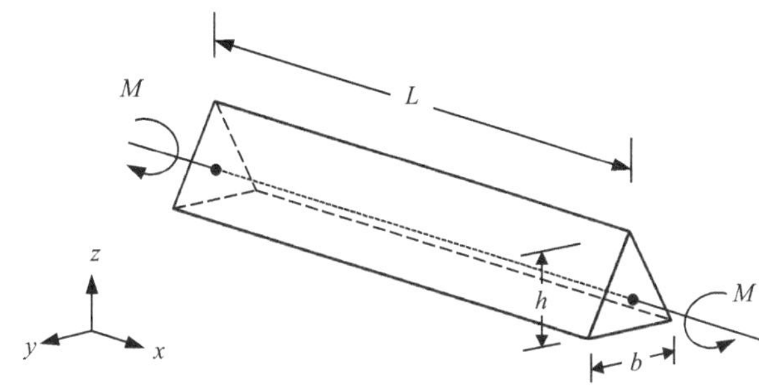

Fig. 3.9 Schematic representation of triangular torsional nanorod

$$k_i = \frac{(2i+1)\pi}{b} \tag{3.P3.8.2}$$

3.9. Compute the first five modes nondimensional frequencies of nonlocal torsional free vibration of triangular (equilateral) clamped-free and clamped-clamped nanorods under parameters as follows. Discuss the effect of nonlocal parameter and boundary conditions on differences between results (Fig. 3.9).

Nanorod length: $L = 20$ nm, edge of equilateral triangular cross-section: $b = 2$ nm, modulus of elasticity: $E = 1$ TPa, mass of unit volume: $\rho = 2300$ kg/m^3, Poisson's ratio: $\upsilon = 0.19$.

Compute according to three different nondimensional nonlocal parameters: $\alpha = 0$, 0.1, 0.2.

Compute the nondimensional frequencies as: $\overline{\omega} = \omega L\sqrt{\rho/G}$.

Note that: Polar moment of inertia of elliptical nanorod: $I_P = \sqrt{3}b^4/48$.

Warping function of equilateral triangular cross-section is given as follows [19]:

$$\psi(y,z) = \frac{1}{6\eta} y\left(3z^2 - y^2\right) \tag{3.P3.9.1}$$

where:

$$\eta = \frac{b\sqrt{3}}{6} \tag{3.P3.9.2}$$

References

1. J.S. Rao, *Advanced Theory of Vibration* (Wiley, New York, 1992)
2. Z. Deng, D. Chen, F. Tang, J. Ren, A.J. Muscat, Synthesis and purple-blue emission of antimony trioxide single-crystalline nanobelts with elliptical cross section. Nano Res. **2**, 151–160 (2009)
3. S.Y. Bae, H.W. Seo, J. Park, H. Yang, H. Kim, S. Kim, Triangular gallium nitride nanorods. Appl. Phys. Lett. **82**(25), 4564–4566 (2003)

4. X. Zhang, X. Zhang, W. Shi, X. Meng, C. Lee, S. Lee, Single-crystal organic microtubes with a rectangular cross section. Angew. Chem. **46**(9), 1525–1528 (2007)
5. S.P. Timoshenko, J.N. Goodier, *Theory of Elasticity* (McGraw-Hill, New York, 1956)
6. J.M. Gere, Torsional vibrations of beams of thin walled open section. J. Appl. Mech. **21**(4), 381–387 (1954)
7. S.P. Timoshenko, Theory of bending, torsion and buckling of thin-walled member of open cross-section. J. Franklin Inst. **239**(4), 249–268 (1945)
8. M. Arda, M. Aydogdu, Torsional statics and dynamics of nanotubes embedded in an elastic medium. Compos. Struct. **114**, 80–91 (2014)
9. C. Li, Torsional vibration of carbon nanotubes: comparison of two nonlocal models and a semi-continuum model. Int. J. Mech. Sci. **82**, 25–31 (2014)
10. C.W. Lim, C. Li, J.L. Yu, Free torsional vibration of nanotubes based on nonlocal stress theory. J. Sound Vib. **331**(12), 2798–2808 (2012)
11. T. Murmu, S. Adhikari, C.Y. Wang, Torsional vibration of carbon nanotube–buckyball systems based on nonlocal elasticity theory. Phys. E **43**(6), 1276–1280 (2011)
12. J.A. Loya, J. Aranda-Ruiz, J. Fernández-Sáez, Torsion of cracked nanorods using a nonlocal elasticity model. J. Phys. D. Appl. Phys. **47**, 115304 (2014)
13. M.Ö. Yayli, On the torsional vibrations of restrained nanotubes embedded in an elastic medium model. J. Braz. Soc. Mech. Sci. Eng. **40**, 419 (2018)
14. M.Ö. Yayli, S.Y. Kandemir, A.E. Çerçevik, Torsional vibration of cracked carbon nanotubes with torsional restraints using Eringen's nonlocal differential. J. Low Freq. Noise Vibr. Act. Contr. **38**(1), 70–87 (2019)
15. Ö. Civalek, B. Uzun, M.Ö. Yayli, Torsional vibrations of functionally graded restrained nanotubes. Eur. Phys. J. Plus **137**, 113 (2022)
16. C.W. Lim, M.Z. Islam, G. Zhang, A nonlocal finite element method for torsional statics and dynamics of circular nanostructures. Int. J. Mech. Sci. **94-95**, 232–243 (2015)
17. H.M. Numanoğlu, Ö. Civalek, On the torsional vibration of nanorods surrounded by elastic matrix via nonlocal FEM. Int. J. Mech. Sci. **161-162**, 105076 (2019)
18. F. Khosravi, S.A. Hosseini, B.A. Hamidi, R. Dimitri, F. Tornabene, Nonlocal torsional vibration of elliptical nanorods with different boundary conditions. Vibration **3**(3), 189–203 (2020)
19. F. Khosravi, S.A. Hosseini, B.A. Hamidi, On torsional vibrations of triangular nanowire. Thin Walled Struct. **148**, 106591 (2020)
20. F. Khosravi, S.A. Hosseini, B.A. Hamidi, Analytical investigation on free torsional vibrations of noncircular nanorods. J. Braz. Soc. Mech. Sci. Eng. **42**, 514 (2020)

Transverse Vibration of Nonlocal Beams

4

4.1 Bending Elements

Bending members are straight rods that act as bears in structural engineering to meet vertical, lateral, and axial loads. Usually, since it is exposed to vertical loads, mechanical analyses that consider only the vertical direction are common. In addition to the internal effects of axial and torsion rods, they are characterized by at least one of the internal effects of shear force and bending moment.

In the macrosense, bending elements are used in building carcasses, bridge decks, steel roofs, and reinforced concrete foundations, etc. In the micro/nanosense, when it is considered that one-directional and longitudinal materials such as carbon nanotubes, silica carbide nanotubes, boron nitride nanotubes, nanowires, and nanorods used in NEMS/MEMS organizations such as microcantilever, microbridge, nanobridge, nanobiosensor, nanomotor-mass system have an intensive utilization, it can be stated that it is necessary to use the straight rod model to investigate the mechanical behavior of NEMS/MEMS organizations and to perform their correct designs. Of course, it is known that the straight rod model should not be combined with classical elasticity theory in nano/microscale, but with atomic size-dependent elasticity theories.

Different theories have been developed to state the mechanical behavior of beams, two of the most widely known are the Euler-Bernoulli beam theory [1], which neglects the deformations due to shear effects, and the Timoshenko beam theory [2, 3], which considers these deformations. Apart from these, higher-order shear deformation beam formulations such as parabolic [4], trigonometric [5], hyperbolic [6], exponential [7] can be sampled. It is also possible to mention approaches that consider only rotational inertia effects, not higher-order terms like Rayleigh beam theory [7]. A similar situation in axial rods can be

Öm. Civalek et al., *Mechanical Behavior and Vibration of Nano-Scaled Rods, Beams, and Frames*, Synthesis Lectures on Engineering, Science, and Technology,
https://doi.org/10.1007/978-3-032-12023-6_4

Table 4.1 Analogies between formulations bending rods (beams) and axial rods

Beam formulation	Axial rod formulation	Reason of analogy
Euler-Bernoulli beam theory	Simple rod theory	Only principal effects
Timoshenko beam theory	Love-bishop (or bishop) rod theory	Shear and mass inertias
Rayleigh beam theory	Rayleigh (or love) rod theory	Only mass inertia

expressed for the Rayleigh rod theory, which can be formulated with the neglect of lateral inertia.

In the chapter on axial rods, we emphasized that the simple rod only considers deformations in the axial direction, namely, it neglects secondary effects such as Poisson's, and that the Love-Bishop rod theory considers the effects of lateral (shear) and mass inertia. If things explained in the above paragraph are considered, an analogy as given in Table 4.1 between the bending and axial rod theories can be constituted [8].

In this chapter, free vibration analyses with nonlocal elasticity of nano-scaled bending elements are examined. Firstly, a general literature review on the subject is given. Then, the equations of motion of the nonlocal free vibration of Euler-Bernoulli and Timoshenko beams, which are the most common beam theories, are obtained. By considering the different environmental factors in NEMS/MEMS organizations, their effects on the equations of motion are also formulated. To investigate the mechanical behavior of nano-scaled beams under these effects, several examples are examined.

4.2 Recent Contributions

It can be stated that the vibration behavior of nanobeams has been examined under various external effects within the framework of nonlocal elasticity. This subchapter presents a literature summary on this subject.

The first studies focusing on this subject are on the free vibration behavior of single-walled carbon nanotubes (SWCNT) and double-walled carbon nanotubes (DWCNT) through nonlocal Euler-Bernoulli beam theory (NL-EBBT) [10] and nonlocal Timoshenko beam theory (NL-TBT) [11]. Additionally, in these studies, the van der Waals interaction between the walls has also been used for DWCNT. Following this, the vibration of simply supported nanobeams has been studied via four different beam theories (Euler-Bernoulli, Timoshenko, Reddy, and Levinson) [12].

On the other hand, the free vibration of tip-mass attached nanobeams (nanotube-based micromass sensor structure) has been investigated with NL-TBT [13]. The large-amplitude thermal vibration of multi-walled carbon nanotubes (MWCNT) resting on elastic foundation have also been studied by NL-TBT [14]. A new nonlocal shear deformable beam theory has been developed for the free vibration of nanobeams [15–17]. The crack effect on the free vibration of nanobeams based on NL-EBBT has been studied [18, 19]. Nonlinear vibration of a DWCNT in a thermo-elastic environment by way of NL-EBBT and van der

Waals interaction has been searched [20]. A differential quadrature solution for the free vibration of SWCNT resting on elastic foundation has been presented based on NL-TBT [21]. Finite element analysis has been introduced for the free vibration of NL-EBBT [22]. Finite element solutions have been given for thermal [23, 24] and thermo-elastic [25] vibrations of nanobeams formulated by using NL-TBT. Moreover, finite element analysis has also been carried out for thermal [26] and thermo-elastic vibrations of nanobeams with NL-EBBT [27]. In addition to these, the differential quadrature method has been applied to four different beam theories in the free vibration of nanobeams [28]. A finite element method has been developed for the nonlocal vibration of a higher-order shear deformable curved nanobeam [29].

The free vibration of tapered nanobeam that has mass and elastic spring attachments has been investigated with NL-EBBT and differential quadrature [30]. Additionally, free vibration analysis of curved NL-EBBT nanobeams is presented via a two-phase nonlocal integral model [31]. A shear locking-free finite element formulation has been applied to the vibration of NL-TBT nanobeams based on a two-phase nonlocal model [32]. Meanwhile, there are different studies for nonlocal free vibration of functionally graded nanobeams [33–38]. Finally, studies aimed at eliminating the paradoxes detected about the nonlocal vibration of nanobeams can also be added [39–42].

4.3 Nonlocal Vibration of Euler-Bernoulli Beam

In the Euler-Bernoulli beam theory (EBBT), sections that are perpendicular to the rod axis and are planar before bending, remain perpendicular and planar after bending, too. However, deformations due to shear stresses during bending are neglected. This beam theory is known to be useful for beams that are not too high, not too short span, and where singular loads do not stand out much compared to other loads.

In this subsection, the equation of motion of free vibration based on the nonlocal elasticity theory of small scaled Euler-Bernoulli beams will be obtained separately according to the equilibrium approach and Hamilton's Principle. Let's address the equilibrium approach first.

4.3.1 Equation of Motion: Equilibrium Approach

The beam element model is a rod with homogeneous, constant cross-section, and straight axis as shown in Fig. 4.1a. The beam element rests on a double-parameter elastic foundation and is exposed to a thermal environment. The equilibrium equation for beams is given as

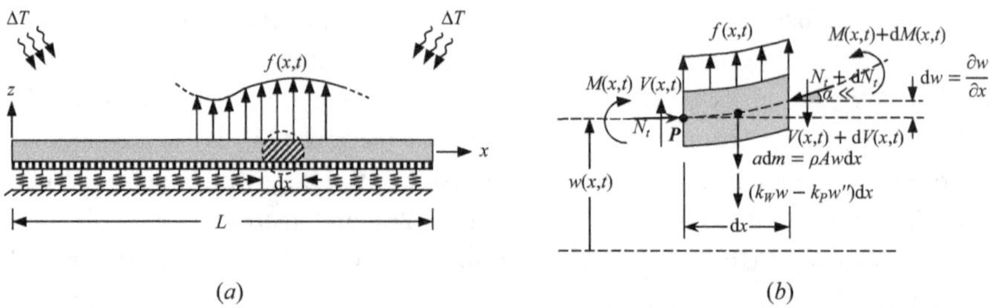

Fig. 4.1 Free vibration of nanobeam resting on thermo-elastic medium. (**a**) Beam model. (**b**) Free body diagram

$$\sum F_{\text{net}} = \rho A \mathrm{d}x \frac{\partial^2 w}{\partial t^2} \tag{4.1}$$

where $w = w(x, t)$ is the transverse displacement of beam. ρ is the mass of unit volume and A is the area of cross-section. t states the time. In bending elements, if there is no axial loading, shear force $V = V(x, t)$ and bending moment $M = M(x, t)$ occur as an internal effect. According to this, the vertical equilibrium of all internal forces occurring in the differential element with length dx, which will perform a transverse displacement w in the beam, is expressed from the free body diagram given in Fig. 4.1b as follows:

$$V - (V + \mathrm{d}V) + \left(k_W w - k_P \frac{\partial^2 w}{\partial x^2}\right) \mathrm{d}x - f\,\mathrm{d}x = -\rho A \mathrm{d}x \frac{\partial^2 w}{\partial t^2} \tag{4.2}$$

where $f = f(x, t)$ is transverse dynamic excitation.

If the ordinary-partial differential relation for the shear force internal effect is substituted into Eq. (4.2):

$$\frac{\partial V}{\partial x} = \rho A \frac{\partial^2 w}{\partial t^2} - f + k_W w - k_P \frac{\partial^2 w}{\partial x^2} \tag{4.3}$$

where k_W and k_P are the stiffnesses of Winkler (linear) and Pasternak (shear) elastic foundations, respectively. On the other hand, the moment equilibrium according to point P in the free body diagram can be written as follows:

$$(M + \mathrm{d}M) - M - (V + \mathrm{d}V)\mathrm{d}x + \left(k_W w - k_P \frac{\partial^2 w}{\partial x^2}\right) \mathrm{d}x \frac{\mathrm{d}x}{2} - N_t \frac{\partial w}{\partial x} \mathrm{d}x - f \mathrm{d}x \frac{\mathrm{d}x}{2} = 0 \tag{4.4}$$

where N_t is the axial load due to ambient temperature and is calculated as follows:

$$N_t = E\int_A \alpha_t \Delta T dA = EA\alpha_t \Delta T \tag{4.5}$$

where ΔT is defined as temperature change of ambient. If $\Delta T < 0$, an axial tensile force occurs, and if $\Delta T > 0$, an axial compressive force occurs. Additionally, α_t explains the coefficient of thermal expansion.

While editing Eq. (4.4), since the higher-order differential terms from the product will be $(dx)^2 \cong 0$ and $dVdx \cong 0$, related terms are vanished. If the partial-ordinary differential relation is used for the bending moment in the remainder expression:

$$\frac{\partial M}{\partial x} = V + N_t \frac{\partial w}{\partial x} \tag{4.6}$$

Substituting Eq. (4.3) into Eq. (4.6) yields:

$$\frac{\partial^2 M}{\partial x^2} = k_W w - k_P \frac{\partial^2 w}{\partial x^2} + N_t \frac{\partial^2 w}{\partial x^2} + \rho A \frac{\partial^2 w}{\partial t^2} - f \tag{4.7}$$

To obtain the nonlocal equation of motion, the following expression is given first in relation to the nonlocal axial stress equation:

$$\left(1 - (e_0 a)^2 \frac{\partial^2}{\partial x^2}\right) \int_A \sigma_{xx} y dA = \int_A s_{xx} y dA \tag{4.8}$$

in which, below definitions are valid:

$$\int_A \sigma_{xx} y dA = M_{\mathrm{nl}} = M, \quad \int_A s_{xx} y dA = M_{\mathrm{c}} = -EI \frac{\partial^2 w}{\partial x^2} \tag{4.9}$$

where M_{nl} and M_{c} are nonlocal and classical bending moment, respectively. E is elasticity modulus and I is moment of inertia. The derivation of the classical bending moment is given in the relevant appendix. By using Eq. (4.7) into the first derivation of Eq. (4.8), the nonlocal motion of equation is attained as follows:

$$\begin{aligned} &\left[-EI - (e_0 a)^2 k_P + (e_0 a)^2 N_t\right] \frac{\partial^4 w}{\partial x^4} + \left[k_P + (e_0 a)^2 k_W - N_t\right] \frac{\partial^2 w}{\partial x^2} - k_W w \\ &- \rho A \frac{\partial^2 w}{\partial t^2} + (e_0 a)^2 \rho A \frac{\partial^4 w}{\partial x^2 \partial t^2} + f - (e_0 a)^2 \frac{\partial^2 f}{\partial x} = 0 \end{aligned} \tag{4.10}$$

To determine the mechanical behavior of the nonlocal nanobeam elements, the internal effects of the bending moment and shear force that will occur in these elements should also be determined. Therefore, another operation that should be done according to the equilibrium approach, the nonlocal bending moment and nonlocal shear force expressions should be obtained. First, to obtain the bending moment, Eq. (4.8) can be rewritten as follows:

$$M-(e_0a)^2\frac{\partial^2 M}{\partial x^2}=-EI\frac{\partial^2 w}{\partial x^2} \tag{4.11}$$

if Eq. (4.7) is used above expression, the bending moment is as follows:

$$M=\left[-EI-(e_0a)^2k_P+(e_0a)^2N_t\right]\frac{\partial^2 w}{\partial x^2}+(e_0a)^2k_Ww+(e_0a)^2\rho A\frac{\partial^2 w}{\partial t^2}-(e_0a)^2f \tag{4.12}$$

On the other hand, Eq. (4.12) can used in Eq. (4.6) to obtain the shear force internal effect:

$$V=\left[-EI-(e_0a)^2k_P+(e_0a)^2N_t\right]\frac{\partial^3 w}{\partial x^3}+\left[(e_0a)^2k_W-N_t\right]\frac{\partial w}{\partial x}+(e_0a)^2\rho A\frac{\partial^3 w}{\partial x\partial t^2}-(e_0a)^2\frac{\partial f}{\partial x} \tag{4.13}$$

However, it should be emphasized that there is a shortcoming here. Shear force should contain the shear layer [9]:

$$V_T(x,t)=V(x,t)+V_P(x,t) \tag{4.14}$$

where $V_T(x, t)$ is the contribution of the shear layer to the total shear force and is defined as follows:

$$V_P(x,t)=k_P\frac{\partial w}{\partial x} \tag{4.15}$$

As a result, the total shear force is given as follows:

$$\begin{aligned}V_T=&\left[-EI-(e_0a)^2k_P+(e_0a)^2N_t\right]\frac{\partial^3 w}{\partial x^3}+\left[k_P+(e_0a)^2k_W-N_t\right]\frac{\partial w}{\partial x}\\&+(e_0a)^2\rho A\frac{\partial^3 w}{\partial x\partial t^2}-(e_0a)^2\frac{\partial f}{\partial x}\end{aligned} \tag{4.16}$$

4.3.2 Equation of Motion: Hamilton's Principle

The equilibrium approach provides a great convenience especially in nonlocal axial nanorod and nonlocal torsional nanorod problems. However, as seen in the previous subchapter, while shear force is obtained, the inability to directly add the shear layer is a weakness of the equilibrium approach. Such weaknesses do not appear in algebra of variation. Also, it will be seen later that the using of the algebra of variation is important to prevent situations that can cause errors in more complex beam problems involving higher-order shear terms.

It was emphasized in the previous chapters that energy equations should be established for the application of variational algebra. Before energy equations are obtained, the following displacement components of beam element should be written as follows:

$$u(x,t)=u^{*}-z\frac{\partial w}{\partial x},\, v(x,t)=0,\, w(x,t)=w \tag{4.17}$$

where $u*$ is pure axial displacement and w is transverse displacement. The displacement components to the deformations are passed as follows:

$$\varepsilon_{xx}=\frac{1}{2}\left(\frac{\partial u}{\partial x}+\frac{\partial u}{\partial x}\right)=\frac{\partial u}{\partial x}=-z\frac{\partial^2 w}{\partial x^2},\, \varepsilon_{yy}=\varepsilon_{zz}=\gamma_{xy}=\gamma_{yz}=\gamma_{xz}=0 \tag{4.18}$$

where ε_{xx}, ε_{yy}, and ε_{zz} represent axial strain components and γ_{xy}, γ_{yz}, and γ_{xz} refers to shear strains. Since only transverse vibration will be examined, $u* = 0$ is taken.

The expressions for strain energy U, kinetic energy K, and work potential energy W for the nanobeam element can be defined as follows:

$$\begin{aligned} U=&\frac{1}{2}\int_V\left(\sigma_{xx}\varepsilon_{xx}+\sigma_{yy}\varepsilon_{yy}+\sigma_{zz}\varepsilon_{zz}+2\tau_{xy}\gamma_{xy}+2\tau_{xz}\gamma_{xz}+2\tau_{yz}\gamma_{yz}\right)\mathrm{d}V+\frac{1}{2}\int_0^L k_W w^2\mathrm{d}x \\ &+\frac{1}{2}\int_0^L k_P\left(\frac{\partial w}{\partial x}\right)^2\mathrm{d}x+\frac{1}{2}\sum_{i=1}^{n}k_{a,i}[w(x_i,t)]^2 \\ =&\frac{1}{2}\int_0^L\left(-M\frac{\partial^2 w}{\partial x^2}\right)\mathrm{d}x+\frac{1}{2}\int_0^L\left[k_W w^2+k_P\left(\frac{\partial w}{\partial x}\right)^2\right]\mathrm{d}x+\frac{1}{2}k_a[w(0,t)]^2 \\ &+\frac{1}{2}k_a[w(L,t)]^2 \end{aligned} \tag{4.19}$$

$$T = \frac{1}{2}\int_V \rho\left[\left(\frac{\partial u}{\partial t}\right)^2 + \left(\frac{\partial v}{\partial t}\right)^2 + \left(\frac{\partial w}{\partial t}\right)^2\right]dV + \frac{1}{2}\sum_{i=1}^{n} m_{a,i}\left(\frac{\partial w(x_i,t)}{\partial t}\right)^2$$
$$= \frac{1}{2}\int_0^L \rho A\left(\frac{\partial w}{\partial t}\right)^2 dx + \frac{1}{2}m_a\left(\frac{\partial w(0,t)}{\partial t}\right)^2 + \frac{1}{2}m_a\left(\frac{\partial w(L,t)}{\partial t}\right)^2 \tag{4.20}$$

$$W = \int_0^L fw dx + \int_0^L N_T\left(\frac{\partial w}{\partial x}\right)^2 dx \tag{4.21}$$

where in Eq. (4.19), from the terms seen for the first time, $k_{a,\,i}$ is the stiffness of generalized point spring attachment. Similarly, $m_{a,\,i}$ is the mass of generalized point attachment. x_i is the location of attachment. In the numerical applications of this chapter, since the attachments will be located at the ends of nanobeam, Eqs. (4.19) and (4.20) are arranged as seen in the equations.

According to Hamilton's Principle, the first variation of the potential energy of the nanobeam in the time interval (t_1, t_2) is equal to 0:

$$\delta\Pi = \delta\int_{t_1}^{t_2}[U - (T + W)]dt = 0 \tag{4.22}$$

Variations of energy expressions can be written as follows:

$$\int_{t_1}^{t_2}\delta U dt = \int_{t_1}^{t_2}\int_0^L\left(M\delta\frac{\partial^2 w}{\partial x^2} + k_W w\delta w + k_P\frac{\partial w}{\partial x}\delta\frac{\partial w}{\partial x}\right)dxdt + \int_{t_1}^{t_2}k_a w(0,t)\delta w(0,t)dt$$
$$+ \int_{t_1}^{t_2}k_a w(L,t)\delta w(L,t)dt \tag{4.23}$$

$$\int_{t_1}^{t_2}\delta T dt = \int_{t_1}^{t_2}\int_0^L\rho A\frac{\partial w}{\partial t}\delta\frac{\partial w}{\partial t}dxdt + \int_{t_1}^{t_2}m_a\frac{\partial w(0,t)}{\partial t}\delta\frac{\partial w(0,t)}{\partial t}dt$$
$$+ \int_{t_1}^{t_2}m_a\frac{\partial w(L,t)}{\partial t}\delta\frac{\partial w(L,t)}{\partial t}dt \tag{4.24}$$

$$\int_{t_1}^{t_2} \delta W \mathrm{d}t = \int_{t_1}^{t_2}\int_0^L f\delta w \mathrm{d}x\mathrm{d}t + \int_{t_1}^{t_2}\int_0^L N_t \frac{\partial w}{\partial x}\delta\left(\frac{\partial w}{\partial x}\right)\mathrm{d}x\mathrm{d}t \tag{4.25}$$

Variation expressions seen in Eqs. (4.20)–(4.22) are calculated as follows:

$$\begin{aligned}
&\int_0^L M\delta\frac{\partial^2 w}{\partial x^2}\mathrm{d}x = M\delta\frac{\partial w}{\partial x}\bigg|_0^L - \frac{\partial M}{\partial x}\delta w\bigg|_0^L + \int_0^L \frac{\partial^2 M}{\partial x^2}\delta w \mathrm{d}x,\\
&\int_0^L k_P\frac{\partial w}{\partial x}\delta\frac{\partial w}{\partial x}\mathrm{d}x = k_P\frac{\partial w}{\partial x}\delta w\bigg|_0^L - \int_0^L k_P\frac{\partial^2 w}{\partial x^2}\delta w \mathrm{d}x,\\
&\int_{t_1}^{t_2} \rho A\frac{\partial w}{\partial t}\delta\frac{\partial w}{\partial t}\mathrm{d}t = \rho A\frac{\partial w}{\partial t}\delta w\bigg|_{t_1}^{t_2} - \int_{t_1}^{t_2}\rho A\frac{\partial^2 w}{\partial t^2}\delta w \mathrm{d}t,\\
&\int_{t_1}^{t_2} m_a\frac{\partial w(0,t)}{\partial t}\delta\frac{\partial w(0,t)}{\partial t}\mathrm{d}t = m_a\frac{\partial w(0,t)}{\partial t}\delta w(0,t)\bigg|_{t_1}^{t_2} - \int_{t_1}^{t_2} m_a\frac{\partial^2 w(0,t)}{\partial t^2}\delta w(0,t)\mathrm{d}t,\\
&\int_{t_1}^{t_2} m_a\frac{\partial w(L,t)}{\partial t}\delta\frac{\partial w(L,t)}{\partial t}\mathrm{d}t = m_a\frac{\partial w(L,t)}{\partial t}\delta w(L,t)\bigg|_{t_1}^{t_2} - \int_{t_1}^{t_2} m_a\frac{\partial^2 w(L,t)}{\partial t^2}\delta w(L,t)\mathrm{d}t,\\
&\int_0^L N_t\frac{\partial w}{\partial x}\delta\frac{\partial w}{\partial x}\mathrm{d}x = N_t\frac{\partial w}{\partial x}\delta w\bigg|_0^L - \int_0^L N_t\frac{\partial^2 w}{\partial x^2}\delta w \mathrm{d}x
\end{aligned} \tag{4.26}$$

If the expressions consisted after Eq. (4.26) is replaced into Eq. (4.23)–(4.25) are used in Eq. (4.22), the following expression is reached

$$\begin{aligned}
&\int_{t_1}^{t_2}\int_0^L\left(-\frac{\partial^2 M}{\partial x^2} + k_w w - k_P\frac{\partial^2 w}{\partial x^2} + \rho A\frac{\partial^2 w}{\partial t^2} - f + N_t\frac{\partial^2 w}{\partial x^2}\right)\delta w \mathrm{d}x\mathrm{d}t + \int_{t_1}^{t_2} - M\delta\frac{\partial w}{\partial x}\bigg|_0^L \mathrm{d}t\\
&+\int_{t_1}^{t_2}\left(\frac{\partial M}{\partial x} + k_P\frac{\partial w}{\partial x} - N_T\frac{\partial w}{\partial x}\right)\delta w\bigg|_0^L \mathrm{d}t + \int_{t_1}^{t_2}\left(k_a w(0,t) + m_a\frac{\partial^2 w}{\partial t^2}(0,t)\right)\delta w(0,t)\mathrm{d}t\\
&+\int_{t_1}^{t_2}\left(k_a w(L,t) + m_a\frac{\partial^2 w}{\partial t^2}(L,t)\right)\delta w(L,t)\mathrm{d}t + \int_0^L\left(-\rho A\frac{\partial w}{\partial t}\right)\delta w\bigg|_{t_1}^{t_2}\mathrm{d}x\\
&+\left(-m_a\frac{\partial w}{\partial t}(0,t)\right)\delta w(0,t)\bigg|_{t_1}^{t_2} + \left(-m_a\frac{\partial w}{\partial t}(L,t)\right)\delta w(L,t)\bigg|_{t_1}^{t_2} = 0
\end{aligned} \tag{4.27}$$

where all parts of the expression should be equal to zero. First, in the case of $\delta w \neq 0$ in the double integral, the following expression can be obtained:

$$-\frac{\partial^2 M}{\partial x^2}+k_w w-k_P\frac{\partial^2 w}{\partial x^2}+\rho A\frac{\partial^2 w}{\partial t^2}-f+N_t\frac{\partial^2 w}{\partial x^2}=0 \tag{4.28}$$

It should be noted that this equation is the same as Eq. (4.7). If Eq. (4.28) is substituted into Eq. (4.8), the equation of motion of nonlocal transverse vibration is derived as given in Eq. (4.10). Additionally, if Eq. (4.28) is substituted into Eq. (4.11), the nonlocal bending moment internal effect can be obtained as given in Eq. (4.12). On the other hand, if the attachments are neglected for now, the other part that should be equal to zero in Eq. (4.27) is as follows:

$$V=\frac{\partial M}{\partial x}+k_P\frac{\partial w}{\partial x}-N_T\frac{\partial w}{\partial x} \tag{4.29}$$

It should be noted that the force of shear layer, which cannot be obtained directly in Eq. (4.16), exists here. After the nonlocal bending moment is obtained, if it is substituted into Eq. (4.29), the nonlocal shear force internal effect can be defined as given in Eq. (4.16). Internal effects in the case of attachments will be discussed later.

4.3.3 Solution of Equation of Motion

The analytical solution of the equation of motion is performed with the separation of variables defined as follows [31]:

$$w(x,t)=W(x)T(t)=W(x)\sin(\omega t-\theta) \tag{4.30}$$

where $w(x, t)$ is the vibration motion, $W(x)$ is the mode shape and $T(t)$ is the harmonic function. θ denotes the phase angle. Since the research subject of this chapter is modal analysis, its solution will be investigated. In addition, $f = 0$ should be taken in Eq. (4.10) since free vibration solution will be carried out. Eq. (4.30) is replaced into Eq. (4.10) and the expressions $\sin(\omega t - \theta)$ are simplified. The resulting equation is written as follows:

$$A_1\frac{\mathrm{d}^4 W(x)}{\mathrm{d}x^4}+A_2\frac{\mathrm{d}^2 W(x)}{\mathrm{d}x^2}+A_3 W(x)=0 \tag{4.31}$$

where

$$A_1 = -EI - (e_0a)^2 k_P + (e_0a)^2 N_t, A_2 = k_P + (e_0a)^2 k_W - N_t - (e_0a)^2 \rho A\omega^2, A_3 = -k_W + \rho A\omega^2 \tag{4.32}$$

On the other hand, bending moment and shear forces, which are mechanical boundary conditions, should also be defined. For now, ignore the attachments. If Eq. (4.30) is applied to Eqs. (4.12) and (4.16), the bending moment and shear force can be rewritten as follows, respectively:

$$M(x) = A_1 \frac{\mathrm{d}^2 W(x)}{\mathrm{d}x^2} + B_1 W(x) \tag{4.33}$$

$$V(x) = A_1 \frac{\mathrm{d}^3 W(x)}{\mathrm{d}x^3} + B_2 \frac{\mathrm{d}W(x)}{\mathrm{d}x} \tag{4.34}$$

in which, below definitions are valid:

$$B_1 = -(e_0a)^2 \rho A\omega^2 + (e_0a)^2 k_W, B_2 = -(e_0a)^2 \rho A\omega^2 + (e_0a)^2 k_W + k_P - N_t \tag{4.35}$$

We proceed with the solution of equation of motion. Since Eq. (4.31) is a homogeneous differential equation with constant coefficients, the solution is searched as follows:

$$W(x) = Ce^{kx} \tag{4.36}$$

If Eq. (4.36) and its derivatives are substituted in Eq. (4.31) and the expressions Ce^{kx} are simplified:

$$k^2 = K \tag{4.37}$$

If a mathematical transformation is used, the following equation is obtained:

$$A_1 K^2 + A_2 K + A_3 = 0 \tag{4.38}$$

For the solution of this equation, the discriminant is defined by $\Delta = {A_2}^2 - 4A_1A_3$ and is assumed as positive. Also, assuming no outside effects ($k_W = k_P = N_t = 0$), $A_1 < 0, A_2 < 0$, and $A_3 > 0$ become. Therefore, the sum of roots is positive and product of roots is negative in Eq. (4.37). Also, the roots of the equation occur as follows:

$$K_1 = \frac{-A_2 + \sqrt{{A_2}^2 - 4A_1A_3}}{2A_1}, K_2 = \frac{-A_2 - \sqrt{{A_2}^2 - 4A_1A_3}}{2A_1} \tag{4.39}$$

where $K_1 < 0$ and $K_2 > 0$. From Eq. (4.37)

$$k_{11} = \mathbf{i}\sqrt{-K_1}, k_{12} = -\mathbf{i}\sqrt{-K_1}, k_{21} = \sqrt{K_2}, k_{22} = -\sqrt{K_2} \tag{4.40}$$

Therefore, solution is:

$$W(x) = C_1 e^{\mathrm{i}\sqrt{-K_1}x} + C_2 e^{-\mathrm{i}\sqrt{-K_1}x} + C_3 e^{\sqrt{K_2}x} + C_4 e^{-\sqrt{K_2}x} \tag{4.41}$$

where C_i $(i = 1, 2, 3, 4)$ are constant numbers. The following transformations can be employed to arrange Eq. (4.41)

$$e^{\mathbf{i}\theta} = \cos\theta + \mathbf{i}\sin\theta \tag{4.42}$$

$$e^{-\mathbf{i}\theta} = \cos\theta - \mathbf{i}\sin\theta \tag{4.43}$$

$$e^{\theta} = \cosh\theta + \sinh\theta \tag{4.44}$$

$$e^{-\theta} = \cosh\theta - \sinh\theta \tag{4.45}$$

If Eqs. (4.42)–(4.45) are replaced into Eq. (4.41), the resulting of solution is as follows

$$W(x) = C_1 \cos\left(\sqrt{-K_1}x\right) + C_2 \sin\left(\sqrt{-K_1}x\right) + C_3 \cosh\left(\sqrt{K_2}x\right) + C_4 \sinh\left(\sqrt{K_2}x\right) \tag{4.46}$$

4.3.4 Boundary Conditions of Nanobeams

The nonlocal frequency equations of nanobeam structures explain the free vibration behavior. In order to obtain the frequency equation, the boundary conditions should be known. Considering different one-dimensional nano/micromaterials, it is conceivable that continuous nanobeam models of these may have different boundary conditions. Here, common end types (Simply supported, fixed supported, free, and guided supported ends) and dynamic end types (mass and spring attached ends) are examined. Subsequently, nonlocal vibration behavior of nanobeams with different boundary conditions is investigated by means of various examples.

1. **Simply supported end (S):** Since there will not become both displacement and bending moment, boundary conditions for a nanobeam with simply supported at both ends are as follows:

$$W = 0, \quad A_1 W'' + B_1 W = 0 \tag{4.47}$$

These expressions are:

$$W = 0, \quad W'' = 0 \tag{4.48}$$

2. **Fixed end (C):** Such type ends do not take displacement and rotation:

$$W = 0, \quad W' = 0 \tag{4.49}$$

3. **Free end (F):** There are no bending moments and shear forces at the free ends.

$$A_1 W'' + B_1 W = 0, \quad A_1 W''' + B_2 W' = 0 \tag{4.50}$$

4. **Guided ends (G):** Guided ends do not rotate but have freedom of motion in the direction of the shear force. Therefore, the shear force is not observed. So, the boundary conditions are written as follows:

$$W' = 0, \quad A_1 W''' + B_2 W' = 0 \tag{4.51}$$

As a result, the shear force boundary condition is arranged as follows:

$$W''' = 0 \tag{4.52}$$

5. **Attached ends (MA and SA):** The end attachment affects the shear force boundary condition at the end which it is attached (mechanical boundary condition). Internally actuated rotary spring or rotary mass attachment. If there are attachments of rotational spring and rotational mass that can take internal effects in the direction of bending freedom, the bending moment is also treated as a mechanical boundary condition, but such attachment models are excluded from this chapter. Equation (4.27) should be examined to establish the mechanical boundary condition. Suppose there are two attachments at the end of nanobeam element. First, the following expression can be written for $\delta w(L,t) \neq 0$ (right end of nanobeam)

$$\left(\frac{\partial M}{\partial x} + k_P \frac{\partial w}{\partial x} - N_T \frac{\partial w}{\partial x}\right)\delta w(L,t) - \left(k_a w(L,t) + m_a \frac{\partial^2 w}{\partial t^2}(L,t)\right)\delta w(L,t) = 0 \tag{4.53}$$

If the variation of displacement at the right end is simplified and remember that the left part will be equal to the shear force of nanobeam:

$$V(L,t) = k_a w(L,t) + m_a \frac{\partial^2 w}{\partial t^2}(L,t) \tag{4.54}$$

By using Eq. (4.34), the mechanical boundary condition for the coexistence of both attachments at the right end can be constituted as follows:

$$A_1 W'''(L) + B_2 W'(L) = \left(k_a - \omega^2 m_a\right) W(L) \tag{4.55}$$

On the other hand, by considering Eq. (4.27), the following expression is given for $\delta w(0,t) \neq 0$ (left end of nanobeam):

$$-\left(\frac{\partial M}{\partial x} + k_P \frac{\partial w}{\partial x} - N_T \frac{\partial w}{\partial x}\right)\delta w(0,t) - \left(k_a w(0,t) + m_a \frac{\partial^2 w}{\partial t^2}(0,t)\right)\delta w(0,t) = 0 \tag{4.56}$$

Similar to Eq. (4.54):

$$V(0,t) = -k_a w(0,t) - m_a \frac{\partial^2 w}{\partial t^2}(0,t) \tag{4.57}$$

The mechanical boundary condition is defined for left end of nanobeam as follows:

$$A_1 W'''(0) + B_2 W'(0) = -\left(k_a - \omega^2 m_a\right) W(0) \tag{4.58}$$

As a result, the mechanical boundary conditions for the different attached ends are summarized as follows:

$$\begin{aligned}
&\text{Mass attachment at } x = 0 : A_1 W''' + B_2 W' = \omega^2 m_a W,\\
&\text{Elastic spring attachment at} x = 0 : A_1 W''' + B_2 W' = -k_a W,\\
&\text{Mass attachment at} x = L : A_1 W''' + B_2 W' = -\omega^2 m_a W,\\
&\text{Elastic spring attachment at} x = L : A_1 W''' + B_2 W' = k_a W.
\end{aligned} \tag{4.59}$$

In addition to the mechanical boundary conditions, Eq. (4.33), which is the bending moment equation, should be written at the same end. Of course, it should be noted that the frequency equations of nanobeams with attachments cannot be attained based on manual solution because the complexity of boundary conditions and the high size of coefficient determinant cause a serious difficulty.

Example 4.1 Determine the frequency equation of nonlocal free vibration of simply supported nanobeam embedded in the thermo-elastic medium shown in Fig. 4.2.

Solution. Based on Eq. (4.48), the boundary conditions for simply supported ends are given as follows:

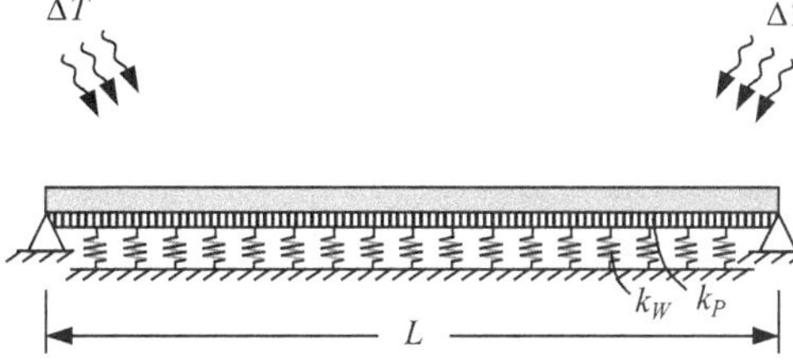

Fig. 4.2 Simply supported nanobeam embedded in the thermo-elastic medium

$$W(0)=0, W''(0)=0, W(L)=0, W''(L)=0 \tag{4.E4.1.1}$$

If the leftmost expression is applied to Eq. (4.46)

$$C_1 \cos 0 + C_2 \sin 0 + C_3 \cosh 0 + C_4 \sinh 0 = 0 \tag{4.E4.1.2}$$

If the second condition is applied to Eq. (4.46)

$$C_1 K_1 \cos 0 + C_2 K_1 \sin 0 + C_3 K_2 \cosh 0 + C_4 K_2 \sinh 0 = 0 \tag{4.E4.1.3}$$

From the common solution of Eqs. (4.E4.1.2) and (4.E4.1.3), the following expression can be given:

$$C_1 = C_3 = 0 \tag{4.E4.1.4}$$

The third and fourth conditions are substituted into the remainder of equation of motion, the following expressions are written:

$$C_2 \sin\left(\sqrt{-K_1}L\right) + C_4 \sinh\left(\sqrt{K_2}L\right) = 0 \tag{4.E4.1.5}$$

$$C_2 K_1 \sin\left(\sqrt{-K_1}L\right) + C_4 K_2 \sinh\left(\sqrt{K_2}L\right) = 0 \tag{4.E4.1.6}$$

These equations are presented in a matrix form as follows:

$$\begin{bmatrix} \sin\left(\sqrt{-K_1}L\right) & \sinh\left(\sqrt{K_2}L\right) \\ K_1 \sin\left(\sqrt{-K_1}L\right) & K_2 \sinh\left(\sqrt{K_2}L\right) \end{bmatrix} \begin{Bmatrix} C_2 \\ C_4 \end{Bmatrix} = 0 \tag{4.E4.1.7}$$

The determinant of coefficients matrix can be computed as follows:

$$(K_1 - K_2) \sin\left(\sqrt{-K_1}L\right) \sinh\left(\sqrt{K_2}L\right) = 0 \tag{4.E4.1.8}$$

This equation means the frequency equation. An analysis can be made here: For $(K_1 - K_2) = 0$ two roots should be equal. Because the sum of roots and the product of roots of characteristic equation of Eq. (4.38) are negative, there are two roots with different signs. Therefore, this cannot happen. On the other hand, for $\sinh\left(\sqrt{K_2}L\right) = 0$, $K_2 = 0$ should become. It was just concluded that there are a negative and a positive root. Therefore, the solution of following equation, which is the last alternative, should be investigated:

$$\sin\left(\sqrt{-K_1}L\right) = 0 \tag{4.E4.1.9}$$

The periodic solution of Eq. (4.E4.1.9) is expressed as follows:

$$-K_1 = \left(\frac{n\pi}{L}\right)^2 \tag{4.E4.1.10}$$

where the integer defined by $n = 1, 2, 3, \ldots$ is defined mode number of vibration. Substituting Eq. (4.E4.1.10) into Eq. (4.39) yields the following expression:

$$\frac{A_2 - \sqrt{{A_2}^2 - 4A_1A_3}}{2A_1} = \left(\frac{n\pi}{L}\right)^2 \tag{4.E4.1.11}$$

After several mathematical operations, the following equation is obtained:

$$A_1\left(\frac{n\pi}{L}\right)^4 - A_2\left(\frac{n\pi}{L}\right)^2 + A_3 = 0 \tag{4.E4.1.12}$$

If Eq. (4.32) are replaced into Eq. (4.E4.1.12), the frequency can be obtained as follows:

$$\omega^2 = \frac{\left[EI + (e_0a)^2k_P - (e_0a)^2N_t\right]\left(\frac{n\pi}{L}\right)^4 + \left[k_P + (e_0a)^2k_W - N_t\right]\left(\frac{n\pi}{L}\right)^2 + k_W}{(e_0a)^2\rho A\left(\frac{n\pi}{L}\right)^2 + \rho A} \tag{4.E4.1.13}$$

With a series expansion that provides boundary conditions [25], it can be stated that the frequency calculation of nanobeam is appropriate for the manual solution. The following series provides the boundary conditions defined in Eq. (4.E4.1.1) for the simply supported nanobeam:

$$w(x,t) = \sum_{n=1}^{\infty} w_n \sin\left(\frac{n\pi x}{L}\right)\sin(\omega t - \theta) \tag{4.E4.1.14}$$

where w_n states an unknown coefficient. Eq. (4.E4.1.1) and its derivatives are substituted into Eq. (4.10):

$$\begin{aligned}&\left[-EI-(e_0a)^2k_P+(e_0a)^2N_t\right]\left(\frac{n\pi}{L}\right)^4\sum_{n=1}^{\infty}w_n\sin\left(\frac{n\pi x}{L}\right)\sin(\omega t-\theta)\\&-\left[k_P+(e_0a)^2k_W-N_t-(e_0a)^2\rho A\omega^2\right]\left(\frac{n\pi}{L}\right)^2\sum_{n=1}^{\infty}w_n\sin\left(\frac{n\pi x}{L}\right)\sin(\omega t-\theta)\\&+(-k_W+\rho A\omega^2)\sum_{n=1}^{\infty}w_n\sin\left(\frac{n\pi x}{L}\right)\sin(\omega t-\theta)=0\end{aligned}\tag{4.E4.1.15}$$

If the series expansions are simplified, Eq. (4.E4.1.12) is obtained. Thus, the frequency equation can be calculated without the need to use the separation of variables method. However, there is no appropriate series expansion for each nanobeam element.

Several numerical results about natural frequency calculation are presented in Example 4.5.

Example 4.2 Determine the frequency equation of nonlocal free vibration of cantilever nanobeam embedded in the thermo-elastic medium shown in Fig. 4.3.

Solution. The boundary conditions of cantilever nanobeam are as follows:

$$W(0)=0, W'(0)=0, A_1W''(L)+B_1W(L)=0, A_1W'''(L)+B_2W'(L)=0 \tag{4.E4.2.1}$$

Substituting these expressions into Eq. (4.46) yields the following expressions, respectively:

$$C_1+C_3=0 \tag{4.E4.2.2}$$

$$C_2\sqrt{-K_1}+C_4\sqrt{K_2}=0 \tag{4.E4.2.3}$$

$$\begin{aligned}&C_1(-A_1K_1+B_1)\cos\left(\sqrt{-K_1}L\right)+C_2(-A_1K_1+B_1)\sin\left(\sqrt{-K_1}L\right)+\\&C_3(A_1K_2+B_1)\cosh\left(\sqrt{K_2}L\right)+C_4(A_1K_2+B_1)\sinh\left(\sqrt{K_2}L\right)=0\end{aligned}\tag{4.E4.2.4}$$

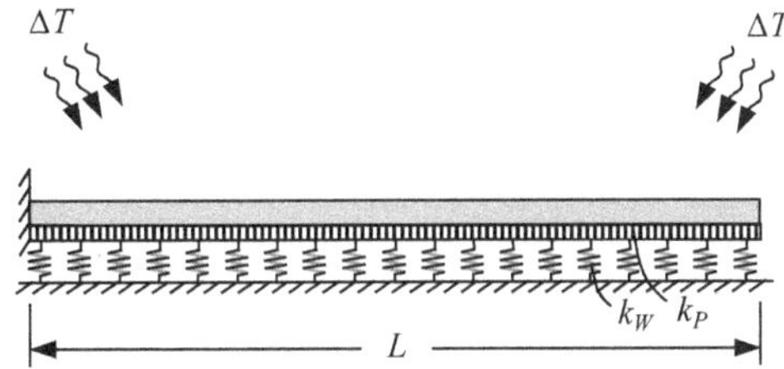

Fig. 4.3. Cantilever nanobeam embedded in the thermo-elastic medium

$$C_1\left(-A_1\sqrt{-K_1}+B_1\sqrt{-K_1{}^3}\right)\sin\left(\sqrt{-K_1}L\right)+C_2\left(A_1\sqrt{-K_1}-B_1\sqrt{-K_1{}^3}\right)\cos\left(\sqrt{-K_1}L\right)+$$
$$C_3\left(A_1\sqrt{K_2}+B_1\sqrt{K_2{}^3}\right)\sinh\left(\sqrt{K_2}L\right)+C_4\left(A_1\sqrt{K_2}+B_1\sqrt{K_2{}^3}\right)\cosh\left(\sqrt{K_2}L\right)=0 \tag{4.E4.2.5}$$

Above equations are represented with a system of equation such as $[a]\{C\}$

$$\begin{bmatrix} a_{11} & a_{12} & a_{13} & a_{14} \\ a_{21} & a_{22} & a_{23} & a_{24} \\ a_{31} & a_{32} & a_{33} & a_{34} \\ a_{41} & a_{42} & a_{43} & a_{44} \end{bmatrix} \begin{Bmatrix} C_1 \\ C_2 \\ C_3 \\ C_4 \end{Bmatrix} = 0 \tag{4.E4.2.6}$$

where

$$\begin{aligned}
&a_{12}=1, a_{12}=0, a_{13}=1, a_{14}=0, a_{21}=0, a_{22}=\sqrt{-K_1}, a_{23}=0, a_{24}=\sqrt{-K_1},\\
&a_{31}=(-A_1K_1+B_1)\cos\left(\sqrt{-K_1}L\right), a_{32}=(-A_1K_1+B_1)\sin\left(\sqrt{-K_1}L\right),\\
&a_{33}=(A_1K_2+B_1)\cosh\left(\sqrt{K_2}L\right), a_{34}=(A_1K_2+B_1)\sinh\left(\sqrt{K_2}L\right),\\
&a_{41}=\left(-A_1\sqrt{-K_1}+B_1\sqrt{-K_1{}^3}\right)\sin\left(\sqrt{-K_1}L\right), a_{42}=\left(A_1\sqrt{-K_1}-B_1\sqrt{-K_1{}^3}\right)\cos\left(\sqrt{-K_1}L\right),\\
&a_{43}=\left(A_1\sqrt{K_2}+B_1\sqrt{K_2{}^3}\right)\sinh\left(\sqrt{K_2}L\right), a_{44}=\left(A_1\sqrt{K_2}+B_1\sqrt{K_2{}^3}\right)\cosh\left(\sqrt{K_2}L\right).
\end{aligned} \tag{4.E4.2.7}$$

In the system of linear equations defined by Eq. (4.E4.2.6), the determinant of the matrix $[a]$ gives the frequency equation. Since to obtain manual solution of frequency is difficult, a solution can be performed with program codes on the computer. Within the scope of this chapter, computer codes for analytical solution are written by symbolic programming.

Several numerical results about natural frequency calculation are presented in Example 4.7.

Example 4.3 Determine the frequency equation of nonlocal free vibration of simply supported-transverse spring attached nanobeam embedded in the thermo-elastic medium shown in Fig. 4.4.

Solution. The boundary conditions of this nanobeam model can be written as follows:

$$\begin{aligned}
&W(0)=0,\ W''(0)=0,\\
&A_1W''(L)+B_1W(L)=0, A_1W'''(L)+B_2W'(L)=k_aW(L).
\end{aligned} \tag{4.E4.3.1}$$

If the first and second boundary conditions are substituted into Eq. (4.46), respectively, a solution as in Eq. (4.E4.1.4) is reached. Thus, the following expressions can be constituted for the third and fourth boundary conditions, respectively:

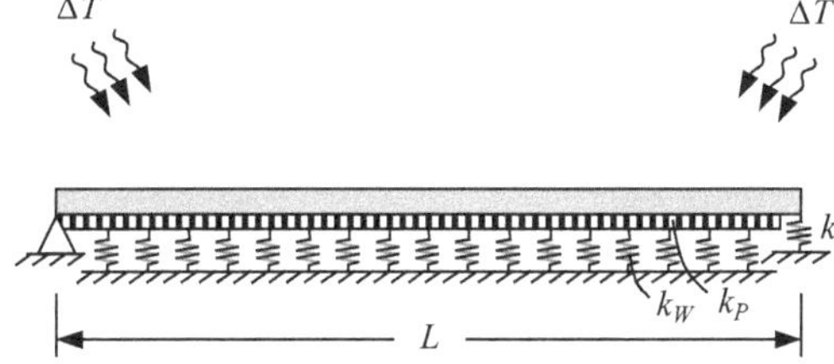

Fig. 4.4 Simply supported-transverse spring attached nanobeam embedded in the thermo-elastic medium

$$C_2(-A_1K_1+B_1)\sin\left(\sqrt{-K_1}L\right)+C_4(A_1K_2+B_1)\sinh\left(\sqrt{K_2}L\right)=0 \tag{4.E4.3.2}$$

$$C_2\left(A_1\sqrt{-K_1}-B_1\sqrt{-K_1}^3\right)\cos\left(\sqrt{-K_1}L\right)+C_4\left(A_1\sqrt{K_2}+B_1\sqrt{K_2}^3\right)\cosh\left(\sqrt{K_2}L\right)=\\ k_a\left[C_2\sin\left(\sqrt{-K_1}L\right)+C_4\sinh\left(\sqrt{K_2}L\right)\right] \tag{4.E4.3.3}$$

Eqs. (4.E4.3.2) and (4.E4.3.3) can be expressed with a system of equations symbolized by $[a]\{C\}$ as follows:

$$\begin{bmatrix} a_{11} & a_{12} \\ a_{21} & a_{22} \end{bmatrix}\begin{Bmatrix} C_2 \\ C_4 \end{Bmatrix}=0 \tag{4.E4.3.4}$$

In which, the inputs of the matrix $[a]$ are as follows:

$$\begin{aligned} &a_{11}=(-A_1K_1+B_1)\sin\left(\sqrt{-K_1}L\right), a_{12}=(A_1K_2+B_1)\sinh\left(\sqrt{K_2}L\right),\\ &a_{21}=\left(A_1\sqrt{-K_1}-B_1\sqrt{-K_1}^3\right)\cos\left(\sqrt{-K_1}L\right)-k_a\sin\left(\sqrt{-K_1}L\right),\\ &a_{22}=\left(A_1\sqrt{K_2}+B_1\sqrt{K_2}^3\right)\cosh\left(\sqrt{K_2}L\right)-k_a\sinh\left(\sqrt{K_2}L\right). \end{aligned} \tag{4.E4.3.5}$$

As mentioned in Example 4.2, the determinant of $[a]$ matrix given in Eq. (4.E4.3.4) presents the frequencies of this nanobeam model. Since manual calculations are difficult, frequency values can be obtained by symbolic programming on the computer.

Some numerical results about natural frequency calculation are presented in Example 4.7.

Example 4.4 Determine the frequency equation of nonlocal free vibration of fixed supported-transverse mass attached nan0obeam embedded in the thermo-elastic medium shown in Fig. 4.5.

Solution. The boundary conditions of this nanobeam model are presented as follows:

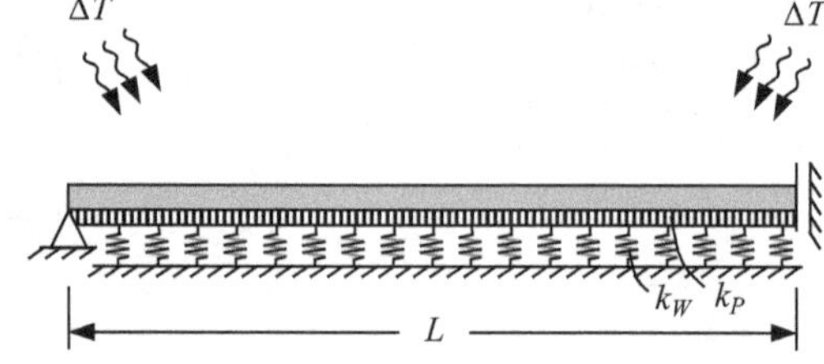

Fig. 4.5 Simply supported-guided supported nanobeam embedded in the thermo-elastic medium

$$\begin{aligned} &W(0)=0, W'(0)=0 \\ &A_1 W''(L)+B_1 W(L)=0, A_1 W'''(L)+B_2 W'(L)=-\omega^2 m_a W(L) \end{aligned} \tag{4.E4.4.1}$$

By using Eq. (4.E4.4.1) into Eq. (4.46), the following equations is obtained:

$$C_1+C_3=0 \tag{4.E4.4.2}$$

$$C_2\sqrt{-K_1}+C_4\sqrt{K_2}=0 \tag{4.E4.4.3}$$

$$\begin{aligned} &C_1(-A_1K_1+B_1)\cos\left(\sqrt{-K_1}L\right)+C_2(-A_1K_1+B_1)\sin\left(\sqrt{-K_1}L\right)+ \\ &C_3(A_1K_2+B_1)\cosh\left(\sqrt{K_2}L\right)+C_4(A_1K_2+B_1)\sinh\left(\sqrt{K_2}L\right)=0 \end{aligned} \tag{4.E4.4.4}$$

$$\begin{aligned} &C_1\left(-A_1\sqrt{-K_1}+B_1\sqrt{-K_1}^3\right)\sin\left(\sqrt{-K_1}L\right)+C_2\left(A_1\sqrt{-K_1}-B_1\sqrt{-K_1}^3\right)\cos\left(\sqrt{-K_1}L\right)+ \\ &C_3\left(A_1\sqrt{K_2}+B_1\sqrt{K_2}^3\right)\sinh\left(\sqrt{K_2}L\right)+C_4\left(A_1\sqrt{K_2}+B_1\sqrt{K_2}^3\right)\cosh\left(\sqrt{K_2}L\right)= \\ &-\omega^2 m_a\left[C_1\cos\left(\sqrt{-K_1}L\right)+C_2\sin\left(\sqrt{-K_1}L\right)+C_3\cosh\left(\sqrt{K_2}L\right)+C_4\sinh\left(\sqrt{K_2}L\right)\right] \end{aligned} \tag{4.E4.4.5}$$

If Eqs. (4.E4.4.2)–(4.E4.4.5) are expressed in a similar way to Eq. (4.E4.2.6), the inputs of $[a]$ matrix are as follows:

$$\begin{aligned} &a_{11}=1, a_{12}=0, a_{13}=1, a_{14}=0, a_{21}=0, a_{22}=\sqrt{-K_1}, a_{23}=0, a_{24}=\sqrt{-K_1}, \\ &a_{31}=(-A_1K_1+B_1)\cos\left(\sqrt{-K_1}L\right), a_{32}=(-A_1K_1+B_1)\sin\left(\sqrt{-K_1}L\right), \\ &a_{33}=(A_1K_2+B_1)\cosh\left(\sqrt{K_2}L\right), a_{34}=(A_1K_2+B_1)\sinh\left(\sqrt{K_2}L\right), \\ &a_{41}=\left(-A_1\sqrt{-K_1}+B_1\sqrt{-K_1}^3\right)\sin\left(\sqrt{-K_1}L\right)+\omega^2 m_a\cos\left(\sqrt{-K_1}L\right), \\ &a_{42}=\left(A_1\sqrt{-K_1}-B_1\sqrt{-K_1}^3\right)\cos\left(\sqrt{-K_1}L\right)+\omega^2 m_a\sin\left(\sqrt{-K_1}L\right), \\ &a_{43}=\left(A_1\sqrt{K_2}+B_1\sqrt{K_2}^3\right)\sinh\left(\sqrt{K_2}L\right)+\omega^2 m_a\cosh\left(\sqrt{K_2}L\right), \\ &a_{44}=\left(A_1\sqrt{K_2}+B_1\sqrt{K_2}^3\right)\cosh\left(\sqrt{K_2}L\right)+\omega^2 m_a\sinh\left(\sqrt{K_2}L\right). \end{aligned} \tag{4.E4.4.6}$$

Again, in the computation of $[a]$ matrix, since the manual operations are complicated, frequency values can be obtained by symbolic programming on the computer.

Several numerical results about natural frequency calculation are tabulated in Example 4.8.

4.4 Nonlocal Vibration of Timoshenko Beam

For many cases of bending members, mechanical analyses based on normal stress and the resulting deformations yield sufficiently accurate results. However, shear stresses are also important for some cases of these elements. For example, in cases such as high height-to-thickness ratio, short spans, exposure to shear forces only, shear stresses can become significant and should be considered. Unless other higher-order terms are mentioned, all formulations that consider such stresses are known as first-order shear deformation beam theory or Timoshenko beam theory (TBT).

In this subchapter, the shear effect in the nonlocal transverse vibration of nanobeams will be investigated. Equation of motion and mechanical boundary conditions will be obtained from Hamilton's Principle and applications for different end types will be given.

4.4.1 Derivation of Equation of Motion

The motion components of the Timoshenko beam are defined as follows:

$$u(x,t) = -z\varphi = -z\frac{\partial w_b}{\partial x}, v(x,t) = 0, w(x,t) = w_b + w_s \tag{4.60}$$

where unlike EBBT, φ represents rotation of beam section, w_b is bending displacement and w_s is shear displacement. Strain components are expressed as follows:

$$\begin{aligned} &\varepsilon_{xx} = \frac{1}{2}\left(\frac{\partial u}{\partial x} + \frac{\partial u}{\partial x}\right) = \frac{\partial u}{\partial x} = -z\frac{\partial \phi}{\partial x}, \varepsilon_{xz} = \frac{1}{2}\left(\frac{\partial u}{\partial z} + \frac{\partial w}{\partial x}\right) = \frac{1}{2}\left(\frac{\partial w}{\partial x} - \phi\right), \\ &\varepsilon_{yy} = \varepsilon_{zz} = \varepsilon_{xy} = \varepsilon_{yz} = 0 \end{aligned} \tag{4.61}$$

Also, the stress expressions are given as follows:

$$\sigma_{xx} = E\varepsilon_{xx} = -Ez\frac{\partial \varphi}{\partial x}, \sigma_{xz} = G\varepsilon_{xz} = \frac{1}{2}G\left(\frac{\partial w}{\partial x} - \phi\right), \sigma_{yy} = \sigma_{zz} = \sigma_{xy} = \sigma_{yz} = 0 \tag{4.62}$$

The energy expressions for the beam element can be constituted as follows:

$$
\begin{aligned}
U &= \frac{1}{2}\int_V \left(\sigma_{xx}\varepsilon_{xx} + 2\sigma_{xy}\varepsilon_{xy}\right)\mathrm{d}V + \frac{1}{2}\int_0^L k_W w^2 \mathrm{d}x + \frac{1}{2}\int_0^L k_P\left(\frac{\partial w}{\partial x}\right)^2 \mathrm{d}x \\
&\quad + \frac{1}{2}\sum_{i=1}^{n} k_{a,i}[w(x_i,t)]^2 \\
&= \int_0^L \left[-M\frac{\partial \varphi}{\partial x} + V\left(\frac{\partial w}{\partial x} - \varphi\right)\right]\mathrm{d}x + \frac{1}{2}\int_0^L \left[k_W w^2 + k_P\left(\frac{\partial w}{\partial x}\right)^2\right]\mathrm{d}x \\
&\quad + \frac{1}{2}k_a[w(0,t)]^2 + \frac{1}{2}k_a[w(L,t)]^2
\end{aligned}
\tag{4.63}
$$

$$
\begin{aligned}
T &= \frac{1}{2}\int_V \rho\left[\left(\frac{\partial u}{\partial t}\right)^2 + \left(\frac{\partial v}{\partial t}\right)^2 + \left(\frac{\partial w}{\partial t}\right)^2\right]\mathrm{d}V + \frac{1}{2}\sum_{i=1}^{n} m_{a,i}\left(\frac{\partial w(x_i,t)}{\partial t}\right)^2 \\
&= \frac{1}{2}\int_0^L \left[\rho A\left(\frac{\partial w}{\partial t}\right)^2 + \rho I\left(\frac{\partial \varphi}{\partial t}\right)^2\right]\mathrm{d}x + \frac{1}{2}m_a\left(\frac{\partial w(0,t)}{\partial t}\right)^2 + \frac{1}{2}m_a\left(\frac{\partial w(L,t)}{\partial t}\right)^2
\end{aligned}
\tag{4.64}
$$

$$
W = \int_0^L f w \mathrm{d}x + \int_0^L N_T\left(\frac{\partial w}{\partial x}\right)^2 \mathrm{d}x \tag{4.65}
$$

Definitions of most of expressions seen here are known from EBBT. Unlike EBBT, second-order mass inertia ρI is included in the kinetic energy equation, where ρ is the mass of unit volume and I is the moment of inertia

$$
I = \int_A z^2 \mathrm{d}A \tag{4.66}
$$

To apply Hamilton's equation given with Eq. (4.22), variations of energy expressions should be stated:

$$
\begin{aligned}
\int_{t_1}^{t_2} \delta U \mathrm{d}t &= \int_{t_1}^{t_2}\int_0^L \left(-M\delta\frac{\partial \varphi}{\partial x} + V\delta\frac{\partial w}{\partial x} - V\delta\varphi + k_W w\delta w + k_P\frac{\partial w}{\partial x}\delta\frac{\partial w}{\partial x}\right)\mathrm{d}x\mathrm{d}t \\
&\quad + \int_{t_1}^{t_2} k_a w(0,t)\delta w(0,t)\mathrm{d}t + \int_{t_1}^{t_2} k_a w(L,t)\delta w(L,t)\mathrm{d}t
\end{aligned}
\tag{4.67}
$$

$$\int_{t_1}^{t_2} \delta T \mathrm{d}t = \int_{t_1}^{t_2} \int_{0}^{L} \left(\rho A \frac{\partial w}{\partial t} \delta \frac{\partial w}{\partial t} + \rho I \frac{\partial \phi}{\partial t} \delta \frac{\partial \phi}{\partial t} \right) \mathrm{d}x \mathrm{d}t + \int_{t_1}^{t_2} m_a \frac{\partial w(0,t)}{\partial t} \delta \frac{\partial w(0,t)}{\partial t} \mathrm{d}t + \int_{t_1}^{t_2} m_a \frac{\partial w(L,t)}{\partial t} \delta \frac{\partial w(L,t)}{\partial t} \mathrm{d}t \tag{4.68}$$

$$\int_{t_1}^{t_2} \delta W \mathrm{d}t = \int_{t_1}^{t_2} \int_{0}^{L} f \delta w \mathrm{d}x \mathrm{d}t + \int_{t_1}^{t_2} \int_{0}^{L} N_t \frac{\partial w}{\partial x} \delta \left(\frac{\partial w}{\partial x} \right) \mathrm{d}x \mathrm{d}t \tag{4.69}$$

The variation expressions seen in Eqs. (4.67)–(4.69) are calculated as follows:

$$\begin{aligned}
&\int_0^L M \delta \frac{\partial \phi}{\partial x} \mathrm{d}x = M \delta \phi|_0^L - \int_0^L \frac{\partial M}{\partial x} \delta \phi \mathrm{d}x, \quad \int_0^L V \delta \frac{\partial w}{\partial x} \mathrm{d}x = V \delta w|_0^L - \int_0^L \frac{\partial V}{\partial x} \delta w \mathrm{d}x, \\
&\int_0^L k_P \frac{\partial w}{\partial x} \delta \frac{\partial w}{\partial x} \mathrm{d}x = k_P \frac{\partial w}{\partial x} \delta w \bigg|_0^L - \int_0^L k_P \frac{\partial^2 w}{\partial x^2} \delta w \mathrm{d}x, \\
&\int_{t_1}^{t_2} \rho A \frac{\partial w}{\partial t} \delta \frac{\partial w}{\partial t} \mathrm{d}t = \rho A \frac{\partial w}{\partial t} \delta w \bigg|_{t_1}^{t_2} - \int_{t_1}^{t_2} \rho A \frac{\partial^2 w}{\partial t^2} \delta w \mathrm{d}t, \\
&\int_{t_1}^{t_1} \rho I \frac{\partial \phi}{\partial t} \delta \frac{\partial \phi}{\partial t} \mathrm{d}x = \rho I \frac{\partial \phi}{\partial t} \delta \phi \bigg|_{t_1}^{t_2} - \int_{t_1}^{t_2} \rho I \frac{\partial^2 \phi}{\partial t^2} \delta \phi \mathrm{d}t, \\
&\int_{t_1}^{t_2} m_a \frac{\partial w(0,t)}{\partial t} \delta \frac{\partial w(0,t)}{\partial t} \mathrm{d}t = m_a \frac{\partial w(0,t)}{\partial t} \delta w(0,t) \bigg|_{t_1}^{t_2} - \int_{t_1}^{t_2} m_a \frac{\partial^2 w(0,t)}{\partial t^2} \delta w(0,t) \mathrm{d}t, \\
&\int_{t_1}^{t_2} m_a \frac{\partial w(L,t)}{\partial t} \delta \frac{\partial w(L,t)}{\partial t} \mathrm{d}t = m_a \frac{\partial w(L,t)}{\partial t} \delta w(L,t) \bigg|_{t_1}^{t_2} - \int_{t_1}^{t_2} m_a \frac{\partial^2 w(L,t)}{\partial t^2} \delta w(L,t) \mathrm{d}t, \\
&\int_0^L N_t \frac{\partial w}{\partial x} \delta \frac{\partial w}{\partial x} \mathrm{d}x = N_t \frac{\partial w}{\partial x} \delta w \bigg|_0^L - \int_0^L N_t \frac{\partial^2 w}{\partial x^2} \delta w \mathrm{d}x
\end{aligned} \tag{4.70}$$

If these expressions are substituted into Eq. (4.22), without going into details this time, the following expressions can be obtained for the variations $\delta\varphi \neq 0$ and $\delta w \neq 0$, respectively [25]:

$$\delta\varphi \neq 0: \quad \frac{\partial M}{\partial x} - V = -\rho I \frac{\partial^2 \varphi}{\partial t^2} \tag{4.71}$$

$$\delta w \neq 0: \quad \frac{\partial V}{\partial x} = k_W w + \left(N_T - k_p\right) \frac{\partial^2 w}{\partial x^2} - f + \rho A \frac{\partial^2 w}{\partial t^2} \tag{4.72}$$

By using these two expressions, the nonlocal equation of motion can be derived. First, the following equations are written for different nonlocal stress components [25, 26]:

$$M - (e_0 a)^2 \frac{\partial^2 M}{\partial x^2} = -EI \frac{\partial \varphi}{\partial x} = -EI \frac{\partial^2 w_b}{\partial x^2} \tag{4.73}$$

$$V - (e_0 a)^2 \frac{\partial^2 V}{\partial x^2} = kGA \left(\frac{\partial w}{\partial x} - \varphi \right) = kGA \frac{\partial w_s}{\partial x} \tag{4.74}$$

where k indicates the shear correction factor. First, Eq. (4.74) is substituted into the first derivative of Eq. (4.73) with respect to x and then, second derivative of the bending moment is defined. By substituting the second derivative of the bending moment into Eq. (4.73)

$$\begin{aligned} M = & -EI \frac{\partial^2 w_b}{\partial x^2} + (e_0 a)^2 k_w (w_b + w_s) + (e_0 a)^2 \left(N_t - k_p\right) \frac{\partial^2 (w_b + w_s)}{\partial x^2} - (e_0 a)^2 f \\ & + (e_0 a)^2 \rho A \frac{\partial^2 (w_b + w_s)}{\partial t^2} - (e_0 a)^2 \rho I \frac{\partial^4 (w_b + w_s)}{\partial x^2 \partial t^2} \end{aligned} \tag{4.75}$$

On the other hand, by substituting the derivative of Eq. (4.74) with respect to x into Eq. (4.72), shear force is described:

$$\begin{aligned} V = & (e_0 a)^2 k_w \frac{\partial (w_b + w_s)}{\partial x} + (e_0 a)^2 \left(N_t - k_p\right) \frac{\partial^3 (w_b + w_s)}{\partial x^3} - (e_0 a)^2 \frac{\partial f}{\partial x} \\ & + (e_0 a)^2 \rho A \frac{\partial^3 (w_b + w_s)}{\partial x \partial t^2} + kGA \frac{\partial w_s}{\partial x} \end{aligned} \tag{4.76}$$

It should be noted that the bending moment and shear force equations include both bending and shear deformations. This result shows that the equation of motion will depend on both shear and bending deformations. It is necessary that the differential equation depends on a single differentiable function so that the analytical solution can be easily obtained. Therefore, a relation between bending and shear displacements should be derived to arrange the equation of motion. If Eqs. (4.75) and (4.76) are written in Eq. (4.71), the following expression is obtained as follows [16]:

$$w_s = -\frac{1}{kGA}\left(EI\frac{\partial^2 w_b}{\partial x^2} - \rho I\frac{\partial^2 w_b}{\partial t^2} + (e_0a)^2\rho I\frac{\partial^4 w_b}{\partial x^2\partial t^2}\right) \tag{4.77}$$

Substituted Eq. (4.77) into Eqs. (4.75) and (4.76) yields the bending moment and shear force equations, respectively:

$$\begin{aligned}M = {} & (e_0a)^2k_Ww_b - \frac{(e_0a)^2k_WEI}{kGA}\frac{\partial^2 w_b}{\partial x^2} + \frac{(e_0a)^2k_W\rho I}{kGA}\frac{\partial^2 w_b}{\partial t^2} - \frac{(e_0a)^4k_W\rho I}{kG}\frac{\partial^4 w_b}{\partial x^2\partial t^2}\\ & +(e_0a)^2(N_T-k_P)\frac{\partial^2 w_b}{\partial x^2} - \frac{(e_0a)^2(N_T-k_P)EI}{kGA}\frac{\partial^4 w_b}{\partial x^4} + \frac{(e_0a)^2(N_T-k_P)\rho I}{kGA}\frac{\partial^4 w_b}{\partial x^2\partial t^2}\\ & -\frac{(e_0a)^4(N_T-k_P)\rho I}{kGA}\frac{\partial^6 w_b}{\partial x^4\partial t^2} - (e_0a)^2f + (e_0a)^2\rho A\frac{\partial^2 w_b}{\partial t^2} - \frac{(e_0a)^2\rho EI}{kG}\frac{\partial^4 w_b}{\partial x^2\partial t^2}\\ & +\frac{(e_0a)^2\rho^2 I}{kG}\frac{\partial^4 w_b}{\partial t^4} + \frac{(e_0a)^4\rho^2 I}{kG}\frac{\partial^6 w_b}{\partial x^4\partial t^2} - (e_0a)^2\rho I\frac{\partial^4 w_b}{\partial x^2\partial t^2} - EI\frac{\partial^2 w_b}{\partial x^2}\end{aligned} \tag{4.78}$$

$$\begin{aligned}V = {} & (e_0a)^2k_W\frac{\partial w_b}{\partial x} - \frac{(e_0a)^2k_WEI}{kGA}\frac{\partial^3 w_b}{\partial x^3} + \frac{(e_0a)^2k_W\rho I}{kGA}\frac{\partial^3 w_b}{\partial x\partial t^2} - \frac{(e_0a)^4k_W\rho I}{kG}\frac{\partial^5 w_b}{\partial x^3\partial t^2}\\ & +(e_0a)^2(N_T-k_P)\frac{\partial^3 w_b}{\partial x^3} - \frac{(e_0a)^2(N_T-k_P)EI}{kGA}\frac{\partial^5 w_b}{\partial x^5} + \frac{(e_0a)^2(N_T-k_P)\rho I}{kGA}\frac{\partial^5 w_b}{\partial x^3\partial t^2}\\ & -\frac{(e_0a)^4(N_T-k_P)\rho I}{kGA}\frac{\partial^7 w_b}{\partial x^5\partial t^2} - (e_0a)^2\frac{\partial f}{\partial x} + (e_0a)^2\rho A\frac{\partial^3 w_b}{\partial x\partial t^2} - \frac{(e_0a)^2\rho EI}{kG}\frac{\partial^5 w_b}{\partial x^3\partial t^2}\\ & +\frac{(e_0a)^2\rho^2 I}{kG}\frac{\partial^5 w_b}{\partial x\partial t^4} + \frac{(e_0a)^4\rho^2 I}{kG}\frac{\partial^7 w_b}{\partial x^3\partial t^4} - EI\frac{\partial^3 w_b}{\partial x^2} + \rho I\frac{\partial^3 w_b}{\partial x\partial t^2} - (e_0a)^2\rho I\frac{\partial^5 w_b}{\partial x^3\partial t^2}\end{aligned} \tag{4.79}$$

If Eqs. (4.78) and (4.79) are substituted into Eq. (4.71), the equation of motion occurs as follows [25]:

$$
\begin{aligned}
&-\frac{(e_0a)^2(N_T-k_P)EI}{kGA}\frac{\partial^6 w_b}{\partial x^6}+(e_0a)^2(N_T-k_P)\frac{\partial^4 w_b}{\partial x^4}\\
&-\frac{\left[(e_0a)^2k_W-(N_T-k_P)+kGA\right]EI}{kGA}\frac{\partial^4 w_b}{\partial x^4}+\left[(e_0a)^2k_W-(N_T-k_P)+kGA\right]\frac{\partial^2 w_b}{\partial x^2}\\
&+\frac{k_W EI}{kGA}\frac{\partial^2 w_b}{\partial x^2}-kGA\frac{\partial^2 w_b}{\partial x^2}-\frac{\rho^2 I}{kG}\frac{\partial^4 w_b}{\partial t^4}-\rho A\frac{\partial^2 w_b}{\partial t^2}-\frac{k_W\rho I}{kGA}\frac{\partial^2 w_b}{\partial t^2}\\
&-\frac{(e_0a)^4(N_T-k_P)\rho I}{kGA}\frac{\partial^8 w_b}{\partial x^6\partial t^2}-\frac{(e_0a)^2\rho EI}{kG}\frac{\partial^6 w_b}{\partial x^4\partial t^2}+\frac{(e_0a)^2(N_T-k_P)\rho I}{kGA}\frac{\partial^6 w_b}{\partial x^4\partial t^2}\\
&-\frac{\left[(e_0a)^2k_W-(N_T-k_P)+kGA\right](e_0a)^2\rho I}{kGA}\frac{\partial^6 w_b}{\partial x^4\partial t^2}+\frac{\rho EI}{kG}\frac{\partial^4 w_b}{\partial x^2\partial t^2}+(e_0a)^2\rho A\frac{\partial^4 w_b}{\partial x^2\partial t^2}\\
&+\frac{\left[(e_0a)^2k_W-(N_T-k_P)+kGA\right]\rho I}{kGA}\frac{\partial^4 w_b}{\partial x^2\partial t^2}+\frac{(e_0a)^2k_w\rho I}{kGA}\frac{\partial^4 w_b}{\partial x^2\partial t^2}-\frac{(e_0a)^4\rho^2 I}{kG}\frac{\partial^8 w_b}{\partial x^4\partial t^4}\\
&+2\frac{(e_0a)^2\rho^2 I}{kG}\frac{\partial^6 w_b}{\partial x^2\partial t^4}-k_W w_b-(e_0a)^2\frac{\partial^2 f}{\partial x^2}+f=0
\end{aligned}
\tag{4.80}
$$

Also, internal effects of bending moment and shear force should also be obtained, but something should be known here. Unlike the EBBT, since it does not include both the force of shear layer and the axial force of thermal environment, Eq. (4.76) cannot reflect the shear force boundary condition. This is not because of a lack of variational calculus, but rather shear force expression obtained for the variation of displacements of end w while Eq. (4.22) for the nonlocal Timoshenko beam is written. We neglect attachments for now and write the equation for the w variation:

$$
V=(N_t-k_P)\frac{\partial w}{\partial x} \tag{4.81}
$$

If this expression is substituted into Eq. (4.76),

$$
\begin{aligned}
&\left[k_p+(e_0a)^2k_w-N_t\right]\frac{\partial(w_b+w_s)}{\partial x}+(e_0a)^2(N_t-k_p)\frac{\partial^3(w_b+w_s)}{\partial x^3}-(e_0a)^2\frac{\partial f}{\partial x}\\
&+(e_0a)^2\rho A\frac{\partial^3(w_b+w_s)}{\partial x\partial t^2}+kGA\frac{\partial w_s}{\partial x}=0
\end{aligned}
\tag{4.82}
$$

This is the mechanical boundary condition for without case of end attachment. If Eq. (4.77) is substituted into Eq. (4.82), the shear force boundary condition will be exactly as follows:

$$
\begin{aligned}
V=&(e_0a)^2k_W\frac{\partial w_b}{\partial x}-\frac{(e_0a)^2k_WEI}{kGA}\frac{\partial^3 w_b}{\partial x^3}+\frac{(e_0a)^2k_W\rho I}{kGA}\frac{\partial^3 w_b}{\partial x\partial t^2}-\frac{(e_0a)^4k_W\rho I}{kG}\frac{\partial^5 w_b}{\partial x^3\partial t^2}\\
&+(e_0a)^2(N_T-k_P)\frac{\partial^3 w_b}{\partial x^3}-\frac{(e_0a)^2(N_T-k_P)EI}{kGA}\frac{\partial^5 w_b}{\partial x^5}+\frac{2(e_0a)^2(N_T-k_P)\rho I}{kGA}\frac{\partial^5 w_b}{\partial x^3\partial t^2}\\
&-\frac{(e_0a)^4(N_T-k_P)\rho I}{kGA}\frac{\partial^7 w_b}{\partial x^5\partial t^2}-(e_0a)^2\frac{\partial f}{\partial x}+(e_0a)^2\rho A\frac{\partial^3 w_b}{\partial x\partial t^2}-\frac{(e_0a)^2\rho EI}{kG}\frac{\partial^5 w_b}{\partial x^3\partial t^2}\\
&+\frac{(e_0a)^2\rho^2 I}{kG}\frac{\partial^5 w_b}{\partial x\partial t^4}+\frac{(e_0a)^4\rho^2 I}{kG}\frac{\partial^7 w_b}{\partial x^3\partial t^4}-EI\frac{\partial^3 w_b}{\partial x^3}+\rho I\frac{\partial^3 w_b}{\partial x\partial t^2}-(e_0a)^2\rho I\frac{\partial^5 w_b}{\partial x^3\partial t^2}\\
&-(N_T-k_P)\frac{\partial w_b}{\partial x}+\frac{(N_T-k_P)EI}{kGA}\frac{\partial^3 w_b}{\partial x^3}-\frac{(N_T-k_P)\rho I}{kGA}\frac{\partial^3 w_b}{\partial x\partial t^2}
\end{aligned}
\tag{4.83}
$$

The bending moment boundary condition is defined in Eq. (4.78). By considering the shear force in Eq. (4.83), the generalized boundary condition for attachments is given as follows:

$$
\begin{aligned}
V(0,t)&=k_a(w_b+w_s)(0,t)+m_a\frac{\partial^2(w_b+w_s)}{\partial t^2}(0,t),\\
V(L,t)&=-k_a(w_b+w_s)(L,t)-m_a\frac{\partial^2(w_b+w_s)}{\partial t^2}(L,t)
\end{aligned}
\tag{4.84}
$$

4.4.2 Solution of Equation of Motion

Similar to EBBT, the solution of equation of motion and the definition of boundary conditions can be performed according to Eq. (4.30). For this, the following template of the separation of variable is defined

$$w_b(x,t)=W(x)T(t)=W(x)\sin(\omega t-\theta) \tag{4.85}$$

First, $f=0$ is taken because of the free vibration solution. In this case, Eq. (4.80) is transformed into the following homogeneous and ordinary differential equation form:

$$A_1\frac{\mathrm{d}^6W(x)}{\mathrm{d}x^6}+A_2\frac{\mathrm{d}^4W(x)}{\mathrm{d}x^4}+A_3\frac{\mathrm{d}^2W(x)}{\mathrm{d}x^2}+A_4W(x)=0 \tag{4.86}$$

where the coefficients A_i $(i=1,2,3,4)$ are specified as follows [25]:

$$
\begin{aligned}
A_1 &= -\frac{(e_0a)^2(N_T-k_P)EI}{kGA}+\frac{(e_0a)^4(N_T-k_P)\rho I}{kGA}\omega^2,\\
A_2 &= -\frac{\left[(e_0a)^2k_W-(N_T-k_P)+kGA\right]EI}{kGA}+(e_0a)^2(N_T-k_P)+\left[\frac{(e_0a)^2\rho EI}{kG}\right.\\
&\quad\left.-\frac{(e_0a)^2(N_T-k_P)\rho I}{kGA}+\frac{\left[(e_0a)^2k_W-(N_T-k_P)+kGA\right](e_0a)^2\rho I}{kGA}\right]\omega^2\\
&\quad+\left[-\frac{(e_0a)^4\rho^2 I}{kG}\right]\omega^4,\\
A_3 &= \frac{k_W EI}{kGA}+(e_0a)^2k_W-(N_T-k_P)+\left[-\frac{\rho EI}{kG}-(e_0a)^2\rho A\right.\\
&\quad\left.-\frac{\left[(e_0a)^2k_W-(N_T-k_P)+kGA\right]\rho I}{kGA}-\frac{(e_0a)^2k_W\rho I}{kGA}\right]\omega^2\\
&\quad+\left[\frac{2(e_0a)^2\rho^2 I}{kG}\right]\omega^4,\\
A_4 &= -k_W+\left(\rho A+\frac{k_W\rho I}{kGA}\right)\omega^2+\left(-\frac{\rho^2 I}{kG}\right)\omega^4.
\end{aligned}
\tag{4.87}
$$

It is difficult to obtain an analytical solution as in Eq. (4.41) since the characteristic polynomial contains third order and contains a constant number such as A_4.

On the other hand, it should be expressed that bending moment and shear forces are formulated. First, If Eq. (4.85) applied to Eq. (4.75), following expressions can be obtained:

$$M = B_1\frac{\mathrm{d}^4 W}{\mathrm{d}x^4}+B_2\frac{\mathrm{d}^2 W}{\mathrm{d}x^2}+B_3 W \tag{4.88}$$

$$V = B_4\frac{\mathrm{d}^5 W}{\mathrm{d}x^5}+B_5\frac{\mathrm{d}^3 W}{\mathrm{d}x^3}+B_6\frac{\mathrm{d}W}{\mathrm{d}x} \tag{4.89}$$

where B_i $(i = 1-6)$ are given as follows

$$
\begin{aligned}
B_1 &= -\frac{(e_0a)^2(N_T-k_P)EI}{kGA}+\left(\frac{(e_0a)^4(N_T-k_P)\rho I}{kGA}-\frac{(e_0a)^4\rho^2 I}{kG}\right)\omega^2,\\
B_2 &= -EI-\frac{(e_0a)^2k_W EI}{kGA}+(e_0a)^2(N_T-k_P)+\left(\frac{(e_0a)^4k_W\rho I}{kG}-\frac{(e_0a)^2(N_T-k_P)\rho I}{kGA}\right.\\
&\quad\left.+\frac{(e_0a)^2\rho EI}{kG}+(e_0a)^2\rho I\right)\omega^2,\\
B_3 &= (e_0a)^2k_W+\left(-(e_0a)^2\rho A-\frac{(e_0a)^2k_W\rho I}{kGA}\right)\omega^2+\left(\frac{(e_0a)^2\rho^2 I}{kG}\right)\omega^4,\\
B_4 &= -\frac{(e_0a)^2(N_T-k_P)EI}{kGA}+\left(\frac{(e_0a)^4(N_T-k_P)\rho I}{kGA}\right)\omega^2,\\
B_5 &= -\frac{(e_0a)^2k_W EI}{kGA}+(e_0a)^2(N_T-k_P)-EI+\left(\frac{(N_T-k_P)EI}{kGA}+\frac{(e_0a)^4k_W\rho I}{kG}\right.\\
&\quad\left.-\frac{2(e_0a)^2(N_T-k_P)\rho I}{kGA}+\frac{(e_0a)^2\rho EI}{kG}+(e_0a)^2\rho I\right)\omega^2+\left(\frac{(e_0a)^4\rho^2 I}{kG}\right)\omega^4,\\
B_6 &= (e_0a)^2k_W-(N_T-k_P)+\left(-\frac{(e_0a)^2k_W\rho I}{kGA}-(e_0a)^2\rho A-\rho I+\frac{(N_T-k_P)\rho I}{kGA}\right)\omega^2\\
&\quad+\frac{(e_0a)^2\rho^2 I}{kG}\omega^4.
\end{aligned}
\tag{4.90}
$$

4.4.3 Boundary Conditions of Nanobeams

Unlike EBBT, the analytical solution of Eq. (4.85) is quite difficult due to the degree of characteristic polynomial and is not appropriate for manual solution. However, in this nonlocal nanobeam model, a hand solution can only be done with series expansions that provide boundary conditions for S-S and S-G nanobeams. The different end types are examined as follows.

1. **Simply supported end:** Since there will not become both displacement and bending moment, boundary conditions for a nanobeam with simply supported at both ends are as follows

$$
W=0, B_1W^{(4)}+B_2W''+B_3W=0 \tag{4.91}
$$

Substituting the displacement conditions into the bending moment conditions, the bending moment conditions are arranged as follows:

$$
B_1W^{(4)}+B_2W''=0 \tag{4.92}
$$

2. **Clamped end:** Such type ends do not take displacement and rotation

$$W = 0, W' = 0 \tag{4.93}$$

3. **Free end:** There are not bending moments and shear forces at the free ends

$$B_1 W^{(4)} + B_2 W'' + B_3 W = 0, B_4 W^{(5)} + B_5 W''' + B_6 W' = 0 \tag{4.94}$$

4. **Guided end:** There are not rotation and shear forces at the free ends

$$W' = 0, B_4 W^{(5)} + B_5 W''' + B_6 W' = 0 \tag{4.95}$$

If the rotation condition is substituted into the shear force condition

$$B_4 W^{(5)} + B_5 W''' = 0 \tag{4.96}$$

5. **Attached end:** Firstly, Eqs. (4.77) and (4.84) and are used in the shear force derived in Eq. (4.89) to constitute the mechanical boundary condition at the attached end. Thus, the shear force boundary condition for $x = 0$ and $x = L$ boundaries can be written as follows, respectively:

$$\begin{aligned} &B_4 W^{(5)}(0) + B_5 W'''(0) + B_6 W'(0) = D_1 W''(0) + D_2 W(0), \\ &B_4 W^{(5)}(L) + B_5 W'''(L) + B_6 W'(L) = -D_1 W''(L) - D_2 W(L) \end{aligned} \tag{4.97}$$

where

$$\begin{aligned} D_1 &= -\frac{k_a EI}{kGA} + \frac{(e_0 a)^2 k_a \rho I}{kGA} + \left(\frac{m_a EI}{kGA}\right)\omega^2 + \left(-\frac{(e_0 a)^2 m_a \rho I}{kGA}\right)\omega^4, \\ D_2 &= k_a + \left(-\frac{k_a \rho I}{kGA} - m_a\right)\omega^2 + \left(\frac{m_a \rho I}{kGA}\right)\omega^4. \end{aligned} \tag{4.98}$$

Example 4.5 Determine the frequency equation of nonlocal free vibration of simply supported nanobeam embedded in the thermo-elastic medium shown in Fig. 4.2 according to nonlocal TBT.

Solution. Considering the Eq. (4.92), the boundary conditions are written as:

$$\begin{aligned} &W(0) = 0, B_1 W^{(4)}(0) + B_2 W''(0) = 0, \\ &W(L) = 0, B_1 W^{(4)}(L) + B_2 W''(L) = 0 \end{aligned} \tag{4.E4.5.1}$$

In addition to the difficulty of derivation of analytical solution via Eqs. (4.86), these boundary conditions make the solution further impossible. However, the manual solution is possible with the following function that provides the boundary conditions:

$$w_b(x,t) = \sum_{n=1}^{\infty} w_n \sin\left(\frac{n\pi x}{L}\right) \sin(\omega t - \theta) \tag{4.E4.5.2}$$

Substituting this expression into Eq. (4.86), we can obtain the following expression:

$$-A_1\left(\frac{n\pi}{L}\right)^6 + A_2\left(\frac{n\pi}{L}\right)^4 - A_3\left(\frac{n\pi}{L}\right)^2 + A_4 = 0 \tag{4.E4.5.3}$$

As a result of several mathematical operations, the frequency equation occurs as a second-order linear equation based on the square of square of frequency. It is preferred not to go into the details of this. In this case, the solution is defined as follows:

$$\omega^2 = \frac{-C_2 + \sqrt{C_2{}^2 - 4C_1C_3}}{2C_1} \tag{4.E4.5.4}$$

where

$$C_1 = c_1\left(\frac{n\pi}{L}\right)^4 - c_2\left(\frac{n\pi}{L}\right)^2 + c_3, C_2 = -c_4\left(\frac{n\pi}{L}\right)^6 + c_5\left(\frac{n\pi}{L}\right)^4 - c_6\left(\frac{n\pi}{L}\right)^2 + c_7,$$
$$C_3 = -c_8\left(\frac{n\pi}{L}\right)^6 + c_9\left(\frac{n\pi}{L}\right)^4 - c_{10}\left(\frac{n\pi}{L}\right)^2 + c_{11} \tag{4.E4.5.5}$$

where the coefficients c_i $(i = 1 - 11)$ are written as follows:

$$c_1 = -\frac{(e_0a)^4\rho^2 I}{kG}, c_2 = \frac{2(e_0a)^2\rho^2 I}{kG}, c_3 = -\frac{\rho^2 I}{kG}, c_4 = \frac{(e_0a)^4(N_t - k_P)\rho I}{kGA},$$
$$c_5 = \frac{(e_0a)^2\rho EI}{kG} - \frac{(e_0a)^2(N_t - k_P)\rho I}{kGA} + \frac{\left[(e_0a)^2 k_W - (N_t - k_P)\right](e_0a)^2\rho I}{kGA} + (e_0a)^2\rho I,$$
$$c_6 = -\frac{\rho EI}{kG} - (e_0a)^2\rho A - \frac{\left[2(e_0a)^2 k_W - (N_t - k_P)\right]\rho I}{kGA} - \rho I, c_7 = \rho A + \frac{k_W\rho I}{kGA},$$
$$c_8 = -\frac{(e_0a)^2(N_t - k_P)EI}{kGA}, c_9 = -\frac{\left[(e_0a)^2 k_W - (N_t - k_P)\right]EI}{kGA} - EI + (e_0a)^2(N_t - k_P),$$
$$c_{10} = \frac{k_W EI}{kGA} + (e_0a)^2 k_W - (N_t - k_P), c_{11} = -k_W c_{11} = -k_W. \tag{4.E.4.5.6}$$

In Table 4.2, the first five nonlocal nondimensional frequency parameters of simply supported nanobeams are calculated for different nonlocal beam theories. Thermo-elastic medium is neglected in the computations ($k_P = k_W = N_T = 0$). The nondimensionalization of frequencies is performed as follows:

$$\overline{\omega} = \omega_i L^2 \sqrt{\frac{\rho A}{EI}} \tag{4.E4.5.7}$$

where i is mode number and ω_i is natural frequency. $\overline{\omega}_i$ defines nondimensional frequency. Also α defines nondimensional nonlocal parameter:

$$\alpha = \frac{e_0 a}{L} \tag{4.E4.5.8}$$

The following parameters are used to calculate the results in Table 4.2: Nanobeam length: $L = 20$ nm, height and thickness of rectangular cross-section: $h = 1$ nm and $b = 2$ nm, modulus of elasticity: $E = 1$ TPa, Poisson's ratio: $\upsilon = 0.19$, shear modulus: $G = E/2(1 + \upsilon) = 0.42$ TPa, mass of unit volume: $\rho = 2300$ kg/m^3, and shear correction factor: $k = 5/6$.

When the results obtained in Table 4.2 are examined, the classical nondimensional frequencies decrease under the atomic parameter and this is a very important inference. On the other hand, as the mode number increases, the nondimensional frequencies decrease more than classical elasticity. Finally, it is understood that the nondimensional frequencies decrease more in the shear deformable beam theory. The shear effect is more pronounced in higher modes.

Example 4.6 Determine the frequency equation of nonlocal free vibration of simply supported-guided supported nanobeam embedded in the thermo-elastic medium shown in Fig. 4.5 according to nonlocal TBT.

Solution. For a nonlocal Timoshenko nanobeam, the following series expansion provides the boundary conditions of S-G nanobeam:

$$w_b(x,t) = \sum_{n=1}^{\infty} w_n \sin\left(\frac{(2n-1)\pi x}{2L}\right) \sin(\omega t - \theta) \tag{4.E4.6.1}$$

Using this solution into Eq. (4.86) yields the following equation:

Table 4.2 Comparisons of nondimensional frequencies of simply supported nanobeams

Mode	EBBT			TBT		
Number	$\alpha = 0$	$\alpha = 0.1$	$\alpha = 0.2$	$\alpha = 0$	$\alpha = 0.1$	$\alpha = 0.2$
1	9.86960	9.41588	8.35692	9.83077	9.37883	8.32404
2	39.47842	33.42768	24.58230	38.87044	32.91288	24.20373
3	88.82644	64.64141	41.62849	85.85397	62.47827	40.23544
4	157.91367	98.32921	58.38034	148.94153	92.74246	55.06336
5	246.74011	132.50666	74.83985	226.00756	121.37268	68.55136

$$-A_1\left(\frac{(2n-1)\pi}{2L}\right)^6+A_2\left(\frac{(2n-1)\pi}{2L}\right)^4-A_3\left(\frac{(2n-1)\pi}{2L}\right)^2+A_4=0 \quad (4.\text{E}4.6.2)$$

Similar to Eq. (4.E4.5.3), the expressions in Eq. (4.E4.6.2) are as follows for S-G condition:

$$\begin{aligned}
C_1&=c_1\left(\frac{(2n-1)\pi}{2L}\right)^4-c_2\left(\frac{(2n-1)\pi}{2L}\right)^2+c_3,\\
C_2&=-c_4\left(\frac{(2n-1)\pi}{2L}\right)^6+c_5\left(\frac{(2n-1)\pi}{2L}\right)^4-c_6\left(\frac{(2n-1)\pi}{2L}\right)^2+c_7,\\
C_3&=-c_8\left(\frac{(2n-1)\pi}{2L}\right)^6+c_9\left(\frac{(2n-1)\pi}{2L}\right)^4-c_{10}\left(\frac{(2n-1)\pi}{2L}\right)^2+c_{11}.
\end{aligned} \quad (4.\text{E}4.6.3)$$

coefficients c_i $(i=1-11)$ are as given in Eq. (4.E4.5.6).

Example 4.7 Determine the frequency equation of nonlocal free vibration of cantilever nanobeam without thermo-elastic environment shown in Fig. 4.6 according to nonlocal TBT.

Solution. In Eq. (4.86), the thermo-elastic environment is neglected with $k_P=k_W=N_T=0$. Thus, $A_1=0$ in Eq. (4.87). Equation (4.86) becomes:

$$A_2\frac{d^4W(x)}{dx^4}+A_3\frac{d^2W(x)}{dx^2}+A_4W(x)=0 \quad (4.\text{E}4.7.1)$$

where

$$\begin{aligned}
A_2&=-EI+(e_0a)^2(N_T-k_P)+\left[\frac{(e_0a)^2\rho EI}{kG}+(e_0a)^2\rho I\right]\omega^2+\left[-\frac{(e_0a)^4\rho^2 I}{kG}\right]\omega^4,\\
A_3&=\left[-\frac{\rho EI}{kG}-(e_0a)^2\rho A-\rho I\right]\omega^2+\left[\frac{2(e_0a)^2\rho^2 I}{kG}\right]\omega^4,\\
A_3&=\left[-\frac{\rho EI}{kG}-(e_0a)^2\rho A-\rho I\right]\omega^2+\left[\frac{2(e_0a)^2\rho^2 I}{kG}\right]\omega^4,
\end{aligned} \quad (4.\text{E}4.7.2)$$

Since Eq. (4.E4.7.1) has a form like Eq. (4.31), the solution is as in Eq. (4.46)

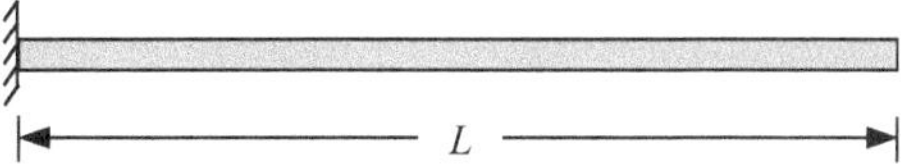

Fig. 4.6 Cantilever nanobeam

$$W(x) = C_1 \cos\left(\sqrt{-K_1}x\right) + C_2 \sin\left(\sqrt{-K_1}x\right) + C_3 \cosh\left(\sqrt{K_2}x\right) + C_4 \sinh\left(\sqrt{K_2}x\right) \tag{4.E4.7.3}$$

Where below definitions are valid:

$$K_1 = \frac{-A_3 + \sqrt{{A_3}^2 - 4A_2A_4}}{2A_2}, K_2 = \frac{-A_3 - \sqrt{{A_3}^2 - 4A_2A_4}}{2A_2} \tag{4.E4.7.4}$$

Boundary conditions of this beam model are given as follows:

$$\begin{aligned} &W(0) = 0, W'(0) = 0, B_1 W^{(4)}(L) + B_2 W''(L) + B_3 W(L) = 0, B_4 W^{(5)}(L) \\ &+ B_5 W'''(L) + B_6 W'(L) = 0 \end{aligned} \tag{4.E4.7.5}$$

By considering Eq. (4.90)

$$\begin{aligned} &B_1 = \left(-\frac{(e_0a)^4\rho^2 I}{kG}\right)\omega^2, B_2 = -EI + \left(\frac{(e_0a)^2\rho EI}{kG} + (e_0a)^2\rho I\right)\omega^2, \\ &B_3 = -(e_0a)^2\rho A\omega^2 + \left(\frac{(e_0a)^2\rho^2 I}{kG}\right)\omega^4, B_4 = 0, \\ &B_5 = -EI + \left(\frac{(e_0a)^2\rho EI}{kG} + (e_0a)^2\rho I\right)\omega^2 + \left(\frac{(e_0a)^4\rho^2 I}{kG}\right)\omega^4, \\ &B_6 = \left(-(e_0a)^2\rho A - \rho I\right)\omega^2 + \frac{(e_0a)^2\rho^2 I}{kG}\omega^4 \end{aligned} \tag{4.E4.7.6}$$

If the above expressions are substituted into Eq. (4.E4.7.3), the following equations are obtained, respectively:

$$C_1 + C_3 = 0 \tag{4.E4.7.7}$$

$$C_2\sqrt{-K_1} + C_4\sqrt{K_2} = 0 \tag{4.E4.7.8}$$

$$\begin{aligned} &C_1\left(B_1{K_1}^2 - B_2K_1 + B_3\right)\cos\left(\sqrt{-K_1}L\right) + C_2\left(B_1{K_1}^2 - B_2K_1 + B_3\right)\sin\left(\sqrt{-K_1}L\right) + \\ &C_3\left(B_1{K_2}^2 + B_2K_2 + B_3\right)\cosh\left(\sqrt{K_2}L\right) + C_4\left(B_1{K_2}^2 + B_2K_2 + B_3\right)\sinh\left(\sqrt{K_2}L\right) = 0 \end{aligned} \tag{4.E4.7.9}$$

$$C_1\left(B_5\sqrt{-K_1}^3-B_6\sqrt{-K_1}\right)\sin\left(\sqrt{-K_1}L\right)+C_2\left(-B_5\sqrt{-K_1}^3+B_6\sqrt{-K_1}\right)\cos\left(\sqrt{-K_1}L\right)$$
$$C_3\left(B_5\sqrt{K_2}^3+B_6\sqrt{K_2}\right)\sinh\left(\sqrt{K_2}L\right)+C_4\left(B_5\sqrt{K_2}^3+B_6\sqrt{K_2}\right)\cosh\left(\sqrt{K_2}L\right)=0 \tag{4.E4.7.10}$$

If Eqs. (4.E4.7.7)–(4.E4.7.10) are expressed in a similar way to Eq. (4.E4.2.6), the inputs of $[a]$ matrix are as follows:

$$\begin{aligned}
&a_{11}=1, a_{12}=0, a_{13}=1, a_{14}=0, a_{21}=0, a_{22}=\sqrt{-K_1}, a_{23}=0, a_{24}=\sqrt{-K_1},\\
&a_{31}=\left(B_1K_1^2-B_2K_1+B_3\right)\cos\left(\sqrt{-K_1}L\right), a_{32}=\left(B_1K_1^2-B_2K_1+B_3\right)\sin\left(\sqrt{-K_1}L\right),\\
&a_{33}=\left(B_1K_2^2+B_2K_2+B_3\right)\cosh\left(\sqrt{K_2}L\right), a_{34}=\left(B_1K_2^2+B_2K_2+B_3\right)\sinh\left(\sqrt{K_2}L\right),\\
&a_{41}=\left(B_5\sqrt{-K_1}^3-B_6\sqrt{-K_1}\right)\sin\left(\sqrt{-K_1}L\right), a_{42}=\left(-B_5\sqrt{-K_1}^3+B_6\sqrt{-K_1}\right)\cos\left(\sqrt{-K_1}L\right),\\
&a_{43}=\left(B_5\sqrt{K_2}^3+B_6\sqrt{K_2}\right)\sinh\left(\sqrt{K_2}L\right), a_{44}=\left(B_5\sqrt{K_2}^3+B_6\sqrt{K_2}\right)\cosh\left(\sqrt{K_2}L\right).
\end{aligned} \tag{4.E4.7.11}$$

By considering the parameters used in the analysis in Table 4.2, the first five modes nondimensional frequencies of cantilever nanobeams are listed in Table 4.3 according to different nonlocal beam theories. Firstly, it is determined that the fundamental mode nondimensional frequencies of both nonlocal nanobeam types rise in response to the increment of atomic parameter. This situation is contrary to the conclusions about the effect of nonlocal elasticity on structure stiffness. Because it was determined that the nonlocal parameter reduces the frequencies both in the previous chapters and in the simply supported nanobeam example. In the scientific literature, this discrepancy of nonlocal elasticity is defined as a paradox and this is a suspicious situation. There are different approaches based on analytical analysis to resolve the paradox [39–42]. Additionally, in this book, an analysis that rules out the paradox of nonlocal cantilever nanobeams is the subject of the seventh chapter. It should be noted that the paradox that comes to the forefront in the fundamental mode is not observed in other modes.

On the other hand, the nonlocal parameter is understood to be quite effective in higher modes. Moreover, the frequencies of shear deformable nanobeams are more reduced by the

Table 4.3 Comparisons of nondimensional frequencies of cantilever nanobeams

Mode	EBBT			TBT		
Number	$\alpha=0$	$\alpha=0.1$	$\alpha=0.2$	$\alpha=0$	$\alpha=0.1$	$\alpha=0.2$
1	3.51602	3.53128	3.57940	3.51521	3.53018	3.57733
2	22.03449	20.67960	17.57601	21.87521	20.51140	17.40707
3	61.69721	51.06382	36.81305	60.41308	49.82984	35.78182
4	120.90192	85.68971	54.19456	115.96414	81.64190	51.44191
5	199.85953	121.36150	71.94097	186.75043	112.27783	66.20461

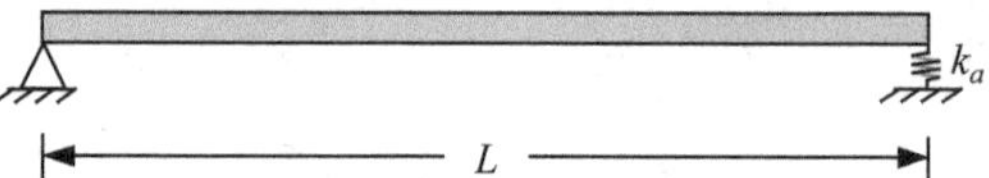

Fig. 4.7 Simply supported-transverse spring attached

nonlocal parameter. As the mode number increases, nonlocal parameter more decreases the frequencies. All these conclusions once again reveal the effect of nonlocal parameter on the mechanical behavior of nano-scaled continuous systems.

Example 4.8 Determine the frequency equation of nonlocal free vibration of simply supported-transverse spring attached nanobeam without thermo-elastic environment shown in Fig. 4.7 according to nonlocal TBT.

Solution. The boundary conditions can be written as follows:

$$\begin{aligned} &W(0)=0, B_1W^{(4)}(0)+B_2W''(0)=0\\ &B_1W^{(4)}(L)+B_2W''(L)+B_3W(L)=0,\\ &B_5W'''(L)+B_6W'(L)=-D_1W''(L)-D_2W(L)\end{aligned} \tag{4.E4.8.1}$$

In which, $B_4=0$ and $m_a=0$ are reminded. By considering Eq. (4.98)

$$D_1=-\frac{k_aEI}{kGA}+\frac{(e_0a)^2k_a\rho I}{kGA}, D_2=k_a+\left(-\frac{k_a\rho I}{kGA}\right)\omega^2 \tag{4.E4.8.2}$$

Applying boundary conditions into Eq. (4.E4.7.3) gives, respectively:

$$C_1+C_3=0 \tag{4.E4.8.3}$$

$$C_1\left(B_1{K_1}^2-B_2K_1\right)\cos\left(\sqrt{-K_1}L\right)+C_3\left(B_1{K_2}^2+B_2K_2\right)\cosh\left(\sqrt{K_2}L\right)=0 \tag{4.E4.8.4}$$

$$\begin{aligned} &C_1\left(B_1{K_1}^2-B_2K_1+B_3\right)\cos\left(\sqrt{-K_1}L\right)+C_2\left(B_1{K_1}^2-B_2K_1+B_3\right)\sin\left(\sqrt{-K_1}L\right)\\ &+C_3\left(B_1{K_2}^2+B_2K_2+B_3\right)\cosh\left(\sqrt{K_2}L\right)+C_4\left(B_1{K_2}^2+B_2K_2+B_3\right)\sinh\left(\sqrt{K_2}L\right)=0\end{aligned} \tag{4.E4.8.5}$$

$$\begin{aligned} &C_1\left(B_5\sqrt{-K_1}^3-B_6\sqrt{-K_1}\right)\sin\left(\sqrt{-K_1}L\right)+C_2\left(-B_5\sqrt{-K_1}^3+B_6\sqrt{-K_1}\right)\cos\left(\sqrt{-K_1}L\right)\\ &+C_3\left(B_5\sqrt{K_2}^3+B_6\sqrt{K_2}\right)\sinh\left(\sqrt{K_2}L\right)+C_4\left(B_5\sqrt{K_2}^3+B_6\sqrt{K_2}\right)\cosh\left(\sqrt{K_2}L\right)=\\ &\qquad C_1(D_1K_1-D_2)\cos\left(\sqrt{-K_1}L\right)+C_2(D_1K_1-D_2)\sin\left(\sqrt{-K_1}L\right)\\ &\qquad +C_3(-D_1K_2-D_2)\cosh\left(\sqrt{K_2}L\right)+C_4(-D_1K_2-D_2)\sinh\left(\sqrt{K_2}L\right)\end{aligned} \tag{4.E4.8.6}$$

Considering Eq. (4.E4.2.6), the inputs of [a] matrix are obtained as follows:

$$a_{11}=1, a_{12}=0, a_{13}=1, a_{14}=0, a_{21}=\left(B_1{K_1}^2-B_2K_1\right)\cos\left(\sqrt{-K_1}L\right), a_{22}=0,$$
$$a_{23}=\left(B_1{K_2}^2+B_2K_2\right)\cosh\left(\sqrt{K_2}L\right), a_{24}=0, a_{31}=\left(B_1{K_1}^2-B_2K_1+B_3\right)\cos\left(\sqrt{-K_1}L\right),$$
$$a_{32}=\left(B_1{K_1}^2-B_2K_1+B_3\right)\sin\left(\sqrt{-K_1}L\right), a_{33}=\left(B_1{K_2}^2+B_2K_2+B_3\right)\cosh\left(\sqrt{K_2}L\right),$$
$$a_{34}=\left(B_1{K_2}^2+B_2K_2+B_3\right)\sinh\left(\sqrt{K_2}L\right),$$
$$a_{41}=\left(B_5\sqrt{-{K_1}^3}-B_6\sqrt{-K_1}\right)\sin\left(\sqrt{-K_1}L\right)-(D_1K_1-D_2)\cos\left(\sqrt{-K_1}L\right),$$
$$a_{42}=\left(-B_5\sqrt{-{K_1}^3}+B_6\sqrt{-K_1}\right)\cos\left(\sqrt{-K_1}L\right)-(D_1K_1-D_2)\sin\left(\sqrt{-K_1}L\right),$$
$$a_{43}=\left(B_5\sqrt{{K_2}^3}+B_6\sqrt{K_2}\right)\sinh\left(\sqrt{K_2}L\right)+(D_1K_2+D_2)\cosh\left(\sqrt{K_2}L\right),$$
$$a_{44}=\left(B_5\sqrt{{K_2}^3}+B_6\sqrt{K_2}\right)\cosh\left(\sqrt{K_2}L\right)+(D_1K_2+D_2)\sinh\left(\sqrt{K_2}L\right). \tag{4.E4.8.7}$$

In Table 4.4, the first five modes nonlocal nondimensional frequencies of simply supported-transverse spring attached nanobeams formulated via two different nonlocal beam theories are calculated for different values of atomic parameter and attachment stiffness. In the computations, the parameters in Table 4.2 are employed. In addition to this, the following parameter (attachment ratio) defines the ratio of attachment stiffness to beam stiffness:

Table 4.4 Comparisons of nondimensional frequencies of simply supported-transverse spring attached nanobeams

	Mode	EBBT			TBT		
β_k	Number	$\alpha=0$	$\alpha=0.1$	$\alpha=0.2$	$\alpha=0$	$\alpha=0.1$	$\alpha=0.2$
0.01	1	0.17319	0.17319	0.17318	0.17313	0.17313	0.17313
	2	15.41950	14.60051	12.74711	15.32407	14.50712	12.66014
	3	49.96526	41.79513	30.33741	48.99371	40.94507	29.68663
	4	104.24789	74.85198	47.89250	100.18228	71.78951	45.85109
	5	178.26984	109.59491	64.92040	166.95271	102.30037	60.44824
1	1	1.71561	1.71389	1.70877	1.71499	1.71328	1.70816
	2	15.54868	14.72987	12.87853	15.45285	14.63607	12.79107
	3	50.00496	41.83548	30.38228	49.03303	40.98498	29.73089
	4	104.26690	74.87206	47.91768	100.20092	71.80914	45.87563
	5	178.28095	109.60738	64.93777	166.96348	102.31239	60.46527
100	1	8.93199	8.59710	7.77811	8.90489	8.57070	7.75343
	2	26.50407	24.75222	20.81391	26.33978	24.58364	20.63077
	3	54.57120	46.39903	34.87376	53.57326	45.52045	34.14939
	4	106.30112	77.04688	50.55748	102.20528	73.95277	48.46710
	5	179.43015	110.91581	66.74515	168.08207	103.58610	62.27362

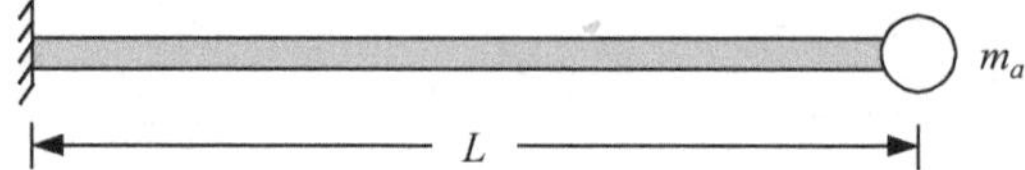

Fig. 4.8 Fixed supported-transverse mass attached nanobeam

$$\beta_k = \frac{k_a}{k_{beam}} = \frac{k_a L^3}{EI} \tag{4.E4.8.8}$$

When the results are examined, it is seen that the frequencies rise with the increment of attachment ratio. It can be stated here that as the attachment ratio increases, the attached end will behave as a simply supported end. Additionally, it appears that the nonlocal parameter reduces the frequencies of Timoshenko nanobeams more than the frequencies of Euler-Bernoulli nanobeams. Moreover, the nonlocal parameter is more effective in reducing the frequencies of nanobeams with low attachment ratio. Therefore, it is understood that the atomic size effect is a factor that must be taken into consideration in semi-rigid mechanical modeling of nanobeams.

Example 4.9 Determine the frequency equation of nonlocal free vibration of fixed supported-transverse mass attached nanobeam without thermo-elastic environment shown in Fig. 4.8 according to nonlocal TBT.

Solution. The boundary conditions are presented as

$$\begin{gathered} W(0)=0, W'(0)=0, B_1 W^{(4)}(L) + B_2 W''(L) + B_3 W(L) = 0 \\ B_5 W'''(L) + B_6 W'(L) = -D_1 W''(L) - D_2 W(L) \end{gathered} \tag{4.E4.9.1}$$

Where

$$D_1 = \left(\frac{m_a EI}{kGA}\right)\omega^2 + \left(-\frac{(e_0 a)^2 m_a \rho I}{kGA}\right)\omega^4, D_2 = -m_a\omega^2 + \left(\frac{m_a \rho I}{kGA}\right)\omega^4 \tag{4.E4.9.2}$$

If the boundary conditions are substituted into Eq. (4.E4.7.3), Eq. (4.E4.7.7), Eq. (4.E4.7.8), Eq. (4.E4.7.9), and Eq. (4.E4.8.6) are reached, respectively. In this case, the inputs of $[a]$ matrix in Eq. (4.E4.2.6) are as follows:

$$
\begin{aligned}
&a_{11}=1, a_{12}=0, a_{13}=1, a_{14}=0, a_{21}=0, a_{22}=\sqrt{-K_1}, a_{23}=0, a_{24}=\sqrt{K_2},\\
&a_{31}=\left(B_1{K_1}^2-B_2K_1+B_3\right)\cos\left(\sqrt{-K_1}L\right), a_{32}=\left(B_1{K_1}^2-B_2K_1+B_3\right)\sin\left(\sqrt{-K_1}L\right),\\
&a_{33}=\left(B_1{K_2}^2+B_2K_2+B_3\right)\cosh\left(\sqrt{K_2}L\right), a_{34}=\left(B_1{K_2}^2+B_2K_2+B_3\right)\sinh\left(\sqrt{K_2}L\right),\\
&a_{41}=\left(B_5\sqrt{-{K_1}^3}-B_6\sqrt{-K_1}\right)\sin\left(\sqrt{-K_1}L\right)-(D_1K_1-D_2)\cos\left(\sqrt{-K_1}L\right),\\
&a_{42}=\left(-B_5\sqrt{-{K_1}^3}+B_6\sqrt{-K_1}\right)\cos\left(\sqrt{-K_1}L\right)-(D_1K_1-D_2)\sin\left(\sqrt{-K_1}L\right),\\
&a_{43}=\left(B_5\sqrt{{K_2}^3}+B_6\sqrt{K_2}\right)\sinh\left(\sqrt{K_2}L\right)+(D_1K_2+D_2)\cosh\left(\sqrt{K_2}L\right),\\
&a_{44}=\left(B_5\sqrt{{K_2}^3}+B_6\sqrt{K_2}\right)\cosh\left(\sqrt{K_2}L\right)+(D_1K_2+D_2)\sinh\left(\sqrt{K_2}L\right).
\end{aligned}
\tag{4.E4.9.3}
$$

Table 4.5 lists the frequencies of fixed supported-transverse mass attached nanobeams according to the variations of nonlocal parameter and attachment ratio. The parameters that lead to the calculations in Table 4.2 are also considered in Table 4.5. Here the attachment ratio is expressed as the ratio of attachment mass to beam mass:

$$
\beta_m=\frac{m_a}{m_{beam}}=\frac{m_a}{\rho AL} \tag{4.E4.8.8}
$$

where since the mass attachment reduces the stiffness, it should be said that the mass attachment reduces the frequencies. Additionally, the fundamental mode frequency of

Table 4.5 Comparisons of nondimensional frequencies of fixed supported-transverse mass attached nanobeams

	Mode	EBBT			TBT		
β_m	number	$\alpha=0$	$\alpha=0.1$	$\alpha=0.2$	$\alpha=0$	$\alpha=0.1$	$\alpha=0.2$
0.01	1	3.44766	3.46233	3.50851	3.44691	3.46130	3.50656
	2	21.61996	20.31260	17.30656	21.46838	20.15214	17.14428
	3	60.56997	50.26443	36.33121	59.34364	49.07824	35.33039
	4	118.75743	84.50823	53.58826	114.02832	80.58900	50.87826
	5	196.41632	119.84661	71.15617	183.82746	110.98953	65.45698
1	1	1.55730	1.55909	1.56455	1.55725	1.55902	1.56439
	2	16.25009	15.37594	13.38275	16.17172	15.29469	13.30016
	3	50.89584	42.57391	30.91409	50.03866	41.77426	30.26733
	4	105.19828	75.52671	48.29347	101.47687	72.55529	46.26776
	5	179.23202	110.17950	65.27546	168.65338	103.01775	60.84662
100	1	0.17300	0.17300	0.17301	0.17300	0.17300	0.17301
	2	15.42765	14.60803	12.75300	15.35697	14.53544	12.67976
	3	49.97488	41.80310	30.34314	49.14237	41.03372	29.72712
	4	104.25769	74.85886	47.89643	100.58780	71.94803	45.91785
	5	178.27973	109.60084	64.92387	167.78529	102.53178	60.57222

nanobeam more decreases at very large increments of attachment ratio. On the other hand, the paradox of nonlocal parameter in the fundamental mode again appears. It is also true here that the frequencies of shear deformable nanobeams more decrease due to the nonlocal parameter. As a result, it is necessary to investigate a more accurate analysis about the nonlocality that is understood to be important in nanosystems, due to the existence of paradox.

Problems

4.1. Determine the frequency equation of nonlocal free vibration of nanobeams without the thermo-elastic medium given in Fig. 4.9 according to nonlocal EBBT.

4.2. Compute the first five mode nondimensional frequencies of nanobeams in Problem 4.1 according to nonlocal EBBT under parameters as follow. Discuss the effects of nonlocal parameter and boundary conditions on differences between results.

 Nanobeam length, $L = 20$ nm, height and thickness of rectangular cross-section, $h = 1$ nm and $b = 2$ nm, modulus of elasticity, $E = 1$ TPa, mass of unit volume, $\rho = 2300$ kg/m^3, stiffness ratio for transverse spring attachment $\beta_k = 1$.

 Compute according to three different nondimensional nonlocal parameters: $\alpha = 0$, 0.1, 0.2.

4.3. Compute the nondimensional frequencies as: $\overline{\omega} = \omega L^2 \sqrt{\rho A / EI}$.

4.4. Determine the frequency equation of nonlocal free vibration of nanobeams embedded in the thermo-elastic medium given in Fig. 4.10 according to nonlocal EBBT.

4.5. Examine how the thermo-elastic medium influences the frequency equations of nanobeams in Problem 4.3 according to nonlocal EBBT.

4.6. Determine the frequency equations of nonlocal free vibration of nanobeams without the thermo-elastic medium given in Fig. 4.9 according to nonlocal TBT.

4.7. Compute the first five mode nondimensional frequencies of nanobeams in Problem 4.1 according to nonlocal TBT under parameters as follow. Considering the results of Problem 4.2, discuss the effects of shear deformation, nonlocal parameter, and boundary conditions on differences between results.

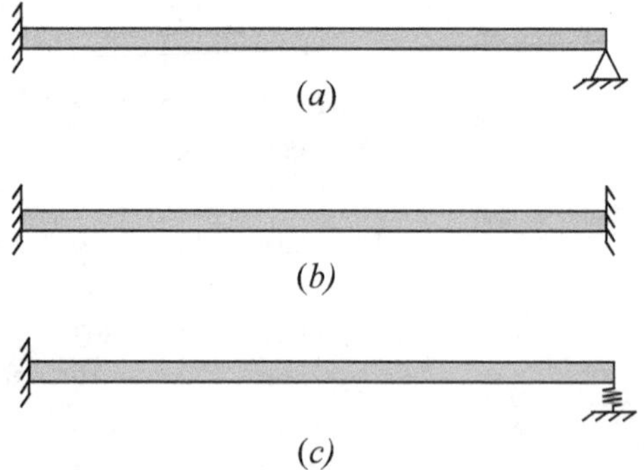

Fig. 4.9 Nanobeams with different boundary conditions. (**a**) Fixed supported-simply supported (**b**) Fixed supported at both ends (**c**) Fixed supported-transverse spring attached

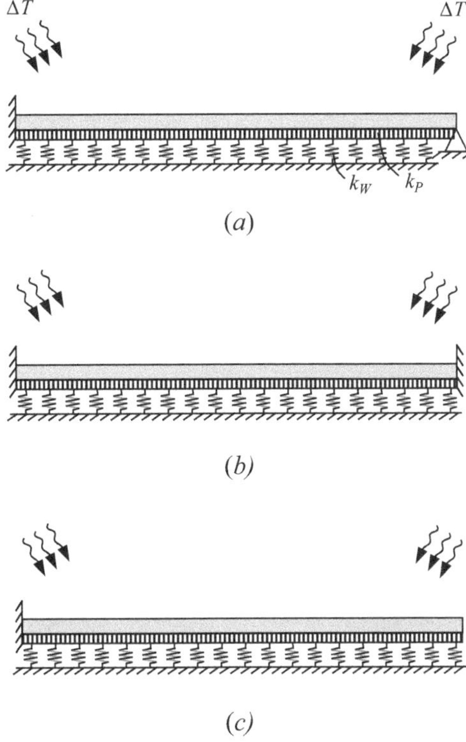

Fig. 4.10 Nanobeams with different boundary conditions embedded thermo-elastic medium. (**a**) Fixed supported-simply supported (**b**) Fixed supported at both ends (**c**) Cantilever

Nanobeam length, $L = 20$ nm, height and thickness of rectangular cross-section, $h = 1$ nm and $b = 2$ nm, modulus of elasticity, $E = 1$ TPa, mass of unit volume, $\rho = 2300$ kg/m^3, stiffness ratio for transverse spring attachment $\beta_k = 1$, shear correction factor, $k = 5/6$.

Compute according to three different nondimensional nonlocal parameters: $\alpha = 0$, 0.1, 0.2.

4.8. Compute the nondimensional frequencies as: $\overline{\omega} = \omega L^2 \sqrt{\rho A / EI}$.

4.9. Determine the frequency equations of nonlocal free vibration of nanobeams given in Fig. 4.10 according to nonlocal TBT and examine how the thermo-elastic medium influences the frequency equations.

References

1. S.S. Rao, Chapter 1: Introduction: Basic concepts and terminology, in *Vibration of Continuous Systems*, (Wiley, Hoboken, NJ, 2019), pp. 1–30

2. S.P. Timoshenko, On the correction for shear of the differential equation for transverse vibrations of prismatic bars. Philos. Mag. Ser. **6**(41), 744–746 (1921)
3. S.P. Timoshenko, On the transverse vibrations of bars of uniform cross sections. Philos. Mag. Ser. **6**(43), 125–131 (1922)
4. M. Levinson, A new rectangular beam theory. J. Sound Vib. **74**(1), 81–87 (1981)
5. M. Touratier, An efficient standard plate theory. Int. J. Eng. Sci. **29**(8), 901–906 (1991)
6. K.P. Soldatos, A transverse shear deformation theory for homogeneous monoclinic plates. Acta Mech. **94**(3–4), 195–220 (1992)
7. J.W.S. Rayleigh, *The Theory of Sound* (Dover, New York, 1945)
8. M. Aydogdu, A new shear deformation theory for laminated composite plates. Compos. Struct. **89**(1), 94–101 (2009)
9. H.M. Numanoğlu, Dynamic Analyses of Nano Scaled Continuous and Discrete Structures Based on Nonlocal Finite Element Method, MSc Thesis, Akdeniz University, Antalya, 2019. (In Turkish)
10. Q. Wang, V.K. Varadan, Vibration of carbon nanotubes studied using nonlocal continuum mechanics. Smart Mater. Struct. **15**(2), 659 (2006)
11. P. Lu, H.P. Lee, C. Lu, P.Q. Zhang, Application of nonlocal beam models for carbon nanotubes. Int. J. Solids Struct. **44**(16), 5289–5300 (2007)
12. J.N. Reddy, Nonlocal theories for bending, buckling and vibration of beams. Int. J. Eng. Sci. **45**(2–8), 288–307 (2007)
13. Z.B. Shen, X.F. Li, L.P. Sheng, G.J. Tang, Transverse vibration of nanotube-based micro-mass sensor via nonlocal Timoshenko beam theory. Comput. Mater. Sci. **53**(1), 340–346 (2012)
14. R. Ansari, H. Ramezannezhad, Nonlocal Timoshenko beam model for the large-amplitude vibrations of embedded multiwalled carbon nanotubes including thermal effects. Phys. E **43**(6), 1171–1178 (2011)
15. H.T. Thai, A nonlocal beam theory for bending, buckling, and vibration of nanobeams. Int. J. Eng. Sci. **52**, 56–64 (2012)
16. S. Thai, H.T. Thai, T.P. Vo, V.I. Patel, A simple shear deformation theory for nonlocal beams. Compos. Struct. **183**, 262–270 (2018)
17. M. Aydoğdu, A general nonlocal beam theory: Its application to nanobeam bending, buckling and vibration. Phys. E **41**(9), 1651–1655 (2009)
18. H. Darban, R. Luciano, M. Basista, Free transverse vibrations of cracked nanobeams using a nonlocal elasticity model. Int. J. Eng. Sci. **177**, 103703 (2022)
19. H. Roostai, M. Haghpanahi, Vibration of nanobeams of different boundary conditions with multiple cracks based on nonlocal elasticity theory. Appl. Math. Model. **38**(3), 1159–1169 (2014)
20. R. Ansari, H. Ramezannezhad, R. Gholami, Nonlocal beam theory for nonlinear vibrations of embedded multiwalled carbon nanotubes in thermal environment. Nonlinear Dynam. **67**, 2241–2254 (2012)
21. C.P. Wu, W.W. Lai, Free vibration of an embedded single-walled carbon nanotube with various boundary conditions using the RMVT-based nonlocal Timoshenko beam theory and DQ method. Phys. E **68**, 8–21 (2015)
22. M. Eltaher, A.E. Alshorbagy, F.F. Mahmoud, Vibration analysis of Euler–Bernoulli nanobeams by using finite element method. Appl. Math. Model. **37**(7), 4787–4797 (2013)
23. S.C. Pradhan, U. Mandal, Finite element analysis of CNTs based on nonlocal elasticity and Timoshenko beam theory including thermal effect. Phys. E **68**, 8–21 (2015)
24. H.M. Numanoğlu, H. Ersoy, B. Akgöz, Ö. Civalek, A new eigenvalue problem solver for thermo-mechanical vibration of Timoshenko nanobeams by an innovative nonlocal finite element method. Math. Meth. Appl. Sci. **45**(5), 2592–2614 (2022)

25. H.M. Numanoğlu, H. Ersoy, O. Civalek, A.J.M. Ferreria, Derivation of nonlocal FEM formulation for thermo-elastic Timoshenko beams on elastic matrix. Compos. Struct. **473**, 114292 (2021)
26. H.M. Numanoğlu, Thermal vibration of zinc oxide nanowires by using nonlocal finite element method. Int. J. Eng. Appl. Sci. **12**(3), 99–110 (2020)
27. Ç. Demir, Ö. Civalek, A new nonlocal FEM via Hermitian cubic shape functions for thermal vibration of nano beams surrounded by an elastic matrix. Compos. Struct. **168**, 872–884 (2017)
28. L. Behera, S. Chakraverty, Application of differential quadrature method in free vibration analysis of nanobeams based on various nonlocal theories. Comput. Math. Appl. **69**(12), 1444–1462 (2015)
29. M. Ganapathi, T. Merzouki, O. Polit, Vibration study of curved nanobeams based on nonlocal higher-order shear deformation theory using finite element approach. Compos. Struct. **184**, 821–838 (2018)
30. M.A. De Rosa, M. Lippiello, E. Babilio, C. Ceraldi, Nonlocal vibration analysis of a nonuniform carbon nanotube with elastic constraints and an attached mass. Materials **14**(13), 3445 (2021)
31. P. Zhang, H. Qing, Free vibration analysis of Euler-Bernoulli curved beams using two-phase nonlocal integral models. J. Vib. Control. **28**(19–20), 2861–2878 (2022)
32. M. Fakher, S. Hosseini-Hashemi, Vibration of two-phase local/nonlocal Timoshenko nanobeams with an efficient shear-locking-free finite-element model and exact solution. Eng. Comput. **38**, 231–245 (2022)
33. M.A. Eltaher, S.A. Emam, F.F. Mahmoud, Free vibration analysis of functionally graded size-dependent nanobeams. Appl. Math. Comput. **218**(14), 7406–7420 (2012)
34. O. Rahmani, O. Pedram, Analysis and modeling the size effect on vibration of functionally graded nanobeams based on nonlocal Timoshenko beam theory. Int. J. Eng. Sci. **77**, 55–70 (2014)
35. R. Nazemnezhad, S. Hosseini-Hashemi, Nonlocal nonlinear free vibration of functionally graded nanobeams. Compos. Struct. **110**, 192–199 (2014)
36. F. Ebrahimi, E. Salari, Nonlocal thermo-mechanical vibration analysis of functionally graded nanobeams in thermal environment. Acta Astronaut. **113**, 29–50 (2015)
37. F. Ebrahimi, M.R. Barati, A nonlocal higher-order shear deformation beam theory for vibration analysis of size-dependent functionally graded nanobeams. Arab. J. Sci. Eng. **41**, 1679–1690 (2016)
38. I. Esen, C. Özarpa, M.A. Eltaher, Free vibration of a cracked FG microbeam embedded in an elastic matrix and exposed to magnetic field in a thermal environment. Compos. Struct. **261**, 113552 (2021)
39. K.G. Eptaimeros, C.C. Koutsoumaris, G.J. Tsamasphyros, Nonlocal integral approach to the dynamical response of nanobeams. Int. J. Mech. Sci. **115-116**, 68–80 (2016)
40. N. Challamel, Z. Zhang, C.M. Wang, J.N. Reddy, Q. Wang, T. Michelitsch, B. Collet, On nonconservativeness of Eringen's nonlocal elasticity in beam mechanics: Correction from a discrete-based approach. Arch. Appl. Mech. **84**, 1275–1292 (2014)
41. A. Apuzzo, R. Barretta, R. Luciano, F.M. de Sciarra, R. Penna, Free vibrations of Bernoulli-Euler nano-beams by the stress-driven nonlocal integral model. Compos. Part B **123**, 105–111 (2017)
42. M. Tuna, M. Kırca, Exact solution of Eringen's nonlocal integral model for vibration and buckling of Euler–Bernoulli beam. Int. J. Eng. Sci. **107**, 54–67 (2016)

5 Introduction to the Finite Element Method

5.1 Introduction

Engineering disciplines state a physical problem mathematically and work on its solution. Every physical problem may not exhibit the same characteristics. If we want to mention about the mechanics, it is known that systems sometimes contain or may contain nonlinear behavior, even if those are not in the majority. According to another example, in addition to external load, elastic foundation, thermal environment, piezo, electric, magnetic, or optical effects may also affect the system. Even without these, the case of external loading may be excessively complex. Or the structure may not be in a simple form, may be layered composite or graded composite. In other words, mechanical problems are branched out according to the field of occupation. The analytical methods that we use in most simple mechanical analyses can fail us in complex problems because the power of these methods is quite limited indeed. This is a problem, but not being able to obtain a solution is a more serious problem since the engineering function cannot be performed because the engineer is the person who interprets the numerical findings and converts it output (product) under certain criteria.

5.2 Recent Contributions

Numerical (approximate) methods have been developed and are still being developed to overcome a physical problem. Today, many numerical methods are alternatives to analytical methods. Finite Element Method (FEM) that is one of these is a quite strong mathematical processing procedure. It can be stated that the history of FEM dates to the 1940s, where the development of FEM is briefly mentioned. Courant [1] investigated an interpolation

Öm. Civalek et al., *Mechanical Behavior and Vibration of Nano-Scaled Rods, Beams, and Frames*, Synthesis Lectures on Engineering, Science, and Technology,
https://doi.org/10.1007/978-3-032-12023-6_5

method for an element examined with triangular subelements. In the 1950s, stress analysis of aircraft wings was studied in the aerospace industry [2]. Energy formulations and matrix algebra were presented in the study by Argyris [3]. Also, Turner et al. [4] explained the fundamental principles of finite element analysis for the first time without using the concept of “finite element,” this concept was first used by Clough [5] in 1960. Additionally, Adini and Clough [6] studied the finite element analyses of some shell structures. Clough and Tocher [7] gave the stiffness matrix of the statics of plates. Zienkiewicz and Cheung [8] presented the first book about FEM. Moreover, the first book on nonlinear finite element analysis was published by Oden [9]. Furthermore, we can mention that in addition to this theoretical development, computer programs such as NASTRAN in 1965, ANSYS in 1971, and ABAQUS in 1978 emerged [2, 10]. On the other hand, the development of computer technology in the 1980s paved the way for the adoption of FEM by some industries starting from the 1990s.

5.3 Fundamental Concepts

5.3.1 Definition of Finite Element

FEM is a method that examined a physical system (whole) by dividing it into subelements (units) whose physical and geometric properties and boundaries are known, and whose mechanical properties can be formulated according to a certain procedure, called finite elements. The reason why it is called “finite” here is that the system assumed to consist of infinite number of points is discretized into a finite number of elements. The meaning by a certain procedure is shape functions that will be introduced later. Generally, the method starts from the interaction between finite elements and compounds the whole system in the light of concepts called global matrices, which will be explained in more detail later. Sets of linear algebraic equations formed as a result of this combination are solved by methods known from matrices algebra. Thus, the solution of system is completed. However, in some problems, the analysis results obtained from the whole may need to be applied to the units. For example, whether discrete or continuous, if the vibration frequency or buckling load of a structure is demanded, it is sufficient to calculate the roots of eigenvalue problem arising from the matrices expressing the whole system. However, for static problem analysis of a truss or frame structures, the internal forces of rod members (discrete member) can be calculated by first obtaining the displacements from the global matrices and then applying these displacements to a series of matrix equations. Meanwhile, in discrete systems operating with simple and one-dimensional elements, the finite elements are the discrete members themselves. On the other hand, a continuous system can be one finite element itself or can be modeled by dividing into many finite elements.

5.3.2 Process of Finite Element Analysis

Although the finite element formulation used in the analysis of different structures is different, the fundamental stages of analysis are essentially the same. The general process of finite element method is briefly introduced below.

1. **Constituting the finite element mesh of system:** In fact, before this step, there is the choice of finite element type to be used. According to this, after choosing the minimum finite element type that can represent the problem, if the structure (physical system) is discrete, all its members, if the structure is continuous, it divided by the chosen finite number of elements, are connected to each other through the finite element boundaries. At this step, which finite elements are connected to each other in which boundaries, node and element numbers of finite elements and element orientations should be determined. All these components are unignorable parts of finite element mesh. Accurate organization of the finite element mesh affects the step of derivation of global matrices of system.
2. **Choosing the interpolation function:** Interpolation (shape) functions represent the characteristics of finite element such as mechanical displacement and rotation. A finite element can have many different shape functions. The important thing here is the choosing of minimum shape function appropriate for the type of problem. For example, two-node linear finite element is appropriate in simple axial rod theory, but this element cannot be employed in axial rods. For such axial rods, minimum-three node linear finite element should be used.
3. **Computing the finite element matrices:** Each finite element has physical and mechanical properties. Considering these properties, the related matrices of each finite element are calculated for related mechanical problem. For example, in the vibration problem, stiffness and mass matrices of each finite element in its own axis, that is, in the local axis, should be derived.
4. **Computing the global matrices of system:** In order to provide the representation of each finite element in the main structure, firstly, the transformation of system on the global axis is performed. In some cases, transformation is unnecessary. If necessary, the transformed finite element matrices are summed not directly, but in the direction of freedoms of each finite element. Thus, global matrices are prepared. There are some properties that global matrices should provide and these are expressed later.
5. **Applying the boundary conditions:** All global matrices required for the analysis of problem are reduced under all global geometric constraints (boundary conditions) of the system. The obligation of reduction arises from the degrees of freedom. We can explain this with an example as: In a cantilever beam element, the first finite element is probably assigned to the element with fixed (or clamped) support. The first end of this finite element represents the clamped support point. In this case, after the global matrices are obtained, since the fixed support is rigid against displacement and rotation, the freedoms

in the direction of these deformations are eliminated by row and column from the spherical stiffness matrix.

6. **Analyzing the system:** If the global matrices are ready, the solution of mechanical problem expressing the system is provided by methods known from linear algebra and vector analysis. For example, the equilibrium (Hooke) equation to be constituted in the linear static analysis problem is solved by the substitution method in the global stiffness matrix and the displacements can be computed. In another example, after the global elastic stiffness and global geometric stiffness matrices are determined in the buckling problem, they are substituted into the relevant eigenvalue problem and the eigenvalue problem is solved either electronically or by known numerical methods such as Newton-Raphson and Stodola.
7. **Analyzing the elements:** Solution of system is an output of the finite element analysis process, but it is not the only output. The results of mechanical solution are used in the matrix equations of each finite element and the solution of finite element is performed. For example, in the linear static analysis of frame element, parameters such as displacement, section rotation and support reactions of the system are obtained as a result of the analysis of system. Then, the displacements and section rotations are transformed to calculate the internal forces of sections represented by the ends of each finite member. The transformed deformations are substituted into the matrix displacement equation of bending finite element and the internal forces are calculated. Thus, the analysis process with finite elements is completed.

5.3.3 Finite Element Modeling

Since the finite element method is a numerical method, it is expected that the results obtained present approximate results to the analytical method. Of course, the convergence of these results can be adjusted under certain procedures. What is meant by the concept of a certain procedure is the choosing of finite elements that have equal sizes and high number as far as possible. Figure 5.1 shows a circular plate uniformly pressurized on circumference and its finite element models. While one finite element model uses small triangular finite elements, the other uses more elements.

It is not exact that the mechanical analysis results of circular plate under mechanical forces can be obtained by analytical methods. However, it is known that an approximate result can be reached with the using of a finite element formulation whose algorithm is coded rightly (unless there are exceptional cases that we cannot foresee). Another exact and very important point that we should emphasize for the finite element method is explained as: If we increase the number of finite elements in the mechanical system, we obtain more accurate results. In this context, the improved finite element model shown in Fig. 5.1c is more advantageous than the model given in Fig. 5.1b. At this point, we should emphasize that the using of incorrect numerical results causes to designs that are contrary to the principles of engineering.

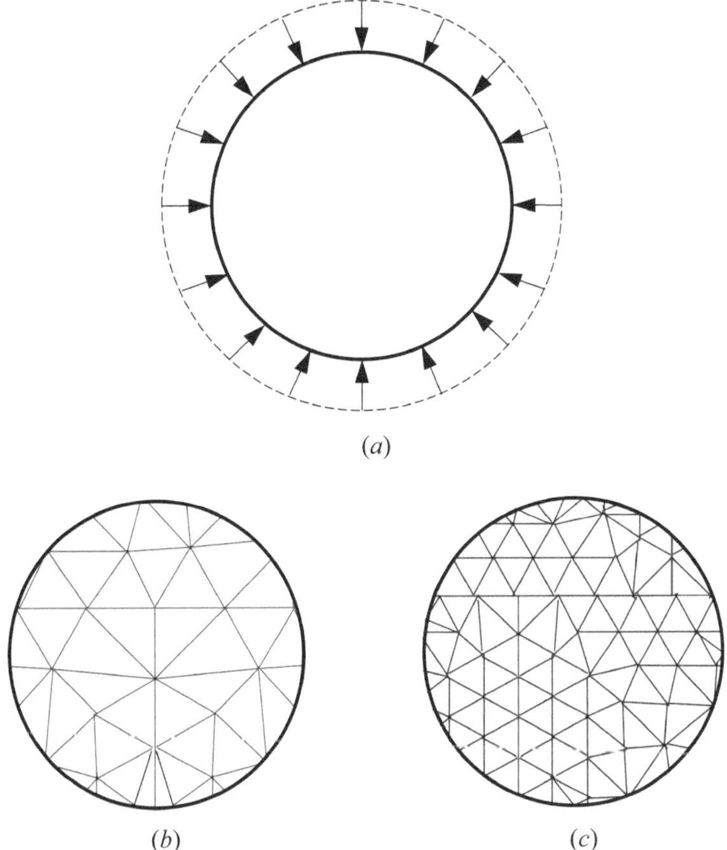

Fig. 5.1 (**a**) Circular plate pressurized on circumference. (**b**) Finite element model (less sensitive). (**c**) Improved finite element model (more sensitive)

Finite elements can vary according to their geometry. A quite general classification can be mentioned such as linear, planar, cubic, prismatic. Linear elements are elements with at least two nodes. Planar, cubic, and prismatic elements are elements that have at least one node at their corners. Planar elements can be triangular and quadrilateral, or have more sides, and moreover, they can be irregular polygons. Three-dimensional (spatial) finite elements can also be of irregular shape. It should be noted that the sensitivity increases in finite element analysis performed with more nodes.

As it is a practical and catchy example, a reinforced concrete carcass system with two-storey and two-span and finite element model is presented in Fig. 5.2.

Due to the scope of this book, mechanical analysis of nano-scaled continuous and discrete systems are examined in the following chapters. In this context, simple finite element models of beams, axial rods, torsional rods, trusses, and frames are also depicted in

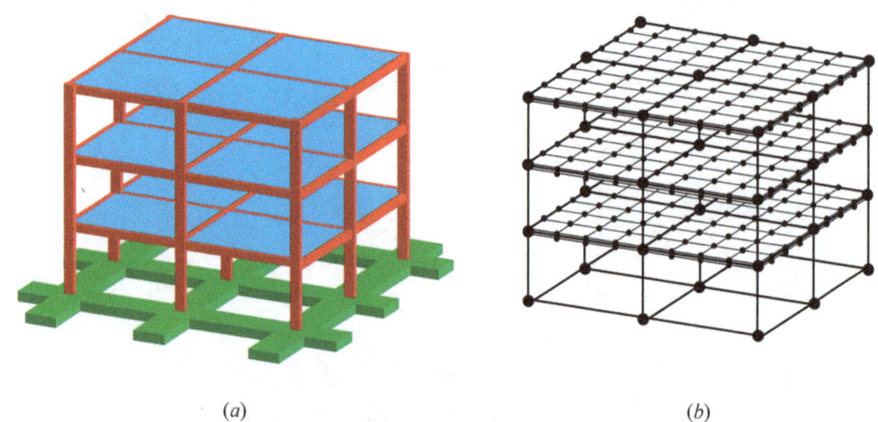

(a) (b)

Fig. 5.2 (**a**) Reinforced concrete structural system (Physical system). (**b**) Finite element model of physical system

Fig. 5.3 in order to give foreknowledge about what will be done in the chapters related to finite element analysis.

5.4 Application Areas

With the understanding that FEM can overcome many problems, its application areas have also started to be clear. FEM is used in many fields such as machinery, automotive, construction, aerospace, electromagnetics, biomedical, medicine, dentistry, and industrial design. A few specific examples from these can be exemplified as measurement of heat conduction in device design, determination of stresses in the aircraft wings, calculation of forces between living tissue and dental structure, durability and vibration tests at the design stage of automobiles, the permeability analysis of soils, stability analysis, the forces that exposed to body in the viscous flow field is and perhaps the most intensive using field, structural analysis can be referred.

There are many computer programs developed based on the finite element method. Examples of these programs are explained as NASTRAN, LS-DYNA, PATRAN, HYPERMESH, ABAQUS, ANSYS, SAP2000, and PLAXIS.

5.5 A General Investigation of Theoretical Background

While the general processing stages of finite element analysis of structure is introduced, it was emphasized that one of them is the choosing of shape functions. In fact, it can be stated that these functions constitute the basis of finite element algorithms. While giving explanation about the relevant stage, it was said that the choosing of appropriate finite element

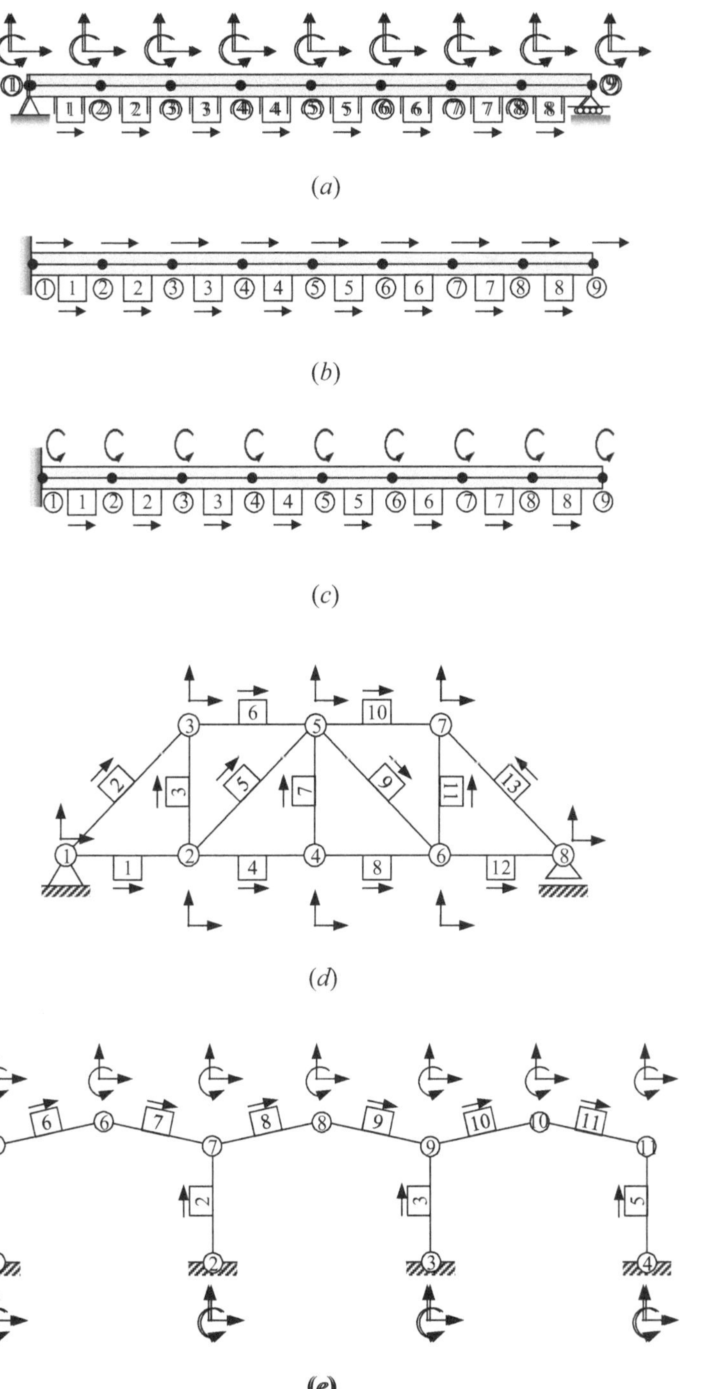

Fig. 5.3 Finite element models of different continuous and discrete structures. (**a**) Axial rod. (**b**) Torsional rod (**c**) Bending rod (Beam). (**d**) Truss. (**e**) Frame

and its shape function for the structure and behavior of system is important. For this reason, shape functions of continuous systems such as axial rods and bending rods are given here to constitute a theoretical background.

5.5.1 Shape Functions of Axial and Torsional Elements

Firstly, it should be mentioned what the axial rod means. Continuous elements that can only strain in the axial direction, in other words, under mechanical forces perpendicular to the cross-sectional area, are called axial rods. Modeling of axial rods with linear finite elements without thickness can be done. This finite element must have minimum two nodes. According to this, consider the two-node (named as i and j) linear finite element depending on the global x axis depicted in Fig. 5.4. The total axial motion of this finite element is defined as follows:

$$u^e = c_1 + c_2 x \tag{5.1}$$

where the unknowns c_1 and c_2 can be determined by the following boundary conditions of finite element:

$$u^e(x_i) = u_i, \quad u^e(x_j) = u_j \tag{5.2}$$

Here x_i and x_j are node coordinates, u_i and u_j denote node displacements. For the geometry of rod finite element, the following expression can be written as follows:

$$x_j - x_i = l_e \tag{5.3}$$

where l_e means the finite element length. By using Eq. (5.2) into Eq. (5.1), the following two equations with the two unknowns can be constituted:

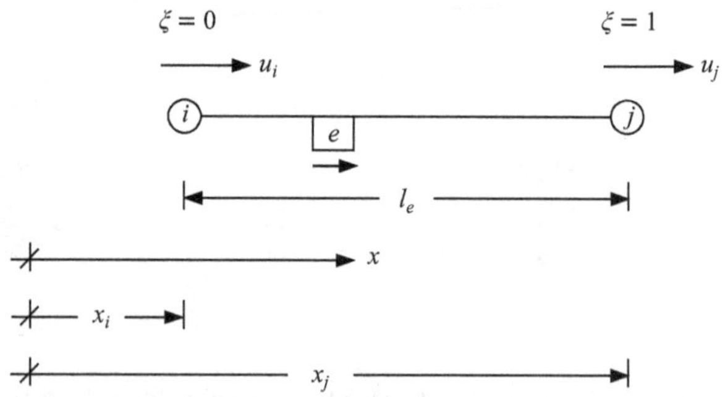

Fig. 5.4 Two-node linear axial finite element

$$u_i = c_1 + c_2 x_i, u_j = c_1 + c_2 x_j \tag{5.4}$$

If equations in Eq. (5.4) are solved side-by-side using Eq. (5.3) and some mathematical manipulations is performed, the following expression is obtained:

$$c_1 = \frac{u_i x_j - u_j x_i}{l_e}, c_2 = \frac{u_j - u_i}{l_e} \tag{5.5}$$

These expressions should be substituted into Eq. (5.1):

$$u^e = \left(\frac{u_i x_j - u_j x_i}{l_e}\right) + \left(\frac{u_j - u_i}{l_e}\right)x \tag{5.6}$$

To further edit this expression, the nondimensionalization parameter is used. Accordingly, a nondimensionalization parameter ξ that changes the nondimensional displacements of finite element between 0 and 1 is defined as follows:

$$\xi = \frac{x - x_i}{l_e} = \frac{x - x_j + l_e}{l_e} \tag{5.7}$$

Now let's go back a step and write Eq. (5.1) in terms of the node displacement components of the finite element as follows:

$$u^e = \left(\frac{x_j - x}{l_e}\right)u_i + \left(\frac{x - x_i}{l_e}\right)u_j \tag{5.8}$$

If Eq. (5.1) is given with $u^e = \phi_i u_i + \phi_j u_j$:

$$\phi_i = \frac{x_j - x}{l_e} = \frac{l_e + x_i - x}{l_e} = 1 - \xi, \quad \phi_j = \frac{x - x_i}{l_e} = \xi \tag{5.9}$$

where ϕ_i and ϕ_j expressions are known as shape functions of axial rod finite element. Meanwhile, the total motion of axial rod finite element is also formulated as:

$$u^e = \begin{bmatrix}\varphi_i & \varphi_j\end{bmatrix}\begin{Bmatrix} u_i \\ u_j \end{Bmatrix} = [\varphi]\{d\} \tag{5.10}$$

where $[\phi]$ is the shape function matrix and $\{d\}$ is the end freedom vector.

In the finite element analysis of axial nanorods, it will be seen that the using of a two-node finite element will be insufficient. The reason for this will be explained when appropriate. In such a case, it is necessary to change the finite element type. The three-node

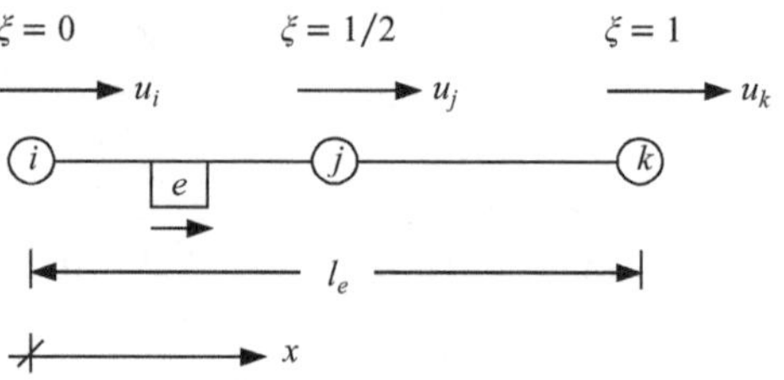

Fig. 5.5 Three-node linear axial finite element

linear finite element to be used this time and its freedoms are depicted in Fig. 5.5. Where the linear axis bases on the i node. The function representing the total axial motion of finite element can be stated as:

$$u^e = c_1 + c_2 x + c_3 x^2 \tag{5.11}$$

Since the finite element has three boundary conditions, the following expressions are defined:

$$u_i = c_1 + c_2 x_i + c_3 {x_i}^2, u_j = c_1 + c_2 x_j + c_3 {x_j}^2, u_k = c_1 + c_2 x_k + c_3 {x_k}^2 \tag{5.12}$$

The following equations should be used as compatibility equations in editing operations:

$$x_k - x_i = l_e, \ \ x_k - x_j = x_j - x_i = \frac{l_e}{2} \tag{5.13}$$

If this set of equations is subjected to some intermediate mathematical operations such as addition and subtraction, first the following parameter is obtained:

$$c_3 = \frac{2}{l_e^2}\left(u_i - 2u_j + u_k\right) \tag{5.14}$$

By using this expression, other constants are also revealed as follows:

$$\begin{aligned} c_1 &= u_k - \frac{u_i x_k}{l_e} + 4\frac{u_j x_k}{l_e} - 3\frac{u_k {x_k}^2}{l_e^2} + 2\frac{u_i {x_k}^2}{l_e^2} - 4\frac{u_j {x_k}^2}{l_e^2} + 2\frac{u_k {x_k}^2}{l_e^2}, \\ c_2 &= \frac{u_i}{l_e} - 4\frac{u_j}{l_e} + 3\frac{u_k}{l_e} - 4\frac{u_i x_k}{l_e^2} + 8\frac{u_j x_k}{l_e^2} - 4\frac{u_k x_k}{l_e^2}. \end{aligned} \tag{5.15}$$

If above expressions are substituted into Eq. (5.11), the total axial motion is explained as follows:

$$u^e = \left(-\frac{x_k}{l_e} + 2\frac{x_k^2}{l_e^2} + \frac{x}{l_e} - 4\frac{x_k x}{l_e^2} + 2\frac{x^2}{l_e^2} \right) u_i + \left(4\frac{x_k}{l_e} - 4\frac{x_k^2}{l_e^2} - 4\frac{x}{l_e} + 8\frac{x_k x}{l_e^2} - 4\frac{x^2}{l_e^2} \right) u_j + \left(-3\frac{x_k}{l_e} + 2\frac{x_k^2}{l_e^2} + 3\frac{x}{l_e} - 4\frac{x_k x}{l_e^2} + 2\frac{x^2}{l_e^2} + 1 \right) u_k \tag{5.16}$$

Since the origin of axis is at the i node, if $x_k = L$ and the nondimensional displacement $\xi = x/l_e$ are employed here:

$$u^e = (2\xi^2 - 3\xi + 1)u_i + (-4\xi^2 + 4\xi)u_j + (2\xi^2 - \xi)u_k \tag{5.17}$$

where the total axial motion is written as $u^e = \varphi_i u_i + \varphi_j u_j + \varphi_k u_k$, the shape functions are presented as follows:

$$\varphi_i = 2\xi^2 - 3\xi + 1, \quad \varphi_j = -4\xi^2 + 4\xi, \quad \varphi_k = 2\xi^2 - \xi \tag{5.18}$$

5.5.2 Shape Functions of Bending Elements

Bending rods, unlike axial bars, are the elements that work under bending and shear effects and deform in the direction of these effects. Of course, these elements can also work under axial forces in addition to bending and shear, but the property that characterizes these types of elements is the presence of at least one of the bending and shear. Modeling of bending elements can be performed through two-node linear finite elements isolated from axial effects depicted in Fig. 5.6. The equation characterizing the total motion of this finite element that has four different freedoms at its nodes is denoted as follows:

$$w^e = c_1 + c_2 x + c_3 x^2 + c_4 x^3 \tag{5.19}$$

Boundary conditions for the four different freedoms in this element are given as follows:

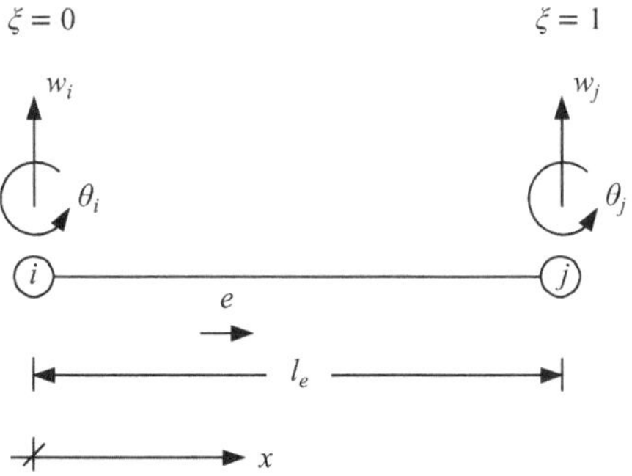

Fig. 5.6 Two-node linear bending finite element

$$w^e(x_i) = w_i, \quad \frac{dw^e}{dx}(x_i) = -\theta_i, \quad w^e(x_j) = w_j, \quad \frac{dw^e}{dx}(x_j) = -\theta_j \tag{5.20}$$

where w and θ means the freedoms of displacement and rotation, respectively.

c_i ($i = 1 - 4$) constants that are calculated as a result of applying the boundary conditions specified in Eq. (5.20) can be given as follows:

$$c_2 = -\theta_i, \quad c_2 = \theta_j, \quad c_3 = -3\frac{w_i}{l_e^2} + 3\frac{w_j}{l_e^2} + 2\frac{\theta_i}{l_e} + \frac{\theta_j}{l_e}, \quad c_4 = 2\frac{w_i}{l_e^3} - 2\frac{w_j}{l_e^3} - \frac{\theta_i}{l_e^2} - \frac{\theta_j}{l_e^2} \tag{5.21}$$

By using above expressions into Eq. (5.19), the following equation can be written:

$$w^e = \varphi_i w_i + \phi_i \theta_i + \varphi_j w_j + \phi_j \theta_j \tag{5.22}$$

where if $\xi = x/l_e$ is considered, shape functions of bending finite element are reached as follows:

$$\varphi_i = 1 - 3\xi^2 + 2\xi^3, \quad \phi_i = L(-\xi + 2\xi^2 - \xi^3), \quad \varphi_j = 3\xi^2 - 2\xi^3, \phi_j = L(\xi^2 - \xi^3) \tag{5.23}$$

Finally, total motion of finite element is described in the matrix form as follows:

$$w^e = \begin{bmatrix} \varphi_i & \phi_i & \varphi_j & \phi_j \end{bmatrix} \begin{Bmatrix} w_i \\ \theta_i \\ w_j \\ \theta_j \end{Bmatrix} = [\varphi]\{d\} \tag{5.24}$$

where $[\phi]$ denotes shape function matrix and $\{d\}$ means end freedom vector.

5.5.3 Stiffness and Mass Matrices

In the previous sections, the fundamentals of finite element method were given. From here on, the mentality of finite elements is mentioned. The finite element method is a numerical method that discretizes a structure that has infinite number of points into subdomains that interact with each other through nodes and called as finite elements. In Sect. 5.3.2, while the steps of finite element method are explained, we talked about some matrices of these subelements that need to be constituted to use in the solution of related problem. In the finite element method, first these matrices are computed for each element and summed at the end. However, this matrix addition is not linear algebraic, that is, not an addition like input-to-input. It is an addition constituted in line with the freedom represented by each input of matrices in the structure evaluated as a whole. These sum matrices are subjected to the linear algebraic operation required by the problem and the output of finite element method is attained.

The finite element analysis of vibration problem that will be examined in the following chapters can be summarized as follows: After the mass and stiffness matrices of each element are calculated, these are summed in the direction of node freedoms. The matrices obtained as a result of this addition are defined as global matrices. The mathematical meaning of vibration problem in the physical sense is a standard eigenvalue problem, and the linear algebraic operation which the global matrices in this eigenvalue problem are subjected yields the natural vibration frequencies of structure. By employing these frequencies, the amplitude vector of structure, that is, the mode shapes can be determined. We will mention about stiffness and mass matrices from here on.

We stated in Chap. 1 that the undamped and free vibration problem constitutes a second order homogeneous and ordinary differential equation. In fact, it can be stated that the finite element method discretizes this homogeneous differential equation that is partial for continuous systems, such as the separation of variable briefly mentioned in Chap. 1 into ordinary differential equation, but of course the other principles of method are different from the separation of variable. Equation (1.37) introduces the final version of discretized differential equation. Of course, to perform the finite element analysis, the stiffness and mass matrices seen in this equation should be constituted.

The stiffness matrix of an element specifies the distribution of its stiffness properties that are continuous throughout the element at discretized points for that element. Similarly, the mass matrix denotes the distribution of continuous mass of that element at discretized points. These distributions are expressed by different compositions of product of related mechanical stiffness expression in the stiffness matrix and the unit volume, cross-section area, and length parameters in the mass matrix.

There are different methods for derivation of stiffness and mass matrix. Direct (equilibrium) method, energy methods, and variational method refer some of them. We will close this section by giving some examples on the energy equations for the classical, that is, macro elasticity. First, let's mention the mass and stiffness matrix of axial rod. According to this, firstly, the stiffness matrix of simple axial rod that does not consider the shear effects is derived. Without going into the details of derivation, we can write the internal strain energy of this rod as follows:

$$U = \frac{1}{2} \int_{0}^{L} AE \left(\frac{\mathrm{d}u^e}{\mathrm{d}x} \right)^2 \mathrm{d}x \tag{5.25}$$

where derivative of axial motion of rod is formulated as follows:

$$\frac{du^e}{dx} = \frac{d\varphi_i}{dx} u_i + \frac{d\varphi_j}{dx} u_j = \frac{u_j - u_i}{L} \quad (5.26)$$

By utilizing Eq. (5.26) into Eq. (5.25):

$$U = \frac{AE}{L} \left(u_i^2 - 2u_i u_j + u_j^2 \right) \quad (5.27)$$

According to a fundamental that needs to know in energy methods, the derivative of internal strain energy with respect to displacement gives the internal force on the rod (Castigliano's I. Theory). Thus, the following equations can be constituted in the two-node element:

$$\frac{dU}{du_i} = f_i = \frac{AE}{L} (u_i - u_j), \quad \frac{dU}{du_j} = f_j = \frac{AE}{L} (-u_i + u_j) \quad (5.28)$$

Since the matrix form of Hooke's law is $\{f\} = [k]\{d\}$, these equations can be combined as follows:

$$\begin{Bmatrix} f_i \\ f_j \end{Bmatrix} = \frac{AE}{L} \begin{bmatrix} 1 & -1 \\ -1 & 1 \end{bmatrix} \begin{Bmatrix} u_i \\ u_j \end{Bmatrix} \quad (5.29)$$

where $\{f\}$ defines the load vector and $\{d\}$ defines the displacement vector. $[k]$ is known as the axial stiffness matrix:

$$[k] = \frac{AE}{L} \begin{bmatrix} 1 & -1 \\ -1 & 1 \end{bmatrix} \quad (5.30)$$

The second example is the energy expression-based derivation of mass matrix for axial rods without shear effect. According to this, considering the work-energy principle, the kinetic energy of body during motion can be stated with the following integral:

$$T = \frac{1}{2} \int_0^V (\rho \ddot{u}^e) u^e dV \quad (5.31)$$

where V is the volume of rod and ρ is the unit volume of mass. Also, $\ddot{u}^e$ define the acceleration of motion and is written as:

$$\frac{\mathrm{d}^2u^e}{\mathrm{d}t^2} = \varphi_i \frac{\mathrm{d}^2u_i}{\mathrm{d}t^2} + \varphi_j \frac{\mathrm{d}^2u_j}{\mathrm{d}t^2} \tag{5.32}$$

If the axial motion and its second derivative according to the time in Eq. (5.32) are substituted into Eq. (5.31):

$$T = \frac{1}{2}\int_0^L \rho A\left[\left(1-\frac{x}{L}\right)^2 \ddot{u}_i u_i + \left(1-\frac{x}{L}\right)\left(\frac{x}{L}\right)\ddot{u}_i u_j + \left(1-\frac{x}{L}\right)\left(\frac{x}{L}\right)u_i \ddot{u}_j + \left(\frac{x}{L}\right)^2 \ddot{u}_j u_j\right]\mathrm{d}x \tag{5.33}$$

Derivatives of energy according to the node displacements are calculated as follows:

$$\frac{\mathrm{d}T}{\mathrm{d}u_i} = f_i = \frac{\rho AL}{6}(2\ddot{u}_i + \ddot{u}_j), \quad \frac{\mathrm{d}T}{\mathrm{d}u_i} = f_j = \frac{\rho AL}{6}(\ddot{u}_i + 2\ddot{u}_j) \tag{5.34}$$

These equations can be expressed as $\{f\} = [m]\{\ddot{d}\}$:

$$\begin{Bmatrix} f_i \\ f_j \end{Bmatrix} = \frac{\rho AL}{6}\begin{bmatrix} 2 & 1 \\ 1 & 2 \end{bmatrix}\begin{Bmatrix} \ddot{u}_i \\ \ddot{u}_j \end{Bmatrix} \tag{5.35}$$

where $[m]$ indicates the mass matrix of axial rod:

$$[m] = \frac{\rho AL}{6}\begin{bmatrix} 2 & 1 \\ 1 & 2 \end{bmatrix} \tag{5.36}$$

Another example is presented for the stiffness matrix of axial rods with shear (or Poisson's) effect (Love-Bishop axial rods). The internal strain energy is written as follows:

$$U = \frac{1}{2}\int_0^L \left[AE\left(\frac{\mathrm{d}u^e}{\mathrm{d}x}\right)^2 + \upsilon^2 GI_p\left(\frac{\mathrm{d}^2u^e}{\mathrm{d}x^2}\right)^2\right]\mathrm{d}x \tag{5.37}$$

Note that in the shear inertia term ($\upsilon^2 GI_p$) seen in Eq. (5.37), since the second order derivative of motion is taken according to the position, the two-node linear finite element cannot be useful. The fact that reveals the necessity of choosing a linear axial finite element with three nodes at least. In this type of finite element, the first and second derivatives of motion expression are obtained respectively as follows:

$$\frac{\mathrm{d}u^e}{\mathrm{d}x} = \frac{\mathrm{d}\varphi_i}{\mathrm{d}x}u_i + \frac{\mathrm{d}\varphi_j}{\mathrm{d}x}u_j + \frac{\mathrm{d}\varphi_k}{\mathrm{d}x}u_k = \left(\frac{4x}{L^2} - \frac{3}{L}\right)u_i + \left(-\frac{8x}{L^2} + \frac{4}{L}\right)u_j + \left(\frac{4x}{L^2} - \frac{1}{L}\right)u_k \quad (5.38)$$

$$\frac{\mathrm{d}^2u^e}{\mathrm{d}x^2} = \frac{\mathrm{d}^2\varphi_i}{\mathrm{d}x^2}u_i + \frac{\mathrm{d}^2\varphi_j}{\mathrm{d}x^2}u_j + \frac{\mathrm{d}^2\varphi_k}{\mathrm{d}x^2}u_k = \frac{4}{L^2}u_i - \frac{8}{L^2}u_j + \frac{4}{L^2}u_k \quad (5.39)$$

Derivatives of internal strain energy according to all components are computed as follows:

$$\begin{aligned}\frac{\partial U}{\partial u_i} = \int_0^L \Bigg[& EA\left[\left(\frac{4x}{L^2} - \frac{3}{L}\right)^2 u_i + \left(\frac{4x}{L^2} - \frac{3}{L}\right)\left(-\frac{8x}{L^2} + \frac{4}{L}\right)u_j + \left(\frac{4x}{L^2} - \frac{3}{L}\right)\left(\frac{4x}{L^2} - \frac{1}{L}\right)u_k\right] \\ & +v^2GI_p\left(\frac{16}{L^4}u_i - \frac{32}{L^4}u_j + \frac{16}{L^4}u_k\right)\Bigg]\mathrm{d}x \\ = {} & \frac{EA}{L}\left(\frac{7}{3}u_i - \frac{8}{3}u_j + \frac{1}{3}u_i\right) + \frac{v^2GI_p}{L^3}(16u_i - 32u_j + 16u_i)\end{aligned} \quad (5.40)$$

$$\begin{aligned}\frac{\partial U}{\partial u_j} = \int_0^L \Bigg[& EA\left[\left(\frac{4x}{L^2} - \frac{3}{L}\right)\left(-\frac{8x}{L^2} + \frac{4}{L}\right)u_i + \left(-\frac{8x}{L^2} + \frac{4}{L}\right)^2 u_j + \left(-\frac{8x}{L^2} + \frac{4}{L}\right)\left(\frac{4x}{L^2} - \frac{1}{L}\right)u_k\right] \\ & +v^2GI_p\left(-\frac{32}{L^4}u_i + \frac{64}{L^4}u_j - \frac{32}{L^4}u_k\right)\Bigg]\mathrm{d}x \\ = {} & \frac{EA}{L}\left(-\frac{8}{3}u_i + \frac{16}{3}u_j - \frac{8}{3}u_i\right) + \frac{v^2GI_p}{L^3}(-32u_i + 64u_j - 32u_i)\end{aligned} \quad (5.41)$$

$$\begin{aligned}\frac{\partial U}{\partial u_k} = \int_0^L \Bigg[& EA\left[\left(\frac{4x}{L^2} - \frac{3}{L}\right)\left(\frac{4x}{L^2} - \frac{1}{L}\right)u_i + \left(-\frac{8x}{L^2} + \frac{4}{L}\right)\left(\frac{4x}{L^2} - \frac{1}{L}\right)u_j + \left(\frac{4x}{L^2} - \frac{1}{L}\right)^2 u_k\right] \\ & +v^2GI_p\left(\frac{16}{L^4}u_i - \frac{32}{L^4}u_j + \frac{16}{L^4}u_k\right)\Bigg]\mathrm{d}x \\ = {} & \frac{EA}{L}\left(\frac{1}{3}u_i - \frac{8}{3}u_j + \frac{7}{3}u_i\right) + \frac{v^2GI_p}{L^3}(16u_i - 32u_j + 16u_i)\end{aligned} \quad (5.42)$$

Considering $\{f\} = [k]\{d\}$, the results of Eqs. (5.40)–(5.42) are symbolized in matrix form as follows:

$$[k] = \begin{bmatrix} k_{11} & k_{12} & k_{13} \\ k_{21} & k_{22} & k_{23} \\ k_{31} & k_{32} & k_{33} \end{bmatrix} \tag{5.43}$$

Inputs of this matrix are given as follows:

$$\begin{aligned} & k_{11} = k_{33} = \frac{7}{3}\frac{EA}{L} + 16\frac{\upsilon^2 GI_p}{L^3}, \quad k_{12} = k_{21} = k_{23} = k_{32} = -\frac{8}{3}\frac{EA}{L} - 32\frac{\upsilon^2 GI_p}{L^3}, \\ & k_{13} = k_{31} = \frac{1}{3}\frac{EA}{L} + 16\frac{\upsilon^2 GI_p}{L^3}, \quad k_{22} = \frac{16}{3}\frac{EA}{L} + 64\frac{\upsilon^2 GI_p}{L^3} \end{aligned} \tag{5.44}$$

Example 5.1 Determine the natural frequencies of classical axial free vibration of clamped-clamped nanorod shown in Fig. 2.2 (Chap. 2) with FEM. Use four finite elements in the analysis as in Fig. 5.7. Physical and mechanical parameters of structure are: Nanobeam length: $L = 20$ nm, diameter of circular cross-section: $d = 1$ nm, modulus of elasticity: $E = 1$ TPa, mass of unit volume: $\rho = 2300$ kg/m^3, and axial medium stiffness: $k_M = 0$ N/m^2.

Solution. Firstly, we know $e_0 a = 0$ for classical vibration. Therefore, nonlocality is neglected in the analysis. Additionally, axial elastic medium is not considered for simplicity in this example. These are already the contents of Chap. 6 in the future.

Equation (1.38) is symbolized for the physical system in the finite element analysis as follows:

$$\det\left(\sum_{e=1}^{n}[k]_e - \omega_i^2 \sum_{e=1}^{n}[m]_e\right) = 0 \tag{5.E5.1.1}$$

If you remember here, $[k]_e$ and $[m]_e$ represent the stiffness and mass matrices of finite element (FE), respectively, and ω_i the natural frequency of ith vibration mode. n means the number of FEs in the system. Of course, as it is mentioned before, global matrices are obtained by combining the related matrices. However, it should be remembered that the related matrices are summed not according to input-to-input, but according to the common freedoms in the FE mesh. Based on this, $\sum_{e=1}^{n}[k]_e$ and $\sum_{e=1}^{n}[m]_e$ denotes the global stiffness and global mass matrices of system, respectively.

The FE mesh of clamped axial rod (physical system) given in Fig. 5.7a and the FEs of this mesh are presented in Fig. 5.7b, c, respectively. According to this, since the physical system is divided into $n = 4$ FEs, the existences of five nodes and four FEs in the FE mesh are understood. To avoid complexity, FE nodes are symbolized with thin and italic numbers, node numbers of system are symbolized with thin numbers, and FE numbers are symbolized with bold numbers. Also, the lengths of FEs are equal to:

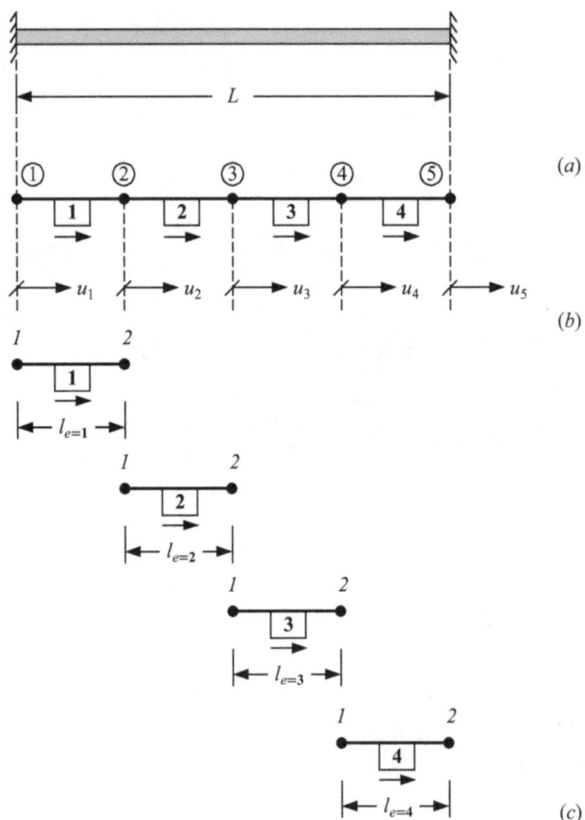

Fig. 5.7 FEM analysis of clamped axial nanorod. (**a**) Physical system. (**b**) FEM model with four elements (Finite element mesh). (**c**) Finite elements

$$l_{e=1} = l_{e=2} = l_{e=3} = l_{e=4} = l = \frac{L}{4} = \frac{20}{4} = 5 \text{ nm} \tag{5.E5.1.2}$$

where it can be stated that the FE mesh has five freedoms, since five nodes in the FE mesh have only axial freedom of movement. Therefore, global stiffness and global mass matrices are 5 × 5 in size. The matrices defining the stiffness and mass of FEs are summed according to node numbers. This emphasizes the accumulation of stiffness and mass at nodes that overlap at boundaries of FEs (e.g., node 2 is common at the boundaries of elements 1 and 2). Now we can constitute the mass and stiffness matrices of four FEs. Let's start with the stiffness matrix first. Eq. (5.30) can be symbolized as follows for a FE by considering Fig. 5.7c:

$$[k]_e = \frac{A_e E_e}{l_e}\left[\begin{array}{c|c} 1 & -1 \\ \hline -1 & 1 \end{array}\right] \begin{array}{l} \textit{FE node numbers: } 1 \mid 2 \text{ (columns)};\ 1,\ 2 \text{ (rows)} \end{array} \tag{5.E5.1.3}$$

Since the system has homogeneous material and a constant cross-section, it should be noted that the physical and mechanical parameters of all elements are equal in itself:

$$E_{e=1} = E_{e=2} = E_{e=3} = E_{e=4} = E = 1 \ \text{TPa} \tag{5.E5.1.4}$$

$$\rho_{e=1} = \rho_{e=2} = \rho_{e=3} = \rho_{e=4} = \rho = 2300 \ \frac{\text{kg}}{\text{m}^3} \tag{5.E5.1.5}$$

As a result, the stiffness matrix of all FEs is equal. Of course, this is also valid for the mass matrices of all FEs.

The analysis process is carried out with the SI (International System of Units) unit system. Then, all the physical parameters affecting the problem should be expressed in units of meter (m), kilogram (kg), and second (s) for length, mass, and time, respectively. The natural frequencies we want to reach are already defined in unit of Hz = 1/s. According to this:

$$\begin{aligned} &E = 1 \ \text{TPa} = 1 \times 10^{12} \frac{\text{N}}{\text{m}^2} = 10^{12} \frac{\text{kg}}{\text{ms}^2}, \ \rho = 2300 \ \frac{\text{kg}}{\text{m}^3}, \ l = 5 \ \text{nm} = 5 \times 10^{-9}\text{m}, \\ &A = \frac{\pi d^2}{4} = \frac{\pi \times (1 \ \text{nm})^2}{4} = \frac{\pi (1 \ \times 10^{-9} \ \text{m})^2}{4} = 0.785 \times 10^{-18} \ \text{m}^2. \end{aligned} \tag{5.E5.1.6}$$

For now, we write the stiffness matrices of FEs without using the above values:

$$[k]_1 = [k]_2 = [k]_3 = [k]_4 = \frac{AE}{l}\begin{bmatrix} 1 & -1 \\ -1 & 1 \end{bmatrix} \tag{5.E5.1.7}$$

On the other hand, mass matrix of FE can be expressed by considering Eq. (5.36) as follows:

$$[m]_e = \frac{\rho_e A_e l_e}{6}\left[\begin{array}{c|c} 2 & 1 \\ \hline 1 & 2 \end{array}\right] \begin{array}{l} \textit{FE node numbers: } 1 \mid 2 \text{ (columns)};\ 1,\ 2 \text{ (rows)} \end{array} \tag{5.E5.1.8}$$

The mass matrices of FEs are given below:

$$[m]_1 = [m]_2 = [m]_3 = [m]_4 = \frac{\rho A l}{6}\begin{bmatrix} 2 & 1 \\ 1 & 2 \end{bmatrix} \tag{5.E5.1.9}$$

According to Fig. 5.7b, it can be noted that the stiffness and mass matrices can be combined as follows:

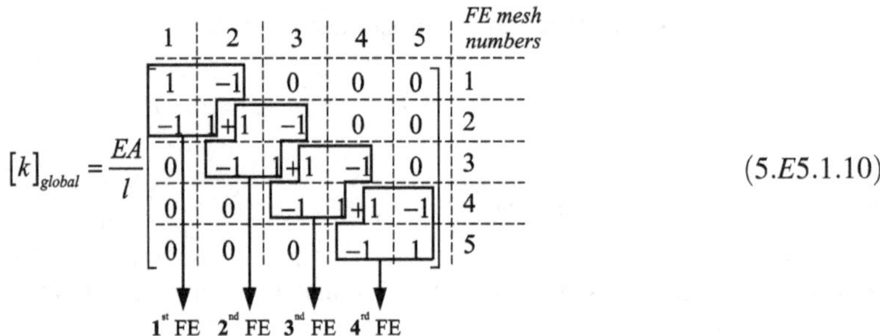

(5.E5.1.10)

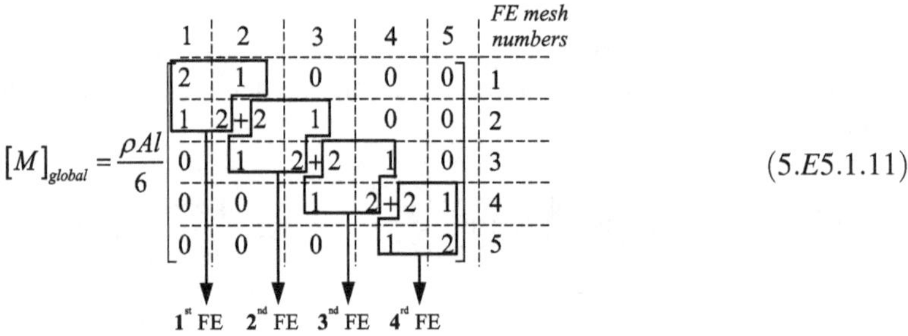

(5.E5.1.11)

By using Eq. (5.E5.1.6) into Eqs. (5.E5.1.10) and (5.E5.1.11), global stiffness and mass matrices are calculated as follows. After the calculation, the global matrices should be reduced. When Fig. 5.7a is examined, the clamped ends representing the geometric boundary condition correspond to nodes 1 and 5. Therefore, the reduced global matrices are obtained by eliminating the first and fifth rows and columns from the global matrices:

Eliminated (column 1), Eliminated (column 5), Result of reduction, Eliminated (row 1), Eliminated (row 5)

$$[k]^{*}_{global} = \begin{bmatrix} 157 & -157 & 0 & 0 & 0 \\ -157 & 314 & -157 & 0 & 0 \\ 0 & -157 & 314 & -157 & 0 \\ 0 & 0 & -157 & 314 & -157 \\ 0 & 0 & 0 & -157 & 157 \end{bmatrix} \frac{\text{kg}}{\text{s}^2} \tag{5.E5.1.12}$$

Eliminated *Result of reduction* *Eliminated*

$$[m]^{*}_{global} = \begin{bmatrix} 3.01\times10^{-24} & 1.505\times10^{-24} & 0 & 0 & 0 \\ 1.505\times10^{-24} & 6.02\times10^{-24} & 1.505\times10^{-24} & 0 & 0 \\ 0 & 1.505\times10^{-24} & 6.02\times10^{-24} & 1.505\times10^{-24} & 0 \\ 0 & 0 & 1.505\times10^{-24} & 6.02\times10^{-24} & 1.505\times10^{-24} \\ 0 & 0 & 0 & 1.505\times10^{-24} & 3.01\times10^{-24} \end{bmatrix} \text{kg} \quad \begin{matrix} \textit{Eliminated} \\ \\ \\ \\ \textit{Eliminated} \end{matrix} \tag{5.E5.1.13}$$

The eigenvalue equation can be constructed based on Eq. (5.E5.1.1) as follows:

$$\det\left(\begin{bmatrix} 314 & -157 & 0 \\ -157 & 314 & -157 \\ 0 & -157 & 314 \end{bmatrix} - \omega_i^2 \times \begin{bmatrix} 6.02\times10^{-24} & 1.505\times10^{-24} & 0 \\ 1.505\times10^{-24} & 6.02\times10^{-24} & 1.505\times10^{-24} \\ 0 & 1.505\times10^{-24} & 6.02\times10^{-24} \end{bmatrix}\right) = 0 \tag{5.E5.1.14}$$

The above eigenvalue equation has six roots and these are as follows:

$$\begin{aligned} &\omega_1 = 3.360 \text{ THz}, \quad \omega_1 = 7.222 \text{ THz}, \quad \omega_1 = 11.736 \text{ THz}, \\ &\omega_4 = -3.360 \text{ THz}, \quad \omega_5 = -7.222 \text{ THz}, \quad \omega_6 = -11.740 \text{ THz} \end{aligned} \tag{5.E5.1.15}$$

where natural frequencies with negative values are physically meaningless. Therefore, the first three frequency values are the fundamental (first), second and third mode natural frequencies, respectively.

If we look closely, the frequency is calculated as much as the size of matrix expressing the reduced eigenvalue problem. In other words, the frequency is obtained as many as the number of freedoms remaining after reduction. As a result, the using of four FEs and the reduction process resulting from boundary conditions give a maximum of three frequency values in this example. Therefore, to obtain frequencies of higher modes is possible by utilizing a larger number of FEs. Now, we compare FEM results and the analytical results obtained with Eq. (2.E2.1.7) (Chap. 2) within the scope of nondimensional frequency calculation. Eq. (2.E2.7.16) is used for nondimensional frequency calculation. In addition to the results obtained in this analysis, comparisons including analysis results with higher number of FEs through a specially prepared computer program are presented in Table 5.1.

As it can be seen in Table 5.1, it is observed that the FEM results generally approach the analytical (exact) results. However, the convergence of higher modes decreases for analyses with low number of FEs. It is also understood that to increase the convergence of higher modes, the finite element number should be increased. For $n = 4$ FEs, the difference between solution manually and computer programming arises due to precision

Table 5.1 Comparisons of nondimensional frequencies of clamped nanorods for classical axial free vibration

Mode	Analytical		FEM results			
Number	Eq. (2.E2. 7.16)	Solution manually	Program code (MATLAB)			
		$n = 4$	$n = 4$	$n = 8$	$n = 20$	$n = 100$
1	3.14159	3.222	3.22283	3.16182	3.14482	3.14172
2	6.28318	6.9272	6.92820	6.44566	6.30905	6.28422
3	9.42478	11.2570	11.25861	9.97439	9.51221	9.42827
4	12.56637	–	–	13.85641	12.77397	12.57464
5	15.70796	–	–	18.11880	16.11416	15.72412

of solution manually. Therefore, although the solution manually for some simple problems in FEM analysis is possible, computer programming provides better results.

5.6 Finite Element Method in Nonlocal Vibration Analysis

In this chapter, we introduced general information about FEM so far. However, general information was given through macromechanics. FEM that is used in many macro-level physical problems has also provided very successful results in nonlocal elasticity problems. The next chapters of book examine the nonlocal dynamic analysis of continuous structures such as rods and beams and discrete structures such as trusses and frames with FEM. Selected studies from the scientific literature on nonlocal free vibration applications of FEM are explained. In order to prepare for next chapters, the relation between FEM and nonlocal vibration is briefly mentioned in this section.

If you remember, we emphasized that the fundamental of FEM analysis is given by Eq. (1.42) in the first chapter. In this analysis, Eq. (1.43) symbolizes the general solution. If the mechanical case that is subject to a finite element analysis is an undamped free vibration in the macro sense, only two concepts emerge: Stiffness and mass matrices. After the global matrices constituted in the system are reduced according to the boundary conditions, the related eigenvalue solution gives the natural frequencies of system. On the other hand, the most general standard eigenvalue equation of a nonlocal undamped free vibration is presented as follows:

$$\det\left[\sum[K] - \omega^2 \sum([M_c] + [M_{nl}])\right] = 0 \tag{5.45}$$

where M_c denotes the classical mass matrix without the nonlocal effect and M_{nl} means the mass matrix with the nonlocal effect. While nonlocal mass is $M_{nl} = 0$ in macro-level problems, it should be considered as $M_{nl} \neq 0$ in atomic-level problems.

In case you have noticed, no definition regarding nonlocality has been made for the stiffness matrix. In the following sections, it will be understood that the stiffness matrix will be affected only by nonlocal environmental factors at the atomic scale. If no environmental factor affects the nonlocal problem, that is, if the problem is studied in simplest form, it can be said that the atomic parameter yields only an additional mass effect in the problem. As a result, since the mass increase occurs against the stiffness that is constant, we know that the natural frequencies of the decrease, based on the knowledge of fundamental structural dynamics. It should certainly be noted that the results obtained in the second, third, and fourth chapters, which concern the analytical solutions of nonlocal vibration analysis, support this conclusion.

Problems

5.1. Determine the mass matrix of Love-Bishop axial rods by using the three-node shape function and following equation:

$$T=\frac{1}{2}\int_{0}^{L}\left[(\rho A\ddot{u}^{e})u^{e}+\upsilon^{2}\rho I_{p}\ddot{u}'^{e}u'^{e}\right]\mathrm{d}x \tag{5.P5.1.1}$$

where u^{e} is the total displacement of axial finite element, ρ is the unit volume mass, υ is the Poisson's ratio, A is the cross-sectional area, I_p is the polar moment of inertia, and L is the finite element length. Also, following expressions can be defined:

$$\ddot{u}^{e}=\frac{\partial^{2}u^{e}}{\partial t^{2}},\quad \ddot{u}'^{e}=\frac{\partial^{3}u^{e}}{\partial x\partial t^{2}} \tag{5.P5.1.2}$$

5.2. Determine the stiffness and mass matrices of two-node bending finite element. The necessary equations for this are formulated as follows:

$$T=\frac{1}{2}\int_{0}^{L}(\rho A\ddot{w}^{e})w^{e}\mathrm{d}V,\quad U=\frac{1}{2}\int_{0}^{L}\left(EIw''^{e}\right)\mathrm{d}x \tag{5.P5.2.1}$$

where w^{e} is the total displacement of bending finite element, ρ is the unit volume mass, E is the modulus of elasticity, A is the cross-sectional area, I is the bending moment of inertia, and L is the finite element length.

5.3. Determine the stiffness and mass matrices of three-node bending finite element given in Fig. 5.8 according to expressions defined in Problem 5.2.

5.4. Determine the stiffness and mass matrices of four-node simple axial rod given in Fig. 5.9.

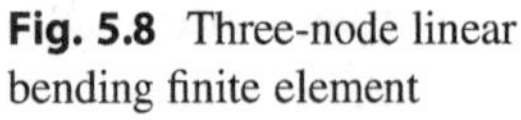

Fig. 5.8 Three-node linear bending finite element

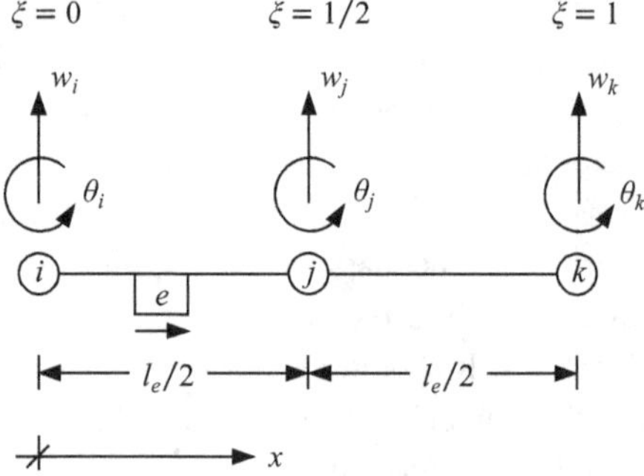

Fig. 5.9 Four-node linear axial finite element

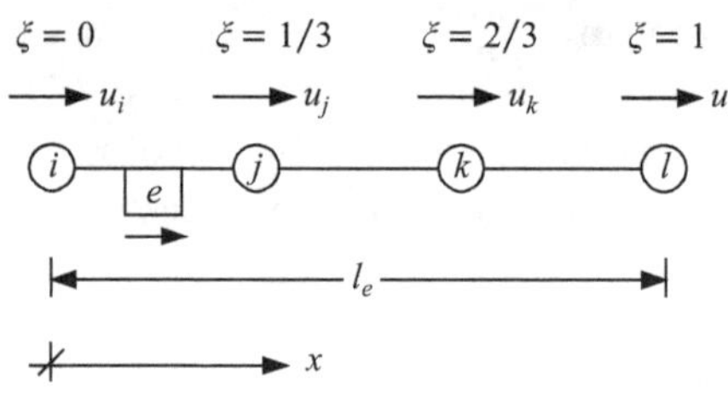

Fig. 5.10 A two-node and four-freedom linear axial finite element

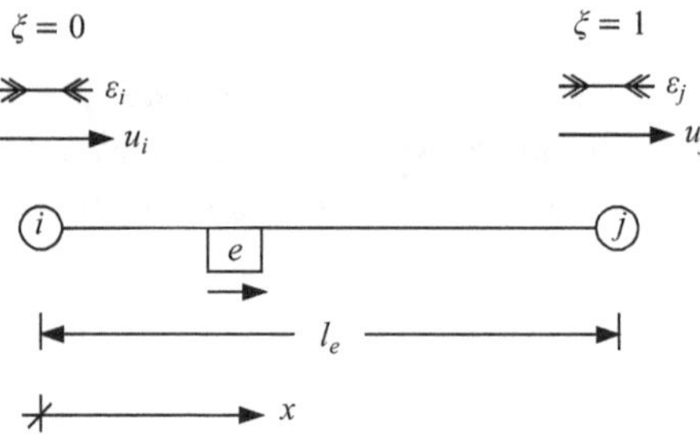

5.5. Axial rods can also be analyzed with a two-node higher-order finite element which includes not only displacements but also strains in Fig. 5.10.

Determine the stiffness and mass matrices of this finite element via the total displacement formulated by Eq. (5.P5.5.1), boundary conditions formulated by Eq. (5.P5.5.2), and related equations given by Eqs. (5.31) and (5.37). For simplicity, consider the simple axial rod. Compare the resulting matrices with the two-node finite element matrices and discuss the differences between the matrices.

$$u^e = c_1 + c_2 x + c_3 x^2 + c_4 x^3 \tag{5.P5.5.1}$$

$$u^e(x_i) = u_i, \quad \frac{\mathrm{d}u^e}{\mathrm{d}x}(x_i) = -\varepsilon_i, \quad u^e(x_j) = u_j, \quad \frac{\mathrm{d}u^e}{\mathrm{d}x}(x_j) = -\varepsilon_j \tag{5.P5.5.2}$$

where u is axial displacement and ε is axial strain.

5.6. Consider the classical free vibration of a simple axial rod with clamped-free boundary condition referring to Example 2.7 (Chap. 2). According to this, calculate the nondimensional free vibration frequencies by using two, three, and four finite elements, respectively, under the parameters used in the calculation of results given in Table 2.1. For simplicity, use the two-node linear finite element. Compare the results with the analytical results given in Table 2.1. Discuss the importance of finite element number in the analysis. *Hint*: The first row and column of global matrices are eliminated since the first node in the finite element mesh corresponds to the clamped support. Also, since the solution manually is difficult to in the analysis with high number of finite elements, a solution in the computer environment is recommended.
5.7. Perform the analyses in Problem 5.6 with a three-node finite element. Compare the results obtained with the analytical results and the two-node finite element results. Discuss the importance of finite element model in analysis.
5.8. Determine the necessary finite element matrices for the analysis of free torsional vibration of circular nanoshaft by using a linear two-node finite element. Compare the finite element formulation with the vibration of simple axial rod. For simplicity, neglect any external influences such as elastic medium. Use Eq. (3.E3.5.8) (Chap. 3) for nondimensional frequency calculations.
5.9. By using the derivations in Problem 5.8, perform the analyses similar to Problem 5.6 for the circular nanorod (clamped-clamped boundary condition). Analytical results are given in Table 3.1. *Hint*: Clamped ends correspond to the first and $(N + 1)$th nodes on the condition that the number of finite elements is N.
5.10. Consider the classical free vibration of a simply supported Euler-Bernoulli beam referring to Example 4.1 (Chap. 4). According to this, calculate the nondimensional free vibration frequencies by using two, three, and four finite elements, respectively, under the parameters used in the calculation of results given in Table 4.2. For simplicity, use the two-node linear finite element. Compare the results with the analytical results given in Table 4.2. Discuss the importance of finite element number in the analysis. *Hint*: Since the simply supported ends correspond to the first and $(N + 1)$th nodes on the condition that the number of finite elements is N, the first and $(2N + 1)$th row and column of global matrices are eliminated. Also, since the solution manually is difficult to in the analysis with high number of finite elements, a solution in the computer environment is recommended.
5.11. Explain mathematically how the finite element analysis can be performed for the classical axial vibration of clamped axial elastic spring attached simple axial rod given in Fig. 5.11.

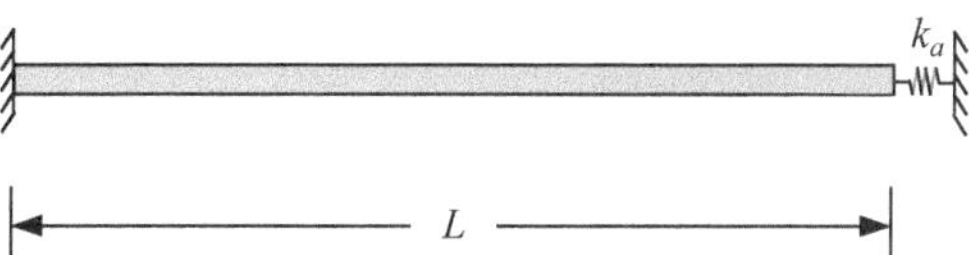

Fig. 5.11 Clamped-spring attached axial rod

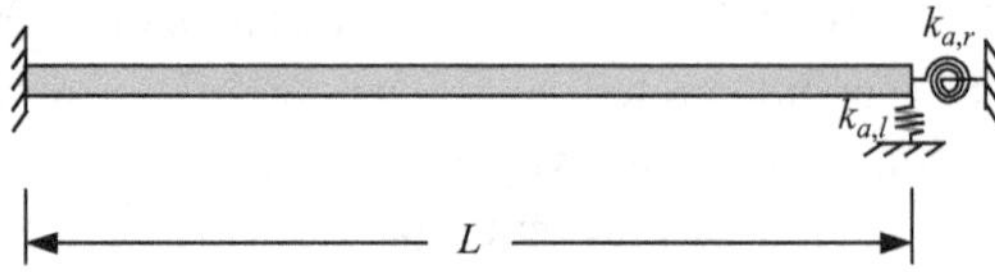

Fig. 5.12 Clamped-transverse and rotational springs attached beam

5.12. Explain mathematically how the finite element analysis can be performed for the classical transverse vibration of clamped-transverse and rotational elastic springs attached Euler-Bernoulli beam given in Fig. 5.12.

References

1. R. Courant, Variational methods for the solution of problems of equilibrium and vibrations. Bull. Am. Math. Soc. **49**(1), 1–23 (1943)
2. R.D. Cook, D.S. Malkus, M.E. Plesha, *Concepts and Applications of Finite Element Analysis* (Wiley, Hoboken, NJ, 1989)
3. J.H. Argyris, *Energy Theorems and Structural Analysis* (Aircraft Eng., Butterworths's Scientific Publications, London, 1954)
4. M.J. Turner, R.W. Clough, H.C. Martin, L.J. Topp, Stiffness and deflection analysis of complex structures. J. Aeronaut. Sci. **23**(9), 805–823 (1956)
5. R.W. Clough, The finite element method, in *Plane Stress Analysis, 2nd ASCE Conference on Electronic Computation, Pittsburgh, 8-10 September*, (1960)
6. A. Adini, R.W. Clough, Analysis of Plate Bending by the Finite Element Method, NSF Report, Grant G7337, 1960
7. R.W. Clough, J.L. Tocher, Finite element stiffness matrices for the analysis of plate bending, in *Proceedings of the Conference on Matrix Methods in Structural Mechanics, Ohio, 26–-28 October*, (1965), pp. 515–545
8. O.C. Zienkiewicz, Y.K. Cheung, *The Finite Element in Structural and Continuum Mechanics* (McGraw-Hill Publishing, London, 1967)
9. J.T. Oden, *Finite Elements of Nonlinear Continua* (McGraw-Hill Publishing, New York, 1972)
10. Anonymous, History of finite element method. https://www.fea-academy.com/index.php/component/content/article/27-blog/fea-generalities/73-fem-history?Itemid=101

Applications of NL-FEM for Nanorods 6

6.1 Introduction

It is stated in Chap. 1 that some nano-scaled materials in NEMS organizations can be examples of one-dimensional and continuous rod models. In addition to this, from the research in Chaps. 2 and 3, we understand that the atomic effect based on nonlocal elasticity is a very important concept in the mechanical behavior of axial and torsional nanorod models. As it is known, accurate analysis of mechanical behavior is necessary in order to carry out the process of exact design in accordance with the principles of engineering.

Investigating the mechanical behavior of axial and torsional nanorods through analytical methods is a fundamental approach. As seen in the related chapters, the analytical method helps to learn mechanical behavior simply, especially when the boundary conditions are not complex. However, there are some situations where analytical methods may be ineffective or completely useless. For example, after a certain point, it becomes difficult to analyze the nonlocal free axial and torsional dynamics of nanorod with an axial elastic spring attached end that models the partially loose firm layer mentioned in the presentation of NEMS models. According to this, the outputs of vibration of higher modes cannot be achieved. Additionally, high values of the atomic parameter make it difficult to obtain frequency. Moreover, if the elastic medium is included in the problem, it is almost impossible to obtain the result of nonlocal vibration even in fundamental modes. On the other hand, the analytical solution is insufficient even in the classical elasticity due to additional terms when the shear stresses of axial rods are considered. These problems already occur even though the analytical method is programmed on the computer. According to a special case, we could not attain the analytical solution of torsion of nanorods with noncircular cross-section due to the missing boundary condition in

Öm. Civalek et al., *Mechanical Behavior and Vibration of Nano-Scaled Rods, Beams, and Frames*, Synthesis Lectures on Engineering, Science, and Technology,
https://doi.org/10.1007/978-3-032-12023-6_6

Chap. 2. As a result, analytical methods greatly weaken the progress of engineering analysis from a certain point onward.

The most well-known way to purify such problems in engineering mechanics is to develop a numerical solution by abandoning the analytical solution. Numerical solution methods are known as approximate solutions because these provide solutions that have a certain difference from the exact result. To reduce this difference considerably depends on generally increase of parameter used according to the nature of numerical method. For example, increasing the number of finite elements in the finite element method, increasing the number of functions in the Rayleigh-Ritz method, increasing the number of grids in the differential quadrature method, and increasing the number of terms in the differential transformation method can be specified. So, as stated in the previous chapter, the approximate solution is much better than no solution.

Starting from this chapter in the book, the primary objective will be to reach more accurate analysis processes of vibration by examining finite element analyses of the nonlocal vibration behavior of various nanostructures. First, in this chapter, we formulated the free axial and torsional vibrations of nanorods via the nonlocal finite element method (NL-FEM). Accordingly, two fundamental topics are investigated: Nonlocal axial vibration and nonlocal torsional vibration. While the first part introduces NL-FEM models of simple and Love-Bishop rods, the second part is interested in the application of NL-FEM to the free torsional vibration of circular and elliptical nanorods. Different numerical applications of NL-FEM formulations developed by considering external effects such as elastic environment and tip attachment in nanorods are presented, and research on the atomic size-dependent behavior of nanorods is completed.

6.2 Recent Contributions

Studies on the using of NL-FEM on nonlocal axial and torsional vibrations of nanorods have fallen within the scientific literature. A summary of this topic is given below.

The studies in which NL-FEM was first developed are about the axial free vibration of nanorods [1, 2]. Additionally, it can be mentioned that the NL-FEM formulated for nonlocal axial and torsional vibration of microtubules succeeds by considering the analytical solution [3]. Axial vibration of variable cross-sectional nanorods that have material change that is modeled with the trigonometric functions has also been investigated [4]. The effects of different parameters such as mechanical crack in axial nanorods [5] and elastic axial spring attachment [6] for nonlocal free vibration have been investigated with some NL-FEM formulations. In addition to these studies dealing with simple axial rods, it can be stated that the nonlocal free dynamics of Poisson's effect nanorods embedded in elastic medium have been studied with a higher-order NL-FEM [7]. Moreover, NL-FEM has been examined for the free vibration of elastic spring attached nanorods according to Poisson's effect [8].

On the other hand, nonlocal torsional vibration of nanorods with general boundary conditions has been explained [9]. Additionally, the free torsional vibration of nanorods medium has been studied under the effects of elastic spring attachment and elastic medium [10].

Also, the finite element formulation of nonlocal strain gradient elasticity theory that is a higher-order nonlocal theory model is introduced in the atomic size-dependent axial vibration of nanorods [11]. It can be added that spectral finite element formulations have been proposed for axial nanorod [12] and torsional nanorod [13] about the nonlocal vibration analyses of nanorods.

6.3 Nonlocal Axial Vibration

Before starting, we remember that Fig. 5.3a is an example of a finite element mesh of an axial rod. In this section, finite element procedures that provide the solution for the nonlocal free vibrations of simple and Love-Bishop nanorods with different tip attachments and fully embedded in the elastic foundation are formulated. Let's first examine simple (without shear effect) nanorods.

6.3.1 Simple Nanorod

6.3.1.1 NL-FEM Expression

It is stated that the FEM research is based on Eq. (1.42). From this, the NL-FEM procedure of free vibration of simple axial nanorods can be defined. According to this, by considering Eq. (2.6), the average weighted residue equation is written as follows:

$$\begin{aligned} \mathrm{I} = \int_0^{l_e} h\Bigg(& EA\frac{\partial^2 u}{\partial x^2} - \rho A\frac{\partial^2 u}{\partial t^2} - k_M u + (e_0 a)^2 k_M \frac{\partial^2 u}{\partial x^2} + (e_0 a)^2 \rho A \frac{\partial^4 u}{\partial x^2 \partial t^2} \\ & - (e_0 a)^2 \frac{\partial^2 q}{\partial x^2} + q \Bigg) \mathrm{d}x \end{aligned} \tag{6.1}$$

where l_e is the length of finite element. In order to arrange this equation, the kinematic relations of axial finite element should be defined. The total motion of two-node linear axial finite element illustrated in Fig. 5.4 and based on Eq. (5.1), is defined as follows:

$$u = \phi \mathbf{d} \tag{6.2}$$

The following definitions are valid here [5]:

$$\phi = [\phi_i \ \phi_j] = [1-\xi \quad \xi] = \left[1-\frac{x}{L} \quad \frac{x}{L}\right], \quad \mathbf{d} = \begin{Bmatrix} d_i \\ d_j \end{Bmatrix} \tag{6.3}$$

where $\xi = x/L$ is the nondimensional coordinate. u denotes the finite element motion, ϕ means the shape function matrix, and $\mathbf{d}$ expresses the displacement vector of end freedoms.

A fundamental definition to constitute NL-FEM formulation is explained as follows [5]:

$$\frac{\partial u}{\partial x} = D^k u = \mathbf{Bd} \tag{6.4}$$

where D^k is the kinematic operator and is explained as $D^k \phi = \mathbf{B}$.

In Eq. (6.1), it should be stated that all of the expressions separated from each other by addition or subtraction constitute a mathematical finite element component. Therefore, Eq. (6.1) should be divided into seven different integrals, and some of these integrals are rearranged by partial integration [5]:

$$\begin{aligned}
\mathrm{I}_1 &= \int_0^{l_e} EAh\frac{\partial^2 u}{\partial x^2}\mathrm{d}x = EAh\frac{\partial u}{\partial x}\bigg|_0^{l_e} - \int_0^{l_e} EA\frac{\partial h}{\partial x}\frac{\partial u}{\partial x}\mathrm{d}x, \\
\mathrm{I}_2 &= \int_0^{l_e} \rho Ah\frac{\partial^2 u}{\partial t^2}\mathrm{d}x, \quad \mathrm{I}_3 = \int_0^{l_e} k_M hu\mathrm{d}x, \\
\mathrm{I}_4 &= \int_0^{l_e} (e_0a)^2 k_M h\frac{\partial^2 u}{\partial x^2}\mathrm{d}x = (e_0a)^2 k_M h\frac{\partial u}{\partial x}\bigg|_0^{l_e} - \int_0^{l_e} (e_0a)^2 k_M\frac{\partial h}{\partial x}\frac{\partial u}{\partial x}\mathrm{d}x, \\
\mathrm{I}_5 &= \int_0^{l_e} (e_0a)^2\rho Ah\frac{\partial^4 u}{\partial x^2\partial t^2}\mathrm{d}x = (e_0a)^2\rho Ah\frac{\partial^3 u}{\partial x\partial t^2}\bigg|_0^{l_e} - \int_0^{l_e} (e_0a)^2\rho A\frac{\partial h}{\partial x}\frac{\partial^3 u}{\partial x\partial t^2}\mathrm{d}x, \\
\mathrm{I}_6 &= \int_0^{l_e} (e_0a)^2 h\frac{\partial^2 q}{\partial x^2}\mathrm{d}x = (e_0a)^2 h\frac{\partial q}{\partial x}\bigg|_0^{l_e} - \int_0^{l_e} (e_0a)^2\frac{\partial h}{\partial x}\frac{\partial q}{\partial x}\mathrm{d}x, \quad \mathrm{I}_7 = \int_0^{l_e} hq\mathrm{d}x
\end{aligned} \tag{6.5}$$

Using these expressions into Eq. (6.1) gives the following expression:

$$\int_0^{l_e} EA\frac{\partial h}{\partial x}\frac{\partial u}{\partial x}\mathrm{d}x + \int_0^{l_e} (e_0a)^2\rho A\frac{\partial h}{\partial x}\frac{\partial^3 u}{\partial x\partial t^2}\mathrm{d}x + \int_0^{l_e} (e_0a)^2k_M\frac{\partial h}{\partial x}\frac{\partial u}{\partial x}\mathrm{d}x - \int_0^{l_e} (e_0a)^2\frac{\partial h}{\partial x}\frac{\partial f}{\partial x}\mathrm{d}x$$
$$+ \int_0^{l_e} k_M hu\mathrm{d}x + \int_0^{l_e} \rho Ah\frac{\partial^2 u}{\partial t^2}\mathrm{d}x - \int_0^{l_e} hq\mathrm{d}x - EAh\frac{\partial u}{\partial x}\bigg|_0^{l_e} - (e_0a)^2\rho Ah\frac{\partial^3 u}{\partial x\partial t^2}\bigg|_0^{l_e}$$
$$- (e_0a)^2k_Mh\frac{\partial u}{\partial x}\bigg|_0^{l_e} + (e_0a)^2h\frac{\partial q}{\partial x}\bigg|_0^{l_e} = 0 \tag{6.6}$$

In the above equation, only integral expressions indicate the weak formulation of weighted residue, that is, the finite element solution. The following expressions are required to rewrite the related expressions:

$$h = \phi^{\mathrm{T}},\ \mathbf{B} = \mathrm{D}^{\mathrm{k}}\phi,\ \frac{\partial h}{\partial x} = (\phi^{\mathrm{T}})' = \mathbf{B}^{\mathrm{T}},\ \frac{\partial u}{\partial x} = \mathbf{Bd},\ \frac{\partial^2 u}{\partial t^2} = \phi\ddot{\mathbf{d}} \tag{6.7}$$

thus, Eq. (6.6) becomes:

$$\int_0^{l_e} EA(\mathbf{B}^{\mathrm{T}}\mathbf{B})\mathbf{d}\mathrm{d}x + \int_0^{l_e} \rho A(\phi^{\mathrm{T}}\phi)\ddot{\mathbf{d}}\mathrm{d}x + \int_0^{l_e} (e_0a)^2\rho A(\mathbf{B}^{\mathrm{T}}\mathbf{B})\ddot{\mathbf{d}}\mathrm{d}x + \int_0^{l_e} (e_0a)^2k_M(\mathbf{B}^{\mathrm{T}}\mathbf{B})\mathbf{d}\mathrm{d}x$$
$$- \int_0^{l_e} (e_0a)^2\mathbf{B}^{\mathrm{T}}q'\mathrm{d}x + \int_0^{l_e} k_M(\phi^{\mathrm{T}}\phi)\mathbf{d}\mathrm{d}x - \int_0^{l_e} \phi^{\mathrm{T}}q\mathrm{d}x = 0 \tag{6.8}$$

This equation is expressed as follows:

$$(K + K_{M,c} + K_{M,nl})\mathbf{d} + (M_c + M_{nl})\ddot{\mathbf{d}} = \mathbf{q}_c + \mathbf{q}_{nl} \tag{6.9}$$

where $\mathbf{d}$ and $\ddot{\mathbf{d}}$ represent the motion and acceleration of finite element, respectively. Also, above expressions are written as follows [3, 5]:

$$K = \int_0^{l_e} EA(\mathbf{B}^{\mathrm{T}}\mathbf{B})\mathrm{d}x = \int_0^{l_e} EA\begin{Bmatrix}\phi_1' \\ \phi_2'\end{Bmatrix}[\phi_1' \quad \phi_2']\mathrm{d}x = \frac{EA}{l_e}\begin{bmatrix}1 & -1 \\ -1 & 1\end{bmatrix} \tag{6.10}$$

$$K_{M,c}=\int_0^{l_e} k_M(\phi^{\mathrm{T}}\phi)\mathrm{d}x=\int_0^{l_e} k_M\begin{Bmatrix}\phi_1\\ \phi_2\end{Bmatrix}[\phi_1 \quad \phi_2]\mathrm{d}x=\frac{k_M l_e}{6}\begin{bmatrix}2 & 1\\ 1 & 2\end{bmatrix} \tag{6.11}$$

$$K_{M,nl}=\int_0^{l_e} (e_0a)^2 k_M(\mathbf{B}^{\mathrm{T}}\mathbf{B})\mathrm{d}x=\int_0^{l_e} (e_0a)^2 k_M\begin{Bmatrix}\phi_1'\\ \phi_2'\end{Bmatrix}[\phi_1' \quad \phi_2']\mathrm{d}x=\frac{(e_0a)^2 k_M}{l_e}\times\begin{bmatrix}1 & -1\\ -1 & 1\end{bmatrix} \tag{6.12}$$

$$M_c=\int_0^{l_e} \rho A(\phi^{\mathrm{T}}\phi)\mathrm{d}x=\int_0^{l_e} \rho A\begin{Bmatrix}\phi_1\\ \phi_2\end{Bmatrix}[\phi_1 \quad \phi_2]\mathrm{d}x=\frac{\rho A l_e}{6}\begin{bmatrix}2 & 1\\ 1 & 2\end{bmatrix} \tag{6.13}$$

$$M_{nl}=\int_0^{l_e} (e_0a)^2 \rho A(\mathbf{B}^{\mathrm{T}}\mathbf{B})\mathrm{d}x=\int_0^{l_e} (e_0a)^2 \rho A\begin{Bmatrix}\phi_1'\\ \phi_2'\end{Bmatrix}[\phi_1' \quad \phi_2']\mathrm{d}x=\frac{(e_0a)^2 \rho A}{l_e}\times\begin{bmatrix}1 & -1\\ -1 & 1\end{bmatrix} \tag{6.14}$$

$$\mathbf{q}_c=\int_0^{l_e} q\phi^{\mathrm{T}}\mathrm{d}x=\int_0^{l_e} q\begin{Bmatrix}\phi_1\\ \phi_2\end{Bmatrix}\mathrm{d}x=\frac{q l_e}{2}\begin{Bmatrix}1\\ 1\end{Bmatrix} \tag{6.15}$$

$$\mathbf{q}_{nl}=\int_0^{l_e} (e_0a)^2 q'\mathbf{B}^{\mathrm{T}}\mathrm{d}x=\int_0^{l_e} (e_0a)^2 q'\begin{Bmatrix}\phi_1'\\ \phi_2'\end{Bmatrix}\mathrm{d}x=\frac{(e_0a)^2 q'}{l_e}\begin{Bmatrix}-1\\ 1\end{Bmatrix} \tag{6.16}$$

where for finite element, K is the axial stiffness matrix, $K_{M,\,c}$ is the classical stiffness matrix of axial medium, and $K_{M,\,nl}$ is the nonlocal stiffness matrix of axial medium. M_c is the classical mass and M_{nl} is the nonlocal mass matrices. Finally, $\mathbf{q}_c$ and $\mathbf{q}_{nl}$ denote the classical and nonlocal external load vectors, respectively.

Interpretation. Note that the stiffness matrix obtained directly with the weighted residual is the same as the stiffness matrix obtained indirectly and presented with Eq. (5.30). In addition to this, if Eq. (6.9) is examined, when firstly the external factor is assumed as $K_M=K_{M,\,nl}=0$, it is concluded that the total mass matrix increases by combining mass with an additional effect by nonlocal elasticity. On the other hand, the stiffness matrix of axial medium also combines with a nonlocal effect. Another remarkable result is that the elastic stiffness matrix of nanorod is not affected in any way by nonlocal elasticity.

Solution stage. The total mass and stiffness matrices of a finite element are defined as follows:

$$[K]_e = K + K_{M,c} + K_{M,nl} \tag{6.17}$$

$$[M]_e = M_c + M_{nl} \tag{6.18}$$

Additionally, the total excitation vector is written as:

$$\mathbf{q}_e = \mathbf{q}_c + \mathbf{q}_{nl} \tag{6.19}$$

For nonlocal free vibration analysis, it should be considered as $\mathbf{q}_c = \mathbf{q}_{nl} = 0$. Remembering the equation of separation of variable given with Eq. (2.21), we formulate the eigenvalue equation as follows:

$$\det\left[\left(\sum_{e=1}^{n}[K]_e\right)^{*} - \omega_i^2\left(\sum_{e=1}^{n}[M]_e\right)^{*}\right] = 0 \tag{6.20}$$

Of course, it should not be forgotten that global matrices are not obtained via an algebraic addition (input to input). According to this, global matrices are attained as a result of addition in which the same freedoms in the finite element network are considered (in other words, the common freedoms of adjacent finite elements).

On the other hand, the $(*)$ superscript in total matrices refers to global matrices reduced by boundary conditions. If the boundary conditions are geometric, they delete the row and column corresponding to the related freedom. If the boundary condition is mechanical (such as a tip attachment), it is either included in the stiffness or the mass depending on internal effect affected by boundary condition. So, there is no reduction process. As a result, the roots of polynomial arising from the standard eigenvalue problem under reduced global matrices should be searched. These roots represent the natural vibration frequencies of different modes, i.e., ω_i ($\omega_1 < \omega_2 < \omega_3 \cdots$ and $i = 1, 2, 3, \cdots i \in Z^+$). Now let's interpret the reduction process.

By the way, in the example applications, $(\sum[\cdots])^*$ notation is preferred instead of $\left(\sum_{e=1}^{n}[\cdots]_e\right)^*$ notation for simplicity in order to symbolize the global matrices.

6.3.1.2 Application of Boundary Conditions

It is necessary to interpret the general boundary conditions of simple axial rods. According to this, since there will be no displacement at the fully clamped end, this defines the geometric boundary condition specified in the NL-FEM procedure. On the other hand, the dynamic boundary condition explains that no axial force will occur at the free end and axial force will occur at the attached end. By eliminating the rows and columns corresponding to the freedom addressing the geometric boundary in the finite element mesh, global stiffness and global mass matrices are obtained. Let's explain the application discussed in Example 5.1 here once again. Let's assume that we have divided the cantilever axial nanorod into n finite elements ($n > 1$). All global matrices of system are $(n + 1) \times (n + 1)$ sizes.

Additionally, since the fully clamped end coincides with the left node of first finite element, the first row and column of global matrices are deleted. Then the remaining matrices in $n \times n$ sizes are known as reduced global matrices.

However, case of attached end should be evaluated differently. The attached end does not reduce the finite element at the end that it is placed, on the contrary, it supports related finite element. Let's think of a clamped-elastic spring attached nanorod. Only the input 22 of an additional stiffness matrix of 2×2 sizes contains the elastic spring constant of attachment on condition that the right end of last finite element shows the attached end, and this new matrix is included in the total stiffness matrix. Of course, this is also valid for the mass matrix. Now also we consider that the right end has a mass attachment. This time, the value of attachment mass is written only into the input 22 of additional mass matrix. The eigenvalue problem that will be computed as explained in the paragraph above is solved. As a result, these explanations are presented mathematically as follows:

$$[K]_{e,a} = [K]_{e,na} + (K_a) = (K_{ae})_e + (K_{M,c})_e + (K_{M,nl})_e + \begin{bmatrix} 0 & 0 \\ 0 & k_a \end{bmatrix} \tag{6.21}$$

$$[M]_{e,a} = [M]_{e,na} + (M_a) = (M_c)_e + (M_{nl})_e + \begin{bmatrix} 0 & 0 \\ 0 & m_a \end{bmatrix} \tag{6.22}$$

where $K_{e,\,a}$ and $K_{e,\,na}$ define the total stiffness matrices of a finite element with and without spring attachment, respectively. Similarly, $M_{e,\,a}$ and $M_{e,\,na}$ are the total mass matrices of a finite element with and without mass attachment. K_a and M_a are the stiffness and mass matrices of end attachment. Also, some notations are made to avoid confusion. Accordingly, the axial stiffness matrix is symbolized by *ae* subscript. While normal parenthesis symbolizes the matrices of continuous or discrete components of finite elements, square parenthesis addresses additive matrices.

It is necessary to pay attention to what is explained because since these explanations are sufficient, detail about the application of boundary conditions is not given when the following NL-FEM procedures are introduced.

Example 6.1 Determine the natural frequencies of nonlocal axial free vibration of clamped-clamped nanorod in Fig. 6.1. Use three finite elements in the analysis. Nanobeam length: $L = 20$ nm, diameter of circular cross-section $d = 1$ nm, modulus of elasticity: $E = 1$ TPa, and mass of unit volume: $\rho = 2300$ kg/m^3. Use the nonlocal parameter as $e_0 a/L = 0.2$.

Solution. Since there are two geometric boundary conditions in this nanorod, using three two-node finite elements makes it possible to calculate the eigenvalues manually, that is, the natural frequencies.

Let's write all the physical and mechanical parameters of finite elements as follows:

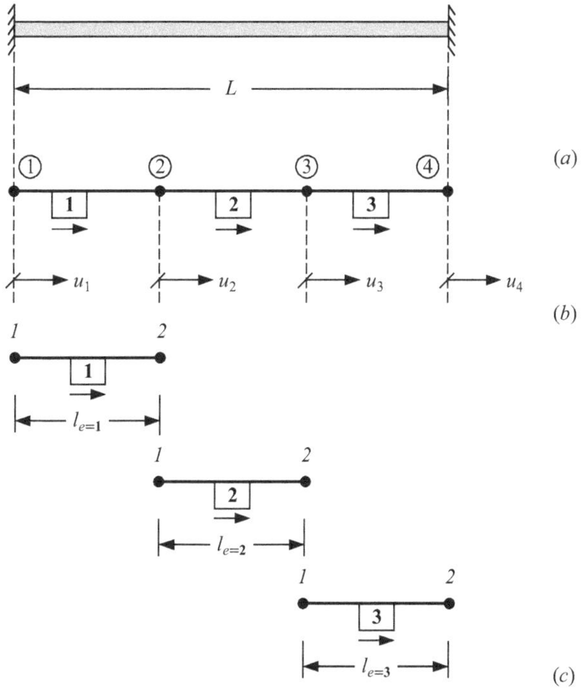

Fig. 6.1 FEM analysis of clamped axial nanorod (**a**) Physical system (**b**) FEM model with three elements (**c**) Finite elements

$$\begin{aligned}
&\text{Modulus of elasticity}: E_{e=1}=E_{e=2}=E_{e=3}=E=1\ \text{TPa}=1\times 10^{12}\ \frac{\text{N}}{\text{m}^2},\\
&\text{Mass of unit volume}: \rho_{e=1}=\rho_{e=2}=\rho_{e=3}=\rho=2300\ \frac{\text{kg}}{\text{m}^3},\\
&\text{Finite element length}: l_{e=1}=l_{e=2}=l_{e=3}=l=\frac{L}{3}=\frac{20}{3}\ \text{nm}\\
&\qquad =6.667\ \text{nm}=6.667\times 10^{-9}\ \text{m},\\
&\text{Cross-section area}: A_{e=1}=A_{e=2}=A_{e=3}=A=\frac{\pi d^2}{4}\\
&\qquad =\frac{\pi\times(1\ \text{nm})^2}{4}=\frac{\pi\left(1\times 10^{-9}\ \text{m}\right)^2}{4}=7.854\times 10^{-19}\ \text{m}^2,\\
&\text{Nonlocal parameter}: (e_0a)_{e=1}=(e_0a)_{e=2}=(e_0a)_{e=3}=e_0a=0.2\times L\\
&\qquad =4\ \text{nm}=4\times 10^{-9}\ \text{m}.
\end{aligned} \tag{6.E6.1.1}$$

Elastic stiffness and classical mass matrices of a two-node axial finite element are as:

$$(K)_e = \frac{A_e E_e}{l_e} \begin{array}{c|c|l} 1 & 2 & FE\ node\ numbers \\ \hline 1 & -1 & 1 \\ -1 & 1 & 2 \end{array} \tag{6.E6.1.2}$$

$$(M_c)_e = \frac{\rho_e A_e l_e}{6} \begin{array}{c|c|l} 1 & 2 & FE\ node\ numbers \\ \hline 2 & 1 & 1 \\ 1 & 2 & 2 \end{array} \tag{6.E6.1.3}$$

Additionally, the nonlocal mass matrix of a finite element can be stated as follows:

$$(M_{nl})_e = \frac{(e_0 a)_e^2 \rho_e A_e}{l_e} \begin{array}{c|c|l} 1 & 2 & FE\ node\ numbers \\ \hline 1 & -1 & 1 \\ -1 & 1 & 2 \end{array} \tag{6.E6.1.4}$$

Thus, the stiffness, classical mass and nonlocal mass matrices of finite elements are written as:

$$(K)_{e=1} = (K)_{e=2} = (K)_{e=3} = \frac{AE}{l} \begin{bmatrix} 1 & -1 \\ -1 & 1 \end{bmatrix} \tag{6.E6.1.5}$$

$$(M_c)_{e=1} = (M_c)_{e=2} = (M_c)_{e=3} = \frac{\rho A l}{6} \begin{bmatrix} 2 & 1 \\ 1 & 2 \end{bmatrix} \tag{6.E6.1.6}$$

$$(M_{nl})_{e=1} = (M_{nl})_{e=2} = (M_{nl})_{e=3} = \frac{(e_0 a)^2 \rho A}{l} \begin{bmatrix} 1 & -1 \\ -1 & 1 \end{bmatrix} \tag{6.E6.1.7}$$

If these matrices are assumed by considering the finite element freedoms in Fig. 6.1c, global matrices are obtained as follows:

$$\sum[K] = \frac{AE}{l} \begin{array}{c|c|c|c|l} 1 & 2 & 3 & 4 & Mesh\ node\ numbers \\ \hline 1 & -1 & 0 & 0 & 1 \\ -1 & 1+1 & -1 & 0 & 2 \\ 0 & -1 & 1+1 & -1 & 3 \\ 0 & 0 & -1 & 1 & 4 \end{array} \tag{6.E6.1.8}$$

$$\sum[M_c]=\frac{\rho A l}{6}\begin{bmatrix} 2 & 1 & 0 & 0 \\ 1 & 2+2 & 1 & 0 \\ 0 & 1 & 2+2 & 1 \\ 0 & 0 & 1 & 2 \end{bmatrix} \quad \begin{matrix}\text{Mesh node numbers: columns } 1,2,3,4;\ \text{rows } 1,2,3,4\end{matrix} \tag{6.E6.1.9}$$

$$\sum[M_{nl}]=\frac{(e_0 a)^2\rho A}{l}\begin{bmatrix} 1 & -1 & 0 & 0 \\ -1 & 1+1 & -1 & 0 \\ 0 & -1 & 1+1 & -1 \\ 0 & 0 & -1 & 1 \end{bmatrix} \quad \begin{matrix}\text{Mesh node numbers: columns } 1,2,3,4;\ \text{rows } 1,2,3,4\end{matrix} \tag{6.E6.1.10}$$

Geometric boundary conditions reduce global matrices as follows:

$$\sum[K]=\begin{bmatrix} 117.804 & -117.804 & 0 & 0 \\ -117.804 & \boxed{\begin{matrix}235.608 & -117.804 \\ -117.804 & 235.608\end{matrix}} & & 0 \\ 0 & & & -117.804 \\ 0 & 0 & -117.804 & 235.608 \end{bmatrix}\frac{\text{kN}}{\text{m}} \tag{6.E6.1.11}$$

(First and last rows and columns: *Eliminated*; boxed part: $\left(\sum[K]\right)^*$)

$$\sum[M_c]=\begin{bmatrix} 4.014\times10^{-24} & 2.007\times10^{-24} & 0 & 0 \\ 2.007\times10^{-24} & \boxed{\begin{matrix}8.029\times10^{-24} & 2.007\times10^{-24} \\ 2.007\times10^{-24} & 8.029\times10^{-24}\end{matrix}} & & 0 \\ 0 & & & -117.804 \\ 0 & 0 & 2.007\times10^{-24} & 4.014\times10^{-24} \end{bmatrix}\text{kg}$$

(First and last rows and columns: *Eliminated*; boxed part: $\left(\sum[M_c]\right)^*$)

(6.E6.1.12)

$$\sum[M_{nl}] = \begin{bmatrix} \cancel{4.335\times10^{-24}} & \cancel{-4.335\times10^{-24}} & \cancel{0} & \cancel{0} \\ -4.335\times10^{-24} & 8.670\times10^{-24} & -4.335\times10^{-24} & 0 \\ 0 & -4.335\times10^{-24} & 8.670\times10^{-24} & -4.335\times10^{-24} \\ \cancel{0} & \cancel{0} & \cancel{-4.335\times10^{-24}} & \cancel{4.335\times10^{-24}} \end{bmatrix} \text{kg}$$

(Columns 1 and 4 and rows 1 and 4: *Eliminated*; the remaining central block is $\left(\sum[M_c]\right)^*$)

(6.*E*6.1.13)

where the reduced total global mass matrix can be expressed as follows:

$$\left(\sum[M]\right)^* = \left(\sum[M_c]\right)^* + \left(\sum[M_{nl}]\right)^* = \begin{bmatrix} 8.029\times10^{-24} & 2.007\times10^{-24} \\ 2.007\times10^{-24} & 8.029\times10^{-24} \end{bmatrix} + \begin{bmatrix} 8.670\times10^{-24} & -4.335\times10^{-24} \\ -4.335\times10^{-24} & 8.670\times10^{-24} \end{bmatrix} = \begin{bmatrix} 1.67\times10^{-23} & -2.328\times10^{-24} \\ -2.328\times10^{-24} & 1.67\times10^{-23} \end{bmatrix} \text{kg} \tag{6.E6.1.14}$$

The eigenvalue equation is formulated below by using Eq. (6.20):

$$\det\left(\begin{bmatrix} 235.608 & -117.804 \\ -117.804 & 235.608 \end{bmatrix} - \omega^2 \times \begin{bmatrix} 1.67\times10^{-23} & -2.328\times10^{-24} \\ -2.328\times10^{-24} & 1.67\times10^{-23} \end{bmatrix}\right) = 0 \tag{6.E6.1.15}$$

This equation can be transformed into the following polynomial:

$$2.735\times10^{-46}\omega^4 - 7.321\times10^{-21}\omega^2 + 41633.347 = 0 \tag{6.E6.1.16}$$

The following expression is used here:

$$\omega^2 = F \tag{6.E6.1.17}$$

So the polynomial of eigenvalue problem can be written:

$$2.735\times10^{-46}F^2 - 7.321\times10^{-21}F + 41633.347 = 0 \tag{6.E6.1.18}$$

The roots of this equation are as follows:

$$F_1 = 0.82 \times 10^{25}\,\text{s}^{-2} \quad \rightarrow \quad \omega_{1,1} = 2.863 \times 10^{12}\,\text{s}^{-1} = 2.863\ \text{THz},$$
$$\omega_{1,2} = -2.863 \times 10^{12}\,\text{s}^{-1} = -2.863\ \text{THz},$$

$$F_2 = 1.857 \times 10^{25}\,\text{s}^{-2} \quad \rightarrow \quad \omega_{2,1} = 4.309 \times 10^{12}\,\text{s}^{-1} = 4.309\ \text{THz},$$
$$\omega_{2,2} = -4.309 \times 10^{12}\,\text{s}^{-1} = -4.309\ \text{THz}.$$

While the smaller positive root of equation is the natural frequency of fundamental mode, the larger one is the natural frequency of second mode. Negative values are physically meaningless. Then, only the natural frequencies of first two modes are calculated with the using of three two-node finite elements in this nanorod. The accuracy of these values is discussed in the next example. According to this, the nondimensional forms of above natural frequencies are presented in Table 6.1 and a discussion on these values is explained.

In all applications after this example, since the polynomials of eigenvalue equation that will occur will be of sixth degree, the roots of the polynomials will be calculated and shared electronically.

Example 6.2 Determine the natural frequencies of nonlocal axial free vibration of clamped-clamped nanorod embedded on elastic foundation in Fig. 2.2. Use four finite elements in the analysis. Physical and mechanical parameters are as in Example 6.1. Firstly, ignore the elastic medium. Then, repeat the analysis for nondimensional stiffness of elastic medium: $K_M = 10$. Nondimensional stiffness of elastic medium is defined as $K_M = k_M L^2 / EA$.

Solution. In the solution of Example 5.1, reduced global stiffness and mass matrices are calculated with Eqs. (5.E5.1.12) and (5.E5.1.13) for classical elasticity. Additionally, in this application, only stiffness matrices due to elastic medium and nonlocal mass matrices are achieved in order to add to reduced global classical stiffness and mass matrices. Therefore, the finite element mesh and node numbers depicted in Fig. 5.7b, c, respectively, are valid.

First, NL-FEM analysis is performed only based on the nonlocal parameter. Like Eq. (6. E6.1.6), the global nonlocal mass matrix is given as follows:

Mesh node numbers: 1, 2, 3, 4, 5 (columns and rows)

$$\sum[M_{nl}] = \frac{(e_0 a)^2 \rho A}{l} \begin{bmatrix} 1 & -1 & 0 & 0 & 0 \\ -1 & 1+1 & -1 & 0 & 0 \\ 0 & -1 & 1+1 & -1 & 0 \\ 0 & 0 & -1 & 1+1 & -1 \\ 0 & 0 & 0 & -1 & 1 \end{bmatrix} \tag{6.E6.2.1}$$

The reduction of global nonlocal mass matrix due to boundary conditions should be written as follows:

Table 6.1 Comparisons of nondimensional frequencies of clamped nanorods for nonlocal axial free vibration ($e_0a/L = 0.2$, $K_M = 0$)

Mode number	Analytical	NL-FEM Results						
	Eq. (2.E2.1.7)	Manual solution		Program code (MATLAB)				
		$n = 3$	$n = 4$	$n = 3$	$n = 4$	$n = 10$	$n = 15$	$n = 20$
1	2.66009	2.746	2.708	2.74625	2.70887	2.66793	2.66358	2.66205
2	3.91239	4.133	4.054	4.13384	4.05442	3.93708	3.92343	3.91862
3	4.41692	–	4.569	–	4.56963	4.45158	4.43264	4.42582
4	4.64576	–	–	–	–	4.68450	4.66376	4.65603
5	4.76445	–	–	–	–	4.80384	4.78345	4.77540

$$\sum[M_{nl}]=\begin{bmatrix} -5.778\times10^{-24} & -5.778\times10^{-24} & 0 & 0 & 0 \\ -5.778\times10^{-24} & 1.156\times10^{-23} & -5.778\times10^{-24} & 0 & 0 \\ 0 & -5.778\times10^{-24} & 1.156\times10^{-23} & -5.778\times10^{-24} & 0 \\ 0 & 0 & -5.778\times10^{-24} & 1.156\times10^{-23} & -5.778\times10^{-24} \\ 0 & 0 & 0 & -5.778\times10^{-24} & -5.778\times10^{-23} \end{bmatrix}\text{kg}$$

Eliminated (first and last rows and columns); the boxed inner block is $\left(\sum[M_{nl}]\right)^*$

(6.E6.2.2)

By using Eqs. (6.E6.2.2) and (5.E5.1.13), the reduced total global mass matrix can be computed as follows:

$$\left(\sum[M]\right)^*=\left(\sum[M_c]\right)^*+\left(\sum[M_{nl}]\right)^*=$$
$$\begin{bmatrix} 6.02\times10^{-24} & 1.505\times10^{-24} & 0 \\ 1.505\times10^{-24} & 6.02\times10^{-24} & 1.505\times10^{-24} \\ 0 & 1.505\times10^{-24} & 6.02\times10^{-24} \end{bmatrix}+\begin{bmatrix} 1.156\times10^{-23} & -5.778\times10^{-24} & 0 \\ -5.778\times10^{-24} & 1.156\times10^{-23} & -5.778\times10^{-24} \\ 0 & -5.778\times10^{-24} & 1.156\times10^{-23} \end{bmatrix}=$$
$$\begin{bmatrix} 1.758\times10^{-23} & -4.273\times10^{-24} & 0 \\ -4.273\times10^{-24} & 1.785\times10^{-23} & -4.273\times10^{-24} \\ 0 & -4.273\times10^{-24} & 1.758\times10^{-23} \end{bmatrix}\text{kg}$$

(6.E6.2.3)

Since the stiffness matrix is not affected by the nonlocal parameter, the eigenvalue problem of nonlocal free vibration is meant as follows through Eqs. (5.E5.1.1), (5.E5.1.12), and (5.E5.1.14):

$$\det\left(\begin{bmatrix} 314 & -157 & 0 \\ -157 & 314 & -157 \\ 0 & -157 & 314 \end{bmatrix}-\omega_i^2\times\begin{bmatrix} 1.758\times10^{-23} & -4.273\times10^{-24} & 0 \\ -4.273\times10^{-24} & 1.758\times10^{-23} & -4.273\times10^{-24} \\ 0 & -4.273\times10^{-24} & 1.758\times10^{-23} \end{bmatrix}\right)=0$$

(6.E6.2.4)

The positive roots of this eigenvalue problem are calculated as follows:

$$\omega_1=2.823\ \text{THz},\quad \omega_2=4.226\ \text{THz},\quad \omega_3=4.764\ \text{THz}.$$

Note that the natural frequency values of first two modes from the above values are lower than those obtained in Example 6.1. On the other hand, since the degree of polynomial resulting from Eq. (6.E6.2.4) is high, it is not easy to calculate the roots manually.

Therefore, the roots of such polynomials should be calculated electronically. MATLAB codes are developed for such solution necessity in this book. According to this, the analytical nonlocal nondimensional frequencies calculated from Eq. (2.E2.1.7) and Eq. (2.E2.7.16) are compared in Table 6.1.

The results in Table 6.1 show that finite element analysis is efficient for nonlocal vibration of axial nanorods. Also, for the finite element analysis of classical vibration (Chap. 5, Table 5.1), it is concluded that the difference between the manual solution and the program code solution decreases in the analysis with nonlocal elasticity. Of course, NL-FEM should be coded on the computer to reduce the difference from the manual solution. Generally, to reduce the difference between finite element analysis and analytical results depends on the increment of number of finite elements. Increment of number of finite elements is already necessary for the convergence of frequencies, especially in higher modes. However, if Table 5.1 and Table 6.1 are examined together, the finite element analysis presents more convergent values to the analytical results in nonlocal elasticity. This conclusion definitely indicates that NL-FEM is of great importance for the mechanical analysis of atomic structures.

Now, the axial elastic medium is also considered in addition to the nonlocal parameter. The stiffness of elastic medium is calculated as follows:

$$\begin{aligned} k_{M,e=1} &= k_{M,e=2} = k_{M,e=3} = k_{M,e=4} = k_M = \frac{K_M EA}{L^2} \\ &= \frac{(10) \times \left(1 \times 10^{12} \, \frac{\mathrm{N}}{\mathrm{m}^2}\right) \times \left(0.785 \times 10^{-18} \mathrm{m}^2\right)}{\left(20 \times 10^{-9} \mathrm{m}\right)^2} = 1.963 \times 10^{10} \, \frac{\mathrm{N}}{\mathrm{m}^2} \end{aligned} \tag{6.E6.2.5}$$

Equations (6.11) and (6.12) should be written for finite elements as follows:

$$\left(K_{M,c}\right)_e = \frac{k_{M,e} l_e}{6} \begin{bmatrix} 2 & 1 \\ 1 & 2 \end{bmatrix} \quad \begin{array}{l} \text{FE node numbers: } 1, 2 \text{ (columns)}; \; 1, 2 \text{ (rows)} \end{array} \tag{6.E6.2.6}$$

$$\left(K_{M,nl}\right)_e = \frac{(e_0 a)_e^2 \, k_{M,e}}{l_e} \begin{bmatrix} 1 & -1 \\ -1 & 1 \end{bmatrix} \quad \begin{array}{l} \text{FE node numbers: } 1, 2 \text{ (columns)}; \; 1, 2 \text{ (rows)} \end{array} \tag{6.E6.2.7}$$

The classical and nonlocal stiffness matrices of axial elastic medium of finite elements are expressed as follows:

$$(K_{M,c})_{e=1} = (K_{M,c})_{e=2} = (K_{M,c})_{e=3} = (K_{M,c})_{e=4} = \frac{k_M l}{6}\begin{bmatrix} 2 & 1 \\ 1 & 2 \end{bmatrix} \tag{6.E6.2.8}$$

$$(K_{M,nl})_{e=1} = (K_{M,nl})_{e=2} = (K_{M,nl})_{e=3} = (K_{M,nl})_{e=4} = \frac{(e_0 a)^2 k_M}{l}\begin{bmatrix} 1 & -1 \\ -1 & 1 \end{bmatrix} \tag{6.E6.2.9}$$

According to Eqs. (6.E6.2.6)-(6.E6.2.9), global classical and nonlocal stiffness matrices are obtained as follows for the axial elastic environment:

$$\sum[K_{M,c}] = \frac{k_M l}{6}\begin{array}{c} \begin{array}{ccccc} 1 & 2 & 3 & 4 & 5 \end{array} \\ \left[\begin{array}{c|c|c|c|c} 2 & 1 & 0 & 0 & 0 \\ \hline 1 & 2+2 & 1 & 0 & 0 \\ \hline 0 & 1 & 2+2 & 1 & 0 \\ \hline 0 & 0 & 1 & 2+2 & 1 \\ \hline 0 & 0 & 0 & 1 & 2 \end{array}\right] \end{array} \begin{array}{c} \textit{Mesh node numbers} \\ 1 \\ 2 \\ 3 \\ 4 \\ 5 \end{array} \tag{6.E6.2.10}$$

$$\sum[K_{M,nl}] = \frac{(e_0 a)^2 k_M}{l}\begin{array}{c} \begin{array}{ccccc} 1 & 2 & 3 & 4 & 5 \end{array} \\ \left[\begin{array}{c|c|c|c|c} 1 & -1 & 0 & 0 & 0 \\ \hline -1 & 1+1 & -1 & 0 & 0 \\ \hline 0 & -1 & 1+1 & -1 & 0 \\ \hline 0 & 0 & -1 & 1+1 & -1 \\ \hline 0 & 0 & 0 & -1 & 1 \end{array}\right] \end{array} \begin{array}{c} \textit{Mesh node numbers} \\ 1 \\ 2 \\ 3 \\ 4 \\ 5 \end{array} \tag{6.E6.2.11}$$

The matrices calculated above are reduced as follows by the boundary conditions:

$$\sum[K_{M,c}] = \begin{bmatrix} \cancel{32.717} & \cancel{16.358} & \cancel{0} & \cancel{0} & \cancel{0} \\ 16.358 & 65.433 & 16.358 & 0 & 0 \\ 0 & 16.358 & 65.433 & 16.358 & 0 \\ 0 & 0 & 16.358 & 65.433 & 16.358 \\ \cancel{0} & \cancel{0} & \cancel{0} & \cancel{16.358} & \cancel{32.717} \end{bmatrix} \frac{\text{N}}{\text{m}} \tag{6.E6.2.12}$$

Eliminated (first column), *Eliminated* (last column), *Eliminated* (first row), *Eliminated* (last row); the remaining inner 3×3 block is $\left(\sum[K_{M,c}]\right)^*$.

$$\sum\left[K_{M,nl}\right]=\begin{bmatrix} 62.816 & -62.816 & 0 & 0 & 0 \\ -62.816 & 125.632 & -62.816 & 0 & 0 \\ 0 & -62.816 & 125.632 & -62.816 & 0 \\ 0 & 0 & -62.816 & 125.632 & -62.816 \\ 0 & 0 & 0 & -62.816 & 62.816 \end{bmatrix}\frac{\text{N}}{\text{m}} \quad (6.E6.2.13)$$

(Eliminated: first and last rows and columns; the remaining central block is $\left(\sum\left[K_{M,c}\right]\right)^{*}$)

By employing Eqs. (5.E5.1.11), (6.E6.2.12), and (6.E6.2.13), the reduced total global stiffness matrix is achieved as follows:

$$\left(\sum[K]\right)^{*}=\left(\sum[K_{ae}]\right)^{*}+\left(\sum[K_{M,c}]\right)^{*}+\left(\sum[K_{M,nl}]\right)^{*}=$$

$$\begin{bmatrix} 314 & -157 & 0 \\ -157 & 314 & -157 \\ 0 & -157 & 314 \end{bmatrix}+\begin{bmatrix} 65.433 & 16.358 & 0 \\ 16.358 & 65.433 & 16.358 \\ 0 & 16.358 & 65.433 \end{bmatrix}+$$

$$\begin{bmatrix} 125.632 & -62.816 & 0 \\ -62.816 & 125.632 & -62.816 \\ 0 & -62.816 & 125.632 \end{bmatrix}=\begin{bmatrix} 505.065 & -203.458 & 0 \\ -203.458 & 505.065 & -203.458 \\ 0 & -203.458 & 505.065 \end{bmatrix}\frac{\text{N}}{\text{m}} \quad (6.E6.2.14)$$

The eigenvalue equation is formulated with the help of Eqs. (6.E6.2.3) and (6.E6.2.14):

$$\det\left(\begin{bmatrix} 505.065 & -203.458 & 0 \\ -203.458 & 505.065 & -203.458 \\ 0 & -203.458 & 505.065 \end{bmatrix}-\omega_i^2\times\begin{bmatrix} 1.785\times10^{-23} & -4.273\times10^{-24} & 0 \\ -4.273\times10^{-24} & 1.785\times10^{-23} & -4.273\times10^{-24} \\ 0 & -4.273\times10^{-24} & 1.785\times10^{-23} \end{bmatrix}\right)=0 \quad (6.E6.2.15)$$

the roots of this equation are given below:

$$\omega_1=4.340\ \text{THz},\ \omega_2=5.360\ \text{THz},\ \omega_3=5.793\ \text{THz}.$$

Table 6.2 Comparisons of nondimensional frequencies of embedded clamped nanorods for nonlocal axial free vibration ($e_0a/L = 0.2$, $K_M = 10$)

Mode number	Analytical	NL-FEM results				
	Eq. (2.E2.1.7)	Manual solution	Program code (MATLAB)			
		$n = 4$	$n = 4$	$n = 10$	$n = 15$	$n = 20$
1	4.13232	4.163	4.16389	4.13737	4.13457	4.13358
2	5.03059	5.141	5.14182	5.04981	5.03918	5.03543
3	5.43224	5.557	5.55712	5.46045	5.44503	5.43948
4	5.61988	–	–	5.65195	5.63477	5.62838
5	5.71839	–	–	5.75125	5.73423	5.72752

It should be noted that in the nonlocal free vibration, the elastic medium increases the frequencies compared to the results computed for the case without elastic medium. A comparison study on the nonlocal nondimensional frequencies of nanorods embedded in elastic medium is presented in Table 6.2.

Example 6.3 Determine the natural frequencies of nonlocal axial free vibration of free-spring attached nanorod embedded on elastic foundation in Fig. 2.5. Use two finite elements in the analysis. Firstly, consider nonlocal elasticity and ignore the elastic medium. Then, repeat the analysis for elastic medium. Physical and mechanical parameters are as in Examples 6.1 and 6.2. In addition to this, attachment ratio is $\beta_k = 5$ and defined as $\beta_k = k_aL/EA$.

Solution. If the finite element mesh given in Fig. 5.7b is considered, three nodes occur in this mesh of nanorod divided into two finite elements. Since the third node is the right end of nanorod, the attachment stiffness should be included in the elastic stiffness matrix of second finite element containing this node. Therefore, the stiffness matrices of related finite elements are calculated as follows:

$$(K)_1 = \frac{EA}{l}\begin{bmatrix} 1 & -1 \\ -1 & 1 \end{bmatrix}, \quad (K)_2 = \frac{EA}{l}\begin{bmatrix} 1 & -1 \\ -1 & 1 \end{bmatrix} + k_a\begin{bmatrix} 0 & 0 \\ 0 & 1 \end{bmatrix} \tag{6.E6.3.1}$$

where the lengths of finite elements and the stiffness of attachment can be written as follows:

$$l_1 = l_2 = l = \frac{L}{2} = \frac{20}{2} = 10 \text{ nm} = 10^{-8}\text{m} \tag{6.E6.3.2}$$

$$k_a = \frac{\beta_k EA}{L} = \frac{(5) \times \left(1 \times 10^{12}\,\frac{\text{N}}{\text{m}^2}\right) \times \left(0.785 \times 10^{-18}\text{m}^2\right)}{\left(20 \times 10^{-9}\text{m}\right)} = 196.25\ \frac{\text{N}}{\text{m}} \tag{6.E6.3.3}$$

The global total stiffness matrix of axial nanorod should be obtained. Since there is no geometric boundary condition, a reduction is not necessary in this matrix and the global matrix is computed as:

$$\sum[K]=\frac{EA}{l}\begin{bmatrix}1&-1&0\\-1&2&-1\\0&-1&1\end{bmatrix}+k_a\begin{bmatrix}0&0&0\\0&0&0\\0&0&1\end{bmatrix}=\begin{bmatrix}78.5&-78.5&0\\-78.5&157&-78.5\\0&-78.5&274.75\end{bmatrix}\frac{\text{N}}{\text{m}} \tag{6.E6.3.4}$$

On the other hand, the total global mass matrix is quickly given as follows:

$$\begin{aligned}\sum[M]&=\frac{\rho Al}{6}\begin{bmatrix}2&1&0\\1&4&1\\0&1&2\end{bmatrix}+\frac{(e_0a)^2\rho A}{l}\begin{bmatrix}1&-1&0\\-1&2&-1\\0&-1&1\end{bmatrix}\\&=\begin{bmatrix}8.907\times10^{-24}&1.204\times10^{-25}&0\\1.204\times10^{-25}&1.781\times10^{-23}&1.204\times10^{-25}\\0&1.204\times10^{-25}&8.907\times10^{-24}\end{bmatrix}\text{kg}\end{aligned} \tag{6.E6.3.5}$$

By means of Eqs. (6.E6.3.4) and (6.E6.3.5), the eigenvalue equation is presented as follows:

$$\det\left(\begin{bmatrix}78.5&-78.5&0\\-78.5&157&-78.5\\0&-78.5&274.75\end{bmatrix}-\omega_i^2\times\begin{bmatrix}8.907\times10^{-24}&1.204\times10^{-25}&0\\1.204\times10^{-25}&1.781\times10^{-23}&1.204\times10^{-25}\\0&1.204\times10^{-25}&8.907\times10^{-24}\end{bmatrix}\right)=0 \tag{6.E6.3.6}$$

The positive roots of Eq. (6.E6.3.3) are as follows:

$$\omega_1=1.362\ \text{THz},\quad \omega_2=3.753\ \text{THz},\quad \omega_3=5.725\ \text{THz}.$$

The nonlocal nondimensional frequencies via Eqs. (2.E2.4.4) and (2.E2.7.16) are compared with NL-FEM results in Table 6.3. The frequency equation of this nanorod is given by Eq. (2.E2.4.4). Since the elastic environment is neglected in this analysis, the definitions explained in Eq. (2.E2.3.4) and (2.E2.3.5) are valid. When the elastic medium is neglected, nonlocal frequencies can be attained within some limits by coding the frequency equation in MATLAB. Computer programming may not provide frequency values under very high values of the nonlocal parameter. Additionally, frequencies of very high modes may not be computed. If elastic ground is considered in analyses carried out with codes,

Table 6.3 Comparisons of nondimensional frequencies of free-spring attached nanorods ($\beta_k = 5$, $e_0a/L = 0.2$, $K_M = 0$)

Mode number	Analytical (MATLAB)	NL-FEM results				
	Eq. (2.E2.4.4)	Manual solution	Program code (MATLAB)			
		$n = 2$	$n = 2$	$n = 5$	$n = 10$	$n = 15$
1	1.28319	1.306	1.30596	1.28679	1.28409	1.28359
2	3.23998	3.600	3.59966	3.30117	3.25525	3.24676
3	4.14645	5.491	5.49140	4.26604	4.17791	4.16053
4	4.52432	–	–	4.65064	4.56328	4.54206
5	4.70204	–	–	4.78861	4.74315	4.72132

obtaining frequency becomes completely impossible. In such cases, the use of NL-FEM provides a great advantage. This will be discussed when appropriate.

The analyses within the scope of this application are repeated for the elastic environment. Accordingly, the stiffnesses of elastic medium of finite elements can be formulated as:

$$(K_{M,c})_1 = (K_{M,c})_2 = \frac{k_M l}{6}\begin{bmatrix} 2 & 1 \\ 1 & 2 \end{bmatrix} \tag{6.E6.3.7}$$

$$(K_{M,nl})_1 = (K_{M,nl})_2 = \frac{(e_0a)^2 k_M}{l}\begin{bmatrix} 1 & -1 \\ -1 & 1 \end{bmatrix} \tag{6.E6.3.8}$$

Using the elastic medium stiffnesses, global matrices are obtained according to the nodes in the finite element mesh as follows:

$$\sum[K_{M,c}] = \frac{k_M l}{6}\begin{bmatrix} 2 & 1 & 0 \\ 1 & 4 & 1 \\ 0 & 1 & 2 \end{bmatrix} = \begin{bmatrix} 65.433 & 32.717 & 0 \\ 32.717 & 130.867 & 32.717 \\ 0 & 32.717 & 65.433 \end{bmatrix}\frac{\text{N}}{\text{m}} \tag{6.E6.3.9}$$

$$\sum[K_{M,nl}] = \frac{(e_0a)^2 k_M}{l}\begin{bmatrix} 1 & -1 & 0 \\ -1 & 2 & -1 \\ 0 & -1 & 1 \end{bmatrix} = \begin{bmatrix} 31.408 & -31.408 & 0 \\ -31.408 & 62.816 & -31.408 \\ 0 & -31.408 & 31.408 \end{bmatrix}\frac{\text{N}}{\text{m}} \tag{6.E6.3.10}$$

If Eqs. (6.E6.3.4), (6.E6.3.9), and (6.E6.3.10) are added, the global stiffness matrix based on the eigenvalue solution is obtained:

Table 6.4 Nonlocal finite element results for nondimensional frequencies of embedded free-spring attached nanorods ($\beta_k = 5$, $e_0a/L = 0.2$, $K_M = 10$)

	Manual solution	Program code (MATLAB)			
Mode number	$n = 2$	$n = 2$	$n = 5$	$n = 10$	$n = 15$
1	3.422	3.42133	3.41406	3.41305	3.41286
2	4.792	4.79141	4.57141	4.53836	4.53227
3	6.337	6.33683	5.31028	5.23975	5.22590
4	–	–	5.62392	5.55189	5.53447
5	–	–	5.73853	5.70066	5.68250

$$\left(\sum[K]\right)^* = \left(\sum[K_{ae}]\right)^* + \left(\sum[K_a]\right)^* + \left(\sum[K_{M,c}]\right)^* + \left(\sum[K_{M,nl}]\right)^*$$
$$= \begin{bmatrix} 175.341 & -77.191 & 0 \\ -77.191 & 350.683 & -77.191 \\ 0 & -77.191 & 175.341 \end{bmatrix} \frac{\text{N}}{\text{m}} \tag{6.E6.3.11}$$

The roots of eigenvalue problem formulated through Eqs. (6.E6.3.5) and (6.E6.3.11) are explained:

$$\omega_1 = 3.568 \text{ THz}, \quad \omega_2 = 4.996 \text{ THz}, \quad \omega_3 = 6.607 \text{ THz}.$$

NL-FEM results about the analysis in question are presented in Table 6.4. It should be noted that if the elastic medium is included in the mechanical analysis in simple nanorods with attachments, computing the frequency through the analytical solution becomes very difficult, despite computer programming. High values of elastic medium or atomic parameter make it impossible to calculate the frequency even in fundamental mode. All the difficulties caused by this approach that is not appropriate for manual solution have been overcome by means of NL-FEM.

6.3.2 Love-Bishop Nanorod

6.3.2.1 NL-FEM Expression

In the analysis of axial nanorods with shear (or Poisson) effect, the NL-FEM procedure is some more crowded due to the higher-order terms. Firstly, the average weighted residual is given as follows [7]:

$$
\begin{aligned}
\mathrm{I}= & \int_0^{l_e} h\left(EA\frac{\partial^2 u}{\partial x^2}+(e_0a)^2k_M\frac{\partial^2 u}{\partial x^2}-\rho A\frac{\partial^2 u}{\partial t^2}-k_M u+q+(e_0a)^2(1+2\upsilon)\rho A\frac{\partial^4 u}{\partial x^2\partial t^2}\right.\\
& \left.+\rho I_P\upsilon^2\frac{\partial^4 u}{\partial x^2\partial t^2}-GI_P\upsilon^2\frac{\partial^4 u}{\partial x^4}-(e_0a)^2\rho I_P\upsilon^2\frac{\partial^6 u}{\partial x^4\partial t^2}-(e_0a)^2\frac{\partial^2 q}{\partial x^2}\right)\mathrm{d}x
\end{aligned}
\tag{6.23}
$$

where ten different integrals can be calculated as follows by partial integration [7]:

$$
\begin{aligned}
\mathrm{I}_1 &= \int_0^{l_e} EAh\frac{\partial^2 u}{\partial x^2}\mathrm{d}x = EAh\frac{\partial u}{\partial x}\Big|_0^{l_e}-\int_0^{l_e} EA\frac{\partial h}{\partial x}\frac{\partial u}{\partial x}\mathrm{d}x,\\
\mathrm{I}_2 &= \int_0^{l_e} (e_0a)^2k_M h\frac{\partial^2 u}{\partial x^2}\mathrm{d}x = (e_0a)^2k_M h\frac{\partial u}{\partial x}\Big|_0^{l_e}-\int_0^{l_e} (e_0a)^2k_M\frac{\partial h}{\partial x}\frac{\partial u}{\partial x}\mathrm{d}x,\ \ \mathrm{I}_3=\int_0^{l_e}\rho Ah\frac{\partial^2 u}{\partial t^2}\mathrm{d}x,\\
\mathrm{I}_4 &= \int_0^{l_e} k_M hu\,\mathrm{d}x,\ \ \mathrm{I}_5=\int_0^{l_e} hq\,\mathrm{d}x,\\
\mathrm{I}_6 &= \int_0^{l_e} (e_0a)^2\rho A(1+2\upsilon)h\frac{\partial^4 u}{\partial x^2\partial t^2}\mathrm{d}x = (e_0a)^2\rho A(1+2\upsilon)h\frac{\partial^3 u}{\partial x\partial t^2}\Big|_0^{l_e}\\
&\quad -\int_0^{l_e} (e_0a)^2\rho A(1+2\upsilon)\frac{\partial h}{\partial x}\frac{\partial^3 u}{\partial x\partial t^2}\mathrm{d}x,\\
\mathrm{I}_7 &= \int_0^{l_e} \rho I_P\upsilon^2 h\frac{\partial^4 u}{\partial x^2\partial t^2}\mathrm{d}x = \rho I_P\upsilon^2 h\frac{\partial^3 u}{\partial x\partial t^2}\Big|_0^{l_e}-\int_0^{l_e}\rho I_P\upsilon^2\frac{\partial h}{\partial x}\frac{\partial^3 u}{\partial x\partial t^2}\mathrm{d}x,\\
\mathrm{I}_8 &= \int_0^{l_e} GI_P\upsilon^2 h\frac{\partial^4 u}{\partial x^4}\mathrm{d}x = GI_P\upsilon^2 h\frac{\partial^3 u}{\partial x^3}\Big|_0^{l_e}-\int_0^{l_e} GI_P\upsilon^2\frac{\partial h}{\partial x}\frac{\partial^3 u}{\partial x^3}\mathrm{d}x,\\
\mathrm{I}_9 &= \int_0^{l_e} (e_0a)^2\rho I_P\upsilon^2 h\frac{\partial^6 u}{\partial x^4\partial t^2}\mathrm{d}x = (e_0a)^2\rho I_P\upsilon^2 h\frac{\partial^6 u}{\partial x^4\partial t^2}\Big|_0^{l_e}-\int_0^{l_e}(e_0a)^2\rho I_P\upsilon^2\frac{\partial h}{\partial x}\frac{\partial^5 u}{\partial x^3\partial t^2}\mathrm{d}x,\\
\mathrm{I}_{10} &= \int_0^{l_e} (e_0a)^2h\frac{\partial^2 q}{\partial x^2}\mathrm{d}x = (e_0a)^2h\frac{\partial q}{\partial x}\Big|_0^{l_e}-\int_0^{l_e}(e_0a)^2\frac{\partial h}{\partial x}\frac{\partial q}{\partial x}\mathrm{d}x.
\end{aligned}
\tag{6.24}
$$

Since the expression defining the motion of finite element is a first-order polynomial in the simple rod, a high number of differentiations according to position are not possible. Since partial integration operations involve equal-order derivatives of weight function and displacement according to position, one partial integration is sufficient. However, Eq. (6.24) includes the third derivative of the shape function I_8 and I_9 with respect to position. This problem can be overcome with the help of two point. The first point specifies that the partial integration process is applied to I_8 and I_9 integrals [7]:

$$I_8 = GI_P\upsilon^2 h\frac{\partial^3 u}{\partial x^3}\bigg|_0^{l_e} - GI_P\upsilon^2\frac{\partial h}{\partial x}\frac{\partial^2 u}{\partial x^2}\bigg|_0^{l_e} + \int_0^{l_e} GI_P\upsilon^2\frac{\partial^2 h}{\partial x^2}\frac{\partial^2 u}{\partial x^2}\mathrm{d}x,$$

$$I_9 = (e_0a)^2\rho I_P\upsilon^2 h\frac{\partial^6 u}{\partial x^4\partial t^2}\bigg|_0^{l_e} - (e_0a)^2\rho I_P\upsilon^2\frac{\partial h}{\partial x}\frac{\partial^5 u}{\partial x^3\partial t^2}\bigg|_0^{l_e} + \int_0^{l_e}(e_0a)^2\rho I_P\upsilon^2\frac{\partial^2 h}{\partial x^2}\frac{\partial^4 u}{\partial x^2\partial t^2}\mathrm{d}x \tag{6.25}$$

where note that the equal-order derivative terms according to position are obtained in the new integration results. However, a second issue should be known at this point. The second-order derivative according to the position prevents obtaining finite element matrices due to the degree of polynomial given by Eq. (5.1). If three-node, in other words, second-order polynomial shape functions are used, this problem is solved. According to this, the following expressions can be written by considering Eq. (5.18):

$$\phi = \begin{Bmatrix}\phi_i\\ \phi_j\\ \phi_k\end{Bmatrix}^{\mathrm{T}} = \begin{Bmatrix}2\xi^2 - 3\xi + 1\\ -4\xi^2 + 4\xi\\ 2\xi^2 - \xi\end{Bmatrix}^{\mathrm{T}}, \quad \mathbf{d} = \begin{Bmatrix}d_i\\ d_j\\ d_k\end{Bmatrix} \tag{6.26}$$

The three-node linear axial finite element is depicted in Fig. 5.5. Now, let's substitute the partial integration results into Eq. (6.23):

$$\begin{aligned}
&EAh\frac{\partial u}{\partial x}\bigg|_0^{l_e} + (e_0a)^2 k_M h\frac{\partial u}{\partial x}\bigg|_0^{l_e} + (e_0a)^2\rho A(1+2\upsilon)h\frac{\partial^3 u}{\partial x\partial t^2}\bigg|_0^{l_e} + \rho I_P\upsilon^2 h\frac{\partial^3 u}{\partial x\partial t^2}\bigg|_0^{l_e}\\
&- GI_P\upsilon^2 h\frac{\partial^3 u}{\partial x^3}\bigg|_0^{l_e} + GI_P\upsilon^2\frac{\partial h}{\partial x}\frac{\partial^2 u}{\partial x^2}\bigg|_0^{l_e} - (e_0a)^2\rho I_P\upsilon^2 h\frac{\partial^6 u}{\partial x^4\partial t^2}\bigg|_0^{l_e} + (e_0a)^2\rho I_P\upsilon^2\frac{\partial h}{\partial x}\frac{\partial^5 u}{\partial x^3\partial t^2}\bigg|_0^{l_e}\\
&- (e_0a)^2 h\frac{\partial q}{\partial x}\bigg|_0^{l_e} - \int_0^{l_e} EA\frac{\partial h}{\partial x}\frac{\partial u}{\partial x}\mathrm{d}x - \int_0^{l_e}(e_0a)^2 k_M\frac{\partial h}{\partial x}\frac{\partial u}{\partial x}\mathrm{d}x - \int_0^{l_e}\rho Ah\frac{\partial^2 u}{\partial t^2}\mathrm{d}x - \int_0^{l_e} k_M hu\mathrm{d}x\\
&+ \int_0^{l_e} hq\mathrm{d}x - \int_0^{l_e}(e_0a)^2\rho A(1+2\upsilon)\frac{\partial h}{\partial x}\frac{\partial^3 u}{\partial x\partial x\partial t^2}\mathrm{d}x - \int_0^{l_e}\rho I_P\upsilon^2\frac{\partial h}{\partial x}\frac{\partial^3 u}{\partial x\partial x\partial t^2}\mathrm{d}x - \int_0^{l_e} GI_P\upsilon^2\frac{\partial^2 h}{\partial x^2}\frac{\partial^2 u}{\partial x^2}\mathrm{d}x\\
&- \int_0^{l_e}(e_0a)^2\rho I_P\upsilon^2\frac{\partial^2 h}{\partial x^2}\frac{\partial^4 u}{\partial x^2\partial x^2\partial t^2}\mathrm{d}x + \int_0^{l_e}(e_0a)^2\frac{\partial h}{\partial x}\frac{\partial q}{\partial x}\mathrm{d}x = 0
\end{aligned} \tag{6.27}$$

where the integral part is taken as follows [7]:

$$\int_0^{l_e} EA\frac{\partial h}{\partial x}\frac{\partial u}{\partial x}\mathrm{d}x + \int_0^{l_e} (e_0a)^2 k_M \frac{\partial h}{\partial x}\frac{\partial u}{\partial x}\mathrm{d}x + \int_0^{l_e} \rho A h\frac{\partial^2 u}{\partial t^2}\mathrm{d}x + \int_0^{l_e} k_M h u\mathrm{d}x - \int_0^{l_e} hq\mathrm{d}x$$
$$+ \int_0^{l_e} (e_0a)^2\rho A(1+2\upsilon)\frac{\partial h}{\partial x}\frac{\partial^3 u}{\partial x\partial t^2}\mathrm{d}x + \int_0^{l_e} \rho I_P\upsilon^2\frac{\partial h}{\partial x}\frac{\partial^3 u}{\partial x\partial t^2}\mathrm{d}x + \int_0^{l_e} GI_P\upsilon^2\frac{\partial^2 h}{\partial x^2}\frac{\partial^2 u}{\partial x^2}\mathrm{d}x$$
$$+ \int_0^{l_e} (e_0a)^2\rho I_P\upsilon^2\frac{\partial^2 h}{\partial x^2}\frac{\partial^4 u}{\partial x^2\partial t^2}\mathrm{d}x - \int_0^{l_e} (e_0a)^2\frac{\partial h}{\partial x}\frac{\partial q}{\partial x}\mathrm{d}x = 0 \tag{6.28}$$

In addition to the expressions in Eq. (6.7), the following equations can be written:

$$\frac{\partial^2 h}{\partial x^2} = \mathbf{B}'^{\mathrm{T}}, \quad \frac{\partial^2 u}{\partial x^2} = \mathbf{B}'\mathbf{d} \tag{6.29}$$

Substituting Eqs. (6.7) and (6.29) into Eq. (6.28) yields the following equation [7]:

$$\int_0^{l_e} EA(\mathbf{B}^{\mathrm{T}}\mathbf{B})\mathbf{d}\mathrm{d}x + \int_0^{l_e} (e_0a)^2 k_M(\mathbf{B}^{\mathrm{T}}\mathbf{B})\mathbf{d}\mathrm{d}x + \int_0^{l_e} \rho A(\phi^{\mathrm{T}}\phi)\ddot{\mathbf{d}}\mathrm{d}x + \int_0^{l_e} k_M(\phi^{\mathrm{T}}\phi)\mathbf{d}\mathrm{d}x$$
$$- \int_0^{l_e} \phi^{\mathrm{T}} q\mathrm{d}x + \int_0^{l_e} (e_0a)^2\rho A(1+2\upsilon)(\mathbf{B}^{\mathrm{T}}\mathbf{B})\ddot{\mathbf{d}}\mathrm{d}x + \int_0^{l_e} \rho I_P\upsilon^2(\mathbf{B}^{\mathrm{T}}\mathbf{B})\ddot{\mathbf{d}}\mathrm{d}x + \int_0^{l_e} GI_P\upsilon^2\left(\mathbf{B}'^{\mathrm{T}}\mathbf{B}'\right)\mathbf{d}\mathrm{d}x$$
$$+ \int_0^{l_e} (e_0a)^2\rho I_P\upsilon^2\left(\mathbf{B}'^{\mathrm{T}}\mathbf{B}'\right)\ddot{\mathbf{d}}\mathrm{d}x - \int_0^{l_e} (e_0a)^2\mathbf{B}^{\mathrm{T}} q'\mathrm{d}x = 0 \tag{6.30}$$

This equation can be transformed into the following definition [7]:

$$(K + K_I + K_{M,c} + K_{M,nl})\mathbf{d} + (M_c + M_{I,c} + M_{nl} + M_{I,nl})\ddot{\mathbf{d}} = \mathbf{f}_c + \mathbf{f}_{nl} \tag{6.31}$$

where K_I is the lateral inertia stiffness matrix, $M_{I,\,c}$ is the classical mass inertia matrix and $M_{I,\,nl}$ is the nonlocal mass inertia matrix. These definitions are denoted as follows [7]:

$$K=\int_0^{l_e} EA\left(\mathbf{B}^{\mathrm{T}}\mathbf{B}\right)\mathrm{d}x=\int_0^{l_e} EA\begin{Bmatrix}\phi_1'\\ \phi_2'\\ \phi_3'\end{Bmatrix}[\phi_1' \quad \phi_2' \quad \phi_3']\mathrm{d}x=\frac{EA}{3l_e}\begin{bmatrix}7 & -8 & 1\\ -8 & 16 & -8\\ 1 & -8 & 7\end{bmatrix} \tag{6.32}$$

$$\begin{aligned} K_I &= \int_0^{l_e} GI_P\upsilon^2\left(\mathbf{B}'^{\mathrm{T}}\mathbf{B}'\right)\mathrm{d}x=\int_0^{l_e} GI_P\upsilon^2\begin{Bmatrix}\phi_1''\\ \phi_2''\\ \phi_3''\end{Bmatrix}[\phi_1'' \quad \phi_2'' \quad \phi_3'']\mathrm{d}x\\ &=\frac{GI_P\upsilon^2}{{l_e}^3}\begin{bmatrix}16 & -32 & 16\\ -32 & 64 & -32\\ 16 & -32 & 16\end{bmatrix}\end{aligned} \tag{6.33}$$

$$K_{M,c}=\int_0^{l_e} k_M\left(\phi^{\mathrm{T}}\phi\right)\mathrm{d}x=\int_0^{l_e} k_M\begin{Bmatrix}\phi_1\\ \phi_2\\ \phi_3\end{Bmatrix}[\phi_1 \quad \phi_2 \quad \phi_3]\mathrm{d}x=\frac{k_M l_e}{30}\begin{bmatrix}4 & 2 & -1\\ 2 & 16 & 2\\ -1 & 2 & 4\end{bmatrix} \tag{6.34}$$

$$\begin{aligned} K_{M,nl} &= \int_0^{l_e} (e_0a)^2 k_M\left(\mathbf{B}^{\mathrm{T}}\mathbf{B}\right)\mathrm{d}x=\int_0^{l_e} (e_0a)^2 k_M\begin{Bmatrix}\phi_1'\\ \phi_2'\\ \phi_3'\end{Bmatrix}[\phi_1' \quad \phi_2' \quad \phi_3']\mathrm{d}x\\ &=\frac{(e_0a)^2 k_M}{3l_e}\begin{bmatrix}7 & -8 & 1\\ -8 & 16 & -8\\ 1 & -8 & 7\end{bmatrix}\end{aligned} \tag{6.35}$$

$$M_c=\int_0^{l_e} \rho A\left(\phi^{\mathrm{T}}\phi\right)\mathrm{d}x=\int_0^{l_e} \rho A\begin{Bmatrix}\phi_1\\ \phi_2\\ \phi_3\end{Bmatrix}[\phi_1 \quad \phi_2 \quad \phi_3]\mathrm{d}x=\frac{\rho A l_e}{30}\begin{bmatrix}4 & 2 & -1\\ 2 & 16 & 2\\ -1 & 2 & 4\end{bmatrix} \tag{6.36}$$

$$M_{I,c}=\int_0^{l_e} \rho I_P\upsilon^2\left(\mathbf{B}^{\mathrm{T}}\mathbf{B}\right)\mathrm{d}x=\int_0^{l_e} \rho I_P\upsilon^2\begin{Bmatrix}\phi_1'\\ \phi_2'\\ \phi_3'\end{Bmatrix}[\phi_1' \quad \phi_2' \quad \phi_3']\mathrm{d}x=\frac{\rho I_P\upsilon^2}{3l_e}\begin{bmatrix}7 & -8 & 1\\ -8 & 16 & -8\\ 1 & -8 & 7\end{bmatrix} \tag{6.37}$$

$$M_{nl} = \int_0^{l_e} (e_0 a)^2 \rho A(1+2\upsilon)\left(\mathbf{B}^{\mathrm{T}}\mathbf{B}\right)\mathrm{d}x = \int_0^{l_e} (e_0 a)^2 \rho A(1+2\upsilon)\begin{Bmatrix} \phi_1{}' \\ \phi_2{}' \\ \phi_3{}' \end{Bmatrix}[\phi_1{}' \quad \phi_2{}' \quad \phi_3{}']\mathrm{d}x$$
$$= \frac{(e_0 a)^2 \rho A(1+2\upsilon)}{3l_e}\begin{bmatrix} 7 & -8 & 1 \\ -8 & 16 & -8 \\ 1 & -8 & 7 \end{bmatrix} \tag{6.38}$$

$$M_{I,nl} = \int_0^{l_e} (e_0 a)^2 \rho I_P \upsilon^2\left(\mathbf{B}'^{\mathrm{T}}\mathbf{B}'\right)\mathrm{d}x = \int_0^{l_e} (e_0 a)^2 \rho I_P \upsilon^2 \begin{Bmatrix} \phi_1{}'' \\ \phi_2{}'' \\ \phi_3{}'' \end{Bmatrix}[\phi_1{}'' \quad \phi_2{}'' \quad \phi_3{}'']\mathrm{d}x$$
$$= \frac{(e_0 a)^2 \rho I_P \upsilon^2}{l_e^3}\begin{bmatrix} 16 & -32 & 16 \\ -32 & 64 & -32 \\ 16 & -32 & 16 \end{bmatrix} \tag{6.39}$$

$$\mathbf{q}_c = \int_0^{l_e} q\phi^{\mathrm{T}}\mathrm{d}x = \int_0^{l_e} q\begin{Bmatrix} \phi_1 \\ \phi_2 \\ \phi_3 \end{Bmatrix}\mathrm{d}x - \frac{ql_e}{6}\begin{Bmatrix} 1 \\ 4 \\ 1 \end{Bmatrix} \tag{6.40}$$

$$\mathbf{q}_{nl} = \int_0^{l_e} (e_0 a)^2 q' \mathbf{B}^{\mathrm{T}}\mathrm{d}x = \int_0^{l_e} (e_0 a)^2 q' \begin{Bmatrix} \phi_1{}' \\ \phi_2{}' \\ \phi_3{}' \end{Bmatrix}\mathrm{d}x = \frac{(e_0 a)^2 q'}{l_e}\begin{Bmatrix} 1 \\ -4 \\ 3 \end{Bmatrix} \tag{6.41}$$

Interpretation. When Eqs. (6.9) and (6.31) are compared, it is understood that a stiffness and a mass matrices are added to the simple nanorod primarily for classical elasticity in the Love-Bishop rod. While the stiffness matrix indicates the lateral inertia, the mass matrix denotes the mass inertia resulting from lateral motion. The frequencies of Love-Bishop rod are calculated to be lower than simple rods because mass inertia generally dominates lateral inertia. On the other hand, in the NL-FEM analysis of nonlocal Love-Bishop nanorod, it is also noted that there is a second nonlocal mass matrix to be added the sum of mass. To add the nonlocal effect to shear deformation that is already important due to their short span and deep section in axial rods reveals that this rod formulation is important in the nanomechanical research. Of course, numerical analyses need to be examined to investigate this importance more comprehensively.

Solution stage. In this nonlocal rod formulation, the total mass matrix, stiffness matrix, and the load vector of a finite element are written as:

$$[K]_e = K + K_I + K_{M,c} + K_{M,nl} \tag{6.42}$$

$$[M]_e = M_c + M_{I,c} + M_{nl} + M_{I,nl} \tag{6.43}$$

$$\mathbf{q}_e = \mathbf{q}_c + \mathbf{q}_{nl} \tag{6.44}$$

According to the free vibration analysis, Eq. (6.20) is constituted and the solution is carried out. The arrangement of global matrices by boundary conditions is mentioned as follows.

6.3.2.2 Application of Boundary Conditions

Since there is only axial freedom at each point in the three-node finite element, if we divide the rod into n finite elements, the rod has $(2n + 1)$ freedom. After applying the boundary conditions to the global matrices, the eigenvalue equation can be solved.

On the other hand, since an end attachment is considered as lumped at the end of axial rod, it should be added to input 33 of related matrix of finite element:

$$[K]_{e,a} = [K]_{e,na} + (K_a) = (K_{ae})_e + (K_I)_e + (K_{M,c})_e + (K_{M,nl})_e + \begin{bmatrix} 0 & 0 & 0 \\ 0 & 0 & 0 \\ 0 & 0 & k_a \end{bmatrix} \tag{6.45}$$

$$[M]_{e,a} = [M]_{e,na} + (M_a) = (M_c)_e + (M_{I,c})_e + (M_{nl})_e + (M_{I,nl})_e + \begin{bmatrix} 0 & 0 & 0 \\ 0 & 0 & 0 \\ 0 & 0 & m_a \end{bmatrix} \tag{6.46}$$

Example 6.4 Determine the natural frequencies of nonlocal axial free vibration of soft-clamped Love-Bishop nanorod in Fig. 6.2. Use two finite elements in the analysis. Physical and mechanical parameters are as in Examples 6.1 and 6.2. Additionally, Poisson's ratio is $\upsilon = 0.19$.

Solution. The finite element information of this rod in Fig. 6.2b, c should be examined. Now the nonlocal parameter is considered and the elastic medium is neglected. Therefore, considering Eqs. (6.32)–(6.39), the related stiffness and mass matrices are formed as follows:

$$(K_{ae})_1 = (K_{ae})_2 = \frac{EA}{3l} \begin{bmatrix} 7 & -8 & 1 \\ -8 & 16 & -8 \\ 1 & -8 & 7 \end{bmatrix} \tag{6.E6.4.1}$$

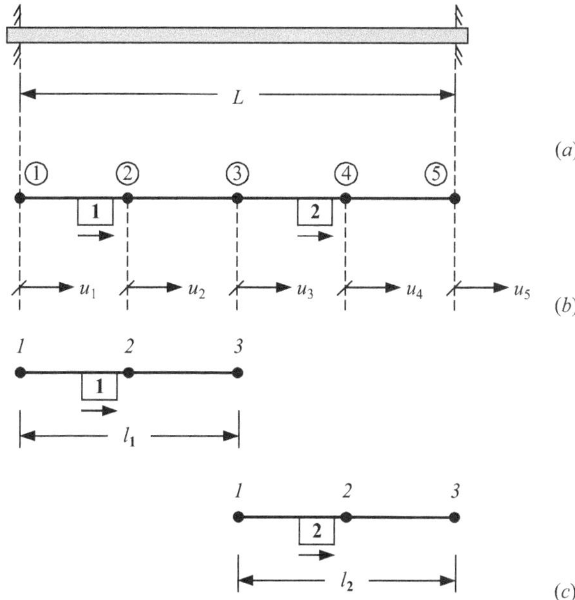

Fig. 6.2 FEM analysis of soft-clamped axial nanorod. (**a**) Physical system. (**b**) FEM model with two elements. (**c**) Finite elements

$$(K_I)_1 = (K_I)_2 = \frac{GI_P v^2}{l^3}\begin{bmatrix} 16 & -32 & 16 \\ -32 & 64 & -32 \\ 16 & -32 & 16 \end{bmatrix} \tag{6.E6.4.2}$$

$$(M_c)_1 = (M_c)_2 = \frac{\rho A l}{30}\begin{bmatrix} 4 & 2 & -1 \\ 2 & 16 & 2 \\ -1 & 2 & 4 \end{bmatrix} \tag{6.E6.4.3}$$

$$(M_{I,c})_1 = (M_{I,c})_2 = \frac{\rho I_P v^2}{3l}\begin{bmatrix} 7 & -8 & 1 \\ -8 & 16 & -8 \\ 1 & -8 & 7 \end{bmatrix} \tag{6.E6.4.4}$$

$$(M_{nl})_1 = (M_{nl})_2 = \frac{(e_0 a)^2 \rho A(1+2v)}{3l}\begin{bmatrix} 7 & -8 & 1 \\ -8 & 16 & -8 \\ 1 & -8 & 7 \end{bmatrix} \tag{6.E6.4.5}$$

$$(M_{I,nl})_1 = (M_{I,nl})_2 = \frac{(e_0 a)^2 \rho I_P v^2}{l^3}\begin{bmatrix} 16 & -32 & 16 \\ -32 & 64 & -32 \\ 16 & -32 & 16 \end{bmatrix} \tag{6.E6.4.6}$$

where in addition to the physical and mechanical parameters given by Eq. (6.E6.1.1), the required parameters for the Love-Bishop rod are given as:

$$G = \frac{E}{2(1+\upsilon)} = \frac{1\ \text{TPa}}{2\times(1+0.19)} = 0.42\ \text{TPa} = 0.42\times 10^{12}\frac{\text{N}}{\text{m}^2} = 4.202 \times 10^{11}\frac{\text{kg}}{\text{ms}^2} \tag{6.E6.4.7}$$

$$I_P = \frac{\pi d^2}{32} = \frac{\pi\times(1\ \text{nm})^4}{32} = \frac{\pi\times\left(1\ \times 10^{-9}\ \text{m}\right)^4}{32} = 9.817\times 10^{-38}\ \text{m}^2 \tag{6.E6.4.8}$$

To assemble matrices written in Eqs. (6.E6.4.1)–(6.E6.4.6) by considering to the freedom numbers of finite element mesh gives the following global matrices. For sample assembly, only the elastic stiffness, lateral stiffness, and classical mass matrices are shown in detail and the others are expressed directly:

Columns/rows labelled by *Mesh node numbers* 1–5:

$$\sum[K_{ae}] = \frac{AE}{3l}\begin{bmatrix} 7 & -8 & 1 & 0 & 0 \\ -8 & 16 & -8 & 0 & 0 \\ 1 & -8 & 7+7 & -8 & 1 \\ 0 & 0 & -8 & 16 & -8 \\ 0 & 0 & 1 & -8 & 7 \end{bmatrix} \tag{6.E6.4.9}$$

$$\sum[K_l] = \frac{GI_p\upsilon^2}{l^3}\begin{bmatrix} 16 & -32 & 16 & 0 & 0 \\ -32 & 64 & -32 & 0 & 0 \\ 16 & -32 & 16+16 & -32 & 16 \\ 0 & 0 & -32 & 64 & -32 \\ 0 & 0 & 16 & -32 & 16 \end{bmatrix} \tag{6.E6.4.10}$$

$$\sum[M_c] = \frac{\rho Al}{30}\begin{bmatrix} 4 & 2 & -1 & 0 & 0 \\ 2 & 16 & 2 & 0 & 0 \\ -1 & 2 & 4+4 & 2 & -1 \\ 0 & 0 & 2 & 16 & 2 \\ 0 & 0 & -1 & 2 & 4 \end{bmatrix} \tag{6.E6.4.11}$$

$$\sum[M_{I,c}] = \frac{\rho I_P v^2}{3l}\begin{bmatrix} 7 & -8 & 1 & 0 & 0 \\ -8 & 16 & -8 & 0 & 0 \\ 1 & -8 & 14 & -8 & 1 \\ 0 & 0 & -8 & 16 & -8 \\ 0 & 0 & 1 & -8 & 7 \end{bmatrix} \tag{6.E6.4.12}$$

$$\sum[M_{nl}] = \frac{(e_0a)^2\rho A(1+2v)}{3l}\begin{bmatrix} 7 & -8 & 1 & 0 & 0 \\ -8 & 16 & -8 & 0 & 0 \\ 1 & -8 & 14 & -8 & 1 \\ 0 & 0 & -8 & 16 & -8 \\ 0 & 0 & 1 & -8 & 7 \end{bmatrix} \tag{6.E6.4.13}$$

$$\sum[M_{I,nl}] = \frac{(e_0a)^2\rho I_P v^2}{l^3}\begin{bmatrix} 16 & -32 & 16 & 0 & 0 \\ -32 & 64 & -32 & 0 & 0 \\ 16 & -32 & 32 & -32 & 16 \\ 0 & 0 & -32 & 64 & -32 \\ 0 & 0 & 16 & -32 & 16 \end{bmatrix} \tag{6.E6.4.14}$$

After the physical and mechanical parameters are written into the global matrices, their first and fifth columns and rows are reduced due to geometric boundary conditions in order for the global matrices to enter the eigenvalue problem. With Eqs. (6.20), (6.42) and (6.43), the reduced global matrices are calculated below:

$$\left(\sum[K]\right)^* = \left(\sum[K_{ae}]\right)^* + \left(\sum[K_I]\right)^* = \begin{bmatrix} 418.762 & -209.381 & 0 \\ -209.381 & 366.381 & -209.381 \\ 0 & -209.381 & 418.762 \end{bmatrix}\frac{\text{N}}{\text{m}} \tag{6.E6.4.15}$$

$$\begin{aligned}\left(\sum[M]\right)^* &= \left(\sum[M_c]\right)^* + \left(\sum[M_{I,c}]\right)^* + \left(\sum[M_{nl}]\right)^* + \left(\sum[M_{I,nl}]\right)^* \\ &= \begin{bmatrix} 3.090\times10^{-23} & -9.433\times10^{-24} & 0 \\ -9.433\times10^{-24} & 2.343\times10^{-23} & -9.433\times10^{-24} \\ 0 & -9.433\times10^{-24} & 3.090\times10^{-23} \end{bmatrix}\text{kg}\end{aligned} \tag{6.E6.4.16}$$

Natural frequencies can be calculated by substituting Eqs. (6.E6.4.15) and (6.E6.4.16) into Eq. (6.20):

Table 6.5 Comparisons of nondimensional frequencies of soft-clamped Love-Bishop nanorods for nonlocal axial free vibration ($e_0a/L = 0.2$, $K_M = 0$)

Mode number	Analytical	NL-FEM results				
	Eq. (2.E2.7.13)	Manual solution	Program code (MATLAB)			
		$n = 2$	$n = 2$	$n = 10$	$n = 15$	$n = 20$
1	2.52756	2.5336	2.53369	2.52757	2.52756	2.52756
2	3.52357	3.5310	3.53080	3.52369	3.52359	3.52357
3	3.87838	3.9845	3.98454	3.87873	3.87844	3.87838
4	4.03013	–	–	4.03082	4.03024	4.03008
5	4.10630	–	–	4.10742	4.10643	4.10608

$$\omega_1 = 2.642 \text{ THz}, \quad \omega_2 = 3.681 \text{ THz}, \quad \omega_3 = 4.154 \text{ THz}.$$

A comparison study of nondimensional frequencies in nonlocal axial vibration of soft-clamped Love-Bishop nanorods is presented in Table 6.5. Here the results of analytical solution through Eq. (2.E2.7.13). According to the three-node finite element model we considered in this problem, the using of two finite elements enabled to be obtained the frequencies of first three modes due to boundary conditions. In Example 6.1, the first two modes can be obtained with three two-node elements, and in Example 6.2, the first three modes can be obtained with four two-node elements. It is concluded from here that more frequencies can be calculated with finite element mesh that have more nodes. Moreover, in the using of high-order finite elements, frequency values are obtained closer to the analytical result since the number of nodes in the system increases. In the related nonlocal vibration analysis, if the fundamental mode is taken into consideration, same values with the analytical result are obtained, especially in NL-FEM analyses containing a high number of elements. However, according to one point that draws attention here, the frequencies of high modes drop below the analytical result as a result of the using of a high number of finite elements ($n = 20$). Generally, appropriate results obtained with the using of $n = 10$ elements.

Now let's consider the elastic medium as well as the nonlocal parameter. In addition to Eqs. (6.E6.4.9)–(6.E6.4.14), the global matrices calculated according to Eqs. (6.34) and (6.35) due to the elastic environment in which the Love-Bishop nanorod is embedded are given as follows:

$$\sum[K_{M,c}] = \frac{k_M l}{30}\begin{bmatrix} 4 & 2 & -1 & 0 & 0 \\ 2 & 16 & 2 & 0 & 0 \\ -1 & 2 & 8 & 2 & -1 \\ 0 & 0 & 2 & 16 & 2 \\ 0 & 0 & -1 & 2 & 4 \end{bmatrix} \tag{6.E6.4.17}$$

$$\sum[K_{M,nl}] = \frac{(e_0a)^2k_M}{3l}\begin{bmatrix} 7 & -8 & 1 & 0 & 0 \\ -8 & 16 & -8 & 0 & 0 \\ 1 & -8 & 14 & -8 & 1 \\ 0 & 0 & -8 & 16 & -8 \\ 0 & 0 & 1 & -8 & 7 \end{bmatrix} \quad (6.E6.4.18)$$

The global stiffness matrix of eigenvalue problem can be obtained by adding the expressions resulting from the reduction of these matrices with Eq. (6.E6.4.15):

$$\left(\sum[K]\right)^* = \left(\sum[K_{ae}]\right)^* + \left(\sum[K_I]\right)^* + \left(\sum[K_{M,c}]\right)^* + \left(\sum[K_{M,nl}]\right)^*$$
$$= \begin{bmatrix} 418.762 & -209.381 & 0 \\ -209.381 & 366.381 & -209.381 \\ 0 & -209.381 & 418.762 \end{bmatrix}\frac{\text{N}}{\text{m}} \quad (6.E6.4.19)$$

if the global matrices given in Eqs. (6.E6.4.16) and (6.E6.4.19) are substituted into Eq. (6.20), the natural frequencies are calculated as follows:

$$\omega_1 = 4.097\ \text{THz},\quad \omega_2 = 4.729\ \text{THz},\quad \omega_3 = 5.050\ \text{THz}.$$

Comparisons of nondimensional frequencies of soft-clamped Love-Bishop nanorods embedded in elastic medium are listed in Table 6.6.

Example 6.5 Determine the natural frequencies of nonlocal axial free vibration of free-mass attached Love-Bishop nanorod in Fig. 2.10. Use only one finite element in the analysis. Physical and mechanical parameters are as in Example 6.1. Neglect the axial elastic medium. In addition to these, attachment ratio is $\beta_m = 5$ and defined as $\beta_m = m_a/\rho AL$.

Table 6.6 Comparisons of nondimensional frequencies of embedded soft-clamped Love-Bishop nanorods for nonlocal axial free vibration ($e_0a/L = 0.2$, $K_M = 10$)

Mode number	Analytical	NL-FEM results				
	Eq. (2.E2.7.13)	Manual solution	Program code (MATLAB)			
		$n = 2$	$n = 2$	$n = 10$	$n = 15$	$n = 20$
1	3.92639	3.923	3.92974	3.92640	3.92639	3.92639
2	4.53046	4.536	4.53519	4.53054	4.53047	4.53045
3	4.76956	4.844	4.84347	4.76980	4.76954	4.76941
4	4.87459	–	–	4.87504	4.87441	4.87393
5	4.92758	–	–	4.92828	4.92698	4.92543

Solution. Since a single finite element is employed and there is not geometric boundary condition, while the global stiffness matrices are stated by Eq. (6.32) and (6.33), Eqs. (6.36)–(6.39) define the global mass matrices. The end mass attachment is expressed in Eq. (6.46).

On the other hand, the attachment mass is calculated as follows:

$$\begin{aligned} m_a &= \beta_m \rho AL = (5) \times \left(2300\frac{\text{kg}}{\text{m}^3}\right) \times \left(0.785 \times 10^{-18}\text{m}^2\right) \times \left(20 \times 10^{-9}\text{m}\right) \\ &= 1.806 \times 10^{-22}\text{kg} \end{aligned} \tag{6.E6.5.1}$$

In consequence, the global matrices for the eigenvalue problem can be written as:

$$\begin{aligned} \left(\sum[K]\right)^* &= \left(\sum[K_{ae}]\right)^* + \left(\sum[K_I]\right)^* \\ &= \frac{EA}{3L}\begin{bmatrix} 7 & -8 & 1 \\ -8 & 16 & -8 \\ 1 & -8 & 7 \end{bmatrix} + \frac{GI_P v^2}{L^3}\begin{bmatrix} 16 & -32 & 16 \\ -32 & 64 & -32 \\ 16 & -32 & 16 \end{bmatrix} \\ &= \begin{bmatrix} 91.586 & -104.673 & 13.086 \\ -104.673 & 209.345 & -104.673 \\ 13.086 & -104.673 & 91.586 \end{bmatrix} \frac{\text{N}}{\text{m}} \end{aligned} \tag{6.E6.5.2}$$

$$\begin{aligned} \left(\sum[M]\right)^* &= \left(\sum[M_c]\right)^* + \left(\sum[M_{I,c}]\right)^* + \left(\sum[M_{nl}]\right)^* + \left(\sum[M_{I,nl}]\right)^* + \left(\sum[M_a]\right)^* \\ &= \frac{\rho AL}{30}\begin{bmatrix} 4 & 2 & -1 \\ 2 & 16 & 2 \\ -1 & 2 & 4 \end{bmatrix} + \frac{\rho I_P v^2}{3L}\begin{bmatrix} 7 & -8 & 1 \\ -8 & 16 & -8 \\ 1 & -8 & 7 \end{bmatrix} + \frac{(e_0 a)^2 \rho A(1+2v)}{3L}\begin{bmatrix} 7 & -8 & 1 \\ -8 & 16 & -8 \\ 1 & -8 & 7 \end{bmatrix} + \\ &\quad \frac{(e_0 a)^2 \rho I_P v^2}{L^3}\begin{bmatrix} 16 & -32 & 16 \\ -32 & 64 & -32 \\ 16 & -32 & 16 \end{bmatrix} + m_a\begin{bmatrix} 0 & 0 & 0 \\ 0 & 0 & 0 \\ 0 & 0 & 1 \end{bmatrix} \\ &= \begin{bmatrix} 9.467\times10^{-24} & -2.910\times10^{-24} & -5.388\times10^{-25} \\ -2.910\times10^{-24} & 2.989\times10^{-23} & -2.910\times10^{-24} \\ -5.388\times10^{-25} & -2.910\times10^{-24} & 1.900\times10^{-22} \end{bmatrix} \text{kg} \end{aligned} \tag{6.E6.5.3}$$

Natural frequencies can be attained by using the above expressions in Eq. (6.20):

$$\omega_1 = 0.0037i \text{ THz}, \quad \omega_2 = 1.644 \text{ THz}, \quad \omega_3 = 3.557 \text{ THz}.$$

It should be noted here that the vibration frequency of first mode is calculated as an imaginary number. Therefore, the nanorod cannot vibrate in the first mode. In this nanorod that does not have a rigid boundary condition, the mass addition, which is a factor that weakens the dynamic response of structure, causes this result.

Also, it is said that the frequencies cannot be obtained by manual calculation when the analytical method-based solution for the free vibration of nanorod that is the subject of this application is predicated on. Unfortunately, the process of reading the frequencies by coding the analytical solution on MATLAB is observed to be quite insufficient. However, the frequencies of mass attached Love-Bishop nanorods could be obtained by NL-FEM analysis. By using a higher number of finite elements in the written program codes, the frequencies of such nanorods can be reached with higher accuracy.

6.4 Nonlocal Torsional Vibration

An example depiction for a finite element mesh of torsional rods is presented in Fig. 5.3b. In this section, we study NL-FEM formulations for the free torsional vibration of nanorods with circular and elliptical cross-section.

6.4.1 Circular Nanorod

6.4.1.1 NL-FEM Expression

It is observed that the analytical investigation of torsional vibration of circular nanorods given in Sect. 3.3 is very similar to the simple axial nanorods in Sect. 2.3. Since the equation of motion is obtained in a similar mathematical structure, the frequency equations of nanorods are also similar for nonlocal free axial and torsional vibrations. The differences arise only from the mechanical and geometric parameters (E and G, A, and I_P) have a part in the topic. It is understood that the finite element procedures of these two nonlocal mechanical analyses are also like each other. However, it is preferred to introduce the formulation so that the behavior can be investigated in detail.

First, considering the equation of motion of nonlocal torsional free vibration defined in Eq. (3.10), the weighted average residual equation is formulated as [10]:

$$\begin{aligned} \mathrm{I} = \int_0^{l_e} h\Bigg(& GI_P \frac{\partial^2 \theta}{\partial x^2} - \rho I_P \frac{\partial^2 \theta}{\partial t^2} - k_T \theta + (e_0 a)^2 k_T \frac{\partial^2 \theta}{\partial x^2} + (e_0 a)^2 \rho I_P \frac{\partial^4 \theta}{\partial x^2 \partial t^2} \\ & - (e_0 a)^2 \frac{\partial^2 m}{\partial x^2} + m \Bigg) \mathrm{d}x \end{aligned} \tag{6.47}$$

Like the two-node linear axial member, the kinematics of torsional element are written below:

$$\theta = \phi \mathbf{d} \tag{6.48}$$

Equation (6.3) is considered for the shape function matrix and the torsional motion vector of end freedoms. Additionally, here the kinematic operator is determined as follows:

$$\frac{\partial \theta}{\partial x} = D^k \theta = \mathbf{B}\mathbf{d} \tag{6.49}$$

where D^k is the kinematic operator and is denoted as $D^k \phi = \mathbf{B}$.

When Eq. (6.5) is taken into consideration, partial integrations for NL-FEM analysis of torsional vibration are obtained as follows [10]:

$$\begin{aligned}
I_1 &= \int_0^{l_e} GI_P h \frac{\partial^2 \theta}{\partial x^2} dx = GI_P h \frac{\partial \theta}{\partial x}\bigg|_0^{l_e} - \int_0^{l_e} GI_P \frac{\partial h}{\partial x}\frac{\partial \theta}{\partial x} dx, \\
I_2 &= \int_0^{l_e} \rho I_P h \frac{\partial^2 \theta}{\partial t^2} dx, \quad I_3 = \int_0^{l_e} k_T h \theta dx, \\
I_4 &= \int_0^{l_e} (e_0 a)^2 k_T h \frac{\partial^2 \theta}{\partial x^2} dx = (e_0 a)^2 k_T h \frac{\partial \theta}{\partial x}\bigg|_0^{l_e} - \int_0^{l_e} (e_0 a)^2 k_T \frac{\partial h}{\partial x}\frac{\partial \theta}{\partial x} dx, \\
I_5 &= \int_0^{l_e} (e_0 a)^2 \rho I_P h \frac{\partial^4 \theta}{\partial x^2 \partial t^2} dx = (e_0 a)^2 \rho I_P h \frac{\partial^3 \theta}{\partial x \partial t^2}\bigg|_0^{l_e} - \int_0^{l_e} (e_0 a)^2 \rho I_P \frac{\partial h}{\partial x}\frac{\partial^3 \theta}{\partial x \partial t^2} dx, \\
I_6 &= \int_0^{l_e} (e_0 a)^2 h \frac{\partial^2 m}{\partial x^2} dx = (e_0 a)^2 h \frac{\partial m}{\partial x}\bigg|_0^{l_e} - \int_0^{l_e} (e_0 a)^2 \frac{\partial h}{\partial x}\frac{\partial m}{\partial x} dx, \quad I_7 = \int_0^{l_e} h m dx.
\end{aligned} \tag{6.50}$$

using these expressions into Eq. (6.1) gives the following equation:

$$\int_0^{l_e} GI_P \frac{\partial h}{\partial x}\frac{\partial \theta}{\partial x}\mathrm{d}x + \int_0^{l_e} (e_0a)^2 \rho I_P \frac{\partial h}{\partial x}\frac{\partial^3 \theta}{\partial x \partial t^2}\mathrm{d}x + \int_0^{l_e} (e_0a)^2 k_T \frac{\partial h}{\partial x}\frac{\partial \theta}{\partial x}\mathrm{d}x$$
$$- \int_0^{l_e} (e_0a)^2 \frac{\partial h}{\partial x}\frac{\partial m}{\partial x}\mathrm{d}x + \int_0^{l_e} k_T h\theta \mathrm{d}x + \int_0^{l_e} \rho I_P h \frac{\partial^2 \theta}{\partial t^2}\mathrm{d}x - \int_0^{l_e} hm\mathrm{d}x - GI_P h \frac{\partial \theta u}{\partial x}\bigg|_0^{l_e}$$
$$- (e_0a)^2 \rho I_P h \frac{\partial^3 \theta}{\partial x \partial t^2}\bigg|_0^{l_e} - (e_0a)^2 k_T h \frac{\partial \theta}{\partial x}\bigg|_0^{l_e} + (e_0a)^2 h \frac{\partial m}{\partial x}\bigg|_0^{l_e} = 0 \tag{6.51}$$

where definitions similar to Eq. (6.7) are presented as follows:

$$h = \phi^{\mathrm{T}}, \;\; \mathbf{B} = \mathrm{D}^{\mathrm{k}}\phi, \;\; \frac{\partial h}{\partial x} = \left(\phi^{\mathrm{T}}\right)' = \mathbf{B}^{\mathrm{T}}, \;\; \frac{\partial \theta}{\partial x} = \mathbf{Bd}, \;\; \frac{\partial^2 \theta}{\partial t^2} = \phi \ddot{\mathbf{d}}. \tag{6.52}$$

By utilizing Eq. (6.52) into the related part of Eq. (6.51), the following expression is obtained:

$$\int_0^{l_e} GI_P\left(\mathbf{B}^{\mathrm{T}}\mathbf{B}\right)\mathbf{d}\mathrm{d}x + \int_0^{l_e} \rho I_P\left(\phi^{\mathrm{T}}\phi\right)\ddot{\mathbf{d}}\mathrm{d}x + \int_0^{l_e} (e_0a)^2 \rho I_P\left(\mathbf{B}^{\mathrm{T}}\mathbf{B}\right)\ddot{\mathbf{d}}\mathrm{d}x + \int_0^{l_e} (e_0a)^2 k_T\left(\mathbf{B}^{\mathrm{T}}\mathbf{B}\right)\mathbf{d}\mathrm{d}x$$
$$- \int_0^{l_e} (e_0a)^2 \mathbf{B}^{\mathrm{T}} m' \mathrm{d}x + \int_0^{l_e} k_T\left(\phi^{\mathrm{T}}\phi\right)\mathbf{d}\mathrm{d}x - \int_0^{l_e} \phi^{\mathrm{T}} m \mathrm{d}x = 0 \tag{6.53}$$

This equation is given with matrix notation as follows:

$$(K + K_{T,c} + K_{T,nl})\mathbf{d} + (I_c + I_{nl})\ddot{\mathbf{d}} = \mathbf{m}_c + \mathbf{m}_{nl} \tag{6.54}$$

In which, K is the torsional stiffness matrix, $K_{T,\ c}$ is the classical stiffness matrix of an elastic rigid medium, and $K_{T,\ nl}$ is the nonlocal stiffness matrix of this medium. Also, I_c is the classical mass and I_{nl} is the nonlocal torsional inertia matrices. On the other hand, $\mathbf{m}_c$ and $\mathbf{m}_{nl}$ mean the classical and nonlocal external torsional excitation vectors, respectively. These expressions are formulated as follows [10]:

$$K = \int_0^{l_e} GI_P(\mathbf{B}^{\mathrm{T}}\mathbf{B})\mathrm{d}x = \int_0^{l_e} GI_P \begin{Bmatrix} \phi_1' \\ \phi_2' \end{Bmatrix} [\phi_1' \quad \phi_2']\mathrm{d}x = \frac{GI_P}{l_e}\begin{bmatrix} 1 & -1 \\ -1 & 1 \end{bmatrix} \tag{6.55}$$

$$K_{T,c} = \int_0^{l_e} k_T(\phi^{\mathrm{T}}\phi)\mathrm{d}x = \int_0^{l_e} k_T \begin{Bmatrix} \phi_1 \\ \phi_2 \end{Bmatrix} [\phi_1 \quad \phi_2]\mathrm{d}x = \frac{k_T l_e}{6}\begin{bmatrix} 2 & 1 \\ 1 & 2 \end{bmatrix} \tag{6.56}$$

$$K_{T,nl} = \int_0^{l_e} (e_0a)^2 k_T(\mathbf{B}^{\mathrm{T}}\mathbf{B})\mathrm{d}x = \int_0^{l_e} (e_0a)^2 k_M \begin{Bmatrix} \phi_1' \\ \phi_2' \end{Bmatrix} [\phi_1' \quad \phi_2']\mathrm{d}x = \frac{(e_0a)^2 k_T}{l_e}\begin{bmatrix} 1 & -1 \\ -1 & 1 \end{bmatrix} \tag{6.57}$$

$$I_c = \int_0^{l_e} \rho I_P(\phi^{\mathrm{T}}\phi)\mathrm{d}x = \int_0^{l_e} \rho I_P \begin{Bmatrix} \phi_1 \\ \phi_2 \end{Bmatrix} [\phi_1 \quad \phi_2]\mathrm{d}x = \frac{\rho I_P l_e}{6}\begin{bmatrix} 2 & 1 \\ 1 & 2 \end{bmatrix} \tag{6.58}$$

$$I_{nl} = \int_0^{l_e} (e_0a)^2 \rho I_P(\mathbf{B}^{\mathrm{T}}\mathbf{B})\mathrm{d}x = \int_0^{l_e} (e_0a)^2 \rho I_P \begin{Bmatrix} \phi_1' \\ \phi_2' \end{Bmatrix} [\phi_1' \quad \phi_2']\mathrm{d}x = \frac{(e_0a)^2 \rho I_P}{l_e}\begin{bmatrix} 1 & -1 \\ -1 & 1 \end{bmatrix} \tag{6.59}$$

$$\mathbf{m}_c = \int_0^{l_e} m\phi^{\mathrm{T}}\mathrm{d}x = \int_0^{l_e} m \begin{Bmatrix} \phi_1 \\ \phi_2 \end{Bmatrix}\mathrm{d}x = \frac{ml_e}{2}\begin{Bmatrix} 1 \\ 1 \end{Bmatrix} \tag{6.60}$$

$$\mathbf{m}_{nl} = \int_0^{l_e} (e_0a)^2 m' \mathbf{B}^{\mathrm{T}}\mathrm{d}x = \int_0^{l_e} (e_0a)^2 m' \begin{Bmatrix} \phi_1' \\ \phi_2' \end{Bmatrix}\mathrm{d}x = \frac{(e_0a)^2 m'}{l_e}\begin{Bmatrix} -1 \\ 1 \end{Bmatrix} \tag{6.61}$$

Interpretation. As stated in the introduction of this section, the NL-FEM matrices shown in Eq. (6.55)–(6.61) are like the NL-FEM matrices of nonlocal simple axial nanorod. Therefore, our interpretations of nonlocal simple axial nanorod are also valid here.

Solution stage. As with the nonlocal simple axial nanorod, the total stiffness and mass matrices of torsional finite element and the external torsional moment vector can be denoted as follows:

$$[K]_e = K + K_{T,c} + K_{T,nl} \tag{6.62}$$

$$[M]_e = I_c + I_{nl} \tag{6.63}$$

$$\mathbf{m}_e = \mathbf{m}_c + \mathbf{m}_{nl} \tag{6.64}$$

Free vibration analysis indicates that $\mathbf{m}_\mathrm{c} = \mathbf{m}_\mathrm{nl} = 0$. Therefore, nonlocal free torsional vibration frequencies can be obtained by solving the eigenvalue equation presented in Eq. (6.20).

6.4.1.2 Application of Boundary Conditions

The effect of geometric boundary conditions on global matrices is naturally as explained in the nonlocal simple axial nanorods. Additionally, the totals of finite element matrices for different add-ons are written as follows:

$$[K]_{e,a} = [K]_{e,na} + (K_a) = (K_{te})_e + (K_{T,c})_e + (K_{T,nl})_e + \begin{bmatrix} 0 & 0 \\ 0 & k_a \end{bmatrix} \tag{6.65}$$

$$[M]_{e,a} = [M]_{e,na} + (I_a) = (I_c)_e + (I_{nl})_e + \begin{bmatrix} 0 & 0 \\ 0 & I_a \end{bmatrix} \tag{6.66}$$

Example 6.6 Determine the natural frequencies of nonlocal torsional free vibration of clamped-free nanoshaft embedded on elastic foundation in Fig. 6.3. Use three finite elements in the analysis. Nanoshaft length: $L = 20$ nm, diameter of circular cross-section $d = 1$nm, modulus of elasticity: $E = 1$ TPa, and mass of unit volume: $\rho = 2300\text{kg/m}^3$. Firstly, ignore the elastic medium and use the nonlocal parameter as $e_0a/L = 0.2$. Then, repeat the analysis for nondimensional stiffness of elastic medium: $K_T = 10$. Nondimensional stiffness of elastic medium is defined as $K_T = k_T L^2/GI_P$.

Solution. The finite element mesh of nanoshaft is depicted in Fig. 6.3b. Here the stiffness, classical torsional inertia, and nonlocal torsional inertia matrices that concern the nonlocal free vibration of any two-node torsional finite element are defined as follows:

$$(K_{te})_e = \frac{G_e I_{P,e}}{l_e}\begin{bmatrix} 1 & -1 \\ -1 & 1 \end{bmatrix} \tag{6.E6.6.1}$$

(column FE node numbers: *1*, *2*; row FE node numbers: *1*, *2*)

$$(I_c)_e = \frac{\rho_e I_{P,e} l_e}{6}\begin{bmatrix} 2 & 1 \\ 1 & 2 \end{bmatrix} \tag{6.E6.6.2}$$

(column FE node numbers: *1*, *2*; row FE node numbers: *1*, *2*)

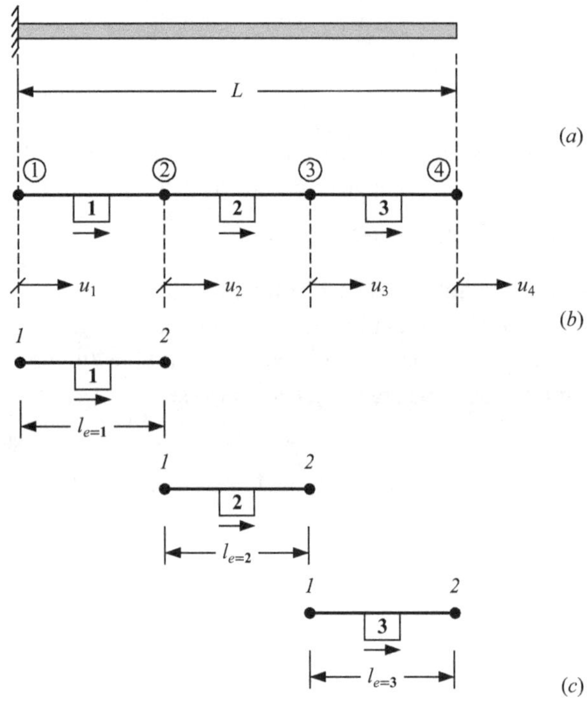

Fig. 6.3 FEM analysis of clamped-free torsional nanoshaft. (**a**) Physical system. (**b**) Finite element mesh. (**c**) Finite elements

$$\left(I_{nl}\right)_e = \frac{\left(e_0 a\right)_e^2 \rho_e I_{P,e}}{l_e} \begin{bmatrix} 1 & -1 \\ -1 & 1 \end{bmatrix} \begin{matrix} 1 \\ 2 \end{matrix} \qquad \text{(columns: 1, 2 — FE node numbers)} \tag{6.E6.6.3}$$

It is known that all physical and mechanical parameters of finite elements in Fig. 6.3c are the same:

$$\begin{aligned}
&\rho_{e=1} = \rho_{e=2} = \rho_{e=3} = \rho = 2300 \ \frac{\text{kg}}{\text{m}^3},\\
&I_{P,e=1} = I_{P,e=2} = I_{P,e=3} = I_P = 9.817 \times 10^{-38} \ \text{m}^2,\\
&G_{e=1} = G_{e=2} = G_{e=3} = G = 4.202 \times 10^{11} \ \frac{\text{kg}}{\text{ms}^2},\\
&l_{e=1} = l_{e=2} = l_{e=3} = l = \frac{L}{3} = \frac{20}{3} \ \text{nm} = 6.667 \times 10^{-9} \ \text{m},\\
&(e_0 a)_{e=1} = (e_0 a)_{e=2} = (e_0 a)_{e=3} = e_0 a = L \times 20 \ \text{nm} = 4 \ \text{nm} = 4 \ \times 10^{-9} \ \text{m}.
\end{aligned} \tag{6.E6.6.4}$$

The global matrices can be combined as follows:

$$\sum[K_{te}] = \frac{GI_P}{l}\begin{bmatrix} 1 & -1 & 0 & 0 \\ -1 & 1+1 & -1 & 0 \\ 0 & -1 & 1+1 & -1 \\ 0 & 0 & -1 & 1 \end{bmatrix} \tag{6.E6.6.5}$$

(Rows and columns labelled with mesh node numbers 1, 2, 3, 4.)

$$\sum[I_c] = \frac{\rho I_P l}{6}\begin{bmatrix} 2 & 1 & 0 & 0 \\ 1 & 2+2 & 1 & 0 \\ 0 & 1 & 2+2 & 1 \\ 0 & 0 & 1 & 2 \end{bmatrix} \tag{6.E6.6.6}$$

(Rows and columns labelled with mesh node numbers 1, 2, 3, 4.)

$$\sum[I_{nl}] = \frac{(e_0 a)^2 \rho I_P}{l}\begin{bmatrix} 1 & -1 & 0 & 0 \\ -1 & 1+1 & -1 & 0 \\ 0 & -1 & 1+1 & -1 \\ 0 & 0 & -1 & 1 \end{bmatrix} \tag{6.E6.6.7}$$

(Rows and columns labelled with mesh node numbers 1, 2, 3, 4.)

As a result of calculating the above expressions, the reduced global matrices are obtained as follows:

$$\sum[K_{te}] = \begin{bmatrix} 6.174\times10^{-18} & -6.174\times10^{-18} & 0 & 0 \\ -6.174\times10^{-18} & 1.235\times10^{-17} & -6.174\times10^{-18} & 0 \\ 0 & -6.174\times10^{-18} & 1.235\times10^{-17} & -6.174\times10^{-18} \\ 0 & 0 & -6.174\times10^{-18} & 6.174\times10^{-18} \end{bmatrix} \mathrm{kg}\frac{\mathrm{m}^2}{\mathrm{s}^2} \tag{6.E6.6.8}$$

(First row and first column: Eliminated. The remaining lower-right 3×3 block is $\left(\sum[K_{te}]\right)^*$.)

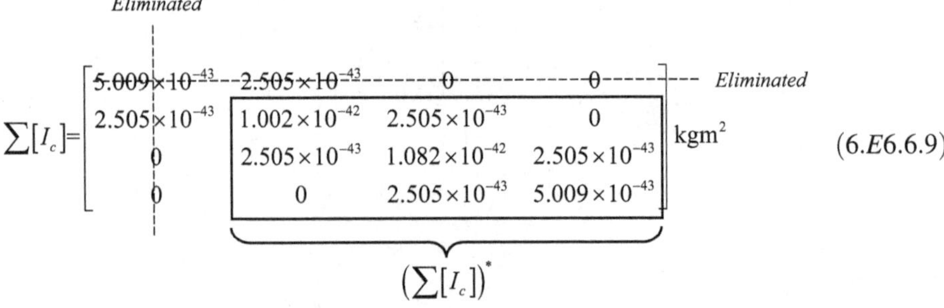

(6.*E*6.6.9)

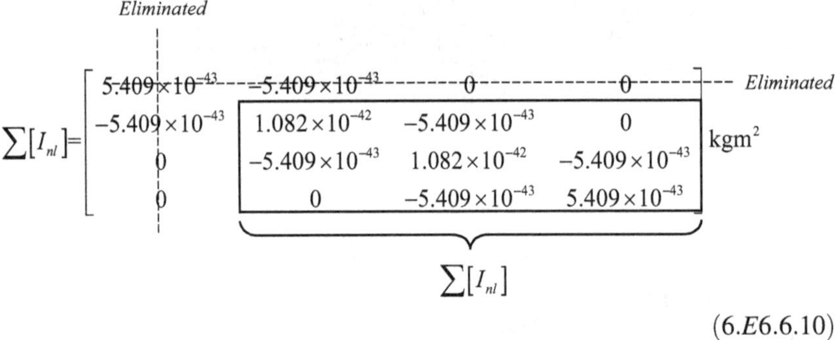

(6.*E*6.6.10)

To formulate the eigenvalue equation in Eq. (6.20), the total mass inertia matrix is calculated according to Eq. (6.63):

$$\left(\sum[M]\right)^* = \left(\sum[I_c]\right)^* + \left(\sum[I_{nl}]\right)^*$$
$$= \begin{bmatrix} 2.084\times10^{-42} & -2.905\times10^{-43} & 0 \\ -2.905\times10^{-43} & 2.084\times10^{-42} & -2.905\times10^{-43} \\ 0 & -2.905\times10^{-43} & 1.042\times10^{-42} \end{bmatrix} \text{kgm}^2 \quad (6.\text{E}6.6.11)$$

If the reduced global matrices are substituted into Eq. (6.20):

$$\det\left(\begin{bmatrix} 1.235\times10^{-17} & -6.174\times10^{-18} & 0 \\ -6.174\times10^{-18} & 1.235\times10^{-17} & -6.174\times10^{-18} \\ 0 & -6.174\times10^{-18} & 6.174\times10^{-18} \end{bmatrix} - \omega_i^2 \times \begin{bmatrix} 2.084\times10^{-42} & -2.905\times10^{-43} & 0 \\ -2.905\times10^{-43} & 2.084\times10^{-42} & -2.905\times10^{-43} \\ 0 & -2.905\times10^{-43} & 1.042\times10^{-42} \end{bmatrix}\right) = 0$$

(6.E6.6.12)

Table 6.7 Comparisons of nondimensional frequencies of clamped-free nanoshafts for nonlocal torsional free vibration ($e_0a/L = 0.2$, $K_T = 0$)

Mode number	Analytical	NL-FEM results				
	Eq. (3.E3.2.5)	Manual solution	Program code (MATLAB)			
		$n = 3$	$n = 3$	$n = 10$	$n = 15$	$n = 20$
1	1.49858	1.515	1.51419	1.49999	1.49921	1.49893
2	3.42933	3.603	3.60288	3.44607	3.43679	3.43353
3	4.21782	4.417	4.41711	4.24842	4.23159	4.22559
4	4.55152	–	–	4.58876	4.56860	4.56122
5	4.71386	–	–	4.75326	4.73248	4.72453

where the positive roots are presented below:

$$\omega_1 = 1.023 \ \text{THz}, \quad \omega_2 = 2.434 \ \text{THz}, \quad \omega_3 = 2.984 \ \text{THz}.$$

Analytical nondimensional frequencies calculated by using Eqs. (3.E3.2.5) and (3.E3.5.8) are compared with NL-FEM results in Table 6.7. Generally, it is understood that the NL-FEM solution also gives appropriate values in the nonlocal free torsional vibration of nanorods. It is reminded that the number of finite elements should be increased for high accuracy of nonlocal natural frequencies.

From here on, let's include the elastic medium in the nonlocal free vibration analysis. Firstly, the stiffness of elastic medium against torsional motion is obtained as follows:

$$k_{T,e=1} = k_{T,e=2} = k_{T,e=3} = k_T = \frac{K_T G I_P}{L^2} = \frac{(10) \times \left(0.42 \times 10^{12} \frac{\text{N}}{\text{m}^2}\right) \times \left(0.098 \times 10^{-36} \text{m}^4\right)}{\left(20 \times 10^{-9}\right)^2} = 1.029 \times 10^{-9} \ \text{N} \tag{6.E6.6.13}$$

Based on Eqs. (6.56) and (6.57), the global stiffness matrices of elastic medium can be written as follows for classical and nonlocal elasticity, respectively:

$$\sum [K_{T,c}] = \frac{k_T l}{6} \begin{bmatrix} 2 & 1 & 0 & 0 \\ 1 & 4 & 1 & 0 \\ 0 & 1 & 4 & 1 \\ 0 & 0 & 1 & 2 \end{bmatrix} \tag{6.E6.6.14}$$

$$\sum[K_{T,nl}] = \frac{(e_0 a)^2 k_T}{l}\begin{bmatrix} 1 & -1 & 0 & 0 \\ -1 & 2 & -1 & 0 \\ 0 & -1 & 2 & -1 \\ 0 & 0 & -1 & 1 \end{bmatrix} \quad \text{(E6.6.15)}$$

After the global matrices are reduced, these are calculated as follows:

$$\left(\sum[K_{T,c}]\right)^* = \begin{bmatrix} 4.574\times10^{-18} & 1.143\times10^{-18} & 0 \\ 1.143\times10^{-18} & 4.546\times10^{-18} & 1.143\times10^{-18} \\ 0 & 1.143\times10^{-18} & 2.287\times10^{-18} \end{bmatrix} \text{kg}\frac{\text{m}^2}{\text{s}^2} \quad \text{(6.E6.6.16)}$$

$$\left(\sum[K_{T,nl}]\right)^* = \begin{bmatrix} 4.939\times10^{-18} & -2.469\times10^{-18} & 0 \\ -2.469\times10^{-18} & 4.939\times10^{-18} & -2.469\times10^{-18} \\ 0 & -2.469\times10^{-18} & 2.469\times10^{-18} \end{bmatrix} \text{kg}\frac{\text{m}^2}{\text{s}^2} \quad \text{(6.E6.6.17)}$$

Thus, the reduced global stiffness matrix is given below:

$$\left(\sum[K]\right)^* = \left(\sum[K_{te}]\right)^* + \left(\sum[K_{T,c}]\right)^* + \left(\sum[K_{T,nl}]\right)^* = \begin{bmatrix} 2.186\times10^{-17} & -7.5\times10^{-18} & 0 \\ -7.5\times10^{-18} & 2.186\times10^{-17} & -7.5\times10^{-18} \\ 0 & -7.5\times10^{-18} & 1.093\times10^{-17} \end{bmatrix} \text{kg}\frac{\text{m}^2}{\text{s}^2} \quad \text{(6.E6.6.18)}$$

Substituting Eqs. (6.E6.6.11) and (6.E6.6.18) into Eq. (6.19) yields the following natural frequencies:

$$\omega_1 = 2.369 \text{ THz}, \quad \omega_2 = 3.239 \text{ THz}, \quad \omega_3 = 3.670 \text{ THz}. \quad \text{(6.E6.6.19)}$$

It should be noted that the elastic medium increases the frequencies compared to the results given by Eq. (6.E6.1.8). A comparison study on the nonlocal nondimensional frequencies of nanorods embedded in elastic medium is listed in Table 6.8.

Example 6.7 Determine the natural frequencies of nonlocal torsional free vibration of free-mass attached nanorod embedded on elastic foundation in Fig. 3.4. Use two finite

Table 6.8 Comparisons of nondimensional frequencies of embedded clamped-free nanoshafts for nonlocal torsional free vibration ($e_0a/L = 0.2$, $K_T = 10$)

Mode nuber	Analytical	NL-FEM results				
	Eq. (3.E3.2.5)	Manual solution	Program code (MATLAB)			
		$n = 3$	$n = 3$	$n = 10$	$n = 15$	$n = 20$
1	3.49939	3.506	3.50610	3.49999	3.49966	3.49954
2	4.66480	4.793	4.79382	4.67711	4.67028	4.66788
3	5.27162	5.432	5.43239	5.29614	5.28264	5.27785
4	5.54223	–	–	5.57286	5.55626	5.55020
5	5.67631	–	–	5.70907	5.69178	5.68517

elements in the analysis. Physical and mechanical parameters are as in Example 6.5. In addition to this, attachment ratio is $\beta_I = 10$ and defined as $\beta_I = I_a/\rho I_P L$.

Solution. In the nanorod divided into two finite elements, since the third node is the right end the nanorod, the mass matrix of second finite element that has this node is added the mass inertia of attachment.

The stiffness and mass matrices of finite elements should be defined as follows, respectively:

$$(K_{te})_1 = (K_{te})_2 = \frac{GI_P}{l}\begin{bmatrix} 1 & -1 \\ -1 & 1 \end{bmatrix} + \frac{k_T l}{6}\begin{bmatrix} 2 & 1 \\ 1 & 2 \end{bmatrix} + \frac{(e_0a)^2 k_T}{l}\begin{bmatrix} 1 & -1 \\ -1 & 1 \end{bmatrix} \tag{6.E6.7.1}$$

$$(M)_1 = \frac{\rho I_P l}{6}\begin{bmatrix} 2 & 1 \\ 1 & 2 \end{bmatrix} + \frac{(e_0a)^2 \rho I_P}{l}\begin{bmatrix} 1 & -1 \\ -1 & 1 \end{bmatrix}, \quad (M)_2 = (M)_1 + I_a\begin{bmatrix} 0 & 0 \\ 0 & 1 \end{bmatrix} \tag{6.E6.7.2}$$

The torsional mass inertia of disk (or rotating disk) attachment is calculated as follows

$$\begin{aligned} I_a &= \beta_I \rho I_P L = (10) \times \left(2300\,\frac{\text{kg}}{\text{m}^3}\right) \times \left(0.098 \times 10^{-36}\text{m}^4\right) \times \left(20 \times 10^{-9}\text{m}\right) \\ &= 4.508 \times 10^{-41}\ \text{kgm}^2 \end{aligned} \tag{6.E6.7.3}$$

On the other hand, since the nanorod does not have a geometric boundary condition, a reduction is not performed and the global stiffness and mass matrices are constituted as follows:

$$\sum[K]=\frac{GI_P}{l}\begin{bmatrix}1&-1&0\\-1&2&-1\\0&-1&1\end{bmatrix}+\frac{k_T l}{6}\begin{bmatrix}2&1&0\\1&4&1\\0&1&2\end{bmatrix}+\frac{(e_0a)^2k_T}{l}\begin{bmatrix}1&-1&0\\-1&2&-1\\0&-1&1\end{bmatrix}$$
$$=\begin{bmatrix}9.192\times10^{-18}&-4.074\times10^{-18}&0\\-4.074\times10^{-18}&1.838\times10^{-17}&-4.074\times10^{-18}\\0&-4.074\times10^{-18}&9.192\times10^{-18}\end{bmatrix}\text{kNm} \tag{6.E6.7.4}$$

$$\sum[M]=\frac{\rho I_P l}{6}\begin{bmatrix}2&1&0\\1&4&1\\0&1&2\end{bmatrix}+\frac{(e_0a)^2\rho I_P}{l}\begin{bmatrix}1&-1&0\\-1&2&-1\\0&-1&1\end{bmatrix}+m_a\begin{bmatrix}0&0&0\\0&0&0\\0&0&1\end{bmatrix}$$
$$=\begin{bmatrix}1.112\times10^{-42}&1.503\times10^{-44}&0\\1.503\times10^{-44}&2.224\times10^{-42}&1.503\times10^{-44}\\0&1.503\times10^{-44}&4.619\times10^{-41}\end{bmatrix}\text{kgm}^2 \tag{6.E6.7.5}$$

Natural frequencies can be calculated by substituting Eqs. (6.E6.7.4) and (6.E6.7.5) into Eq. (6.20). Table 6.9 provides a comparison study for the nondimensional frequencies of first three modes of free torsional mass attached nanorods. Vibration results with and without elastic medium are given separately in the table. Since the natural frequencies of this nanorod cannot be obtained by manual solution, the natural frequencies are reached by coding the frequency equation on MATLAB. Unfortunately, the program code does not yield any results in the vibration analysis where the elastic medium exists. It is seen that this difficulty has been overcome with the NL-FEM. On the other hand, it should be noted that since there is no rigid support condition, the nanostructure cannot show a vibration motion in the fundamental mode when there is no elastic medium.

Table 6.9 Comparisons of nondimensional frequencies of free torsional mass attached nanorods ($\beta_I = 10$, $e_0a/L = 0.2$)

K_T	Mode number	Analytical (MATLAB)	NL-FEM results			
		Eq. (3.E3.3.3)	Manual solution	Program code (MATLAB)		
			$n = 2$	$n = 5$	$n = 10$	$n = 15$
0	1	0.00000	0.000		0.00000	
	2	1.55144	1.587	1.55722	1.55289	1.55208
	3	3.43745	3.742	3.50323	3.45415	3.44489
10	1	–	0.622	0.60498	0.60232	0.60182
	2	–	3.514	3.50622	3.50437	3.50403
	3	–	4.902	4.71676	4.68036	4.67352

6.4.2 Elliptical Nanorod

6.4.2.1 NL-FEM Expression

The NL-FEM approach is formulated for the nonlocal free vibration of elliptical nanorods. Firstly, the following average weighted residue equation can be created based on the equation of motion given in Eq. (3.61):

$$\begin{aligned}\mathrm{I}=&\int_0^{l_e} h\Bigg(GI_\varphi\frac{\partial^2\theta}{\partial x^2}+(e_0a)^2k_T\frac{\partial^2\theta}{\partial x^2}-\rho I_P\frac{\partial^2\theta}{\partial t^2}-k_T\theta-(e_0a)^2\rho I_\psi\frac{\partial^6\theta}{\partial x^4\partial t^2}\\&+\rho I_\psi\frac{\partial^4\theta}{\partial x^2\partial t^2}+(e_0a)^2\rho I_P\frac{\partial^4\theta}{\partial x^2\partial t^2}-(e_0a)^2\frac{\partial^2 m}{\partial x^2}+m\Bigg)\mathrm{d}x=0\end{aligned}\tag{6.67}$$

this equation shows mathematically similarity and mechanically an analogy with Eq. (6.23). However as stated in Chap. 3, since the Saint-Venant's Principle is similar to the Rayleigh rod theory, not the Love-Bishop rod theory, the torsion of this elliptical rod does not contain the additional stiffness term.

The integrals in Eq. (6.67) can be arranged as follows by partial integration:

$$\begin{aligned}
&\mathrm{I}_1=\int_0^{l_e} GI_\varphi h\frac{\partial^2\theta}{\partial x^2}\mathrm{d}x=GI_\varphi h\frac{\partial\theta}{\partial x}\bigg|_0^{l_e}-\int_0^{l_e} GI_\varphi\frac{\partial h}{\partial x}\frac{\partial\theta}{\partial x}\mathrm{d}x,\\
&\mathrm{I}_2=\int_0^{l_e}(e_0a)^2k_Th\frac{\partial^2\theta}{\partial x^2}\mathrm{d}x=(e_0a)^2k_Th\frac{\partial\theta}{\partial x}\bigg|_0^{l_e}-\int_0^{l_e}(e_0a)^2k_T\frac{\partial h}{\partial x}\frac{\partial\theta}{\partial x}\mathrm{d}x,\ \mathrm{I}_3=\int_0^{l_e}\rho I_Ph\frac{\partial^2\theta}{\partial t^2}\mathrm{d}x,\\
&\mathrm{I}_4=\int_0^{l_e}k_Th\theta\mathrm{d}x,\ \mathrm{I}_5=\int_0^{l_e}hm\mathrm{d}x,\\
&\mathrm{I}_6=\int_0^{l_e}(e_0a)^2\rho I_Ph\frac{\partial^4\theta}{\partial x^2\partial t^2}\mathrm{d}x=(e_0a)^2\rho I_Ph\frac{\partial^3\theta}{\partial x\partial t^2}\bigg|_0^{l_e}-\int_0^{l_e}(e_0a)^2\rho I_P\frac{\partial h}{\partial x}\frac{\partial^3\theta}{\partial x\partial t^2}\mathrm{d}x,\\
&\mathrm{I}_7=\int_0^{l_e}\rho I_\psi h\frac{\partial^4\theta}{\partial x^2\partial t^2}\mathrm{d}x=\rho I_\psi h\frac{\partial^3\theta}{\partial x\partial t^2}\bigg|_0^{l_e}-\int_0^{l_e}\rho I_\psi\frac{\partial h}{\partial x}\frac{\partial^3\theta}{\partial x\partial t^2}\mathrm{d}x,\\
&\mathrm{I}_8=\int_0^{l_e}(e_0a)^2\rho I_\psi h\frac{\partial^6\theta}{\partial x^4\partial t^2}\mathrm{d}x=(e_0a)^2\rho I_\psi h\frac{\partial^6\theta}{\partial x^4\partial t^2}\bigg|_0^{l_e}-(e_0a)^2\rho I_\psi\frac{\partial h}{\partial x}\frac{\partial^5\theta}{\partial x^3\partial t^2}\bigg|_0^{l_e}\\
&\qquad+\int_0^{l_e}(e_0a)^2\rho I_\psi\frac{\partial^2 h}{\partial x^2}\frac{\partial^4\theta}{\partial x^2\partial t^2}\mathrm{d}x,\\
&\mathrm{I}_9=\int_0^{l_e}(e_0a)^2h\frac{\partial^2 m}{\partial x^2}\mathrm{d}x=(e_0a)^2h\frac{\partial m}{\partial x}\bigg|_0^{l_e}-\int_0^{l_e}(e_0a)^2\frac{\partial h}{\partial x}\frac{\partial m}{\partial x}\mathrm{d}x.
\end{aligned}\tag{6.68}$$

The partial integration of I_8 includes a higher-order derivative term. For this reason, NL-FEM formulation is required to using of three-node linear torsion element.

The following expression is written by using Eq. (6.68) into Eq. (6.67):

$$\begin{aligned}
&GI_{\varphi}h\frac{\partial\theta}{\partial x}\Big|_0^{l_e} + (e_0a)^2k_Th\frac{\partial\theta}{\partial x}\Big|_0^{l_e} + (e_0a)^2\rho I_Ph\frac{\partial^3\theta}{\partial x\partial t^2}\Big|_0^{l_e} + \rho I_{\psi}h\frac{\partial^3\theta}{\partial x\partial t^2}\Big|_0^{l_e}\\
&-(e_0a)^2\rho I_{\psi}h\frac{\partial^6\theta}{\partial x^4\partial t^2}\Big|_0^{l_e} + (e_0a)^2\rho I_{\psi}\frac{\partial h}{\partial x}\frac{\partial^5\theta}{\partial x^3\partial t^2}\Big|_0^{l_e} - (e_0a)^2h\frac{\partial q}{\partial x}\Big|_0^{l_e}\\
&-\int_0^{l_e} GI_{\varphi}\frac{\partial h}{\partial x}\frac{\partial\theta}{\partial x}\mathrm{d}x - \int_0^{l_e}(e_0a)^2k_T\frac{\partial h}{\partial x}\frac{\partial\theta}{\partial x}\mathrm{d}x - \int_0^{l_e}\rho I_Ph\frac{\partial^2\theta}{\partial t^2}\mathrm{d}x - \int_0^{l_e}k_Mh\theta\mathrm{d}x\\
&+\int_0^{l_e} hm\mathrm{d}x - \int_0^{l_e}(e_0a)^2\rho I_P\frac{\partial h}{\partial x}\frac{\partial^3\theta}{\partial x\partial t^2}\mathrm{d}x - \int_0^{l_e}\rho I_{\psi}\frac{\partial h}{\partial x}\frac{\partial^3\theta}{\partial x\partial t^2}\mathrm{d}x\\
&-\int_0^{l_e}(e_0a)^2\rho I_{\psi}\frac{\partial^2 h}{\partial x^2}\frac{\partial^4\theta}{\partial x^2\partial t^2}\mathrm{d}x + \int_0^{l_e}(e_0a)^2\frac{\partial h}{\partial x}\frac{\partial m}{\partial x}\mathrm{d}x = 0
\end{aligned} \tag{6.69}$$

where the required part for equation of motion of finite element can be defined as follows:

$$\begin{aligned}
&\int_0^{l_e} GI_{\varphi}\frac{\partial h}{\partial x}\frac{\partial\theta}{\partial x}\mathrm{d}x + \int_0^{l_e}(e_0a)^2k_T\frac{\partial h}{\partial x}\frac{\partial\theta}{\partial x}\mathrm{d}x + \int_0^{l_e}\rho I_Ph\frac{\partial^2\theta}{\partial t^2}\mathrm{d}x + \int_0^{l_e}k_Mh\theta\mathrm{d}x - \int_0^{l_e}hm\mathrm{d}x\\
&+\int_0^{l_e}(e_0a)^2\rho I_P\frac{\partial h}{\partial x}\frac{\partial^3\theta}{\partial x\partial t^2}\mathrm{d}x + \int_0^{l_e}\rho I_{\psi}\frac{\partial h}{\partial x}\frac{\partial^3\theta}{\partial x\partial t^2}\mathrm{d}x + \int_0^{l_e}(e_0a)^2\rho I_{\psi}\frac{\partial^2 h}{\partial x^2}\frac{\partial^4\theta}{\partial x^2\partial t^2}\mathrm{d}x\\
&-\int_0^{l_e}(e_0a)^2\frac{\partial h}{\partial x}\frac{\partial m}{\partial x}\mathrm{d}x = 0
\end{aligned} \tag{6.70}$$

By using Eqs. (6.7) and (6.29) into Eq. (6.67), the following expression is obtained:

$$\int_0^{l_e} GI_\varphi(\mathbf{B}^\mathrm{T}\mathbf{B})\mathbf{d}\mathrm{d}x + \int_0^{l_e} (e_0a)^2 k_T(\mathbf{B}^\mathrm{T}\mathbf{B})\mathbf{d}\mathrm{d}x + \int_0^{l_e} \rho I_P(\phi^\mathrm{T}\phi)\ddot{\mathbf{d}}\mathrm{d}x + \int_0^{l_e} k_T(\phi^\mathrm{T}\phi)\mathbf{d}\mathrm{d}x$$
$$- \int_0^{l_e} \phi^\mathrm{T} m\mathrm{d}x + \int_0^{l_e} (e_0a)^2 \rho I_P(\mathbf{B}^\mathrm{T}\mathbf{B})\ddot{\mathbf{d}}\mathrm{d}x + \int_0^{l_e} \rho I_\psi(\mathbf{B}^\mathrm{T}\mathbf{B})\ddot{\mathbf{d}}\mathrm{d}x$$
$$+ \int_0^{l_e} (e_0a)^2 \rho I_\psi\left(\mathbf{B}'^\mathrm{T}\mathbf{B}'\right)\ddot{\mathbf{d}}\mathrm{d}x - \int_0^{l_e} (e_0a)^2 \mathbf{B}^\mathrm{T} m'\mathrm{d}x = 0 \tag{6.71}$$

this equation can be rewritten in the following form:

$$(K + K_{T,c} + K_{T,nl})\mathbf{d} + \left(I_c + I_{\psi,c} + I_{nl} + I_{\psi,nl}\right)\ddot{\mathbf{d}} = \mathbf{m}_c + \mathbf{m}_{nl} \tag{6.72}$$

Here, unlike the circular nanorod, $I_{\psi,c}$ is the classical mass inertia matrix of warping and $I_{\psi,nl}$ is the nonlocal mass inertia matrix of warping. All expressions seen in Eq. (6.72) are formulated as follows:

$$K = \int_0^{l_e} GI_\varphi(\mathbf{B}^\mathrm{T}\mathbf{B})\mathrm{d}x = \int_0^{l_e} GI_\varphi \begin{Bmatrix} \phi_1' \\ \phi_2' \\ \phi_3' \end{Bmatrix} [\phi_1' \quad \phi_2' \quad \phi_3']\mathrm{d}x = \frac{GI_\varphi}{3l_e}\begin{bmatrix} 7 & -8 & 1 \\ -8 & 16 & -8 \\ 1 & -8 & 7 \end{bmatrix} \tag{6.73}$$

$$K_{T,c} = \int_0^{l_e} k_T(\phi^\mathrm{T}\phi)\mathrm{d}x = \int_0^{l_e} k_T \begin{Bmatrix} \phi_1 \\ \phi_2 \\ \phi_3 \end{Bmatrix} [\phi_1 \quad \phi_2 \quad \phi_3]\mathrm{d}x = \frac{k_T l_e}{30}\begin{bmatrix} 4 & 2 & -1 \\ 2 & 16 & 2 \\ -1 & 2 & 4 \end{bmatrix} \tag{6.74}$$

$$K_{T,nl} = \int_0^{l_e} (e_0a)^2 k_T(\mathbf{B}^\mathrm{T}\mathbf{B})\mathrm{d}x = \int_0^{l_e} (e_0a)^2 k_T \begin{Bmatrix} \phi_1' \\ \phi_2' \\ \phi_3' \end{Bmatrix} [\phi_1' \quad \phi_2' \quad \phi_3']\mathrm{d}x$$
$$= \frac{(e_0a)^2 k_T}{3l_e}\begin{bmatrix} 7 & -8 & 1 \\ -8 & 16 & -8 \\ 1 & -8 & 7 \end{bmatrix} \tag{6.75}$$

$$I_c = \int_0^{l_e} \rho I_P (\phi^{\mathrm{T}} \phi) \mathrm{d}x = \int_0^{l_e} \rho I_P \begin{Bmatrix} \phi_1 \\ \phi_2 \\ \phi_3 \end{Bmatrix} [\phi_1 \quad \phi_2 \quad \phi_3] \mathrm{d}x = \frac{\rho I_P l_e}{30} \begin{bmatrix} 4 & 2 & -1 \\ 2 & 16 & 2 \\ -1 & 2 & 4 \end{bmatrix} \tag{6.76}$$

$$I_{\psi,c} = \int_0^{l_e} \rho I_\psi (\mathbf{B}^{\mathrm{T}} \mathbf{B}) \mathrm{d}x = \int_0^{l_e} \rho I_\psi \begin{Bmatrix} \phi_1' \\ \phi_2' \\ \phi_3' \end{Bmatrix} [\phi_1' \quad \phi_2' \quad \phi_3'] \mathrm{d}x = \frac{\rho I_\psi}{3 l_e} \begin{bmatrix} 7 & -8 & 1 \\ -8 & 16 & -8 \\ 1 & -8 & 7 \end{bmatrix} \tag{6.77}$$

$$\begin{aligned} I_{nl} &= \int_0^{l_e} (e_0 a)^2 \rho I_P (\mathbf{B}^{\mathrm{T}} \mathbf{B}) \mathrm{d}x = \int_0^{l_e} (e_0 a)^2 \rho I_P \begin{Bmatrix} \phi_1' \\ \phi_2' \\ \phi_3' \end{Bmatrix} [\phi_1' \quad \phi_2' \quad \phi_3'] \mathrm{d}x \\ &= \frac{(e_0 a)^2 \rho I_P}{3 l_e} \begin{bmatrix} 7 & -8 & 1 \\ -8 & 16 & -8 \\ 1 & -8 & 7 \end{bmatrix} \end{aligned} \tag{6.78}$$

$$\begin{aligned} I_{\psi,nl} &= \int_0^{l_e} (e_0 a)^2 \rho I_\psi \left(\mathbf{B}'^{\mathrm{T}} \mathbf{B}' \right) \mathrm{d}x = \int_0^{l_e} (e_0 a)^2 \rho I_\psi \begin{Bmatrix} \phi_1'' \\ \phi_2'' \\ \phi_3'' \end{Bmatrix} [\phi_1'' \quad \phi_2'' \quad \phi_3''] \mathrm{d}x \\ &= \frac{(e_0 a)^2 \rho I_\psi}{l_e^3} \begin{bmatrix} 16 & -32 & 16 \\ -32 & 64 & -32 \\ 16 & -32 & 16 \end{bmatrix} \end{aligned} \tag{6.79}$$

$$\mathbf{m}_c = \int_0^{l_e} m \phi^{\mathrm{T}} \mathrm{d}x = \int_0^{l_e} m \begin{Bmatrix} \phi_1 \\ \phi_2 \\ \phi_3 \end{Bmatrix} \mathrm{d}x = \frac{m l_e}{6} \begin{Bmatrix} 1 \\ 4 \\ 1 \end{Bmatrix} \tag{6.80}$$

$$\mathbf{m}_{nl} = \int_0^{l_e} (e_0 a)^2 m' \mathbf{B}^{\mathrm{T}} \mathrm{d}x = \int_0^{l_e} (e_0 a)^2 m' \begin{Bmatrix} \phi_1' \\ \phi_2' \\ \phi_3' \end{Bmatrix} \mathrm{d}x = \frac{(e_0 a)^2 m'}{l_e} \begin{Bmatrix} 1 \\ -4 \\ 3 \end{Bmatrix} \tag{6.81}$$

Interpretation. According to the comparison of Eqs. (6.54) and (6.72), it is observed that the elliptic nanorod only changes the classical stiffness matrix and is included in the mass matrix with an additional matrix containing the warping effect. Also, in terms of nonlocal elasticity, it is understood that the warping effect also incorporates a nonlocal mass matrix into the finite element analysis. These conclusions emphasize that the atomic size effect is more important for the torsion of elliptical nanorod.

Solution stage. The related expressions of finite element are stated as follows:

$$[K]_e = K + K_I + K_{M,c} + K_{M,nl} \tag{6.82}$$

$$[M]_e = M_c + M_{I,c} + M_{nl} + M_{I,nl} \tag{6.83}$$

$$\mathbf{m}_e = \mathbf{m}_c + \mathbf{m}_{nl} \tag{6.84}$$

In free vibration analysis, considering that the external excitation is equal to zero, Eq. (6.20) containing global matrices rearranged by boundary conditions is solved and natural frequencies are reached.

6.4.2.2 Application of Boundary Conditions

The application of boundary conditions to global matrices is as introduced in the nonlocal Love-Bishop nanorod. In addition to this, the matrices of finite element comprising the attachment are specified as follows for nonlocal nanorods with end attachment:

$$[K]_{e,a} = [K]_{e,na} + (K_a) = (K_{te})_e + (K_{T,c})_e + (K_{T,nl})_e + \begin{bmatrix} 0 & 0 & 0 \\ 0 & 0 & 0 \\ 0 & 0 & k_a \end{bmatrix} \tag{6.85}$$

$$[M]_{e,a} = [M]_{e,na} + (M_a) = (I_c)_e + (I_{\psi,c})_e + (I_{nl})_e + (I_{\psi,nl})_e + \begin{bmatrix} 0 & 0 & 0 \\ 0 & 0 & 0 \\ 0 & 0 & I_a \end{bmatrix} \tag{6.86}$$

Example 6.8 Determine the natural frequencies of nonlocal torsional free vibration of clamped-clamped elliptical nanoshaft in Fig. 6.4. Use two finite elements in the analysis.

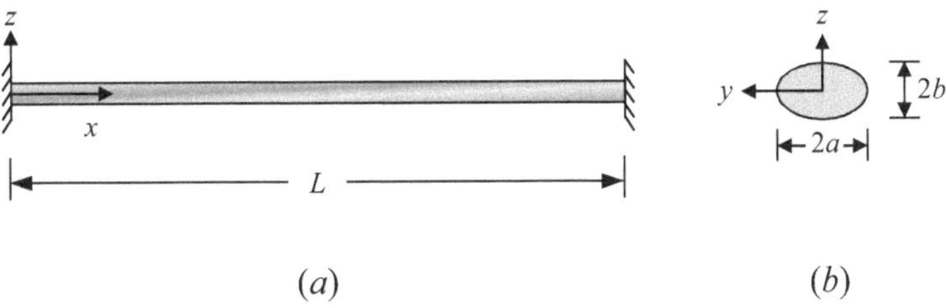

Fig. 6.4 (**a**) Clamped-clamped elliptical nanorod. (**b**) Elliptical cross-section

Nanoshaft length: $L = 20$ nm, horizontal and vertical diameters of elliptical cross-section: $2a = 2$ nm and $2b = 1$ nm, modulus of elasticity: $E = 1$ TPa, and mass of unit volume: $\rho = 2300$ kg/m^3, nondimensional nonlocal parameter: $e_0a/L = 0.2$.

Solution. We know that a three-node finite element should be used for this nanostructure. Additionally, applications of two three-node finite elements are described in Example 6.4. Based on the related example and considering Eqs. (6.73)–(6.79), the global matrices required for NL-FEM analysis can be listed as follows:

$$\sum[K]=\frac{GI_\varphi}{3l}\begin{bmatrix} 7 & -8 & 1 & 0 & 0 \\ -8 & 16 & -8 & 0 & 0 \\ 1 & -8 & 7+7 & -8 & 1 \\ 0 & 0 & -8 & 16 & -8 \\ 0 & 0 & 1 & -8 & 7 \end{bmatrix} \quad (6.E6.8.1)$$

(Columns: 1, 2, 3, 4, 5; rows: 1, 2, 3, 4, 5 — *Mesh node numbers*)

$$\sum[I_c]=\frac{\rho I_p l}{30}\begin{bmatrix} 4 & 2 & -1 & 0 & 0 \\ 2 & 16 & 2 & 0 & 0 \\ -1 & 2 & 4+4 & 2 & -1 \\ 0 & 0 & 2 & 16 & 2 \\ 0 & 0 & -1 & 2 & 4 \end{bmatrix} \quad (6.E6.8.2)$$

(Columns: 1, 2, 3, 4, 5; rows: 1, 2, 3, 4, 5 — *Mesh node numbers*)

$$\sum\left[I_{\psi,c}\right]=\frac{\rho I_\psi}{3l}\begin{bmatrix} 7 & -8 & 1 & 0 & 0 \\ -8 & 16 & -8 & 0 & 0 \\ 1 & -8 & 14 & -8 & 1 \\ 0 & 0 & -8 & 16 & -8 \\ 0 & 0 & 1 & -8 & 7 \end{bmatrix} \quad (6.E6.8.3)$$

$$\sum[I_{nl}]=\frac{(e_0a)^2\rho I_P}{3l}\begin{bmatrix} 7 & -8 & 1 & 0 & 0 \\ -8 & 16 & -8 & 0 & 0 \\ 1 & -8 & 14 & -8 & 1 \\ 0 & 0 & -8 & 16 & -8 \\ 0 & 0 & 1 & -8 & 7 \end{bmatrix} \quad (6.E6.8.4)$$

$$\sum\left[I_{\psi,nl}\right]=\frac{(e_0a)^2\rho I_\psi}{l^3}\begin{bmatrix}16 & -32 & 16 & 0 & 0\\ -32 & 64 & -32 & 0 & 0\\ 16 & -32 & 32 & -32 & 16\\ 0 & 0 & -32 & 64 & -32\\ 0 & 0 & 16 & -32 & 16\end{bmatrix} \tag{6.E6.8.5}$$

where the warping stiffness and mass moments of inertia that are two important parameters are defined by Eqs. (3.E3.5.3) and (3.E3.5.7), respectively. While the stiffness and moment of inertia of warping can be calculated manually, unfortunately, programming must be done on the computer as integral transformations are required in the mass moment of inertia of the distortion. I_y, I_z, and I_P that are the moments of inertia formulated by Eq. (3.E3.5.5) for the stiffness moment of inertia of warping and the multiplication coefficient S_ψ should be computed:

$$\begin{aligned}
S_\psi&=\frac{b^2-a^2}{b^2+a^2}=\frac{(0.5)^2-(1)^2}{(0.5)^2+(1)^2}=-0.6,\\
I_y&=\frac{1}{4}\pi ab^3=\frac{\pi\times(1)\times(0.5)^3}{4}\times\left(10^{-36}\ \mathrm{m}^4\right)=9.817\times10^{-38}\ \mathrm{m}^4,\\
I_z&=\frac{1}{4}\pi a^3b=\frac{\pi\times(1)^3\times(0.5)}{4}\times\left(10^{-36}\ \mathrm{m}^4\right)=3.927\times10^{-37}\ \mathrm{m}^4,\\
I_P&=I_z+I_y=\left(9.817\times10^{-38}\ \mathrm{m}^4\right)+\left(3.927\times10^{-37}\ \mathrm{m}^4\right)=4.909\times10^{-37}\ \mathrm{m}^4.
\end{aligned} \tag{6.E6.8.6}$$

Thus, the stiffness moment of inertia of warping is written as follows:

$$\begin{aligned}
I_\varphi&=\left({S_\psi}^2+1\right)I_P+2S_\psi\left(I_z-I_y\right)\\
&=\left[(-0.6)^2+(1)\right]\times\left(4.909\times10^{-37}\right)+(2)\times(-0.6)\times\left(3.927\times10^{-37}-9.817\times10^{-38}\right)\\
&=3.142\times10^{-37}\mathrm{m}^4
\end{aligned} \tag{6.E6.8.7}$$

Additionally, as a result of the electronic computation, the mass moment of inertia of warping is:

$$
\begin{aligned}
I_{\psi} &= \frac{S_{\psi}{}^{2}}{6}\int_{0}^{2\pi}\left(\frac{ab}{\sqrt{(b\cos\theta)^{2}+(a\sin\theta)^{2}}}\right)^{6}\sin^{2}\theta\cos^{2}\theta d\theta \\
&= \frac{(-0.6)^{2}}{6}\times\left(\int_{0}^{2\pi}\frac{(1)^{6}\times(0.5)^{6}}{\left[(1)^{2}\times\cos^{2}\theta+(0.5)^{2}\times\sin^{2}\theta\right]^{3}}\sin^{2}\theta\cos^{2}\theta d\theta\right)\times 10^{-54}\mathrm{m}^{6} \\
&= 5.89\times10^{-57}\mathrm{m}^{6}
\end{aligned}
\tag{6.E6.8.8}
$$

After reducing the global matrices defined by Eqs. (6.E6.8.1)–(6.E6.8.5) by geometric boundary conditions, the stiffness and mass matrices that should be included in the eigenvalue formulation of nanorod are attained as follows:

$$
\left(\sum[K]_{e}\right)^{*} = \begin{bmatrix} 7.401\times10^{-17} & -3.521\times10^{-17} & 0 \\ -3.521\times10^{-17} & 6.161\times10^{-17} & -3.521\times10^{-17} \\ 0 & -3.521\times10^{-17} & 7.401\times10^{-17} \end{bmatrix}\mathrm{Nm}
\tag{6.E6.8.9}
$$

$$
\begin{aligned}
\left(\sum[M]_{e}\right)^{*} &= \left(\sum[I_{c}]_{e}\right)^{*}+\left(\sum[I_{\psi,c}]_{e}\right)^{*}+\left(\sum[I_{nl}]_{e}\right)^{*}+\left(\sum[I_{\psi,nl}]_{e}\right)^{*} \\
&= \begin{bmatrix} 1.568\times10^{-41} & -4.075\times10^{-42} & 0 \\ -4.075\times10^{-42} & 1.145\times10^{-41} & -4.075\times10^{-42} \\ 0 & -4.075\times10^{-42} & 1.568\times10^{-41} \end{bmatrix}\mathrm{kgm}^{2}
\end{aligned}
\tag{6.E6.8.10}
$$

Natural frequencies can be calculated with the help of Eq. (6.20):

$$
\omega_{1}=1.519\ \mathrm{THz},\quad \omega_{2}=2.173\ \mathrm{THz},\quad \omega_{3}=2.483\ \mathrm{THz}.
$$

In the nonlocal torsional vibration of clamped elliptical nanorods, the nondimensional forms of natural frequencies of first five modes are given in Table 6.10. As will be remembered, analytical results are presented in Example 3.5. NL-FEM generally computes successful values in this atomic size-dependent vibration problem. However, here one topic attracted attention compared to previous NL-FEM analyses. According to this, when considering the results on the using of two finite elements, the difference between the manual and computer solutions is high, especially in the fundamental mode. The result of computer programming is closer to the analytical result. It can be stated that this difference is due to the fact that this problem contains different moment of inertia terms based on

Table 6.10 Comparisons of nondimensional frequencies of clamped elliptical nanoshafts for nonlocal torsional free vibration ($e_0a/L = 0.2$)

Mode number	Analytical	NL-FEM results				
	Eq. (3.E3.2.5)	Manual solution	Program code (MATLAB)			
		$n = 2$	$n = 2$	$n = 5$	$n = 10$	$n = 15$
1	2.1278	2.247	2.1335	2.1279	2.1278	2.1278
2	3.1281	3.215	3.1357	3.1301	3.1281	2.1279
3	3.5288	3.674	3.6574	3.5349	3.5289	3.5276
4	3.7078	–	–	3.7187	3.7074	3.7019
5	3.7975	–	–	3.7971	3.7955	3.7782

various calculations. Because as the number of calculations increases, the deviation of manual calculation increases due to the digit precision used.

On the other hand, it is observed that the frequencies of some modes calculated via NL-FEM drop below the analytical results, especially as the number of finite elements increases. Here excessive decrease in the frequency evokes that some mass matrices form high numbers. This problem can be specifically investigated by examining finite element formulations for classical elasticity and nonlocal elasticity separately, but this research is not covered in the book.

Example 6.9 Determine the natural frequencies of nonlocal torsional free vibration of clamped-mass attached elliptical nanorod embedded on elastic foundation. Use two finite elements in the analysis. Physical and mechanical parameters are as in Example 6.8. In addition to this, attachment ratio: $\beta_I = 10$, nondimensional stiffness of elastic medium: $K_T = 10$. These two parameters are defined in Examples 6.5 and 6.6.

Solution. Since the number of finite elements used is the same as in the previous example, in addition to the global matrices obtained according to Eqs. (6.E6.8.1)–(6. E6.8.5), the stiffness matrices resulting from the elastic environment are assembled in Eqs. (6.E6.9.1) and (6.E6.9.2). Also, the mass matrix resulting from the rotating disk attachment is written as in Eq. (6.E6.9.3):

$$\sum [K_{T,c}] = \frac{k_T l_e}{30} \begin{bmatrix} 4 & 2 & -1 & 0 & 0 \\ 2 & 16 & 2 & 0 & 0 \\ -1 & 2 & 4+4 & 2 & -1 \\ 0 & 0 & 2 & 16 & 2 \\ 0 & 0 & -1 & 2 & 4 \end{bmatrix} \tag{6.E6.9.1}$$

$$\sum[K_{T,nl}] = \frac{(e_0a)^2 k_T}{3l}\begin{bmatrix} 7 & -8 & 1 & 0 & 0 \\ -8 & 16 & -8 & 0 & 0 \\ 1 & -8 & 14 & -8 & 1 \\ 0 & 0 & -8 & 16 & -8 \\ 0 & 0 & 1 & -8 & 7 \end{bmatrix} \tag{6.E6.9.2}$$

$$\sum[I_a] = \begin{bmatrix} 0 & 0 & 0 & 0 & 0 \\ 0 & 0 & 0 & 0 & 0 \\ 0 & 0 & 0 & 0 & 0 \\ 0 & 0 & 0 & 0 & 0 \\ 0 & 0 & 0 & 0 & I_a \end{bmatrix} \tag{6.E6.9.3}$$

where the stiffness of elastic medium is calculated as follows:

$$k_T = \frac{K_T G I_P}{L^2} = \frac{(10)\times\left(4.202\times10^{11}\mathrm{N/m^2}\right)\times\left(4.909\times10^{-37}\,\mathrm{m^4}\right)}{\left(20\times10^{-9}\right)^2\mathrm{m^2}}$$
$$= 5.157\times10^{-9}\ \mathrm{N} \tag{6.E6.9.4}$$

On the other hand, the mass of rotating disk is as:

$$I_a = \beta_I \rho I_P L = (10)\times\left(2300\,\frac{\mathrm{kg}}{\mathrm{m^3}}\right)\times\left(4.909\times10^{-37}\ \mathrm{m^4}\right)\ \times\left(20\times10^{-9}\mathrm{m}\right)$$
$$= 2.258\times10^{-40}\ \mathrm{kgm^2} \tag{6.E6.9.5}$$

After the matrices given by Eqs. (6.E6.8.1)–(6.E6.8.5), (6.E6.9.1) and (6.E6.9.2) is reduced by left end of nanoshaft, the total reduced global stiffness and mass matrices can be achieved below:

$$\left(\sum[K]\right)^* = \left(\sum[K_{te}]\right)^* + \left(\sum[K_{T,c}]\right)^* + \left(\sum[K_{T,nl}]\right)^*$$
$$= \begin{bmatrix} 1.419\times10^{-16} & -5.377\times10^{-16} & 0 & 0 \\ -5.377\times10^{-17} & 1.139\times10^{-16} & -5.377\times10^{-17} & 5.432\times10^{-18} \\ 0 & -5.377\times10^{-17} & 1.419\times10^{-16} & -5.377\times10^{-16} \\ 0 & 5.432\times10^{-18} & -5.377\times10^{-17} & 5.694\times10^{-17} \end{bmatrix}\frac{\mathrm{N}}{\mathrm{m}} \tag{6.E6.9.6}$$

Table 6.11 Comparisons of nondimensional frequencies of clamped-torsional mass attached elliptical nanorods ($\beta_I = 10$, $e_0a/L = 0.2$)

K_T	Mode number	NL-FEM results		
		Manual solution	Program code (MATLAB)	
		$n = 2$	$n = 5$	$n = 10$
0	1	0.248	0.2484	0.2484
	2	2.149	2.1432	2.1430
	3	3.139	3.1331	3.1312
10	1	0.564	0.5628	0.5628
	2	3.818	3.8139	3.8138
	3	4.452	4.4492	4.4477

$$\left(\sum[I]\right)^* = \left(\sum[I_c]\right)^* + \left(\sum[I_{\psi,c}]\right)^* + \left(\sum[I_{nl}]\right)^* + \left(\sum[I_{\psi,nl}]\right)^* + \left(\sum[I_a]\right)^*$$
$$= \begin{bmatrix} 1.568\times10^{-41} & -4.075\times10^{-42} & 0 & 0 \\ -4.075\times10^{-42} & 1.145\times10^{-41} & -4.075\times10^{-42} & 2.297\times10^{-43} \\ 0 & -4.075\times10^{-42} & 1.568\times10^{-41} & -4.075\times10^{-42} \\ 0 & 2.297\times10^{-43} & -4.075\times10^{-42} & 2.315\times10^{-40} \end{bmatrix} \mathrm{kgm}^2 \tag{6.E6.9.7}$$

Natural frequencies are obtained by substituting these two expressions into Eq. (6.20). The nondimensional forms of natural frequencies based on NL-FEM according to Eq. (3.E3.5.8) are compared in Table 6.11. If you remember, in Chap. 3, the natural frequency equations of the attached nanorods are not obtained by analytical solution due to the insufficient of mechanical boundary conditions. Here the using of NL-FEM enables to approximately learn the vibration frequencies of a nanorod with mass attachment and embedded in an elastic medium.

Problems

6.1. Compute the first five modes nondimensional frequencies of axial nanorods given in Fig. 6.5 according to nonlocal SRT under parameters as follows. Use 1, 2 and 3 finite elements, respectively for manual solution. Also, if you have a program code or package program based on finite element analysis, try to obtain the approximate values by using 10, 15, and 20 finite elements. Discuss the effects of nonlocal parameter and boundary conditions on differences between results.

Nanobeam length: $L = 20$ nm, diameter of circular cross-section: $d = 1$ nm, modulus of elasticity: $E = 1$ TPa, Poisson's ratio: $\upsilon = 0.19$, mass of unit

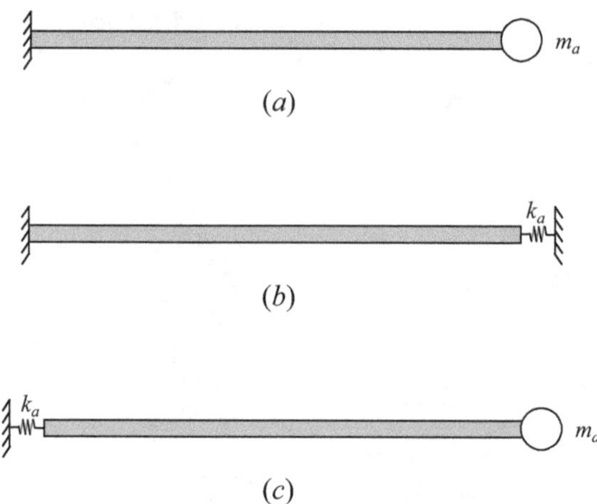

Fig. 6.5 Axial nanorods with different boundary conditions. (**a**) Clamped supported-mass attached (**b**) Clamped-spring attached (**c**) Spring attached-mass attached

volume: $\rho = 2300$ kg/m^3, stiffness ratio for axial spring attachment $\beta_k = 10$, mass ratio for axial spring attachment $\beta_m = 10$.

Compute according to three different nondimensional nonlocal parameters: $\alpha = 0$, 0.1, 0.2.

Compute the mass, stiffness, and nondimensional frequencies of nanorod as: $\beta_k = k_a L/EA$, $\beta_m = m_a/\rho AL$, and $\overline{\omega} = \omega L^2 \sqrt{\rho/E}$.

6.2. Examine the Problem 6.1 by considering also an axial elastic medium according to three different medium stiffnesses: $k_M = 1$ kN/m^2, 10 kN/m^2, 100 kN/m^2. Discuss the effect of nonlocal parameter and medium stiffness on differences between results.

6.3. Compute the first five modes nondimensional frequencies of axial nanorods given in Fig. 6.6 according to nonlocal LBRT under physical and mechanical parameters given in Example 6.5. Use 1, 2 and 3 finite elements, respectively for manual solution. Also, if you have a program code or package program based on finite element analysis, try to obtain the approximated values by using 10, 15, and 20 finite elements. Discuss the effects of nonlocal parameter and boundary conditions on differences between results.

6.4. Examine the Problem 6.3 by considering also an axial elastic medium according to three different medium stiffnesses: $k_M = 1$ kN/m^2, 10 kN/m^2, 100 kN/m^2. Discuss the effect of nonlocal parameter and medium stiffness on differences between results.

Compute the elastic medium stiffness as: $K_M = k_M L^2/EA$.

6.5. Develop the NL-FEM analysis for nonlocal SRT nanorods according to a three-node axial linear element. Then, compute the first five modes nondimensional frequencies of clamped-free and clamped-clamped nanorods by using this NL-FEM procedure. Consider the physical and mechanical parameters given in Problem 6.1. Since the operation volume increases, it is recommended to perform the analysis electronically and examine the convergence of results by using 4, 5 and 6 elements.

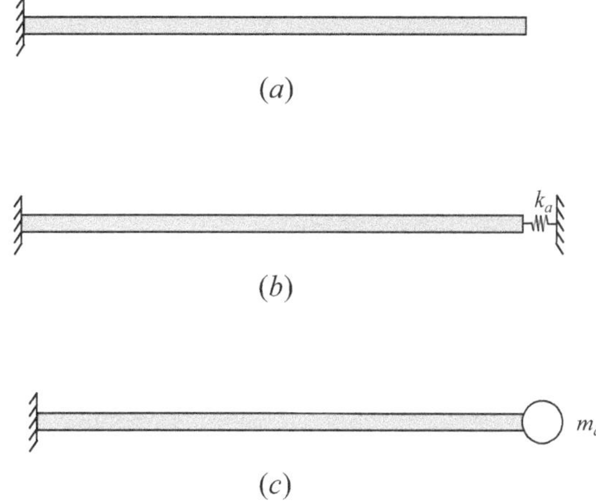

Fig. 6.6 Axial nanorods with different boundary conditions. (**a**) Clamped free. (**b**) Clamped-spring attached. (**c**) Clamped-mass attached

6.6. Compute the first five modes nondimensional torsional frequencies of circular nanorods given in Fig. 3.7 under parameters as follows. Use 1, 2 and 3 finite elements, respectively for manual solution. Also, if you have a program code or package program based on finite element analysis, try to obtain the approximate values by using 10, 15, and 20 finite elements. Discuss the effects of nonlocal parameter and boundary conditions on differences between results.

Nanobeam length: $L = 20$ nm, diameter of circular cross-section: $d = 1$ nm, modulus of elasticity: $E = 1$ TPa, Poisson's ratio: $\upsilon = 0.19$, mass of unit volume: $\rho = 2300$ kg/m^3, stiffness and mass ratios for torsional spring and mass attachments: $\beta_k = 10$ and $\beta_I = 10$.

Compute according to three different nondimensional nonlocal parameters: $\alpha = 0$, 0.1, 0.2.

Compute the mass, stiffness, and nondimensional frequencies of nanorod as: $\beta_r = k_a L/GI_P$, $\beta_I = I_a/\rho I_P L$, and $\overline{\omega} = \omega L^2 \sqrt{\rho/G}$.

6.7. Examine the Problem 6.6 by considering also torsional elastic medium according to two different nondimensional medium stiffnesses: $K_M = 1$ and 20. Discuss the effect of elastic medium on differences between results.

Compute the elastic medium stiffness as: $K_M = k_M L^2/GI_P$.

6.8. Compute the first five modes nondimensional torsional frequencies of elliptical nanorod given in Problem 3.6. It is recommended to perform the analysis electronically and examine the convergence of results by using 4, 5 and 6 elements. Discuss the effects of nonlocal parameter on the differences between results.

6.9. Compute the first five modes nondimensional torsional frequencies of nanorod given in Fig. 3.7 under parameters as follows by considering elliptical cross-section. It is recommended to perform the analysis electronically and examine the convergence of results by using 4, 5 and 6 elements. Discuss the effects of nonlocal parameter on the differences between results.

Nanorod length: $L = 20$ nm, vertical and horizontal radii of elliptical cross-section: $a = 1$ nm, $b = 0.5$ nm, modulus of elasticity: $E = 1$ TPa, mass of unit volume: $\rho = 2300$ kg/m^3, Poisson's ratio: $\upsilon = 0.19$.

Compute according to three different nondimensional nonlocal parameters: $\alpha = 0$, 0.1, 0.2.

Compute the nondimensional frequencies as: $\overline{\omega} = \omega L \sqrt{\rho / G}$.

6.10. Examine the Problem 6.9 by considering also torsional elastic medium according to two different nondimensional medium stiffnesses: $K_M = 1$ and 20. Discuss the effect of elastic medium on differences between results.

Compute the elastic medium stiffness as: $K_M = k_M L^2 / G I_P$.

6.11. Compute the first five modes nondimensional torsional frequencies of nanorod given in Fig. 3.8. It is recommended to perform the analysis electronically and examine the convergence of results by using 4, 5 and 6 elements. Discuss the effects of nonlocal parameter on the differences between results.

References

1. S. Adhikari, T. Murmu, M.A. McCarthy, Dynamic finite element analysis of axially vibrating nonlocal rods. Finite Elem. Anal. Des. **63**, 42–50 (2013)
2. S. Adhikari, T. Murmu, M.A. McCarthy, Frequency domain analysis of nonlocal rods embedded in an elastic medium. Phys. E **59**, 33–40 (2014)
3. Ç. Demir, Ö. Civalek, Torsional and longitudinal frequency and wave response of microtubules based on the nonlocal continuum and nonlocal discrete models. App. Math. Model. **37**(22), 9355–9367 (2013)
4. T.P. Chang, Axial vibration of non-uniform and non-homogeneous nanorods based on nonlocal elasticity theory. Appl. Math Comput. **219**(10), 4933–4941 (2013)
5. H.M. Numanoğlu, Ö. Civalek, Novel size-dependent finite element formulation for modal analysis of cracked nanorods. Mater Today Commun **31**, 103545 (2022)
6. H.M. Numanoğlu, F. Şen, Axial dynamics of elastic deformable supported nonlocal rods using a higher-order nonlocal FEM, in *2nd International Symposium on Characterization, 22–-25 September 2022*, (2022), pp. 222–230
7. Ö. Civalek, H.M. Numanoğlu, Nonlocal finite element analysis for axial vibration of embedded love–bishop nanorods. Int. J. Mech. Sci. **188**, 105939 (2020)
8. H.M. Numanoğlu, K. Mercan, Ö. Civalek, Free vibration analysis of Love-Bishop nanorods with deformable restraints based on nonlocal finite element solution, in *21th National Machine Theory Symposium, 13–15 September 2023*, (2023), pp. 295–305. (In Turkish)
9. C.W. Lim, M.Z. Islam, G. Zhang, A nonlocal finite element method for torsional statics and dynamics of circular nanostructures. Int. J. Mech. Sci. **94-95**, 232–243 (2015)
10. H.M. Numanoğlu, Ö. Civalek, On the torsional vibration of nanorods surrounded by elastic matrix via nonlocal FEM. Int. J. Mech. Sci. **161–162**, 105076 (2019)

11. L. Li, Y. Hu, X. Li, Longitudinal vibration of size-dependent rods via nonlocal strain gradient theory. Int. J. Mech. Sci. **115–116**, 135–144 (2016)
12. S. Narendar, S. Gopalakrishnan, Spectral finite element formulation for nanorods via nonlocal continuum mechanics. J. Appl. Mech. **78**(6), 061018 (2011)
13. S. Narendar, Spectral finite element and nonlocal continuum mechanics based formulation for torsional wave propagation in nanorods. Finite Elem. Anal. Des. **62**, 65–75 (2012)

Applications of NL-FEM for Nanobeams

7

7.1 Introduction

Nano-scaled different materials in the NEMS organizations can be modeled with axial or torsional rods as well as evoke a flexural element (beam) model. As it is known, the fourth chapter focuses entirely on this issue. The nonlocal free vibration behavior of nanobeam structure was investigated with different beam theories.

On the other hand, it is stated in the previous chapter that analytical investigation for nonlocal vibration of axial and torsional nanorods is a fundamental approach and inefficient in most cases. This fact is also evident in the free vibration of nonlocal nanobeams. In the simplest sense, even if there are no environmental external factors in the Euler-Bernoulli nanobeam that neglects shear effects, the nonlocal form of all boundary conditions except for the simply supported nanobeam complicates the manual solution. Additionally, adding Winkler or Pasternak foundations to the problem makes it difficult to obtain frequencies from the analytical solution. Moreover, the thermal effect causes a similar problem. Even if the analytical solution is moved to program codes, it is quite possible that results will not be obtained. Also, the frequency of high modes cannot be calculated in the nanobeams with end attachments, especially for high values of the nonlocal parameter. These problems experienced in the Euler-Bernoulli beam increase in the Timoshenko nanobeam because, as noted in Chap. 4, shear deformation further complicates the process of mechanical analysis. In conclusion, it should definitely be noted that analytical methods are unfortunately poor for nonlocal nanobeams as well as nonlocal nanorods.

In the previous chapter, it is explained in detail that the inefficiencies of analytical method are eliminated by using numerical (approximate) methods. It should be reminded again that the approximate solution obtained by approximate methods is better than the unsolvableness of analytical method, because the engineer only deals with numbers in both

Öm. Civalek et al., *Mechanical Behavior and Vibration of Nano-Scaled Rods, Beams, and Frames*, Synthesis Lectures on Engineering, Science, and Technology,
https://doi.org/10.1007/978-3-032-12023-6_7

the analysis and design processes. So it is useful to know at least a high accuracy numerical result. Based on this idea, various examples of the developed nonlocal finite element method (NL-FEM) on axial and torsional nanorods are mentioned in the previous chapter. Related applications amply emphasize that the NL-FEM approach has greater significance to understand the dynamics of nanorods.

This chapter searches the using of NL-FEM in the free vibration analysis of nano scaled beams. According to this, NL-FEM solutions of Euler-Bernoulli and Timoshenko nanobeams are introduced and several examples about free vibration analyses of nanobeams resting on thermo-elastic environment are examined. Thereby, a more comprehensive investigation on the free vibration behavior of nanobeams is provided.

7.2 Recent Contributions

The vibration behavior of nanobeams has been extensively studied in the scientific literature through the NL-FEM formulation. It can be said that generally NL-FEM analyses of nanobeams are mostly subject to approaches such as Euler-Bernoulli beam theory (EBBT) and Timoshenko beam theory (TBT). In this subchapter, selected studies focusing on the free vibration of nanobeams with NL-FEM are mentioned.

Finite element solution of free vibration of EBBT nanobeams on is a fundamental study that can be given as an example in this regard [1, 2]. Nonlocal vibration of functionally graded nanobeam has been investigated with EBBT [3]. On the other hand, the effects of environmental factors such as various elastic foundation and temperature on the free vibration behavior of nanobeams have been discussed according to EBBT [4, 5] and TBT [6, 7]. Moreover, the nonlocal vibration of Timoshenko nanobeams has been studied by utilizing a high-order finite element [8]. Additionally, high-order shear deformable vibration of curved nanobeams has been expressed [9]. Also, a new NL-FEM for nonlocal free vibration analysis via EBBT and TBT is presented [10].

Apart from these, NL-FEM has been proposed for the vibration behavior of a cracked Euler-Bernoulli nanobeam resting on a thermo-elastic foundation [11]. A NL-FEM closed to shear locking has been developed in the vibration analysis with a two-phase local/nonlocal model of Timoshenko nanobeams [12]. In addition to these, NL-FEM analysis of the free vibration of nanobeams based on TBT has been carried out by utilizing the nonlocal integral model [13].

7.3 Nonlocal Vibration of Euler-Bernoulli Nanobeam

Before clarifying the NL-FEM solutions for free vibration of nanobeams, it should be noted that Fig. 5.3b shows an example for the finite element mesh of beam structure and Fig. 5.6 shows the bending finite element.

7.3.1 NL-FEM Expression

With Eqs. (1.42) and (4.10), the average weighted residual equation of the Euler-Bernoulli nanobeam can be written as follows:

$$\begin{aligned} \mathrm{I} = \int_0^{l_e} h\Bigg(& EI\frac{\partial^4 w}{\partial x^4} - k_P\frac{\partial^2 w}{\partial x^2} + N_t\frac{\partial^2 w}{\partial x^2} + k_W w + \rho A\frac{\partial^2 w}{\partial t^2} + (e_0 a)^2 k_P\frac{\partial^4 w}{\partial x^4} \\ & - (e_0 a)^2 N_t\frac{\partial^4 w}{\partial x^4} - (e_0 a)^2 k_W\frac{\partial^2 w}{\partial x^2} - (e_0 a)^2\rho A\frac{\partial^4 w}{\partial x^2 \partial t^2} - f + (e_0 a)^2\frac{\partial^2 f}{\partial x}\Bigg)\mathrm{d}x \end{aligned} \tag{7.1}$$

To edit this equation, it is necessary to know the following definition of motion of finite element:

$$w = \phi\mathbf{d} \tag{7.2}$$

where the shape function matrix ϕ and the end freedom vector $\mathbf{d}$ are given as follows:

$$\phi = \begin{bmatrix} \phi_i \\ \varphi_i \\ \phi_j \\ \varphi_j \end{bmatrix}^{\mathrm{T}} = \begin{bmatrix} 1 - 3\xi^2 + 2\xi^3 \\ l_e(-\xi - 2\xi^2 + \xi^3) \\ 3\xi^2 - 2\xi^3 \\ l_e(-\xi^2 + \xi^3) \end{bmatrix}^{\mathrm{T}}, \quad \mathbf{d} = \begin{Bmatrix} w_i \\ \theta_i \\ w_j \\ \theta_j \end{Bmatrix} \tag{7.3}$$

here l_e defines the finite element length. i and j are the node numbers of the finite element. In the shape function vector, ϕ and φ are shape functions that concern displacement and rotation, respectively. Also, $\xi = x/L$ denotes the dimensionless coordinate. In the end freedom vector, w defines the displacement and θ means the rotation.

On the other hand, the following expression should also be expressed for the flexural element:

$$\frac{\partial w}{\partial x} = \mathrm{D}^{\mathrm{k}} w = \mathbf{B}\mathbf{d} \tag{7.4}$$

where D^{k} is known as the kinematic operator and $\mathrm{D}^{\mathrm{k}}\phi = \mathbf{B}$ can be achieved.

Equation (7.1) can be divided into different integrals written below:

$$
\begin{aligned}
\mathrm{I}_1 &= \int_0^{l_e} EIh\frac{\partial^4 w}{\partial x^4}\mathrm{d}x = EIh\frac{\partial^3 w}{\partial x^3}\bigg|_0^{l_e} - EI\frac{\partial h}{\partial x}\frac{\partial^2 w}{\partial x^2}\bigg|_0^{l_e} + \int_0^{l_e} EI\frac{\partial^2 h}{\partial x^2}\frac{\partial^2 w}{\partial x^2}\mathrm{d}x,\\
\mathrm{I}_2 &= \int_0^{l_e} k_P h\frac{\partial^2 w}{\partial x^2}\mathrm{d}x = k_P h\frac{\partial w}{\partial x}\bigg|_0^{l_e} - \int_0^{l_e} k_P\frac{\partial h}{\partial x}\frac{\partial w}{\partial x}\mathrm{d}x,\\
\mathrm{I}_3 &= \int_0^{l_e} N_t h\frac{\partial^2 w}{\partial x^2}\mathrm{d}x = N_t h\frac{\partial w}{\partial x}\bigg|_0^{l_e} - \int_0^{l_e} N_t\frac{\partial h}{\partial x}\frac{\partial w}{\partial x}\mathrm{d}x,\quad \mathrm{I}_4 = \int_0^{l_e} k_W h w\mathrm{d}x,\\
\mathrm{I}_5 &= \int_0^{l_e} \rho A h\frac{\partial^2 w}{\partial t^2}\mathrm{d}x,\\
\mathrm{I}_6 &= \int_0^{l_e} (e_0a)^2 k_P h\frac{\partial^4 w}{\partial x^4}\mathrm{d}x = (e_0a)^2 k_P h\frac{\partial^3 w}{\partial x^3}\bigg|_0^{l_e} - (e_0a)^2 k_P\frac{\partial h}{\partial x}\frac{\partial^2 w}{\partial x^2}\bigg|_0^{l_e}\\
&\quad + \int_0^{l_e} (e_0a)^2 k_P\frac{\partial^2 h}{\partial x^2}\frac{\partial^2 w}{\partial x^2}\mathrm{d}x,\\
\mathrm{I}_7 &= \int_0^{l_e} (e_0a)^2 N_t h\frac{\partial^4 w}{\partial x^4}\mathrm{d}x = (e_0a)^2 N_t h\frac{\partial^3 w}{\partial x^3}\bigg|_0^{l_e} - (e_0a)^2 N_t\frac{\partial h}{\partial x}\frac{\partial^2 w}{\partial x^2}\bigg|_0^{l_e}\\
&\quad + \int_0^{l_e} (e_0a)^2 N_t\frac{\partial^2 h}{\partial x^2}\frac{\partial^2 w}{\partial x^2}\mathrm{d}x,\\
\mathrm{I}_8 &= \int_0^{l_e} (e_0a)^2 k_W h\frac{\partial^2 w}{\partial x^2}\mathrm{d}x = (e_0a)^2 k_W h\frac{\partial w}{\partial x}\bigg|_0^{l_e} - \int_0^{l_e} (e_0a)^2 k_W\frac{\partial h}{\partial x}\frac{\partial w}{\partial x}\mathrm{d}x,\\
\mathrm{I}_9 &= \int_0^{l_e} (e_0a)^2 \rho A h\frac{\partial^4 w}{\partial x^2\partial t^2}\mathrm{d}x = (e_0a)^2 \rho A h\frac{\partial^3 w}{\partial x\partial t^2}\bigg|_0^{l_e}\\
&\quad - \int_0^{l_e} (e_0a)^2 \rho A\frac{\partial h}{\partial x}\frac{\partial^3 w}{\partial x\partial t^2}\mathrm{d}x,\\
\mathrm{I}_{11} &= \int_0^{l_e} (e_0a)^2 h\frac{\partial^2 f}{\partial x^2}\mathrm{d}x = (e_0a)^2 h\frac{\partial f}{\partial x}\bigg|_0^{l_e} - \int_0^{l_e} (e_0a)^2\frac{\partial h}{\partial x}\frac{\partial f}{\partial x}\mathrm{d}x.
\end{aligned}
\tag{7.5}
$$

The weak form of average weighted residue can be decomposed as follows from the equation obtained by substituting Eq. (7.5) into Eq. (7.1):

$$\int_0^{l_e} EI\frac{\partial^2 h}{\partial x^2}\frac{\partial^2 w}{\partial x^2}\mathrm{d}x + \int_0^{l_e} (e_0a)^2 k_P\frac{\partial^2 h}{\partial x^2}\frac{\partial^2 w}{\partial x^2}\mathrm{d}x - \int_0^{l_e} (e_0a)^2 N_t\frac{\partial^2 h}{\partial x^2}\frac{\partial^2 w}{\partial x^2}\mathrm{d}x + \int_0^{l_e} k_P\frac{\partial h}{\partial x}\frac{\partial w}{\partial x}\mathrm{d}x$$
$$+ \int_0^{l_e} (e_0a)^2 k_W\frac{\partial h}{\partial x}\frac{\partial w}{\partial x}\mathrm{d}x - \int_0^{l_e} N_t\frac{\partial h}{\partial x}\frac{\partial w}{\partial x}\mathrm{d}x + \int_0^{l_e} k_W hw\mathrm{d}x - \int_0^{l_e} hf\mathrm{d}x + \int_0^{l_e} \rho Ah\frac{\partial^2 w}{\partial t^2}\mathrm{d}x$$
$$+ \int_0^{l_e} (e_0a)^2\rho A\frac{\partial h}{\partial x}\frac{\partial^3 w}{\partial x\partial t^2}\mathrm{d}x - \int_0^{l_e} (e_0a)^2\frac{\partial h}{\partial x}\frac{\partial f}{\partial x}\mathrm{d}x = 0 \tag{7.6}$$

this equation roughly describes the NL-FEM equation of motion. To arrange the equation, the following expressions should be known:

$$h=\phi^{\mathrm{T}},\ \ \mathbf{B}=\mathrm{D}^{\mathrm{k}}\phi,\ \ \frac{\partial h}{\partial x}=(\phi^{\mathrm{T}})'=\mathbf{B}^{\mathrm{T}},\ \ \frac{\partial w}{\partial x}=\mathbf{Bd},\ \ \frac{\partial^2 w}{\partial x^2}=\mathbf{B}'\mathbf{d},\ \ \frac{\partial^2 w}{\partial t^2}=\phi\ddot{\mathbf{d}}. \tag{7.7}$$

Equation (7.6) is rewritten below with the help of Eq. (7.7):

$$\int_0^{l_e} EI\left(\mathbf{B}'^{\mathrm{T}}\mathbf{B}'\right)\mathbf{d}\mathrm{d}x + \int_0^{l_e} (e_0a)^2 k_P\left(\mathbf{B}'^{\mathrm{T}}\mathbf{B}'\right)\mathbf{d}\mathrm{d}x - \int_0^{l_e} (e_0a)^2 N_t\left(\mathbf{B}'^{\mathrm{T}}\mathbf{B}'\right)\mathbf{d}\mathrm{d}x$$
$$+ \int_0^{l_e} k_P(\mathbf{B}^{\mathrm{T}}\mathbf{B})\mathbf{d}\mathrm{d}x + \int_0^{l_e} (e_0a)^2 k_W(\mathbf{B}^{\mathrm{T}}\mathbf{B})\mathbf{d}\mathrm{d}x - \int_0^{l_e} N_t(\mathbf{B}^{\mathrm{T}}\mathbf{B})\mathbf{d}\mathrm{d}x + \int_0^{l_e} k_W(\phi^{\mathrm{T}}\phi)\mathbf{d}\mathrm{d}x$$
$$- \int_0^{l_e} \phi^{\mathrm{T}} f\mathrm{d}x + \int_0^{l_e} \rho A(\phi^{\mathrm{T}}\phi)\ddot{\mathbf{d}}\mathrm{d}x + \int_0^{l_e} (e_0a)^2\rho A(\mathbf{B}^{\mathrm{T}}\mathbf{B})\ddot{\mathbf{d}}\mathrm{d}x - \int_0^{l_e} (e_0a)^2\mathbf{B}^{\mathrm{T}} f'\mathrm{d}x = 0 \tag{7.8}$$

this equation can be expressed in matrix notation as follows:

$$(K + K_{W,c} + K_{W,nl} + K_{P,c} + K_{P,nl} - K_{T,c} - K_{T,nl})\mathbf{d} + (M_c + M_{nl})\ddot{\mathbf{d}} = \mathbf{f}_c + \mathbf{f}_{nl} \tag{7.9}$$

The definitions seen here are presented as follows [4]:

$$K=\int_0^{l_e} EI\left(\mathbf{B}'^{\mathrm{T}}\mathbf{B}'\right)\mathrm{d}x=\int_0^{l_e} EI\begin{Bmatrix}\phi_i''\\ \varphi_i''\\ \phi_j''\\ \varphi_j''\end{Bmatrix}\left[\phi_i''\ \varphi_i''\ \phi_j''\ \varphi_j''\right]\mathrm{d}x$$
$$=\frac{EI}{l_e^3}\begin{bmatrix}12 & 6l_e & -12 & 6l_e\\ 6l_e & 4l_e^2 & -6l_e & 2l_e^2\\ -12 & -6l_e & 12 & -6l_e\\ 6l_e & 2l_e^2 & -6l_e & 4l_e^2\end{bmatrix} \tag{7.10}$$

$$K_{W,c}=\int_0^{l_e} k_W\left(\phi^{\mathrm{T}}\phi\right)\mathrm{d}x=\int_0^{l_e} k_W\begin{Bmatrix}\phi_i\\ \varphi_i\\ \phi_j\\ \varphi_j\end{Bmatrix}\left[\phi_i\quad \varphi_i\quad \phi_j\quad \varphi_j\right]\mathrm{d}x$$
$$=\frac{k_W l_e}{420}\begin{bmatrix}156 & 22l_e & 54 & -13l_e\\ 22l_e & 4l_e^2 & 13l_e & -3l_e^2\\ 54 & 13l_e & 156 & -22l_e\\ -13l_e & -3l_e^2 & -22l_e & 4l_e^2\end{bmatrix} \tag{7.11}$$

$$K_{W,nl}=\int_0^{l_e} (e_0a)^2k_W\left(\mathbf{B}^{\mathrm{T}}\mathbf{B}\right)\mathrm{d}x=\int_0^{l_e} (e_0a)^2k_W\begin{Bmatrix}\phi_i'\\ \varphi_i'\\ \phi_j'\\ \varphi_j'\end{Bmatrix}\left[\phi_i'\quad \varphi_i'\quad \phi_j'\quad \varphi_j'\right]\mathrm{d}x$$
$$=\frac{(e_0a)^2k_W}{30l_e}\begin{bmatrix}36 & 3l_e & -36 & 3l_e\\ 3l_e & 4l_e^2 & -3l_e & -l_e^2\\ -36 & -3l_e & 36 & -3l_e\\ 3l_e & -l_e^2 & -3l_e & 4l_e^2\end{bmatrix} \tag{7.12}$$

$$K_{P,c} = \int_0^{l_e} k_P(\mathbf{B}^{\mathrm{T}}\mathbf{B})\mathrm{d}x = \int_0^{l_e} k_P \begin{Bmatrix} \phi_i' \\ \varphi_i' \\ \phi_j' \\ \varphi_j' \end{Bmatrix} [\phi_i' \quad \varphi_i' \quad \phi_j' \quad \varphi_j']\mathrm{d}x$$
$$= \frac{k_P}{30l_e}\begin{bmatrix} 36 & 3l_e & -36 & 3l_e \\ 3l_e & 4l_e^2 & -3l_e & -l_e^2 \\ -36 & -3l_e & 36 & -3l_e \\ 3l_e & -l_e^2 & -3l_e & 4l_e^2 \end{bmatrix} \tag{7.13}$$

$$K_{P,nl} = \int_0^{l_e} (e_0a)^2 k_P\left(\mathbf{B}'^{\mathrm{T}}\mathbf{B}'\right)\mathrm{d}x = \int_0^{l_e} (e_0a)^2 k_P \begin{Bmatrix} \phi_i'' \\ \varphi_i'' \\ \phi_j'' \\ \varphi_j'' \end{Bmatrix} [\phi_i'' \quad \varphi_i'' \quad \phi_j'' \quad \varphi_j'']\mathrm{d}x$$
$$= \frac{(e_0a)^2 k_P}{l_e^3}\begin{bmatrix} 12 & 6l_e & -12 & 6l_e \\ 6l_e & 4l_e^2 & -6l_e & 2l_e^2 \\ -12 & -6l_e & 12 & -6l_e \\ 6l_e & 2l_e^2 & -6l_e & 4l_e^2 \end{bmatrix} \tag{7.14}$$

$$K_{T,c} = \int_0^{l_e} N_t(\mathbf{B}^{\mathrm{T}}\mathbf{B})\mathrm{d}x = \int_0^{l_e} N_t \begin{Bmatrix} \phi_i' \\ \varphi_i' \\ \phi_j' \\ \varphi_j' \end{Bmatrix} [\phi_i' \quad \varphi_i' \quad \phi_j' \quad \varphi_j']\mathrm{d}x$$
$$= \frac{N_t}{30l_e}\begin{bmatrix} 36 & 3l_e & -36 & 3l_e \\ 3l_e & 4l_e^2 & -3l_e & -l_e^2 \\ -36 & -3l_e & 36 & -3l_e \\ 3l_e & -l_e^2 & -3l_e & 4l_e^2 \end{bmatrix} \tag{7.15}$$

$$K_{T,nl}=\int_0^{l_e}(e_0a)^2N_t\left(\mathbf{B}'^{\mathrm{T}}\mathbf{B}'\right)\mathrm{d}x=\int_0^{l_e}(e_0a)^2N_t\begin{Bmatrix}\phi_i''\\ \varphi_i''\\ \phi_j''\\ \varphi_j''\end{Bmatrix}\left[\phi_i''\quad \varphi_i''\quad \phi_j''\quad \varphi_j''\right]\mathrm{d}x$$
$$=\frac{(e_0a)^2N_t}{l_e^3}\begin{bmatrix}12 & 6l_e & -12 & 6l_e\\ 6l_e & 4l_e^2 & -6l_e & 2l_e^2\\ -12 & -6l_e & 12 & -6l_e\\ 6l_e & 2l_e^2 & -6l_e & 4l_e^2\end{bmatrix} \tag{7.16}$$

$$M_c=\int_0^{l_e}\rho A(\phi^T\phi)dx=\int_0^{l_e}\rho A\begin{Bmatrix}\phi_i\\ \varphi_i\\ \phi_j\\ \varphi_j\end{Bmatrix}\left[\phi_i\quad \varphi_i\quad \phi_j\quad \varphi_j\right]\mathrm{d}x$$
$$=\frac{\rho A l_e}{420}\begin{bmatrix}156 & 22l_e & 54 & -13l_e\\ 22l_e & 4l_e^2 & 13l_e & -3l_e^2\\ 54 & 13l_e & 156 & -22l_e\\ -13l_e & -3l_e^2 & -22l_e & 4l_e^2\end{bmatrix} \tag{7.17}$$

$$M_{nl}=\int_0^{l_e}(e_0a)^2\rho A\left(\mathbf{B}^{\mathrm{T}}\mathbf{B}\right)\mathrm{d}x=\int_0^{l_e}(e_0a)^2\rho A\begin{Bmatrix}\phi_i'\\ \varphi_i'\\ \phi_j'\\ \varphi_j'\end{Bmatrix}\left[\phi_i'\quad \varphi_i'\quad \phi_j'\quad \varphi_j'\right]\mathrm{d}x$$
$$=\frac{(e_0a)^2\rho A}{30l_e}\begin{bmatrix}36 & 3l_e & -36 & 3l_e\\ 3l_e & 4l_e^2 & -3l_e & -l_e^2\\ -36 & -3l_e & 36 & -3l_e\\ 3l_e & -l_e^2 & -3l_e & 4l_e^2\end{bmatrix} \tag{7.18}$$

$$\mathbf{f}_c=\int_0^{l_e}f\phi^{\mathrm{T}}\mathrm{d}x=\int_0^{l_e}f\begin{Bmatrix}\phi_i\\ \varphi_i\\ \phi_j\\ \varphi_j\end{Bmatrix}\mathrm{d}x=\frac{l_e}{12}f\begin{Bmatrix}6\\ l_e\\ 6\\ -l_e\end{Bmatrix} \tag{7.19}$$

$$\mathbf{f}_{nl} = \int_0^{l_e} (e_0 a)^2 f' \mathbf{B}^{\mathrm{T}} \mathrm{d}x = \int_0^{l_e} (e_0 a)^2 f' \begin{Bmatrix} \phi_i' \\ \varphi_i' \\ \phi_j' \\ \varphi_j' \end{Bmatrix} \mathrm{d}x = (e_0 a)^2 f' \begin{Bmatrix} -1 \\ 0 \\ 1 \\ 0 \end{Bmatrix} \tag{7.20}$$

Here, K is the bending stiffness matrix, $K_{W,\ c}$ is the classical stiffness matrix of Winkler foundation, $K_{W,\ nl}$ is the nonlocal stiffness matrix of Winkler foundation, $K_{P,\ c}$ is the classical stiffness matrix of Pasternak foundation, $K_{P,\ c}$ is the nonlocal stiffness matrix of Pasternak foundation. T_c denotes the classical stiffness matrix due to temperature change and T_{nl} defines the nonlocal stiffness matrix due to temperature change. On the other hand, M_c is known as classical mass and M_{nl} means nonlocal mass matrices. $\mathbf{f}_c$ and $\mathbf{f}_{nl}$ are the classical and nonlocal external force vectors, respectively.

7.3.2 Interpretation

In Eq. (7.9), when the elastic foundation and temperature parameters are zero, an additional effect is observed on the classical mass due to the nonlocal parameter, as in the simple axial nanorod. On the other hand, since the additional effects of all environmental factors (Winkler foundation, Pasternak foundation, and ambient temperature) combine with the nonlocal parameter, additional effects of these emerge. Additionally, while both classical and nonlocal foundations increase stiffness, classical and nonlocal temperatures decrease the total stiffness. Also, the stiffness matrix is not affected by the nonlocal parameter as in the simple nanorod.

7.3.3 Solution Stage

The total mass and stiffness matrices and transverse load vector of nonlocal bending finite element are given as:

$$[K]_e = K + K_{W,c} + K_{W,nl} + K_{P,c} + K_{P,nl} - K_{T,c} - K_{T,nl} \tag{7.21}$$

$$[M]_e = M_c + M_{nl} \tag{7.22}$$

$$\mathbf{f}_e = \mathbf{f}_c + \mathbf{f}_{nl} \tag{7.23}$$

According to free vibration analysis, it is known that $\mathbf{f}_c = \mathbf{f}_{nl} = 0$. Like what is stated in Chap. 6, substituting Eq. (4.30) into Eq. (7.9) and considering reduced global matrices, the following eigenvalue equation is obtained:

$$\det\left[\left(\sum_{e=1}^{n} [K]_e\right)^* - \omega_i^2 \left(\sum_{e=1}^{n} [M]_e\right)^*\right] = 0 \tag{7.24}$$

7.3.4 Application of Boundary Conditions

In the bending elements, it is a geometric boundary condition that the simply supported end does not allow displacement. In the fixed (or clamped) support, the displacement and rotation equal to zero. Apart from these, the attached ends are examined as follows: Firstly, it should be known to which freedom the attachment is included in the nanobeams. For example, an elastic spring attachment should not be included directly in the input expressing the freedom in the vertical direction. It should be noted that the elastic spring attachment can also have rotational freedom (in the bending direction, not in the torsional direction). A similar statement can be said for mass attachment. If you remember, rotational additions are not considered in the fourth chapter. Therefore, we consider that the attachments are in vertical freedom only and lumpy at the right end of finite element. As a result, the total stiffness and mass matrices of attached finite element are expressed as follows:

$$K_{e,a} = K_{e,na} + K_a = K + K_{W,c} + K_{W,nl} + K_{P,c} + K_{P,nl} - K_{T,c} - K_{T,nl} + \begin{bmatrix} 0 & 0 & 0 & 0 \\ 0 & 0 & 0 & 0 \\ 0 & 0 & k_a & 0 \\ 0 & 0 & 0 & 0 \end{bmatrix} \tag{7.25}$$

$$M_{e,a} = M_{e,na} + M_a = M_c + M_{nl} + \begin{bmatrix} 0 & 0 & 0 & 0 \\ 0 & 0 & 0 & 0 \\ 0 & 0 & m_a & 0 \\ 0 & 0 & 0 & 0 \end{bmatrix} \tag{7.26}$$

As can be understood, while the attachment that work with mechanical internal influence in the direction of vertical freedom defines the input 33 of related matrix, the attachment in the direction of rotational freedom is in the input 44. If the attachment is at the left end of finite element, inputs 11 and 22 should be evaluated according to the direction of freedom occupied by the attachment.

Example 7.1 Determine the natural frequencies of nonlocal transverse free vibration of simply supported nanobeam in Fig. 7.1. Use a single finite element in the analysis. Nanobeam length: $L = 20$ nm, height and thickness of rectangular cross-section: $h = 1$ nm and $b = 2$ nm, modulus of elasticity: $E = 1$ TPa, and mass of unit volume: $\rho = 2300\text{kg/m}^3$. Use the nonlocal parameter as $e_0a/L = 0.2$.

Solution. The nodes of single finite element in Fig. 7.1b are also the system node numbers. In this case, in the matrices given by Eq. (7.10) for stiffness and Eqs. (7.17) and (7.18) for mass, it is symbolized that $l_e = l_{e=1} = L$. In addition to this, the freedom numbers (node numbers of finite element mesh) are also depicted. Then, let's write the stiffness, mass, and nonlocal mass matrices of finite element as follows:

Fig. 7.1 NL-FEM analysis of simply supported nanobeam (**a**) Physical system (**b**) FEM model with single element

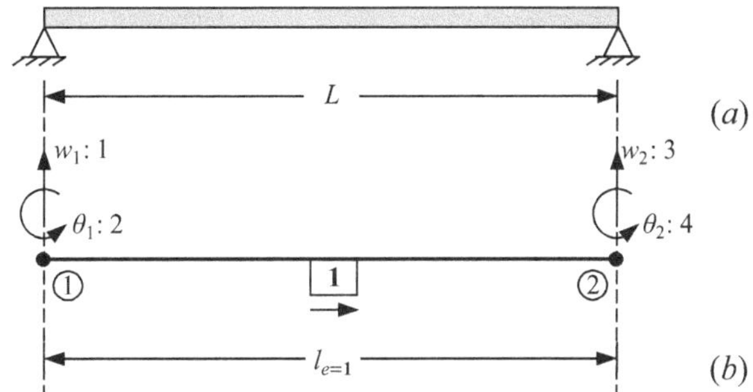

$$K=\frac{EI}{L^3}\begin{bmatrix} 12 & 6L & -12 & 6L \\ 6L & 4L^2 & -6L & 2L^2 \\ -12 & -6L & 12 & -6L \\ 6L & 2L^2 & -6L & 4L^2 \end{bmatrix} \tag{7.E7.1.1}$$

$$M_c=\frac{\rho AL}{420}\begin{bmatrix} 156 & 22L & 54 & -13L \\ 22L & 4L^2 & 13L & -3L^2 \\ 54 & 13L & 156 & -22L \\ -13L & -3L^2 & -22L & 4L^2 \end{bmatrix} \tag{7.E7.1.2}$$

$$M_c=\frac{(e_0a)^2\rho A}{30L}\begin{bmatrix} 36 & 3L & -36 & 3L \\ 3L & 4L^2 & -3L & -L^2 \\ -36 & -3L & 36 & -3L \\ 3L & -L^2 & -3L & 4L^2 \end{bmatrix} \tag{7.E7.1.3}$$

where the geometric and mechanical properties of the beam are given as follows:

$$\begin{aligned} &\text{Finite element length}: L=20\ \text{nm}=2\times10^{-8}\ \text{m}, \\ &\text{Nonlocal parameter}: e_0a=0.2\times L=(0.2)\times(20\ \text{nm})=4\ \times10^{-9}\ \text{m}, \\ &\text{Cross-section area}: A=bh=(1\ \text{nm})\times(2\ \text{nm})=2\times10^{-18}\ \text{m}^2, \\ &\text{Moment of inertia}: I=\frac{bh^3}{12}=\frac{(2\ \text{nm})\times(1\ \text{nm})^3}{12}=1.667\times10^{-37}\ \text{m}^4. \end{aligned} \tag{7.E7.1.4}$$

Vertical displacement does not occur in the simply support and these correspond to freedoms 1 and 3 seen in Fig. 7.1b. That is, rows and columns 1 and 3 should be eliminated in the matrices given by Eqs. (7.E7.1.1)–(7.E7.1.3). According to this, after the stiffness and mass matrices are calculated, they are eliminated as follows:

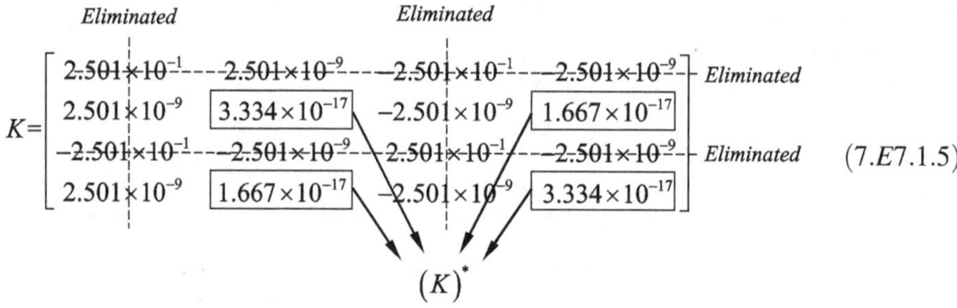

(7.E7.1.5)

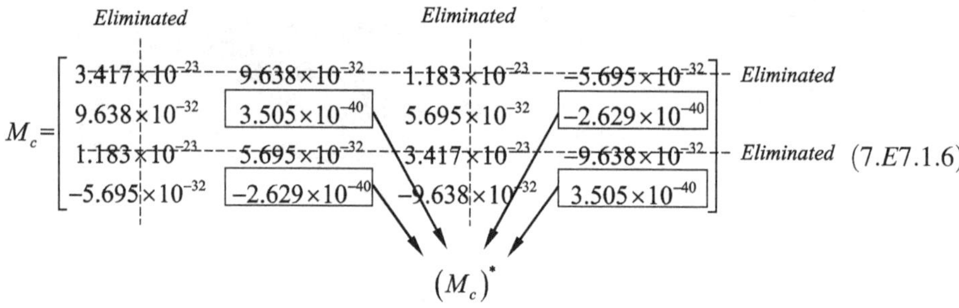

(7.E7.1.6)

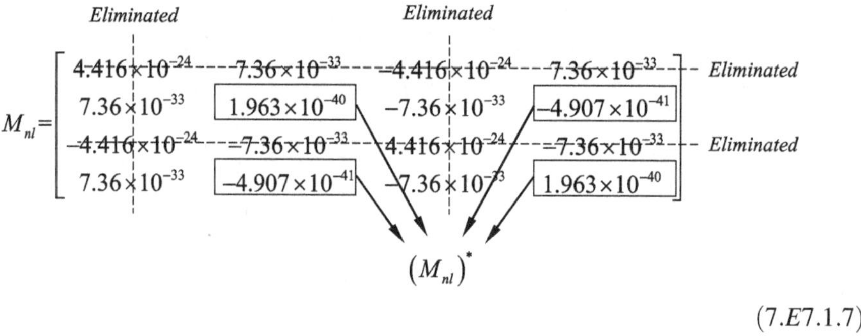

(7.E7.1.7)

where the reduced global mass matrix is determined as follows:

$$M = M_c + M_{nl} = \begin{bmatrix} 5.467 \times 10^{-40} & -3.119 \times 10^{-40} \\ -3.119 \times 10^{-40} & 5.467 \times 10^{-40} \end{bmatrix} \tag{7.E7.1.8}$$

If Eqs. (7.E7.1.5) and (7.E7.1.8) are replaced into Eq. (7.24), eigenvalue equation can be written:

$$\det\left(\begin{bmatrix} 3.334\times10^{-17} & 1.667\times10^{-17} \\ 1.667\times10^{-17} & 3.334\times10^{-17} \end{bmatrix} - \omega^2\times\begin{bmatrix} 5.467\times10^{-40} & -3.119\times10^{-40} \\ -3.119\times10^{-40} & 5.467\times10^{-40} \end{bmatrix}\right)=0 \tag{7.E7.1.9}$$

the polynomial form of this equation is obtained as follows:

$$2.016\times10^{-79}\omega^4 - 4.685\times10^{-56}\omega^2 + 8.337\times10^{-34}=0 \tag{7.E7.1.10}$$

Example 6.1 can be investigated for solution of this equation. The positive roots of equation, that is, the natural frequencies of the and second modes are given below:

$$\omega_1 = 139.347\ \text{GHz} \quad \omega_2 = 461.49\ \text{GHz}$$

Here, it is understood that frequencies of two modes can be obtained by using one finite element in the simply supported nanobeam. Because the matrices with 4 ×4 size have been reduced to 2 ×2 size by two geometric boundary conditions. On the other hand, the using of one finite element enables the manual solution for the simply supported nanobeam. However, since more finite elements will be used in the following examples, the eigenvalue equations which the natural frequencies depend on will be solved electronically.

The convergence study given with Table 7.1 in the next example can be examined about the convergence of above obtained results to the analytical results.

Example 7.2 Determine the natural frequencies of nonlocal transverse free vibration of simply supported nanobeam. Use two finite elements in the analysis. Physical and mechanical parameters are as in Example 7.1.

Solution. The numbers of nodes of finite elements in Fig. 7.2c are symbolized in bold and italics to avoid confusion with the numbers of freedom in Fig. 7.2b. According to this, necessary matrices in the nonlocal vibration analysis of any finite element neglecting the thermo-elastic environment are expressed by Eqs. (7.10), (7.17), and (7.18). In Fig. 7.2b, since the lengths of finite elements are equal, the lengths are denoted by $l_{e=\mathbf{1}} = l_{e=\mathbf{2}} = 1\times10^{-8}$ m $= l$. Considering Fig. 7.2b, c, the related matrices should be combined according to the common freedoms as follows. Here, the grid showing the numbers of finite element mesh is drawn as an example in the global elastic stiffness matrix.

1	2	3	4	5	6	*Mesh node numbers*
12	$6l$	−12	$6l$	0	0	1
$6l$	$4l^2$	$-6l$	$2l^2$	0	0	2
−12	$-6l$	12+12	$-6l+6l$	−12	$6l$	3
$6l$	$2l^2$	$-6l+6l$	$4l^2+4l^2$	$-6l$	$2l^2$	4
0	0	−12	$-6l$	12	$-6l$	5
0	0	$6l$	$2l^2$	$-6l$	$4l^2$	6

$$\sum[K]=\frac{EI}{l^3}\begin{bmatrix} 12 & 6l & -12 & 6l & 0 & 0 \\ 6l & 4l^2 & -6l & 2l^2 & 0 & 0 \\ -12 & -6l & 12+12 & -6l+6l & -12 & 6l \\ 6l & 2l^2 & -6l+6l & 4l^2+4l^2 & -6l & 2l^2 \\ 0 & 0 & -12 & -6l & 12 & -6l \\ 0 & 0 & 6l & 2l^2 & -6l & 4l^2 \end{bmatrix} \tag{7.E7.2.1}$$

Table 7.1 Comparisons of nondimensional frequencies of simply supported Euler-Bernoulli nanobeams for nonlocal transverse free vibration ($e_0a/L = 0.2$, $K_W = K_P = \Delta T = 0$)

Mode	Analytical	NL-FEM results						
Number	Eq. (4.E4.1.13)	Manual solution		Program code (MATLAB)				
		$n = 1$	$n = 2$	$n = 1$	$n = 2$	$n = 10$	$n = 15$	$n = 20$
1	8.35692	9.259	8.391	9.25820	8.38946	8.35698	8.35693	8.35692
2	24.58230	30.666	27.180	30.66429	27.17465	24.58492	24.58282	24.58247
3	41.62849	–	50.435	–	50.43013	41.65040	41.63292	41.62990
4	58.38034		72.280		72.26893	58.47429	58.39965	58.38654
5	74.83985		–		–	75.12216	74.89913	74.85903

$$\sum [M_c] = \frac{\rho A l}{420} \begin{bmatrix} 156 & 22l & 54 & -13l & 0 & 0 \\ 22l & 4l^2 & 13l & -3l^2 & 0 & 0 \\ 54 & 13l & 156+156 & -22l+22l & 54 & -13l \\ -13l & -3l^2 & -22l+22l & 4l^2+4l^2 & 13l & -3l^2 \\ 0 & 0 & 54 & 13l & 156 & -22l \\ 0 & 0 & -13l & -3l^2 & -22l & 4l^2 \end{bmatrix} \tag{7.E7.2.2}$$

$$\sum [M_{nl}] = \frac{(e_0 a)^2 \rho A}{30 l_e} \begin{bmatrix} 36 & 3l & -36 & 3l & 0 & 0 \\ 3l & 4l^2 & -3l & -l^2 & 0 & 0 \\ -36 & -3l & 36+36 & -3l+3l & -36 & 3l \\ 3l & -l^2 & -3l+3l & 4l^2+4l^2 & -3l & -l^2 \\ 0 & 0 & -36 & -3l & 36 & -3l \\ 0 & 0 & 3l & -l^2 & -3l & 4l^2 \end{bmatrix} \tag{7.E7.2.3}$$

After the spherical matrices are calculated, they are reduced as follows due to the first and fifth freedoms in Fig. 7.2b. While dashed lines indicate that the corresponding row and column are reduced, matrix parts enclosed in rectangles represent reduced global matrices. That is,

$$\sum [K] = \frac{EI}{l^3} \begin{bmatrix} 12 & 6l & -12 & 6l & 0 & 0 \\ 6l & 4l^2 & -6l & 2l^2 & 0 & 0 \\ -12 & -6l & 24 & 0 & -12 & 6l \\ 6l & 2l^2 & 0 & 8l^2 & -6l & 2l^2 \\ 0 & 0 & -12 & -6l & 12 & -6l \\ 0 & 0 & 6l & 2l^2 & -6l & 4l^2 \end{bmatrix} \tag{7.E7.2.4}$$

$$\left(\sum [K]\right)^*$$

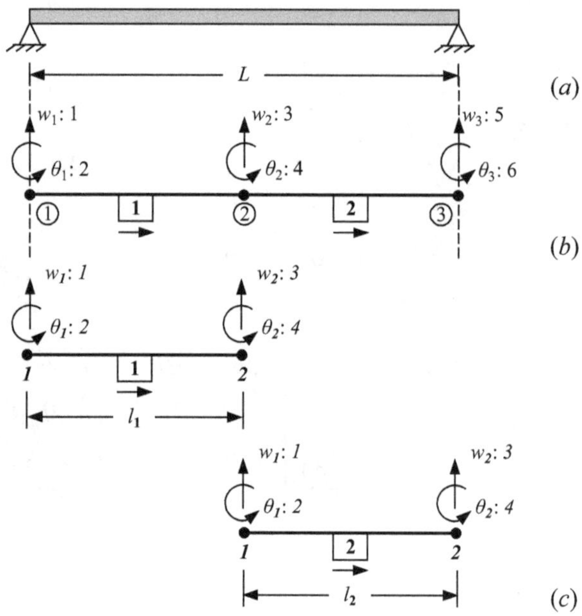

Fig. 7.2 NL-FEM analysis of simply supported nanobeam. (**a**) Physical system. (**b**) FEM model with two elements. (**c**) Finite elements

$$\sum[M_c] = \frac{\rho A l}{420}\begin{bmatrix} 156 & 22l & 54 & -13l & 0 & 0 \\ 22l & 4l^2 & 13l & -3l^2 & 0 & 0 \\ 54 & 13l & 312 & 0 & 54 & -13l \\ -13l & -3l^2 & 0 & 8l^2 & 13l & -3l^2 \\ 0 & 0 & 54 & 13l & 156 & -22l \\ 0 & 0 & -13l & -3l^2 & -22l & 4l^2 \end{bmatrix} \rightarrow \left(\sum[M_c]\right)^* \tag{7.E7.2.5}$$

$$\sum[M_{nl}] = \frac{(e_0a)^2\rho A}{30l_e}\begin{bmatrix} 36 & 3l & -36 & 3l & 0 & 0 \\ 3l & 4l^2 & -3l & -l^2 & 0 & 0 \\ -36 & -3l & 72 & 0 & -36 & 3l \\ 3l & -l^2 & 0 & 8l^2 & -3l & -l^2 \\ 0 & 0 & -36 & -3l & 36 & -3l \\ 0 & 0 & 3l & -l^2 & -3l & 4l^2 \end{bmatrix} \rightarrow \left(\sum[M_{nl}]\right)^* \tag{7.E7.2.6}$$

Global matrices can be calculated by using the related mechanical and physical parameters:

$$\left(\sum[K]\right)^* = \begin{bmatrix} 6.668\times10^{-17} & -1\times10^{-8} & 3.334\times10^{-17} & 0 \\ -1\times10^{-8} & 4 & 0 & 1\times10^{-8} \\ 3.334\times10^{-17} & 0 & 1.334\times10^{-16} & 3.334\times10^{-17} \\ 0 & 1\times10^{-8} & 3.334\times10^{-17} & 6.668\times10^{-17} \end{bmatrix} \tag{7.E7.2.7}$$

$$\left(\sum[M]\right)^* = \left(\sum[M_c]\right)^* + \left(\sum[M_{nl}]\right)^*$$
$$= \begin{bmatrix} 1.419\times10^{-40} & 6.878\times10^{-33} & -5.739\times10^{-41} & 0 \\ 6.878\times10^{-33} & 5.184\times10^{-23} & 0 & -6.878\times10^{-33} \\ -5.739\times10^{-41} & 0 & 2.839\times10^{-40} & -5.739\times10^{-41} \\ 0 & -6.878\times10^{-33} & -5.739\times10^{-41} & 1.419\times10^{-40} \end{bmatrix} \tag{7.E7.2.8}$$

Substituting the reduced spherical matrices into Eq. (7.24) gives the following natural frequencies:

$$\omega_1 = 126.285\ \text{GHz},\quad \omega_2 = 409.051\ \text{GHz},\quad \omega_3 = 759.037\ \text{GHz},\quad \omega_4 = 1087.795\ \text{GHz}.$$

Even though the two finite elements create a solution that is not appropriate for the manual solution, the above values are lower than those obtained as a result of Example 7.1. In other words, as seen in Table 7.1, results that are more convergent to the analytical results are achieved. Also, the solution with two finite elements, the natural frequency values of third and fourth modes that cannot be reached with using of single finite element are also obtained.

According to the nonlocal Euler-Bernoulli beam theory, analytical nondimensional frequencies of first five modes of simply supported nanobeams computed for the value $e_0a/L = 0.2$ are compared with the NL-FEM results in Table 7.1. It is reminded that nondimensional frequencies are calculated with Eq. (4.E4.5.7). Looking at the results, it is generally understood that NL-FEM gives very successful results in this nonlocal free vibration problem. Of course, the number of finite elements that should be used to obtain sensitive results has an importance. In this case, since many elements should be used, the solution cannot be performed manually and the NL-FEM algorithm should be coded on the computer. As a result, it is once again understood that NL-FEM plays an important role in the free vibration behavior of nanobeams.

In the next application, let's examine the thermo-elastic environment in the vibration of nanobeams.

Example 7.3 Determine the natural frequencies of nonlocal transverse free vibration of simply supported nanobeam embedded in thermo-elasti0c medium shown in Fig. 4.2 for two different cases as follows. Use two finite elements in the analysis. Physical and mechanical parameters are as in Example 7.1.

(a) The two-parameter elastic foundation. Nondimensional stiffnesses of Winkler foundation: $K_W = 10$ and Pasternak foundation: $K_P = 10$. Nondimensional stiffness of elastic medium is defined as $K_W = k_W L^4/EI$ and $K_P = k_P L^2/EI$.
(b) The ambient temperature. Temperature change: $\Delta T = 300$ K and thermal expansion: $\alpha_t = 1.6 \times 10^{-6}\ \mathrm{K}^{-1}$.

Solution. (a) According to Eq. (7.21) in the NL-FEM formulation for the free vibration of a nonlocal Euler-Bernoulli nanobeam, the elastic foundation only affects the stiffness matrix and this effect increases the stiffness. Since the classical and nonlocal stiffnesses for Winkler and Pasternak foundations are represented by Eqs. (7.11)–(7.14), similar to obtaining the global matrices in the previous example, the global stiffness matrices for the elastic foundations are obtained as follows in the nanobeams with two finite elements:

$$\sum [K_{W,c}] = \frac{k_W l}{420} \begin{bmatrix} 156 & 22l & 54 & -13l & 0 & 0 \\ 22l & 4l^2 & 13l & -3l^2 & 0 & 0 \\ 54 & 13l & 312 & 0 & 54 & -13l \\ -13l & -3l^2 & 0 & 8l^2 & 13l & -3l^2 \\ 0 & 0 & 54 & 13l & 156 & -22l \\ 0 & 0 & -13l & -3l^2 & -22l & 4l^2 \end{bmatrix} \tag{7.E7.3.1}$$

$$\sum [K_{W,nl}] = \frac{(e_0 a)^2 k_W}{30l} \begin{bmatrix} 36 & 3l & -36 & 3l & 0 & 0 \\ 3l & 4l^2 & -3l & -l^2 & 0 & 0 \\ -36 & -3l & 72 & 0 & -36 & 3l \\ 3l & -l^2 & 0 & 8l^2 & -3l & -l^2 \\ 0 & 0 & -36 & -3l & 36 & -3l \\ 0 & 0 & 3l & -l^2 & -3l & 4l^2 \end{bmatrix} \tag{7.E7.3.2}$$

$$\sum[K_{P,c}] = \frac{k_P}{30l}\begin{bmatrix} 36 & 3l & -36 & 3l & 0 & 0 \\ 3l & 4l^2 & -3l & -l^2 & 0 & 0 \\ -36 & -3l & 72 & 0 & -36 & 3l \\ 3l & -l^2 & 0 & 8l^2 & -3l & -l^2 \\ 0 & 0 & -36 & -3l & 36 & -3l \\ 0 & 0 & 3l & -l^2 & -3l & 4l^2 \end{bmatrix} \tag{7.E7.3.3}$$

$$\sum[K_{P,c}] = \frac{(e_0a)^2 k_P}{l^3}\begin{bmatrix} 12 & 6l & -12 & 6l & 0 & 0 \\ 6l & 4l^2 & -6l & 2l^2 & 0 & 0 \\ -12 & -6l & 24 & 0 & -12 & 6l \\ 6l & 2l^2 & 0 & 8l^2 & -6l & 2l^2 \\ 0 & 0 & -12 & -6l & 12 & -6l \\ 0 & 0 & 6l & 2l^2 & -6l & 4l^2 \end{bmatrix} \tag{7.E7.3.4}$$

here the stiffnesses of elastic foundations are calculated as follows:

$$k_W = K_W\frac{EI}{L^4} = (10)\times\frac{\left(1\times10^{12}\,\frac{\text{N}}{\text{m}^2}\right)\times\left(1.667\times10^{-37}\ \text{m}^4\right)}{\left(20\times10^{-9}\,\text{m}\right)^4} = 1.042\times10^7\ \frac{\text{N}}{\text{m}^2},$$

$$k_P = K_P\frac{EI}{L^2} = (10)\times\frac{\left(1\times10^{12}\,\frac{\text{N}}{\text{m}^2}\right)\times\left(1.667\times10^{-37}\ \text{m}^4\right)}{\left(20\times10^{-9}\,\text{m}\right)^2} = 4.168\times10^{-9}\ \text{N}. \tag{7.E7.3.5}$$

After the first and fifth rows and columns in the global matrices of elastic foundations due to the geometric boundary conditions are deleted, the physical and mechanical parameters are replaced and the reduced global stiffness matrix is calculated as follows by using Eq. (7.E7.2.4):

$$\begin{aligned}\left(\sum[K]\right)^* &= \left(\sum[K_{be}]\right)^* + \left(\sum[K_{W,c}]\right)^* + \left(\sum[K_{W,nl}]\right)^* + \left(\sum[K_{P,c}]\right)^* \\ &\quad + \left(\sum[K_{P,c}]\right)^* \\ &= \begin{bmatrix} 9.923\times10^{-17} & -1.44\times10^{-8} & 4.516\times10^{-17} & 0 \\ -1.44\times10^{-8} & 6.719 & 0 & 1.44\times10^{-8} \\ 4.516\times10^{-17} & 0 & 1.985\times10^{-16} & 4.516\times10^{-17} \\ 0 & 1.44\times10^{-8} & 4.516\times10^{-17} & 9.923\times10^{-17} \end{bmatrix}\end{aligned} \tag{7.E7.3.6}$$

Table 7.2 Comparisons of nondimensional frequencies of simply supported Euler-Bernoulli nanobeams resting on two-parameter elastic foundation for nonlocal transverse free vibration ($e_0a/L = 0.2$, $K_W = K_P = 10$)

Mode	Analytical	NL-FEM results				
Number	Eq. (4.E4.1.13)	Manual solution	Program code (MATLAB)			
		$n = 2$	$n = 2$	$n = 10$	$n = 15$	$n = 20$
1	13.36167	13.39785	13.39089	13.36172	13.36168	13.36167
2	31.76592	34.61142	34.60769	31.76876	31.76649	31.76610
3	51.29518	61.39037	61.38453	51.32008	51.30021	51.29679
4	70.69230	86.84332	86.83049	70.80095	70.71463	70.69947
5	89.87994	–	–	90.20910	89.94906	89.90231

By substituting Eqs. (7.E7.2.8) and (7.E7.3.6) into Eq. (7.24), the natural frequencies of first four vibration modes can be calculated:

$$\omega_1 = 201.634 \text{ GHz}, \quad \omega_2 = 520.893 \text{ GHz}, \quad \omega_3 = 923.909 \text{ GHz}, \quad \omega_4 = 1306.969 \text{ GHz}.$$

A convergence study on the nondimensional natural frequencies is presented in Table 7.2 for the free vibration of simply supported Euler-Bernoulli nanobeams resting on a double-parameter elastic foundation. According to this, it can be easily stated that NL-FEM generally provides results with very high convergence to analytical results.

(b) The classical and nonlocal stiffness matrices due to temperature are given by Eqs. (7.15) and (7.16), respectively. It is understood that the temperature effect has a negative contribution to the total stiffness by examining Eq. (7.25). According to this, let's write the global classical and nonlocal stiffnesses of temperature:

$$\sum [K_{T,c}]_e = \frac{N_t}{30l_e}\begin{bmatrix} 36 & 3l & -36 & 3l & 0 & 0 \\ 3l & 4l^2 & -3l & -l^2 & 0 & 0 \\ -36 & -3l & 72 & 0 & -36 & 3l \\ 3l & -l^2 & 0 & 8l^2 & -3l & -l^2 \\ 0 & 0 & -36 & -3l & 36 & -3l \\ 0 & 0 & 3l & -l^2 & -3l & 4l^2 \end{bmatrix} \tag{7.E7.3.7}$$

$$\sum [K_{T,nl}]_e = \frac{(e_0a)^2 N_t}{l^3} \begin{bmatrix} 12 & 6l & -12 & 6l & 0 & 0 \\ 6l & 4l^2 & -6l & 2l^2 & 0 & 0 \\ -12 & -6l & 24 & 0 & -12 & 6l \\ 6l & 2l^2 & 0 & 8l^2 & -6l & 2l^2 \\ 0 & 0 & -12 & -6l & 12 & -6l \\ 0 & 0 & 6l & 2l^2 & -6l & 4l^2 \end{bmatrix} \quad (7.E7.3.8)$$

where according to Eq. (4.5), the axial force resulting from temperature is:

$$N_t = EA\alpha_t \Delta T = \left(1 \times 10^{12} \frac{\mathrm{N}}{\mathrm{m}^2}\right) \times \left(2 \times 10^{-18}\,\mathrm{m}^2\right) \times \left(1.6 \times 10^{-6}\,\mathrm{K}^{-1}\right) \times (300\,\mathrm{K}) = 9.6 \times 10^{-10}\,\mathrm{N} \quad (7.E7.3.9)$$

After the first and fifth rows and columns of temperature stiffness matrices are reduced by geometric boundary conditions, the reduced global stiffness matrix is calculated below by evaluating these matrices together with Eq. (7.E7.2.4):

$$\left(\sum[K]\right)^* = \left(\sum[K_{be}]\right)^* - \left(\sum[K_{T,c}]\right)^* - \left(\sum[K_{T,nl}]\right)^*$$
$$= \begin{bmatrix} 5.926 \times 10^{-17} & -8.984 \times 10^{-9} & 3.059 \times 10^{-17} & 0 \\ -8.984 \times 10^{-9} & 3.402 & 0 & 8.984 \times 10^{-9} \\ 3.059 \times 10^{-17} & 0 & 1.185 \times 10^{-16} & 3.059 \times 10^{-17} \\ 0 & 8.984 \times 10^{-9} & 3.059 \times 10^{-17} & 5.926 \times 10^{-17} \end{bmatrix} \quad (7.E7.3.10)$$

The natural frequencies calculated from Eqs. (7.E7.2.8), (7.E7.3.10), and (7.24) are given as follows:

$$\omega_1 = 103.875\ \text{GHz},\ \omega_2 = 379.233\ \text{GHz},\ \omega_3 = 716.092\ \text{GHz},\ \omega_4 = 1030.929\ \text{GHz}.$$

The nondimensional frequency parameters of first five modes calculated with NL-FEM for simply supported Euler-Bernoulli nanobeams the temperature effect are presented in Table 7.3.

Example 7.4 Determine the natural frequencies of nonlocal transverse free vibration of cantilever nanobeam shown in Fig. 7.3. Use two finite elements in the analysis. Physical and mechanical parameters are as in Example 7.1.

Solution. According to the NL-FEM solution with two finite element of nonlocal free vibration of a simply supported nanobeam isolated from the thermo-elastic environment, the reduced stiffness, classical mass, and nonlocal mass matrices are presented by Eqs. (7.

Table 7.3 Comparisons of nondimensional frequencies of simply supported Euler-Bernoulli nanobeams under temperature effect for nonlocal transverse free vibration ($e_0a/L = 0.2$, $\Delta T = 300$ K)

Mode	Analytical	NL-FEM results				
Number	Eq. (E4.1.13)	Manual solution	Program code (MATLAB)			
		$n = 2$	$n = 2$	$n = 10$	$n = 15$	$n = 20$
1	6.86284	6.90214	6.89846	6.86290	6.86285	6.86284
2	22.65682	25.19864	25.19839	22.65939	22.65733	22.65698
3	39.09316	47.58169	47.57409	39.11433	39.09744	39.09453
4	55.17637	68.50145	68.49326	55.26660	55.19492	55.18232
5	70.94021	–	–	71.21058	70.99699	70.95859

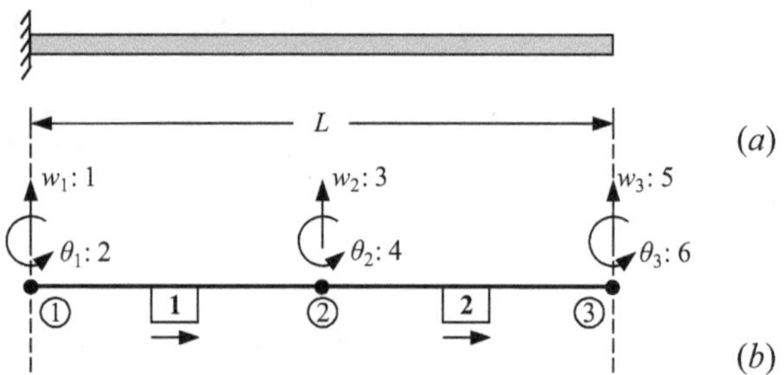

Fig. 7.3 NL-FEM analysis of cantilever nanobeam. (**a**) Physical system. (**b**) FEM model with two elements.

E7.2.1)-(7.E7.2.3), respectively. Considering the finite element mesh in Fig. 7.3b, the freedoms 1 and 2 corresponding to the fixed support are geometric boundary conditions because displacement and rotation do not occur in the fixed support. Then the first and second row and column of global matrices should be reduced.

$$\sum[K] = \frac{EI}{l^3}\begin{bmatrix} 12 & 6l & -12 & 6l & 0 & 0 \\ 6l & 4l^2 & -6l & 2l^2 & 0 & 0 \\ -12 & -6l & 24 & 0 & -12 & 6l \\ 6l & 2l^2 & 0 & 8l^2 & -6l & 2l^2 \\ 0 & 0 & -12 & -6l & 12 & -6l \\ 0 & 0 & 6l & 2l^2 & -6l & 4l^2 \end{bmatrix} \qquad (7.E7.4.1)$$

$$\left(\sum[K]\right)^* = \begin{bmatrix} 24 & 0 & -12 & 6l \\ 0 & 8l^2 & -6l & 2l^2 \\ -12 & -6l & 12 & -6l \\ 6l & 2l^2 & -6l & 4l^2 \end{bmatrix}$$

$$\sum[M_c]=\frac{\rho Al}{420}\begin{bmatrix} 156 & 22l & 54 & -13l & 0 & 0 \\ 22l & 4l^2 & 13l & -3l^2 & 0 & 0 \\ 54 & 13l & 312 & 0 & 54 & -13l \\ -13l & -3l^2 & 0 & 8l^2 & 13l & -3l^2 \\ 0 & 0 & 54 & 13l & 156 & -22l \\ 0 & 0 & -13l & -3l^2 & -22l & 4l^2 \end{bmatrix} \quad (\sum[M_c])^* \text{ = lower-right } 4\times4 \text{ block} \tag{7.E7.4.2}$$

$$\sum[M_{nl}]=\frac{(e_0a)^2\rho A}{30l_e}\begin{bmatrix} 36 & 3l & -36 & 3l & 0 & 0 \\ 3l & 4l^2 & -3l & -l^2 & 0 & 0 \\ -36 & -3l & 72 & 0 & -36 & 3l \\ 3l & -l^2 & 0 & 8l^2 & -3l & -l^2 \\ 0 & 0 & -36 & -3l & 36 & -3l \\ 0 & 0 & 3l & -l^2 & -3l & 4l^2 \end{bmatrix} \quad (\sum[M_{nl}])^* \text{ = lower-right } 4\times4 \text{ block} \tag{7.E7.4.3}$$

where after the related physical and mechanical parameters are written into the reduced global matrices, the stiffness and mass matrices are calculated as follows:

$$\left(\sum[K]\right)^*=\begin{bmatrix} 6.668\times10^{-17} & -1\times10^{-8} & 3.334\times10^{-17} & 0 \\ -1\times10^{-8} & 4 & 0 & 1\times10^{-8} \\ 3.334\times10^{-17} & 0 & 1.334\times10^{-16} & 3.334\times10^{-17} \\ 0 & 1\times10^{-8} & 3.334\times10^{-17} & 6.668\times10^{-17} \end{bmatrix} \tag{7.E7.4.4}$$

$$\left(\sum[M]\right)^*=\left(\sum[M_c]\right)^*+\left(\sum[M_{nl}]\right)^*$$
$$=\begin{bmatrix} 1.419\times10^{-40} & 6.878\times10^{-33} & -5.739\times10^{-41} & 0 \\ 6.878\times10^{-33} & 5.184\times10^{-23} & 0 & -6.878\times10^{-33} \\ -5.739\times10^{-41} & 0 & 2.839\times10^{-40} & -5.739\times10^{-41} \\ 0 & -6.878\times10^{-33} & -5.739\times10^{-41} & 1.419\times10^{-40} \end{bmatrix} \tag{7.E7.4.5}$$

Natural frequencies are calculated by employing the above stiffness and mass matrices into Eq. (7.24):

$$\omega_1 = 48.729 \text{ GHz}, \quad \omega_2 = 220.792 \text{ GHz} \quad \omega_3 = 540.917 \text{ GHz} \quad \omega_4 = 971.168 \text{ GHz}.$$

The nondimensional form (Eq. (4.E4.5.7)) of these values and NL-FEM results with high number of finite element results obtained with the program codes are compared with the analytical results in Table 7.4. Analytical results are added from Table 4.3 (Chap. 4). It should be stated that the NL-FEM results in this table attract attention. The analysis results with a high number of elements drop below the analytical results. However, the results do not decrease much with the increment in the number of finite elements and show a certain convergence. Therefore, these results may raise some doubts about the analytical approach.

On the other hand, another study is performed to see whether NL-FEM gives reasonable results. Accordingly, in the NL-FEM analyses performed by selecting the nondimensional atomic parameter $e_0a/L = 0.1$, the frequency value of fundamental mode is higher than the value obtained for $e_0a/L = 0.2$. As one may remember, in the results given in Table 4.3, the frequencies increase as the atomic parameter increases, and this is a paradoxical result for nonlocal elasticity. Here, it is a very important result that NL-FEM gives reasonable results in terms of the nonlocal structural mechanics. By comparing the results obtained via approaches that eliminate the paradox of nonlocal dynamics of cantilever nanobeams in the scientific literature and the results of NL-FEM, a thesis study [14] showing that NL-FEM is quite reliable is available.

Example 7.5 Determine the natural frequencies of nonlocal transverse free vibration of simply supported-transverse spring attached nanobeam in Fig. 7.4. Use two finite elements in the analysis. Physical and mechanical parameters are as in Example 7.1. In addition to this, attachment ratio is $\beta_k = 1$ and defined as $\beta_k = k_a L^3 / EI$.

Solution. The stiffness matrix of attachment should be constituted by considering Eq. (7.25). Since the attachment is located at the right end of nanobeam in Fig. 7.4b, the stiffness matrix of attachment should be formulated as follows according to Eq. (7.25):

$$(K_a)_{e=2} = \begin{bmatrix} 0 & 0 & 0 & 0 \\ 0 & 0 & 0 & 0 \\ 0 & 0 & k_a & 0 \\ 0 & 0 & 0 & 0 \end{bmatrix} \tag{7.E7.4.1}$$

where it is explained that the reason for the attachment determines only input 33 is the freedom in the vertical direction of right node of second finite element (the third node in the finite element mesh). If there is also a rotational spring attachment in this node, that attachment determines the input 44. After expressing the attachment stiffness globally, it is reduced as follows:

Table 7.4 Comparisons of nondimensional frequencies of cantilever Euler-Bernoulli nanobeams resting on two-parameter elastic foundation for nonlocal transverse free vibration ($K_W = K_P = 10$)

e_0a/L	Mode	Analytical	NL-FEM results				
	Number	Eq. (4.E4.1.13)	Manual solution	Program code (MATLAB)			
			$n = 2$	$n = 2$	$n = 10$	$n = 15$	$n = 20$
0.1	1	3.53128	–	3.43829	3.43681	3.43681	3.43681
	2	20.67960		19.26057	19.13685	19.13647	19.13640
	3	51.06382		54.52445	46.50246	46.49535	46.49412
	4	85.68971		111.01837	78.27291	78.22533	78.21701
	5	121.36150		–	112.05262	111.86281	111.82894
0.2	1	3.57940	3.23787	3.22682	3.22580	3.22580	3.22580
	2	17.57601	14.67080	14.66801	14.57584	14.57561	14.57558
	3	36.81305	35.94195	35.93657	31.38691	31.38206	31.38122
	4	54.19456	64.53057	64.51877	48.31299	48.28162	48.27615
	5	71.94097	–	–	65.29985	65.18213	65.16113

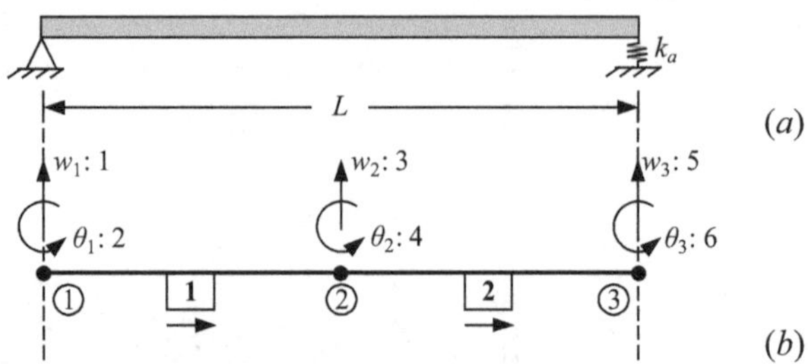

Fig. 7.4 NL-FEM analysis of simply supported-transverse spring attached nanobeam. (**a**) Physical system. (**b**) FEM model with two elements

$$\sum[K_a]=\begin{bmatrix} 0 & 0 & 0 & 0 & 0 & 0 \\ 0 & 0 & 0 & 0 & 0 & 0 \\ 0 & 0 & 0 & 0 & 0 & 0 \\ 0 & 0 & 0 & 0 & 0 & 0 \\ 0 & 0 & 0 & 0 & k_a & 0 \\ 0 & 0 & 0 & 0 & 0 & 0 \end{bmatrix} \quad \underbrace{}_{(\sum[K_a])^*} \tag{7.E7.4.2}$$

so let's calculate the attachment stiffness:

$$k_a=\beta_k\frac{EI}{L^3}=(1)\times\frac{\left(1\times10^{12}\,\frac{\mathrm{N}}{\mathrm{m}^2}\right)\times\left(1.667\times10^{-37}\ \mathrm{m}^4\right)}{\left(20\times10^{-9}\,\mathrm{m}\right)^3}=2.084\times10^{-2}\ \frac{\mathrm{N}}{\mathrm{m}} \tag{7.E7.4.3}$$

On the other hand, if the global stiffness matrix defined in Eq. (7.E7.2.4) is reduced for this nanobeam:

$$\sum[K]=\frac{EI}{l^3}\begin{bmatrix} 12 & 6l & -12 & 6l & 0 & 0 \\ 6l & 4l^2 & -6l & 2l^2 & 0 & 0 \\ -12 & -6l & 24 & 0 & -12 & 6l \\ 6l & 2l^2 & 0 & 8l^2 & -6l & 2l^2 \\ 0 & 0 & -12 & -6l & 12 & -6l \\ 0 & 0 & 6l & 2l^2 & -6l & 4l^2 \end{bmatrix} \quad \underbrace{}_{(\sum[K])^*} \tag{7.E7.4.4}$$

In this NL-FEM analysis, the reduced global stiffness matrix can be calculated as follows:

$$\left(\sum[K]\right)^{*}=\left(\sum[K]\right)^{*}+\left(\sum[K_a]\right)^{*}$$
$$=\begin{bmatrix} 6.668\times10^{-17} & -1\times10^{-8} & 3.334\times10^{-17} & 0 & 0 \\ -1\times10^{-8} & 4.001 & 0 & -2 & 1\times10^{-8} \\ 3.334\times10^{-17} & 0 & 1.334\times10^{-16} & -1\times10^{-8} & 3.334\times10^{-17} \\ 0 & -2 & -1\times10^{-8} & 2.021 & -1\times10^{-8} \\ 0 & 1\times10^{-8} & 3.334\times10^{-17} & -1\times10^{-8} & 6.668\times10^{-17} \end{bmatrix} \tag{7.E7.4.5}$$

If the global classical and nonlocal mass matrices that can be written in Eqs. (7.E7.2.5) and (7.E7.2.6) are reduced in a similar way to the stiffness matrices, the mass matrix required for eigenvalue analysis is calculated as follows:

$$\left(\sum[M]\right)^{*}=\left(\sum[M_c]\right)^{*}+\left(\sum[M_{nl}]\right)^{*}$$
$$=\begin{bmatrix} 1.419\times10^{-40} & 6.878\times10^{-33} & -5.379\times10^{-41} & 0 & 0 \\ 6.878\times10^{-33} & 5.184\times10^{-23} & 0 & -2.918\times10^{-24} & -6.878\times10^{-33} \\ -5.739\times10^{-41} & 0 & 2.839\times10^{-40} & 6.878\times10^{-33} & -5.379\times10^{-41} \\ 0 & -2.918\times10^{-24} & 6.878\times10^{-33} & 2.529\times10^{-23} & -3.146\times10^{-32} \\ 0 & -6.878\times10^{-33} & -5.379\times10^{-41} & -3.146\times10^{-32} & 1.419\times10^{-40} \end{bmatrix} \tag{7.E7.4.6}$$

The natural frequencies of first five modes can be attained by inserting the reduced global stiffness and mass matrices into Eq. (7.24):

$$\omega_1=25.073\,\text{GHz}\ \ \omega_2=163.070\,\text{GHz}\ \ \omega_3=434.359\,\text{GHz}\ \ \omega_4=771.907\,\text{GHz}\ \ \omega_5=1069.725\,\text{GHz}$$

These values and the nondimensional form of values obtained with the program codes based on NL-FEM (Eq. (4.E4.5.7)) and the analytical results through the program codes based on Eqs. (4.E4.2.6) and (4.E4.4.6) are compared in Table 7.5, where it is surprising that the NL-FEM results are less than the analytical results. It is clear from the analytical results that there is not a paradox of nonlocal elasticity in this nanobeam. In addition to this, at high amount of finite element, there is not a rapid decrease in results. In other words, natural frequencies converge to certain values. These inferences may require a revision in the application of boundary conditions to the analytical solution. The fact that the frequencies calculated with FEM are lower than the exact (analytical) results suggests that the analytical solution of nonlocal vibration may be invalid in the actual dynamic behavior of this nanobeam. It should definitely be remembered that the fact that the atomic parameter is a factor that reduces the stiffness of structure increases the importance of learning the mechanical behavior.

Table 7.5 Comparisons of nondimensional frequencies of simply supported-transverse spring attached Euler-Bernoulli nanobeams for nonlocal transverse free vibration ($e_0a/L = 0.2$, $\beta_k = 1$)

Mode	Analytical	NL-FEM results				
Number	(MATLAB)	Manual solution	Program code (MATLAB)			
		$n = 2$	$n = 2$	$n = 10$	$n = 15$	$n = 20$
1	1.70877	1.666	1.62423	1.62407	1.62407	1.62407
2	12.87853	10.835	10.76806	10.72631	10.72624	10.72622
3	30.38228	28.862	28.52468	25.80241	25.80019	25.79981
4	47.91768	51.290	51.25021	42.19099	42.17334	42.17029
5	64.93777	71.079	72.62932	58.73871	58.66392	58.65076

7.4 Nonlocal Vibration of Timoshenko Nanobeam

7.4.1 NL-FEM Expression

In the Timoshenko beam theory, known as the first-order shear deformation beam theory, energy equations are mostly used to constitute the finite element formulation. If you remember, the equation of motion is expressed with only bending displacements by removing the shear displacements in Chap. 4. Therefore, by using this equation of motion, the NL-FEM solution of free vibration can be formulated through finite element kinematics based only on bending deformations. First let's express the weighted average residual according to Eq. (4.80), which is the nonlocal equation of motion of Timoshenko nanobeam [6]:

$$
\begin{aligned}
\mathrm{I}= \int_0^{l_e} h\Bigg(& \frac{(e_0a)^2 N_t EI}{kGA}\frac{\partial^6 w_b}{\partial x^6} - \frac{(e_0a)^2 k_P EI}{kGA}\frac{\partial^6 w_b}{\partial x^6} + EI\frac{\partial^4 w_b}{\partial x^4} - (e_0a)^2 N_T \frac{\partial^4 w_b}{\partial x^4} \\
& + (e_0a)^2 k_P \frac{\partial^4 w_b}{\partial x^4} + \frac{(e_0a)^2 k_W EI}{kGA}\frac{\partial^4 w_b}{\partial x^4} - \frac{N_t EI}{kGA}\frac{\partial^4 w_b}{\partial x^4} + \frac{k_P EI}{kGA}\frac{\partial^4 w_b}{\partial x^4} \\
& - (e_0a)^2 k_W \frac{\partial^2 w_b}{\partial x^2} + N_t \frac{\partial^2 w_b}{\partial x^2} - k_P\frac{\partial^2 w_b}{\partial x^2} - \frac{k_W EI}{kGA}\frac{\partial^2 w_b}{\partial x^2} + \frac{\rho^2 I}{kG}\frac{\partial^4 w_b}{\partial t^4} \\
& + \rho A \frac{\partial^2 w_b}{\partial t^2} + \frac{k_W \rho I}{kGA}\frac{\partial^2 w_b}{\partial t^2} + \frac{(e_0a)^4 N_t \rho I}{kGA}\frac{\partial^8 w_b}{\partial x^6\partial t^2} - \frac{(e_0a)^4 k_P\rho I}{kGA}\frac{\partial^8 w_b}{\partial x^6 \partial t^2} \\
& + \frac{(e_0a)^2\rho EI}{kG}\frac{\partial^6 w_b}{\partial x^4\partial t^2} - 2\frac{(e_0a)^2 N_t\rho I}{kGA}\frac{\partial^6 w_b}{\partial x^4 \partial t^2} + 2\frac{(e_0a)^2 k_P \rho I}{kGA}\frac{\partial^6 w_b}{\partial x^4\partial t^2} \\
& + \frac{(e_0a)^4 k_W\rho I}{kGA}\frac{\partial^6 w_b}{\partial x^4\partial t^2} + (e_0a)^2\rho I \frac{\partial^6 w_b}{\partial x^4 \partial t^2} - \frac{\rho EI}{kG}\frac{\partial^4 w_b}{\partial x^2\partial t^2} - (e_0a)^2 \rho A \frac{\partial^4 w_b}{\partial x^2 \partial t^2} \\
& - 2\frac{(e_0a)^2 k_W \rho I}{kGA}\frac{\partial^4 w_b}{\partial x^2\partial t^2} + \frac{N_t\rho I}{kGA}\frac{\partial^4 w_b}{\partial x^2\partial t^2} - \frac{k_P\rho I}{kGA}\frac{\partial^4 w_b}{\partial x^2 \partial t^2} - \rho I \frac{\partial^4 w_b}{\partial x^2\partial t^2} \\
& + \frac{(e_0a)^4\rho^2 I}{kG}\frac{\partial^8 w_b}{\partial x^4\partial t^4} - 2\frac{(e_0a)^2\rho^2 I}{kG}\frac{\partial^6 w_b}{\partial x^2\partial t^4} + k_W w_b + (e_0a)^2\frac{\partial^2 f}{\partial x^2} - f \Bigg) \mathrm{d}x
\end{aligned}
\tag{7.27}
$$

Additionally, the equation of motion of finite element is defined as follows:

$$w_b = \phi \mathbf{d} \tag{7.28}$$

In order to rearrange Eq. (7.26), the following integral expressions should be written [6]:

$$\mathrm{I}_1 = \int_0^{l_e} \frac{(e_0a)^2 N_t EI}{kGA} h \frac{\partial^6 w_b}{\partial x^6} \mathrm{d}x = \frac{(e_0a)^2 N_t EI}{kGA} h \frac{\partial^5 w_b}{\partial x^5}\bigg|_0^{l_e}$$
$$- \frac{(e_0a)^2 N_t EI}{kGA} \frac{\partial h}{\partial x} \frac{\partial^4 w_b}{\partial x^4}\bigg|_0^{l_e} + \frac{(e_0a)^2 N_t EI}{kGA} \frac{\partial^2 h}{\partial x^2} \frac{\partial^3 w_b}{\partial x^3}\bigg|_0^{l_e} - \int_0^{l_e} \frac{(e_0a)^2 N_t EI}{kGA} \frac{\partial^3 h}{\partial x^3} \frac{\partial^3 w_b}{\partial x^3} \mathrm{d}x,$$

$$\mathrm{I}_2 = \int_0^{l_e} \frac{(e_0a)^2 k_P EI}{kGA} h \frac{\partial^6 w_b}{\partial x^6} \mathrm{d}x = \frac{(e_0a)^2 k_P EI}{kGA} h \frac{\partial^5 w_b}{\partial x^5}\bigg|_0^{l_e}$$
$$- \frac{(e_0a)^2 k_P EI}{kGA} \frac{\partial h}{\partial x} \frac{\partial^4 w_b}{\partial x^4}\bigg|_0^{l_e} + \frac{(e_0a)^2 k_P EI}{kGA} \frac{\partial^2 h}{\partial x^2} \frac{\partial^3 w_b}{\partial x^3}\bigg|_0^{l_e} - \int_0^{l_e} \frac{(e_0a)^2 k_P EI}{kGA} \frac{\partial^3 h}{\partial x^3} \frac{\partial^3 w_b}{\partial x^3} \mathrm{d}x,$$

$$\mathrm{I}_3 = \int_0^{l_e} EIh \frac{\partial^4 w_b}{\partial x^4} \mathrm{d}x = EIh \frac{\partial^3 w_b}{\partial x^3}\bigg|_0^{l_e} - EI \frac{\partial h}{\partial x} \frac{\partial^2 w_b}{\partial x^2}\bigg|_0^{l_e} + \int_0^{l_e} EI \frac{\partial^2 h}{\partial x^2} \frac{\partial^2 w_b}{\partial x^2} \mathrm{d}x,$$

$$\mathrm{I}_4 = \int_0^{l_e} (e_0a)^2 N_t h \frac{\partial^4 w_b}{\partial x^4} \mathrm{d}x = (e_0a)^2 N_t h \frac{\partial^3 w_b}{\partial x^3}\bigg|_0^{l_e} - (e_0a)^2 N_t \frac{\partial h}{\partial x} \frac{\partial^2 w_b}{\partial x^2}\bigg|_0^{l_e} + \int_0^{l_e} (e_0a)^2 N_t \frac{\partial^2 h}{\partial x^2} \frac{\partial^2 w_b}{\partial x^2} \mathrm{d}x,$$

$$\mathrm{I}_5 = \int_0^{l_e} (e_0a)^2 k_P h \frac{\partial^4 w_b}{\partial x^4} \mathrm{d}x = (e_0a)^2 k_P h \frac{\partial^3 w_b}{\partial x^3}\bigg|_0^{l_e} - (e_0a)^2 k_P \frac{\partial h}{\partial x} \frac{\partial^2 w_b}{\partial x^2}\bigg|_0^{l_e} + \int_0^{l_e} (e_0a)^2 k_P \frac{\partial^2 h}{\partial x^2} \frac{\partial^2 w_b}{\partial x^2} \mathrm{d}x,$$

$$\mathrm{I}_6 = \int_0^{l_e} \frac{(e_0a)^2 k_W EI}{kGA} h \frac{\partial^4 w_b}{\partial x^4} \mathrm{d}x = \frac{(e_0a)^2 k_W EI}{kGA} h \frac{\partial^3 w_b}{\partial x^3}\bigg|_0^{l_e} - \frac{(e_0a)^2 k_W EI}{kGA} \frac{\partial h}{\partial x} \frac{\partial^2 w_b}{\partial x^2}\bigg|_0^{l_e}$$
$$+ \int_0^{l_e} \frac{(e_0a)^2 k_W EI}{kGA} \frac{\partial^2 h}{\partial x^2} \frac{\partial^2 w_b}{\partial x^2} \mathrm{d}x,$$

$$\mathrm{I}_7 = \int_0^{l_e} \frac{N_t EI}{kGA} h \frac{\partial^4 w_b}{\partial x^4} \mathrm{d}x = \frac{N_t EI}{kGA} h \frac{\partial^3 w_b}{\partial x^3}\bigg|_0^{l_e} - \frac{N_t EI}{kGA} \frac{\partial h}{\partial x} \frac{\partial^2 w_b}{\partial x^2}\bigg|_0^{l_e} + \int_0^{l_e} \frac{N_t EI}{kGA} \frac{\partial^2 h}{\partial x^2} \frac{\partial^2 w_b}{\partial x^2} \mathrm{d}x,$$

$$\mathrm{I}_8 = \int_0^{l_e} \frac{k_P EI}{kGA} h \frac{\partial^4 w_b}{\partial x^4} \mathrm{d}x = \frac{k_P EI}{kGA} h \frac{\partial^3 w_b}{\partial x^3}\bigg|_0^{l_e} - \frac{k_P EI}{kGA} \frac{\partial h}{\partial x} \frac{\partial^2 w_b}{\partial x^2}\bigg|_0^{l_e} + \int_0^{l_e} \frac{k_P EI}{kGA} \frac{\partial^2 h}{\partial x^2} \frac{\partial^2 w_b}{\partial x^2} \mathrm{d}x,$$

$$\mathrm{I}_9 = \int_0^{l_e} (e_0 a)^2 k_W h \frac{\partial^2 w_b}{\partial x^2} \mathrm{d}x = (e_0 a)^2 k_W h \frac{\partial w_b}{\partial x}\bigg|_0^{l_e} - \int_0^{l_e} (e_0 a)^2 k_W \frac{\partial h}{\partial x} \frac{\partial w_b}{\partial x} \mathrm{d}x,$$

$$\mathrm{I}_{10} = \int_0^{l_e} N_t h \frac{\partial^2 w_b}{\partial x^2} \mathrm{d}x = N_t h \frac{\partial w_b}{\partial x}\bigg|_0^{l_e} - \int_0^{l_e} N_t \frac{\partial h}{\partial x} \frac{\partial w_b}{\partial x} \mathrm{d}x,$$

$$\mathrm{I}_{11} = \int_0^{l_e} k_P h \frac{\partial^2 w_b}{\partial x^2} \mathrm{d}x = k_P h \frac{\partial w_b}{\partial x}\bigg|_0^{l_e} - \int_0^{l_e} k_P \frac{\partial h}{\partial x} \frac{\partial w_b}{\partial x} \mathrm{d}x,$$

$$\mathrm{I}_{12} = \int_0^{l_e} \frac{k_W EI}{kGA} h \frac{\partial^2 w_b}{\partial x^2} \mathrm{d}x = \frac{k_W EI}{kGA} h \frac{\partial w_b}{\partial x}\bigg|_0^{l_e} - \int_0^{l_e} \frac{k_W EI}{kGA} \frac{\partial h}{\partial x} \frac{\partial w_b}{\partial x} \mathrm{d}x,$$

$$\mathrm{I}_{13} = \int_0^{l_e} \frac{\rho^2 I}{kG} h \frac{\partial^4 w_b}{\partial t^4} \mathrm{d}x,\ \mathrm{I}_{14} = \int_0^{l_e} \rho A h \frac{\partial^2 w_b}{\partial t^2} \mathrm{d}x,\ \mathrm{I}_{15} = \int_0^{l_e} \frac{k_W \rho I}{kGA} h \frac{\partial^2 w_b}{\partial t^2} \mathrm{d}x,$$

$$\begin{aligned}\mathrm{I}_{16} = \int_0^{l_e} \frac{(e_0 a)^4 N_t \rho I}{kGA} h \frac{\partial^8 w_b}{\partial x^6 \partial t^2} \mathrm{d}x = \frac{(e_0 a)^4 N_t \rho I}{kGA} h \frac{\partial^7 w_b}{\partial x^5 \partial t^2}\bigg|_0^{l_e} - \frac{(e_0 a)^4 N_t \rho I}{kGA} \frac{\partial h}{\partial x} \frac{\partial^6 w_b}{\partial x^4 \partial t^2}\bigg|_0^{l_e} \\ + \frac{(e_0 a)^4 N_t \rho I}{kGA} \frac{\partial^2 h}{\partial x^2} \frac{\partial^5 w_b}{\partial x^3 \partial t^2}\bigg|_0^{l_e} - \int_0^{l_e} \frac{(e_0 a)^4 N_t \rho I}{kGA} \frac{\partial^3 h}{\partial x^3} \frac{\partial^5 w_b}{\partial x^3 \partial t^2} \mathrm{d}x,\end{aligned}$$

$$\begin{aligned}\mathrm{I}_{17} = \int_0^{l_e} \frac{(e_0 a)^4 k_P \rho I}{kGA} h \frac{\partial^8 w_b}{\partial x^6 \partial t^2} \mathrm{d}x = \frac{(e_0 a)^4 k_P \rho I}{kGA} h \frac{\partial^7 w_b}{\partial x^5 \partial t^2}\bigg|_0^{l_e} - \frac{(e_0 a)^4 k_P \rho I}{kGA} \frac{\partial h}{\partial x} \frac{\partial^6 w_b}{\partial x^4 \partial t^2}\bigg|_0^{l_e} \\ + \frac{(e_0 a)^4 k_P \rho I}{kGA} \frac{\partial^2 h}{\partial x^2} \frac{\partial^5 w_b}{\partial x^3 \partial t^2}\bigg|_0^{l_e} - \int_0^{l_e} \frac{(e_0 a)^4 k_P \rho I}{kGA} \frac{\partial^3 h}{\partial x^3} \frac{\partial^5 w_b}{\partial x^3 \partial t^2} \mathrm{d}x,\end{aligned}$$

$$I_{18} = \int_0^{l_e} \frac{(e_0a)^2\rho EI}{kG} h \frac{\partial^6 w_b}{\partial x^4 \partial t^2} \mathrm{d}x = \frac{(e_0a)^2\rho EI}{kG} h \frac{\partial^5 w_b}{\partial x^3 \partial t^2}\bigg|_0^{l_e} - \frac{(e_0a)^2\rho EI}{kG} \frac{\partial h}{\partial x} \frac{\partial^4 w_b}{\partial x^2 \partial t^2}\bigg|_0^{l_e} + \int_0^{l_e} \frac{(e_0a)^2\rho EI}{kG} \frac{\partial^2 h}{\partial x^2} \frac{\partial^4 w_b}{\partial x^2 \partial t^2} \mathrm{d}x,$$

$$I_{19} = \int_0^{l_e} \frac{2(e_0a)^2 N_t \rho I}{kGA} h \frac{\partial^6 w_b}{\partial x^4 \partial t^2} \mathrm{d}x = \frac{2(e_0a)^2 N_t \rho I}{kGA} h \frac{\partial^5 w_b}{\partial x^3 \partial t^2}\bigg|_0^{l_e} - \frac{2(e_0a)^2 N_t \rho I}{kGA} \frac{\partial h}{\partial x} \frac{\partial^4 w_b}{\partial x^2 \partial t^2}\bigg|_0^{l_e} + \int_0^{l_e} \frac{2(e_0a)^2 N_t \rho I}{kGA} \frac{\partial^2 h}{\partial x^2} \frac{\partial^4 w_b}{\partial x^2 \partial t^2} \mathrm{d}x,$$

$$I_{20} = \int_0^{l_e} \frac{2(e_0a)^2 k_P \rho I}{kGA} h \frac{\partial^6 w_b}{\partial x^4 \partial t^2} \mathrm{d}x = \frac{2(e_0a)^2 k_P \rho I}{kGA} h \frac{\partial^5 w_b}{\partial x^3 \partial t^2}\bigg|_0^{l_e} - \frac{2(e_0a)^2 k_P \rho I}{kGA} \frac{\partial h}{\partial x} \frac{\partial^4 w_b}{\partial x^2 \partial t^2}\bigg|_0^{l_e} + \int_0^{l_e} \frac{2(e_0a)^2 k_P \rho I}{kGA} \frac{\partial^2 h}{\partial x^2} \frac{\partial^4 w_b}{\partial x^2 \partial t^2} \mathrm{d}x,$$

$$I_{21} = \int_0^{l_e} \frac{(e_0a)^4 k_W \rho I}{kGA} h \frac{\partial^6 w_b}{\partial x^4 \partial t^2} \mathrm{d}x = \frac{(e_0a)^4 k_W \rho I}{kGA} h \frac{\partial^5 w_b}{\partial x^3 \partial t^2}\bigg|_0^{l_e} - \frac{(e_0a)^4 k_W \rho I}{kGA} \frac{\partial h}{\partial x} \frac{\partial^4 w_b}{\partial x^2 \partial t^2}\bigg|_0^{l_e} + \int_0^{l_e} \frac{(e_0a)^4 k_W \rho I}{kGA} \frac{\partial^2 h}{\partial x^2} \frac{\partial^4 w_b}{\partial x^2 \partial t^2} \mathrm{d}x,$$

$$I_{22} = \int_0^{l_e} (e_0a)^2 \rho I h \frac{\partial^6 w_b}{\partial x^4 \partial t^2} \mathrm{d}x = (e_0a)^2 \rho I h \frac{\partial^5 w_b}{\partial x^3 \partial t^2}\bigg|_0^{l_e} - (e_0a)^2 \rho I \frac{\partial h}{\partial x} \frac{\partial^4 w_b}{\partial x^2 \partial t^2}\bigg|_0^{l_e} + \int_0^{l_e} (e_0a)^2 \rho I \frac{\partial^2 h}{\partial x^2} \frac{\partial^4 w_b}{\partial x^2 \partial t^2} \mathrm{d}x,$$

$$I_{23} = \int_0^{l_e} \frac{\rho EI}{kG} h \frac{\partial^4 w_b}{\partial x^2 \partial t^2} \mathrm{d}x = \frac{\rho EI}{kG} h \frac{\partial^3 w_b}{\partial x \partial t^2}\bigg|_0^{l_e} - \int_0^{l_e} \frac{\rho EI}{kG} \frac{\partial h}{\partial x} \frac{\partial^3 w_b}{\partial x \partial t^2} \mathrm{d}x,$$

$$\mathrm{I}_{24}=\int_0^{l_e}(e_0a)^2\rho Ah\frac{\partial^4 w_b}{\partial x^2\partial t^2}\mathrm{d}x=(e_0a)^2\rho Ah\frac{\partial^3 w_b}{\partial x\partial t^2}\bigg|_0^{l_e}-\int_0^{l_e}(e_0a)^2\rho A\frac{\partial h}{\partial x}\frac{\partial^3 w_b}{\partial x\partial t^2}\mathrm{d}x,$$

$$\mathrm{I}_{25}=\int_0^{l_e}\frac{2(e_0a)^2k_W\rho I}{kGA}\frac{\partial^4 w_b}{\partial x^2\partial t^2}\mathrm{d}x=\frac{2(e_0a)^2k_W\rho I}{kGA}h\frac{\partial^3 w_b}{\partial x\partial t^2}\bigg|_0^{l_e}-\int_0^{l_e}\frac{2(e_0a)^2k_W\rho I}{kGA}\frac{\partial h}{\partial x}\frac{\partial^3 w_b}{\partial x\partial t^2}\mathrm{d}x,$$

$$\mathrm{I}_{26}=\int_0^{l_e}\frac{N_t\rho I}{kGA}\frac{\partial^4 w_b}{\partial x^2\partial t^2}\mathrm{d}x=\frac{N_t\rho I}{kGA}h\frac{\partial^3 w_b}{\partial x\partial t^2}\bigg|_0^{l_e}-\int_0^{l_e}\frac{N_t\rho I}{kGA}\frac{\partial h}{\partial x}\frac{\partial^3 w_b}{\partial x\partial t^2}\mathrm{d}x,$$

$$\mathrm{I}_{27}=\int_0^{l_e}\frac{k_P\rho I}{kGA}\frac{\partial^4 w_b}{\partial x^2\partial t^2}\mathrm{d}x=\frac{k_P\rho I}{kGA}h\frac{\partial^3 w_b}{\partial x\partial t^2}\bigg|_0^{l_e}-\int_0^{l_e}\frac{k_P\rho I}{kGA}\frac{\partial h}{\partial x}\frac{\partial^3 w_b}{\partial x\partial t^2}\mathrm{d}x,$$

$$\mathrm{I}_{28}=\int_0^{l_e}\rho Ih\frac{\partial^4 w_b}{\partial x^2\partial t^2}\mathrm{d}x=\rho Ih\frac{\partial^3 w_b}{\partial x\partial t^2}\bigg|_0^{l_e}-\int_0^{l_e}\rho I\frac{\partial h}{\partial x}\frac{\partial^3 w_b}{\partial x\partial t^2}\mathrm{d}x,$$

$$\mathrm{I}_{29}=\int_0^{l_e}\frac{(e_0a)^4\rho^2 I}{kG}h\frac{\partial^8 w_b}{\partial x^4\partial t^4}\mathrm{d}x=\frac{(e_0a)^4\rho^2 I}{kG}h\frac{\partial^7 w_b}{\partial x^3\partial t^4}\bigg|_0^{l_e}-\frac{(e_0a)^4\rho^2 I}{kG}\frac{\partial h}{\partial x}\frac{\partial^6 w_b}{\partial x^2\partial t^4}\bigg|_0^{l_e}$$
$$+\int_0^{l_e}\frac{(e_0a)^4\rho^2 I}{kG}\frac{\partial^2 h}{\partial x^2}\frac{\partial^6 w_b}{\partial x^2\partial t^4}\mathrm{d}x,$$

$$\mathrm{I}_{30}=\int_0^{l_e}\frac{2(e_0a)^2\rho^2 I}{kG}h\frac{\partial^6 w_b}{\partial x^2\partial t^4}\mathrm{d}x=\frac{2(e_0a)^2\rho^2 I}{kG}h\frac{\partial^5 w_b}{\partial x\partial t^4}\bigg|_0^{l_e}-\int_0^{l_e}\frac{2(e_0a)^2\rho^2 I}{kG}\frac{\partial h}{\partial x}\frac{\partial^5 w_b}{\partial x\partial t^4}\mathrm{d}x,$$

$$\mathrm{I}_{31}=\int_0^{L}k_W h w_b\mathrm{d}x$$

$$\mathrm{I}_{32}=\int_0^{l_e}(e_0a)^2h\frac{\partial^2 f}{\partial x^2}\mathrm{d}x=(e_0a)^2h\frac{\partial f}{\partial x}\bigg|_0^{l_e}-\int_0^{l_e}(e_0a)^2\frac{\partial h}{\partial x}\frac{\partial f}{\partial x}\mathrm{d}x,$$

$$\mathrm{I}_{33}=\int_0^{l_e}fh\mathrm{d}x \tag{7.29}$$

Weak formulation is achieved by substituting the above integral expressions into Eq. (7.24):

$$\begin{aligned}
\mathrm{I} = & -\int_0^{l_e} \frac{(e_0a)^2 N_t EI}{kGA} \frac{\partial^3 h}{\partial x^3} \frac{\partial^3 w_b}{\partial x^3} \mathrm{d}x + \int_0^{l_e} \frac{(e_0a)^2 k_P EI}{kGA} \frac{\partial^3 h}{\partial x^3} \frac{\partial^3 w_b}{\partial x^3} \mathrm{d}x + \int_0^{l_e} \frac{(e_0a)^2 k_W EI}{kGA} \frac{\partial^2 h}{\partial x^2} \frac{\partial^2 w_b}{\partial x^2} \\
& - \int_0^{l_e} \frac{N_t EI}{kGA} \frac{\partial^2 h}{\partial x^2} \frac{\partial^2 w_b}{\partial x^2} \mathrm{d}x + \int_0^{l_e} \frac{k_P EI}{kGA} \frac{\partial^2 h}{\partial x^2} \frac{\partial^2 w_b}{\partial x^2} \mathrm{d}x + \int_0^{l_e} EI \frac{\partial^2 h}{\partial x^2} \frac{\partial^2 w_b}{\partial x^2} \mathrm{d}x + \int_0^{l_e} \rho A h \frac{\partial^2 w_b}{\partial t^2} \mathrm{d}x \\
& + \int_0^{l_e} \frac{(e_0a)^2 \rho EI}{kG} \frac{\partial^2 h}{\partial x^2} \frac{\partial^4 w_b}{\partial x^2 \partial t^2} \mathrm{d}x + \int_0^{l_e} \frac{\rho EI}{kG} \frac{\partial h}{\partial x} \frac{\partial^3 w_b}{\partial x \partial t^2} \mathrm{d}x + \int_0^{l_e} \frac{k_W EI}{kGA} \frac{\partial h}{\partial x} \frac{\partial w_b}{\partial x} \mathrm{d}x \\
& + \int_0^{l_e} k_W h w_b \mathrm{d}x - \int_0^{l_e} (e_0a)^2 N_t \frac{\partial^2 h}{\partial x^2} \frac{\partial^2 w_b}{\partial x^2} \mathrm{d}x + \int_0^{l_e} (e_0a)^2 k_P \frac{\partial^2 h}{\partial x^2} \frac{\partial^2 w_b}{\partial x^2} \mathrm{d}x \\
& + \int_0^{l_e} (e_0a)^2 k_W \frac{\partial h}{\partial x} \frac{\partial w_b}{\partial x} \mathrm{d}x - \int_0^{l_e} N_t \frac{\partial h}{\partial x} \frac{\partial w_b}{\partial x} \mathrm{d}x + \int_0^{l_e} k_P \frac{\partial h}{\partial x} \frac{\partial w_b}{\partial x} \mathrm{d}x + \int_0^{l_e} (e_0a)^2 \rho A \frac{\partial h}{\partial x} \frac{\partial^3 w_b}{\partial x \partial t^2} \mathrm{d}x \\
& - \int_0^{l_e} f h \mathrm{d}x + \int_0^{l_e} \frac{k_W \rho I}{kGA} h \frac{\partial^2 w_b}{\partial t^2} \mathrm{d}x - \int_0^{l_e} (e_0a)^2 \frac{\partial h}{\partial x} \frac{\partial f}{\partial x} \mathrm{d}x - \int_0^{l_e} \frac{2(e_0a)^2 N_t \rho I}{kGA} \frac{\partial^2 h}{\partial x^2} \frac{\partial^4 w_b}{\partial x^2 \partial t^2} \mathrm{d}x \\
& + \int_0^{l_e} \frac{2(e_0a)^2 k_P \rho I}{kGA} \frac{\partial^2 h}{\partial x^2} \frac{\partial^4 w_b}{\partial x^2 \partial t^2} \mathrm{d}x + \int_0^{l_e} \frac{k_P \rho I}{kGA} \frac{\partial h}{\partial x} \frac{\partial^3 w_b}{\partial x \partial t^2} \mathrm{d}x + \int_0^{l_e} \frac{2(e_0a)^2 \rho^2 I}{kG} \frac{\partial h}{\partial x} \frac{\partial^5 w_b}{\partial x \partial t^4} \mathrm{d}x \\
& + \int_0^{l_e} \frac{2(e_0a)^2 k_W \rho I}{kGA} \frac{\partial h}{\partial x} \frac{\partial^3 w_b}{\partial x \partial t^2} \mathrm{d}x - \int_0^{l_e} \frac{N_t \rho I}{kGA} \frac{\partial h}{\partial x} \frac{\partial^3 w_b}{\partial x \partial t^2} \mathrm{d}x + \int_0^{l_e} \rho I \frac{\partial h}{\partial x} \frac{\partial^3 w_b}{\partial x \partial t^2} \mathrm{d}x \\
& + \int_0^{l_e} \frac{\rho^2 I}{kG} h \frac{\partial^4 w_b}{\partial t^4} \mathrm{d}x + \int_0^{l_e} \frac{(e_0a)^4 k_W \rho I}{kGA} \frac{\partial^2 h}{\partial x^2} \frac{\partial^4 w_b}{\partial x^2 \partial t^2} \mathrm{d}x + \int_0^{l_e} (e_0a)^2 \rho I \frac{\partial^2 h}{\partial x^2} \frac{\partial^4 w_b}{\partial x^2 \partial t^2} \mathrm{d}x \\
& - \int_0^{l_e} \frac{(e_0a)^4 N_t \rho I}{kGA} \frac{\partial^3 h}{\partial x^3} \frac{\partial^5 w_b}{\partial x^3 \partial t^2} \mathrm{d}x + \int_0^{l_e} \frac{(e_0a)^4 k_P \rho I}{kGA} \frac{\partial^3 h}{\partial x^3} \frac{\partial^5 w_b}{\partial x^3 \partial t^2} \mathrm{d}x \\
& + \int_0^{l_e} \frac{(e_0a)^4 \rho^2 I}{kG} \frac{\partial^2 h}{\partial x^2} \frac{\partial^6 w_b}{\partial x^2 \partial t^4} \mathrm{d}x
\end{aligned} \tag{7.30}$$

the following relations can be given to rewritten above equation:

$$
\begin{aligned}
&h=\phi^{\mathrm{T}},\ \ \mathbf{B}=\mathrm{D}^{\mathrm{k}}\phi,\ \ \frac{\partial h}{\partial x}=\left(\phi^{\mathrm{T}}\right)'=\mathbf{B}^{\mathrm{T}},\ \ \frac{\partial w_b}{\partial x}=\mathbf{Bd},\\
&\frac{\partial^2 w_b}{\partial x^2}=\mathbf{B}'\mathbf{d},\ \ \frac{\partial^3 w_b}{\partial x^3}=\mathbf{B}''\mathbf{d},\ \ \frac{\partial^2 w_b}{\partial t^2}=\phi\ddot{\mathbf{d}},\ \ \frac{\partial^4 w_b}{\partial t^4}=\phi\bar{\mathbf{d}}.
\end{aligned}
\tag{7.31}
$$

using Eqs. (7.28) and (7.31) into Eq. (7.30) gives:

$$
\begin{aligned}
\mathrm{I}=&-\int_0^{l_e}\frac{(e_0a)^2N_tEI}{kGA}\left(\mathbf{B}''^{\mathrm{T}}\mathbf{B}''\right)\mathbf{d}\mathrm{d}x+\int_0^{l_e}\frac{(e_0a)^2k_PEI}{kGA}\left(\mathbf{B}''^{\mathrm{T}}\mathbf{B}''\right)\mathbf{d}\mathrm{d}x+\int_0^{l_e}\frac{(e_0a)^2k_WEI}{kGA}\left(\mathbf{B}'^{\mathrm{T}}\mathbf{B}'\right)\mathbf{d}\mathrm{d}x\\
&-\int_0^{l_e}\frac{N_tEI}{kGA}\left(\mathbf{B}'^{\mathrm{T}}\mathbf{B}'\right)\mathbf{d}\mathrm{d}x+\int_0^{l_e}\frac{k_PEI}{kGA}\left(\mathbf{B}'^{\mathrm{T}}\mathbf{B}'\right)\mathbf{d}\mathrm{d}x+\int_0^{l_e}EI\left(\mathbf{B}'^{\mathrm{T}}\mathbf{B}'\right)\mathbf{d}\mathrm{d}x+\int_0^{l_e}\rho A\left(\phi^{\mathrm{T}}\phi\right)\ddot{\mathbf{d}}\mathrm{d}x\\
&+\int_0^{l_e}\frac{(e_0a)^2\rho EI}{kG}\left(\mathbf{B}'^{\mathrm{T}}\mathbf{B}'\right)\ddot{\mathbf{d}}\mathrm{d}x+\int_0^{l_e}\frac{\rho EI}{kG}\left(\mathbf{B}^{\mathrm{T}}\mathbf{B}\right)\ddot{\mathbf{d}}\mathrm{d}x+\int_0^{l_e}\frac{k_WEI}{kGA}\left(\mathbf{B}^{\mathrm{T}}\mathbf{B}\right)\mathbf{d}\mathrm{d}x+\int_0^{l_e}k_W\left(\phi^{\mathrm{T}}\phi\right)\mathbf{d}\mathrm{d}x\\
&-\int_0^{l_e}(e_0a)^2N_t\left(\mathbf{B}'^{\mathrm{T}}\mathbf{B}'\right)\mathbf{d}\mathrm{d}x+\int_0^{l_e}(e_0a)^2k_P\left(\mathbf{B}'^{\mathrm{T}}\mathbf{B}'\right)\mathbf{d}\mathrm{d}x+\int_0^{l_e}(e_0a)^2k_W\left(\mathbf{B}^{\mathrm{T}}\mathbf{B}\right)\mathbf{d}\mathrm{d}x\\
&-\int_0^{l_e}N_t\left(\mathbf{B}^{\mathrm{T}}\mathbf{B}\right)\mathbf{d}\mathrm{d}x+\int_0^{l_e}k_P\left(\mathbf{B}^{\mathrm{T}}\mathbf{B}\right)\mathbf{d}\mathrm{d}x+\int_0^{l_e}(e_0a)^2\rho A\left(\mathbf{B}^{\mathrm{T}}\mathbf{B}\right)\ddot{\mathbf{d}}\mathrm{d}x-\int_0^{l_e}\phi^{\mathrm{T}}f\mathrm{d}x\\
&+\int_0^{l_e}\frac{k_W\rho I}{kGA}\left(\phi^{\mathrm{T}}\phi\right)\ddot{\mathbf{d}}\mathrm{d}x-\int_0^{l_e}(e_0a)^2\mathbf{B}^{\mathrm{T}}f'\mathrm{d}x-\int_0^{l_e}\frac{2(e_0a)^2N_t\rho I}{kGA}\left(\mathbf{B}'^{\mathrm{T}}\mathbf{B}'\right)\ddot{\mathbf{d}}\mathrm{d}x\\
&+\int_0^{l_e}\frac{2(e_0a)^2k_P\rho I}{kGA}\left(\mathbf{B}'^{\mathrm{T}}\mathbf{B}'\right)\ddot{\mathbf{d}}\mathrm{d}x+\int_0^{l_e}\frac{k_P\rho I}{kGA}\left(\mathbf{B}^{\mathrm{T}}\mathbf{B}\right)\mathbf{d}\mathrm{d}x+\int_0^{l_e}\frac{2(e_0a)^2\rho^2I}{kG}\left(\mathbf{B}^{\mathrm{T}}\mathbf{B}\right)\ddot{\ddot{\mathbf{d}}}\mathrm{d}x\\
&+\int_0^{l_e}\frac{2(e_0a)^2k_W\rho I}{kGA}\left(\mathbf{B}^{\mathrm{T}}\mathbf{B}\right)\ddot{\mathbf{d}}\mathrm{d}x-\int_0^{l_e}\frac{N_t\rho I}{kGA}\left(\mathbf{B}'^{\mathrm{T}}\mathbf{B}'\right)\ddot{\mathbf{d}}\mathrm{d}x+\int_0^{l_e}\rho I\left(\mathbf{B}^{\mathrm{T}}\mathbf{B}\right)\ddot{\mathbf{d}}\mathrm{d}x\\
&+\int_0^{l_e}\frac{\rho^2I}{kG}\left(\phi^{\mathrm{T}}\phi\right)\ddot{\ddot{\mathbf{d}}}\mathrm{d}x+\int_0^{l_e}\frac{(e_0a)^4k_W\rho I}{kGA}\left(\mathbf{B}'^{\mathrm{T}}\mathbf{B}'\right)\ddot{\mathbf{d}}\mathrm{d}x+\int_0^{l_e}(e_0a)^2\rho I\left(\mathbf{B}'^{\mathrm{T}}\mathbf{B}'\right)\ddot{\mathbf{d}}\mathrm{d}x\\
&-\int_0^{l_e}\frac{(e_0a)^4N_t\rho I}{kGA}\left(\mathbf{B}''^{\mathrm{T}}\mathbf{B}''\right)\ddot{\mathbf{d}}\mathrm{d}x+\int_0^{l_e}\frac{(e_0a)^4k_P\rho I}{kGA}\left(\mathbf{B}''^{\mathrm{T}}\mathbf{B}''\right)\ddot{\mathbf{d}}\mathrm{d}x\\
&+\int_0^{l_e}\frac{(e_0a)^4\rho^2I}{kG}\left(\mathbf{B}'^{\mathrm{T}}\mathbf{B}'\right)\ddot{\ddot{\mathbf{d}}}\mathrm{d}x=0
\end{aligned}
\tag{7.32}
$$

this equation can be expressed as follows:

$$\left(\sum K\right)\mathbf{d} + \left(\sum M^0\right)\ddot{\mathbf{d}} + \left(\sum M^1\right)\ddddot{\mathbf{d}} = \sum \mathbf{f} \tag{7.33}$$

where $\sum K$ is the total stiffness matrix, $\sum M^0$ is the total simple mass matrix, $\sum M^1$ is the total higher-order mass matrix, and $\sum \mathbf{f}$ is the total load vector. These are defined below [6]:

$$\begin{aligned}\sum K = {} & K + K_{W,c} + K_{W,nl} + K_{P,c} + K_{P,nl} + K_{W,R,c} + K_{W,R,nl} + K_{P,R,c} + K_{P,R,nl} \\ & - K_{T,c} - K_{T,nl} - K_{T,R,c} - K_{T,R,nl}\end{aligned} \tag{7.34}$$

$$\begin{aligned}\sum M^0 = {} & M_{A,c} + M_{A,nl} + M_{I,c} + M_{I,nl} + M_{A,R,c} + M_{A,R,nl} + M_{I,W,c} + M_{I,W,nl} \\ & + M_{I,P,c} + M_{I,P,nl} - M_{I,T,c} - M_{I,T,nl} + M_{I,W,HO-nl} + M_{I,P,HO-nl} \\ & - M_{I,T,HO-nl}\end{aligned} \tag{7.35}$$

$$\sum M^1 = M_{I,R,c} + M_{I,R,nl} + M_{I,R,HO-nl} \tag{7.36}$$

$$\sum \mathbf{f} = \mathbf{f}_c + \mathbf{f}_{nl} \tag{7.37}$$

The definitions in Eqs. (7.34)–(7.37) are formulated as follows [6]:

$$\begin{aligned} K = {} & \int_0^{l_e} EI\left(\mathbf{B}'^{\mathrm{T}}\mathbf{B}'\right)\mathrm{d}x = \int_0^{l_e} EI \begin{Bmatrix} \phi_i'' \\ \varphi_i'' \\ \phi_j'' \\ \varphi_j'' \end{Bmatrix} \left[\phi_i'' \quad \varphi_i'' \quad \phi_j'' \quad \varphi_j''\right]\mathrm{d}x \\ = {} & \frac{EI}{l_e^3}\begin{bmatrix} 12 & 6l_e & -12 & 6l_e \\ 6l_e & 4l_e^2 & -6l_e & 2l_e^2 \\ -12 & -6l_e & 12 & -6l_e \\ 6l_e & 2l_e^2 & -6l_e & 4l_e^2 \end{bmatrix}\end{aligned} \tag{7.38}$$

$$K_{W,c} = \int_0^{l_e} k_W(\phi^T\phi)dx = \int_0^{l_e} k_W \begin{Bmatrix} \phi_i \\ \varphi_i \\ \phi_j \\ \varphi_j \end{Bmatrix} [\phi_i \quad \varphi_i \quad \phi_j \quad \varphi_j] dx$$
$$= \frac{k_W l_e}{420} \begin{bmatrix} 156 & 22l_e & 54 & -13l_e \\ 22l_e & 4l_e^2 & 13l_e & -3l_e^2 \\ 54 & 13l_e & 156 & -22l_e \\ -13l_e & -3l_e^2 & -22l_e & 4l_e^2 \end{bmatrix} \tag{7.39}$$

$$K_{W,nl} = \int_0^{l_e} (e_0a)^2 k_W(\mathbf{B}^T\mathbf{B})dx = \int_0^{l_e} (e_0a)^2 k_W \begin{Bmatrix} \phi_i' \\ \varphi_i' \\ \phi_j' \\ \varphi_j' \end{Bmatrix} [\phi_i' \quad \varphi_i' \quad \phi_j' \quad \varphi_j'] dx$$
$$= \frac{(e_0a)^2 k_W}{30l_e} \begin{bmatrix} 36 & 3l_e & -36 & 3l_e \\ 3l_e & 4l_e^2 & -3l_e & -l_e^2 \\ -36 & -3l_e & 36 & -3l_e \\ 3l_e & -l_e^2 & -3l_e & 4l_e^2 \end{bmatrix} \tag{7.40}$$

$$K_{P,c} = \int_0^{l_e} k_P(\mathbf{B}^T\mathbf{B})dx = \int_0^{l_e} k_P \begin{Bmatrix} \phi_i' \\ \varphi_i' \\ \phi_j' \\ \varphi_j' \end{Bmatrix} [\phi_i' \quad \varphi_i' \quad \phi_j' \quad \varphi_j'] dx$$
$$= \frac{k_P}{30l_e} \begin{bmatrix} 36 & 3l_e & -36 & 3l_e \\ 3l_e & 4l_e^2 & -3l_e & -l_e^2 \\ -36 & -3l_e & 36 & -3l_e \\ 3l_e & -l_e^2 & -3l_e & 4l_e^2 \end{bmatrix} \tag{7.41}$$

$$K_{P,nl} = \int_0^{l_e} (e_0 a)^2 k_P \left(\mathbf{B}'^{\mathrm{T}}\mathbf{B}'\right) \mathrm{d}x = \int_0^{l_e} (e_0 a)^2 k_P \begin{Bmatrix} \phi_i'' \\ \varphi_i'' \\ \phi_j'' \\ \varphi_j'' \end{Bmatrix} \left[\phi_i'' \quad \varphi_i'' \quad \phi_j'' \quad \varphi_j''\right] \mathrm{d}x$$
$$= \frac{(e_0 a)^2 k_P}{l_e^3} \begin{bmatrix} 12 & 6l_e & -12 & 6l_e \\ 6l_e & 4l_e^2 & -6l_e & 2l_e^2 \\ -12 & -6l_e & 12 & -6l_e \\ 6l_e & 2l_e^2 & -6l_e & 4l_e^2 \end{bmatrix} \tag{7.42}$$

$$K_{W,R,c} = \int_0^{l_e} \frac{k_W EI}{kGA} \left(\mathbf{B}^{\mathrm{T}}\mathbf{B}\right) \mathrm{d}x = \int_0^{l_e} \frac{k_W EI}{kGA} \begin{Bmatrix} \phi_i' \\ \varphi_i' \\ \phi_j' \\ \varphi_j' \end{Bmatrix} \left[\phi_i' \quad \varphi_i' \quad \phi_j' \quad \varphi_j'\right] \mathrm{d}x$$
$$= \frac{k_W EI}{30 kGA l_e} \begin{bmatrix} 36 & 3l_e & -36 & 3l_e \\ 3l_e & 4l_e^2 & -3l_e & -l_e^2 \\ -36 & -3l_e & 36 & -3l_e \\ 3l_e & -l_e^2 & -3l_e & 4l_e^2 \end{bmatrix} \tag{7.43}$$

$$K_{W,R,nl} = \int_0^{l_e} \frac{(e_0 a)^2 k_W EI}{kGA} \left(\mathbf{B}'^{\mathrm{T}}\mathbf{B}'\right) \mathrm{d}x = \int_0^{l_e} \frac{(e_0 a)^2 k_W EI}{kGA} \begin{Bmatrix} \phi_i'' \\ \varphi_i'' \\ \phi_j'' \\ \varphi_j'' \end{Bmatrix} \left[\phi_i'' \quad \varphi_i'' \quad \phi_j'' \quad \varphi_j''\right] \mathrm{d}x$$
$$= \frac{(e_0 a)^2 k_W EI}{kGA l_e^3} \begin{bmatrix} 12 & 6l_e & -12 & 6l_e \\ 6l_e & 4l_e^2 & -6l_e & 2l_e^2 \\ -12 & -6l_e & 12 & -6l_e \\ 6l_e & 2l_e^2 & -6l_e & 4l_e^2 \end{bmatrix} \tag{7.44}$$

$$K_{P,R,c}=\int_0^{l_e}\frac{k_PEI}{kGA}\left(\mathbf{B}'^{\mathrm{T}}\mathbf{B}'\right)\mathrm{d}x=\int_0^{l_e}\frac{k_PEI}{kGA}\begin{Bmatrix}\phi_i''\\ \varphi_i''\\ \phi_j''\\ \varphi_j''\end{Bmatrix}\left[\phi_i''\quad \varphi_i''\quad \phi_j''\quad \varphi_j''\right]\mathrm{d}x$$
$$=\frac{k_PEI}{kGAl_e^3}\begin{bmatrix}12 & 6l_e & -12 & 6l_e\\ 6l_e & 4l_e^2 & -6l_e & 2l_e^2\\ -12 & -6l_e & 12 & -6l_e\\ 6l_e & 2l_e^2 & -6l_e & 4l_e^2\end{bmatrix} \tag{7.45}$$

$$K_{P,R,nl}=\int_0^{l_e}\frac{(e_0a)^2k_PEI}{kGA}\left(\mathbf{B}''^{\mathrm{T}}\mathbf{B}''\right)\mathrm{d}x=\int_0^{l_e}\frac{(e_0a)^2k_PEI}{kGA}\begin{Bmatrix}\phi_i'''\\ \varphi_i'''\\ \phi_j'''\\ \varphi_j'''\end{Bmatrix}\left[\phi_i'''\quad \varphi_i'''\quad \phi_j'''\quad \varphi_j'''\right]\mathrm{d}x$$
$$=\frac{36(e_0a)^2k_PEI}{kGAl_e^5}\begin{bmatrix}4 & 2l_e & -4 & 2l_e\\ 2l_e & l_e^2 & -2l_e & l_e^2\\ -4 & -2l_e & 4 & -2l_e\\ 2l_e & l_e^2 & -2l_e & l_e^2\end{bmatrix} \tag{7.46}$$

$$K_{T,c}=\int_0^{l_e}N_t\left(\mathbf{B}^{\mathrm{T}}\mathbf{B}\right)\mathrm{d}x=\int_0^{l_e}N_t\begin{Bmatrix}\phi_i'\\ \varphi_i'\\ \phi_j'\\ \varphi_j'\end{Bmatrix}\left[\phi_i'\quad \varphi_i'\quad \phi_j'\quad \varphi_j'\right]\mathrm{d}x$$
$$=\frac{N_t}{30l_e}\begin{bmatrix}36 & 3l_e & -36 & 3l_e\\ 3l_e & 4l_e^2 & -3l_e & -l_e^2\\ -36 & -3l_e & 36 & -3l_e\\ 3l_e & -l_e^2 & -3l_e & 4l_e^2\end{bmatrix} \tag{7.47}$$

$$K_{T,nl}=\int_0^{l_e}(e_0a)^2N_t\left(\mathbf{B}'^{\mathrm{T}}\mathbf{B}'\right)\mathrm{d}x=\int_0^{l_e}(e_0a)^2N_t\begin{Bmatrix}\phi_i''\\ \varphi_i''\\ \phi_j''\\ \varphi_j''\end{Bmatrix}\left[\phi_i''\quad\varphi_i''\quad\phi_j''\quad\varphi_j''\right]\mathrm{d}x$$
$$=\frac{(e_0a)^2N_t}{l_e^3}\begin{bmatrix}12 & 6l_e & -12 & 6l_e\\ 6l_e & 4l_e^2 & -6l_e & 2l_e^2\\ -12 & -6l_e & 12 & -6l_e\\ 6l_e & 2l_e^2 & -6l_e & 4l_e^2\end{bmatrix} \tag{7.48}$$

$$K_{T,R,c}=\int_0^{l_e}\frac{N_tEI}{kGA}\left(\mathbf{B}'^{\mathrm{T}}\mathbf{B}'\right)\mathrm{d}x=\int_0^{l_e}\frac{N_tEI}{kGA}\begin{Bmatrix}\phi_i''\\ \varphi_i''\\ \phi_j''\\ \varphi_j''\end{Bmatrix}\left[\phi_i''\quad\varphi_i''\quad\phi_j''\quad\varphi_j''\right]\mathrm{d}x$$
$$=\frac{N_tEI}{kGAl_e^3}\begin{bmatrix}12 & 6l_e & -12 & 6l_e\\ 6l_e & 4l_e^2 & -6l_e & 2l_e^2\\ -12 & -6l_e & 12 & -6l_e\\ 6l_e & 2l_e^2 & -6l_e & 4l_e^2\end{bmatrix} \tag{7.49}$$

$$K_{T,R,nl}=\int_0^{l_e}\frac{(e_0a)^2N_TEI}{kGA}\left(\mathbf{B}''^{\mathrm{T}}\mathbf{B}''\right)\mathrm{d}x=\int_0^{l_e}\frac{(e_0a)^2N_tEI}{kGA}\begin{Bmatrix}\phi_i'''\\ \varphi_i'''\\ \phi_j'''\\ \varphi_j'''\end{Bmatrix}\left[\phi_i'''\quad\varphi_i'''\quad\phi_j'''\quad\varphi_j'''\right]\mathrm{d}x$$
$$=\frac{36(e_0a)^2N_tEI}{kGAl_e^5}\begin{bmatrix}4 & 2l_e & -4 & 2l_e\\ 2l_e & l_e^2 & -2l_e & l_e^2\\ -4 & -2l_e & 4 & -2l_e\\ 2l_e & l_e^2 & -2l_e & l_e^2\end{bmatrix} \tag{7.50}$$

$$M_{A,c} = \int_0^{l_e} \rho A(\phi^T \phi) dx = \int_0^{l_e} \rho A \begin{Bmatrix} \phi_i \\ \varphi_i \\ \phi_j \\ \varphi_j \end{Bmatrix} [\phi_i \quad \varphi_i \quad \phi_j \quad \varphi_j] dx$$
$$= \frac{\rho A l_e}{420} \begin{bmatrix} 156 & 22l_e & 54 & -13l_e \\ 22l_e & 4l_e^2 & 13l_e & -3l_e^2 \\ 54 & 13l_e & 156 & -22l_e \\ -13l_e & -3l_e^2 & -22l_e & 4l_e^2 \end{bmatrix} \tag{7.51}$$

$$M_{A,nl} = \int_0^{l_e} (e_0 a)^2 \rho A(\mathbf{B}^T \mathbf{B}) dx = \int_0^{l_e} (e_0 a)^2 \rho A \begin{Bmatrix} \phi_i' \\ \varphi_i' \\ \phi_j' \\ \varphi_j' \end{Bmatrix} [\phi_i' \quad \varphi_i' \quad \phi_j' \quad \varphi_j'] dx$$
$$= \frac{(e_0 a)^2 \rho A}{30 l_e} \begin{bmatrix} 36 & 3l_e & -36 & 3l_e \\ 3l_e & 4l_e^2 & -3l_e & -l_e^2 \\ -36 & -3l_e & 36 & -3l_e \\ 3l_e & -l_e^2 & -3l_e & 4l_e^2 \end{bmatrix} \tag{7.52}$$

$$M_{I,c} = \int_0^{l_e} \rho I(\mathbf{B}^T \mathbf{B}) dx = \int_0^{l_e} \rho I \begin{Bmatrix} \phi_i' \\ \varphi_i' \\ \phi_j' \\ \varphi_j' \end{Bmatrix} [\phi_i' \quad \varphi_i' \quad \phi_j' \quad \varphi_j'] dx$$
$$= \frac{\rho I}{30 l_e} \begin{bmatrix} 36 & 3l_e & -36 & 3l_e \\ 3l_e & 4l_e^2 & -3l_e & -l_e^2 \\ -36 & -3l_e & 36 & -3l_e \\ 3l_e & -l_e^2 & -3l_e & 4l_e^2 \end{bmatrix} \tag{7.53}$$

$$M_{I,nl}=\int_0^{l_e}(e_0a)^2\rho I\left(\mathbf{B}'^{\mathrm{T}}\mathbf{B}'\right)\mathrm{d}x=\int_0^{l_e}(e_0a)^2\rho I\begin{Bmatrix}\phi_i''\\ \varphi_i''\\ \phi_j''\\ \varphi_j''\end{Bmatrix}\left[\phi_i''\quad \varphi_i''\quad \phi_j''\quad \varphi_j''\right]\mathrm{d}x$$
$$=\frac{(e_0a)^2\rho I}{l_e^3}\begin{bmatrix}12 & 6l_e & -12 & 6l_e\\ 6l_e & 4l_e^2 & -6l_e & 2l_e^2\\ -12 & -6l_e & 12 & -6l_e\\ 6l_e & 2l_e^2 & -6l_e & 4l_e^2\end{bmatrix} \tag{7.54}$$

$$M_{A,R,c}=\int_0^{l_e}\frac{\rho EI}{kG}\left(\mathbf{B}^{\mathrm{T}}\mathbf{B}\right)\mathrm{d}x=\int_0^{l_e}\frac{\rho EI}{kG}\begin{Bmatrix}\phi_i'\\ \varphi_i'\\ \phi_j'\\ \varphi_j'\end{Bmatrix}\left[\phi_i'\quad \varphi_i'\quad \phi_j'\quad \varphi_j'\right]\mathrm{d}x$$
$$=\frac{\rho EI}{30kGl_e}\begin{bmatrix}36 & 3l_e & -36 & 3l_e\\ 3l_e & 4l_e^2 & -3l_e & -l_e^2\\ -36 & -3l_e & 36 & -3l_e\\ 3l_e & -l_e^2 & -3l_e & 4l_e^2\end{bmatrix} \tag{7.55}$$

$$M_{A,R,nl}=\int_0^{l_e}\frac{(e_0a)^2\rho EI}{kG}\left(\mathbf{B}'^{\mathrm{T}}\mathbf{B}'\right)\mathrm{d}x=\int_0^{l_e}\frac{(e_0a)^2\rho EI}{kG}\begin{Bmatrix}\phi_i''\\ \varphi_i''\\ \phi_j''\\ \varphi_j''\end{Bmatrix}\left[\phi_i''\quad \varphi_i''\quad \phi_j''\quad \varphi_j''\right]\mathrm{d}x$$
$$=\frac{(e_0a)^2\rho EI}{kGl_e^3}\begin{bmatrix}12 & 6l_e & -12 & 6l_e\\ 6l_e & 4l_e^2 & -6l_e & 2l_e^2\\ -12 & -6l_e & 12 & -6l_e\\ 6l_e & 2l_e^2 & -6l_e & 4l_e^2\end{bmatrix} \tag{7.56}$$

$$M_{I,W,c} = \int_0^{l_e} \frac{k_W \rho I}{kGA} (\phi^{\mathrm{T}} \phi) \mathrm{d}x = \int_0^{l_e} \frac{k_W \rho I}{kGA} \begin{Bmatrix} \phi_i \\ \varphi_i \\ \phi_j \\ \varphi_j \end{Bmatrix} [\phi_i \quad \varphi_i \quad \phi_j \quad \varphi_j] \mathrm{d}x$$
$$= \frac{k_W \rho I l_e}{420 kGA} \begin{bmatrix} 156 & 22l_e & 54 & -13l_e \\ 22l_e & 4{l_e}^2 & 13l_e & -3{l_e}^2 \\ 54 & 13l_e & 156 & -22l_e \\ -13l_e & -3{l_e}^2 & -22l_e & 4{l_e}^2 \end{bmatrix} \tag{7.57}$$

$$M_{I,W,nl} = \int_0^{l_e} \frac{2(e_0 a)^2 k_W \rho I}{kGA} (\mathbf{B}^{\mathrm{T}} \mathbf{B}) \mathrm{d}x = \int_0^{l_e} \frac{2(e_0 a)^2 k_W \rho I}{kGA} \begin{Bmatrix} \phi_i' \\ \varphi_i' \\ \phi_j' \\ \varphi_j' \end{Bmatrix} [\phi_i' \quad \varphi_i' \quad \phi_j' \quad \varphi_j'] \mathrm{d}x$$
$$= \frac{2(e_0 a)^2 k_W \rho I}{30 kGA l_e} \begin{bmatrix} 36 & 3l_e & -36 & 3l_e \\ 3l_e & 4{l_e}^2 & -3l_e & -{l_e}^2 \\ -36 & -3l_e & 36 & -3l_e \\ 3l_e & -{l_e}^2 & -3l_e & 4{l_e}^2 \end{bmatrix} \tag{7.58}$$

$$M_{I,P,c} = \int_0^{l_e} \frac{k_P \rho I}{kGA} (\mathbf{B}^{\mathrm{T}} \mathbf{B}) \mathrm{d}x = \int_0^{l_e} \frac{k_P \rho I}{kGA} \begin{Bmatrix} \phi_i' \\ \varphi_i' \\ \phi_j' \\ \varphi_j' \end{Bmatrix} [\phi_i' \quad \varphi_i' \quad \phi_j' \quad \varphi_j'] \mathrm{d}x$$
$$= \frac{k_P \rho I}{30 kGA l_e} \begin{bmatrix} 36 & 3l_e & -36 & 3l_e \\ 3l_e & 4{l_e}^2 & -3l_e & -{l_e}^2 \\ -36 & -3l_e & 36 & -3l_e \\ 3l_e & -{l_e}^2 & -3l_e & 4{l_e}^2 \end{bmatrix} \tag{7.59}$$

$$
M_{I,P,nl} = \int_0^{l_e} \frac{2(e_0a)^2 k_P \rho I}{kGA}\left(\mathbf{B}'^{\mathrm{T}}\mathbf{B}'\right)\mathrm{d}x = \int_0^{l_e} \frac{2(e_0a)^2 k_P \rho I}{kGA}\begin{Bmatrix} \phi_i'' \\ \varphi_i'' \\ \phi_j'' \\ \varphi_j'' \end{Bmatrix}\left[\phi_i'' \quad \varphi_i'' \quad \phi_j'' \quad \varphi_j''\right]\mathrm{d}x
$$

$$
= \frac{2(e_0a)^2 k_P \rho I}{kGAl_e^3}\begin{bmatrix} 12 & 6l_e & -12 & 6l_e \\ 6l_e & 4l_e^2 & -6l_e & 2l_e^2 \\ -12 & -6l_e & 12 & -6l_e \\ 6l_e & 2l_e^2 & -6l_e & 4l_e^2 \end{bmatrix} \tag{7.60}
$$

$$
M_{I,T,c} = \int_0^{l_e} \frac{N_t \rho I}{kGA}\left(\mathbf{B}^{\mathrm{T}}\mathbf{B}\right)\mathrm{d}x = \int_0^{l_e} \frac{N_t \rho I}{kGA}\begin{Bmatrix} \phi_i' \\ \varphi_i' \\ \phi_j' \\ \varphi_j' \end{Bmatrix}\left[\phi_i' \quad \varphi_i' \quad \phi_j' \quad \varphi_j'\right]\mathrm{d}x
$$

$$
= \frac{N_t \rho I}{30kGAl_e}\begin{bmatrix} 36 & 3l_e & -36 & 3l_e \\ 3l_e & 4l_e^2 & -3l_e & -l_e^2 \\ -36 & -3l_e & 36 & -3l_e \\ 3l_e & -l_e^2 & -3l_e & 4l_e^2 \end{bmatrix} \tag{7.61}
$$

$$
M_{I,T,nl} = \int_0^{l_e} \frac{2(e_0a)^2 N_t \rho I}{kGA}\left(\mathbf{B}'^{\mathrm{T}}\mathbf{B}'\right)\mathrm{d}x = \int_0^{l_e} \frac{2(e_0a)^2 N_t \rho I}{kGA}\begin{Bmatrix} \phi_i'' \\ \varphi_i'' \\ \phi_j'' \\ \varphi_j'' \end{Bmatrix}\left[\phi_i'' \quad \varphi_i'' \quad \phi_j'' \quad \varphi_j''\right]\mathrm{d}x
$$

$$
= \frac{2(e_0a)^2 N_t \rho I}{kGAl_e^3}\begin{bmatrix} 12 & 6l_e & -12 & 6l_e \\ 6l_e & 4l_e^2 & -6l_e & 2l_e^2 \\ -12 & -6l_e & 12 & -6l_e \\ 6l_e & 2l_e^2 & -6l_e & 4l_e^2 \end{bmatrix} \tag{7.62}
$$

$$M_{I,W,HO-nl} = \int_0^{l_e} \frac{(e_0a)^4 k_W \rho I}{kGA}\left(\mathbf{B}'^{\mathrm{T}}\mathbf{B}'\right)\mathrm{d}x = \int_0^{l_e} \frac{(e_0a)^4 k_W \rho I}{kGA}\begin{Bmatrix} \phi_i'' \\ \varphi_i'' \\ \phi_j'' \\ \varphi_j'' \end{Bmatrix}\left[\phi_i'' \quad \varphi_i'' \quad \phi_j'' \quad \varphi_j''\right]\mathrm{d}x$$

$$= \frac{(e_0a)^4 k_W \rho I}{kGAl_e^3}\begin{bmatrix} 12 & 6l_e & -12 & 6l_e \\ 6l_e & 4l_e^2 & -6l_e & 2l_e^2 \\ -12 & -6l_e & 12 & -6l_e \\ 6l_e & 2l_e^2 & -6l_e & 4l_e^2 \end{bmatrix} \tag{7.63}$$

$$M_{I,P,HO-nl} = \int_0^{l_e} \frac{(e_0a)^4 k_P \rho I}{kGA}\left(\mathbf{B}''^{\mathrm{T}}\mathbf{B}''\right)\mathrm{d}x = \int_0^{L} \frac{(e_0a)^4 k_P \rho I}{kGA}\begin{Bmatrix} \phi_i''' \\ \varphi_i''' \\ \phi_j''' \\ \varphi_j''' \end{Bmatrix}\left[\phi_i''' \quad \varphi_i''' \quad \phi_j''' \quad \varphi_j'''\right]\mathrm{d}x$$

$$= \frac{36(e_0a)^4 k_P \rho I}{kGAl_e^5}\begin{bmatrix} 4 & 2l_e & -4 & 2l_e \\ 2l_e & l_e^2 & -2l_e & l_e^2 \\ -4 & -2l_e & 4 & -2l_e \\ 2l_e & l_e^2 & -2l_e & l_e^2 \end{bmatrix} \tag{7.64}$$

$$M_{I,T,HO-nl} = \int_0^{l_e} \frac{(e_0a)^4 N_t \rho I}{kGA}\left(\mathbf{B}''^{\mathrm{T}}\mathbf{B}''\right)\mathrm{d}x = \int_0^{l_e} \frac{(e_0a)^4 N_t \rho I}{kGA}\begin{Bmatrix} \phi_i''' \\ \varphi_i''' \\ \phi_j''' \\ \varphi_j''' \end{Bmatrix}\left[\phi_i''' \quad \varphi_i''' \quad \phi_j''' \quad \varphi_j'''\right]\mathrm{d}x$$

$$= \frac{36(e_0a)^4 N_t \rho I}{kGAl_e^5}\begin{bmatrix} 4 & 2l_e & -4 & 2l_e \\ 2l_e & l_e^2 & -2l_e & l_e^2 \\ -4 & -2l_e & 4 & -2l_e \\ 2l_e & l_e^2 & -2l_e & l_e^2 \end{bmatrix} \tag{7.65}$$

$$M_{I,R,c} = \int_0^{l_e} \frac{\rho^2 I}{kG}(\phi^{\mathrm{T}}\phi)\mathrm{d}x = \int_0^{l_e} \frac{\rho^2 I}{kG} \begin{Bmatrix} \phi_i \\ \varphi_i \\ \phi_j \\ \varphi_j \end{Bmatrix} [\phi_i \quad \varphi_i \quad \phi_j \quad \varphi_j]\mathrm{d}x$$

$$= \frac{\rho^2 I l_e}{420kG} \begin{bmatrix} 156 & 22l_e & 54 & -13l_e \\ 22l_e & 4l_e^2 & 13l_e & -3l_e^2 \\ 54 & 13l_e & 156 & -22l_e \\ -13l_e & -3l_e^2 & -22l_e & 4l_e^2 \end{bmatrix} \tag{7.66}$$

$$M_{I,R,nl} = \int_0^{l_e} \frac{2(e_0a)^2\rho^2 I}{kG}(\mathbf{B}^{\mathrm{T}}\mathbf{B})\mathrm{d}x = \int_0^{l_e} \frac{2(e_0a)^2\rho^2 I}{kG} \begin{Bmatrix} \phi_i' \\ \varphi_i' \\ \phi_j' \\ \varphi_j' \end{Bmatrix} [\phi_i' \quad \varphi_i' \quad \phi_j' \quad \varphi_j']\mathrm{d}x$$

$$= \frac{2(e_0a)^2\rho^2 I}{30kGl_e} \begin{bmatrix} 36 & 3l_e & -36 & 3l_e \\ 3l_e & 4l_e^2 & -3l_e & -l_e^2 \\ -36 & -3l_e & 36 & -3l_e \\ 3l_e & -l_e^2 & -3l_e & 4l_e^2 \end{bmatrix} \tag{7.67}$$

$$M_{I,R,HO-nl} = \int_0^{l_e} \frac{(e_0a)^4\rho^2 I}{kG}\left(\mathbf{B}'^{\mathrm{T}}\mathbf{B}'\right)\mathrm{d}x = \int_0^{l_e} \frac{(e_0a)^4\rho^2 I}{kG} \begin{Bmatrix} \phi_i'' \\ \varphi_i'' \\ \phi_j'' \\ \varphi_j'' \end{Bmatrix} [\phi_i'' \quad \varphi_i'' \quad \phi_j'' \quad \varphi_j'']\mathrm{d}x$$

$$= \frac{(e_0a)^4\rho^2 I}{kGl_e^3} \begin{bmatrix} 12 & 6l_e & -12 & 6l_e \\ 6l_e & 4l_e^2 & -6l_e & 2l_e^2 \\ -12 & -6l_e & 12 & -6l_e \\ 6l_e & 2l_e^2 & -6l_e & 4l_e^2 \end{bmatrix} \tag{7.68}$$

$$\mathbf{f}_c = \int_0^{l_e} \phi^{\mathrm{T}} f \mathrm{d}x = \int_0^{l_e} f \begin{Bmatrix} \phi_i \\ \varphi_i \\ \phi_j \\ \varphi_j \end{Bmatrix} \mathrm{d}x = \frac{l_e}{12} f \begin{Bmatrix} 6 \\ l_e \\ 6 \\ -l_e \end{Bmatrix} \tag{7.69}$$

$$\mathbf{f}_{nl} = \int_0^{l_e} (e_0 a)^2 \mathbf{B}^{\mathrm{T}} f' \mathrm{d}x = \int_0^{l_e} (e_0 a)^2 f' \begin{Bmatrix} \phi_i' \\ \varphi_i' \\ \phi_j' \\ \varphi_j' \end{Bmatrix} \mathrm{d}x = (e_0 a)^2 f' \begin{Bmatrix} -1 \\ 0 \\ 1 \\ 0 \end{Bmatrix} \tag{7.70}$$

where instead of introducing the expressions shown in Eqs. (7.38)–(7.70) individually, it is preferred to make a definition regarding the subscripts of these expressions. Accordingly, at least one of the following definitions is the effective factor in the formation of related stiffness matrix: "*W*" is the Winkler foundation parameter, "*P*" is the Pasternak foundation parameter, "*T*" is the temperature parameter, "*R*" is at least one of the shear stiffness (kGA) and bending stiffness (EI). "*c*" indicates classical and "*nl*" indicates nonlocal elasticity. In addition to these, subscripts of "*A*" and "*I*" in the mass matrices and external force vectors explain the mass integrity and inertia integrity, respectively, in the related mass matrix. In the nonlocal elasticity analysis, in order to address the matrices and vectors containing the factor $(e_0a)^4$ that is a higher-order nonlocal parameter than $(e_0a)^2$, $HO-nl$ subscript is used.

7.4.2 Interpretation

The expressions of NL-FEM analysis for the bending finite element are presented by Eqs. (7.34)–(7.37). First, the external excitations are taken as $\mathbf{f}_c = \mathbf{f}_{nl} = 0$ for the free vibration analysis. When the total stiffness matrix presented in Eq. (7.34) is compared with Eq. (7.21), it is understood that Winker foundation, Pasternak foundation, and thermal effect that are the components of thermo-elastic environment combine with the stiffness terms. Additionally, when Eqs. (7.22) and (7.35) are examined, it is seen that environmental effects include other classical and nonlocal matrices into the analysis due to shear deformation. Moreover, a higher-order mass matrix that is not seen in nonlocal EBBT exists due to the fourth derivative according to time for nonlocal TBT. As a result, nonlocal TBT indicates a more comprehensive NL-FEM analysis than nonlocal EBBT. It is understood that the shear effect has a particular importance in nanobeams, but numerical studies should be carried out to determine the details of this importance.

7.4.3 Solution Stage

For nonlocal free vibration analysis, the global matrices constituted under the influences of stiffness given by Eqs. (7.34)–(7.36) and two different total mass matrices are reduced by boundary conditions. After this, the following eigenvalue equation should be solved by using Eqs. (4.30) and (7.33):

$$\det\left[\left(\sum_{e=1}^{n}[K]_e\right)^{*}-\omega_i^{2}\left(\sum_{e=1}^{n}\left[M^{0}\right]_e\right)^{*}+\omega_i^{4}\left(\sum_{e=1}^{n}\left[M^{1}\right]_e\right)^{*}\right]=0 \tag{7.71}$$

7.4.4 Application of Boundary Conditions

There is no difference between nonlocal TBT and nonlocal EBBT in the application of boundary conditions. However, it should be definitely stated that the point mass attachment should be added with Eq. (7.35) that is total simple mass matrix due to Newton's second law.

Example 7.6 Determine the natural frequencies of nonlocal transverse free vibration of simply supported nanobeam in Fig. 7.1. Use two finite elements in the analysis. Physical and mechanical parameters are as in Example 7.1. In addition to these, shear modulus: $G = 0.42$ TPa, and shear correction factor: $k = 5/6$.

Solution. For the NL-FEM analysis of free vibration in the Timoshenko nanobeam isolated from thermo-elastic medium, the elastic stiffness matrix given by Eq. (7.38), the simple mass components given by Eqs. (7.52)–(7.56), and the higher-order mass components given by Eqs. (7.66)–(7.68) should be used. The reduced elastic stiffness matrix for NL-FEM analysis according to two finite elements is written by Eq. (7.E7.2.4). On the other hand, reduced classical mass (Eq. (7.51)) and nonlocal mass (Eq. (7.52)) matrices, which are simple mass matrices, are computed by Eqs. (7.E7.2.5) and (7.E7.2.6), respectively. Here, other simple mass matrices should be obtained:

$$\sum\left[M_{I,c}\right]=\frac{\rho I}{30l}\begin{bmatrix} 36 & 3l & -36 & 3l & 0 & 0 \\ 3l & 4l^2 & -3l & -l^2 & 0 & 0 \\ -36 & -3l & 72 & 0 & -36 & 3l \\ 3l & -l^2 & 0 & 8l^2 & -3l & -l^2 \\ 0 & 0 & -36 & -3l & 36 & -3l \\ 0 & 0 & 3l & -l^2 & -3l & 4l^2 \end{bmatrix} \rightarrow \left(\sum\left[M_{I,c}\right]\right)^{*} \tag{7.E7.6.1}$$

$$\sum\left[M_{I,nl}\right]=\frac{(e_0a)^2\rho I}{l^3}\begin{bmatrix} 12 & 6l & -12 & 6l & 0 & 0 \\ 6l & 4l^2 & -6l & 2l^2 & 0 & 0 \\ -12 & -6l & 24 & 0 & -12 & 6l \\ 6l & 2l^2 & 0 & 8l^2 & -6l & 2l^2 \\ 0 & 0 & -12 & -6l & 12 & -6l \\ 0 & 0 & 6l & 2l^2 & -6l & 4l^2 \end{bmatrix} \rightarrow \left(\sum\left[M_{I,c}\right]\right)^* \tag{7.E7.6.2}$$

$$\sum\left[M_{A,R,c}\right]=\frac{\rho EI}{30kGl}\begin{bmatrix} 36 & 3l & -36 & 3l & 0 & 0 \\ 3l & 4l^2 & -3l & -l^2 & 0 & 0 \\ -36 & -3l & 72 & 0 & -36 & 3l \\ 3l & -l^2 & 0 & 8l^2 & -3l & -l^2 \\ 0 & 0 & -36 & -3l & 36 & -3l \\ 0 & 0 & 3l & -l^2 & -3l & 4l^2 \end{bmatrix} \rightarrow \left(\sum\left[M_{A,R,c}\right]\right)^* \tag{7.E7.6.3}$$

$$\sum\left[M_{A,R,nl}\right]=\frac{(e_0a)^2\rho EI}{kGl_e^3}\begin{bmatrix} 12 & 6l & -12 & 6l & 0 & 0 \\ 6l & 4l^2 & -6l & 2l^2 & 0 & 0 \\ -12 & -6l & 24 & 0 & -12 & 6l \\ 6l & 2l^2 & 0 & 8l^2 & -6l & 2l^2 \\ 0 & 0 & -12 & -6l & 12 & -6l \\ 0 & 0 & 6l & 2l^2 & -6l & 4l^2 \end{bmatrix} \rightarrow \left(\sum\left[M_{A,R,nl}\right]\right)^* \tag{7.E7.6.4}$$

On the other hand, reduced higher-order global mass matrices can be stated as follows:

$$\sum\left[M_{I,R,c}\right]=\frac{\rho^2 Il}{420kG}\begin{bmatrix} 156 & 22l & 54 & -13l & 0 & 0 \\ 22l & 4l^2 & 13l & -3l^2 & 0 & 0 \\ 54 & 13l & 312 & 0 & 54 & -13l \\ -13l & -3l^2 & 0 & 8l^2 & 13l & -3l^2 \\ 0 & 0 & 54 & 13l & 156 & -22l \\ 0 & 0 & -13l & -3l^2 & -22l & 4l^2 \end{bmatrix} \rightarrow \left(\sum\left[M_{I,R,c}\right]\right)^* \tag{7.E7.6.5}$$

$$\sum\left[M_{I,R,nl}\right]=\frac{(e_0a)^2\rho^2 I}{15kGl_e}\begin{bmatrix} 36 & 3l & -36 & 3l & 0 & 0 \\ 3l & 4l^2 & -3l & -l^2 & 0 & 0 \\ -36 & -3l & 72 & 0 & -36 & 3l \\ 3l & -l^2 & 0 & 8l^2 & -3l & -l^2 \\ 0 & 0 & -36 & -3l & 36 & -3l \\ 0 & 0 & 3l & -l^2 & -3l & 4l^2 \end{bmatrix} \rightarrow \left(\sum\left[M_{I,R,nl}\right]\right)^* \tag{7.E7.6.6}$$

$$\sum\left[M_{I,R,HO-nl}\right]=\frac{(e_0a)^4\rho^2 I}{kGl_e^3}\begin{bmatrix} 12 & 6l & -12 & 6l & 0 & 0 \\ 6l & 4l^2 & -6l & 2l^2 & 0 & 0 \\ -12 & -6l & 24 & 0 & -12 & 6l \\ 6l & 2l^2 & 0 & 8l^2 & -6l & 2l^2 \\ 0 & 0 & -12 & -6l & 12 & -6l \\ 0 & 0 & 6l & 2l^2 & -6l & 4l^2 \end{bmatrix} \rightarrow \left(\sum\left[M_{I,R,HO-nl}\right]\right)^* \tag{7.E7.6.7}$$

If the reduced matrices calculated with Eqs. (7.E7.6.1)-(7.E7.6.4) are added with Eq. (7.E7.2.8), the reduced total simple mass matrix can be achieved:

$$\left(\sum[M^0]\right)^* = \left(\sum[M_c]\right)^* + \left(\sum[M_{nl}]\right)^* + \left(\sum[M_{I,c}]\right)^* + \left(\sum[M_{I,nl}]\right)^*$$
$$\left(\sum[M_{A,R,c}]\right)^* + \left(\sum[M_{A,R,nl}]\right)^*$$
$$= \begin{bmatrix} 1.534\times10^{-40} & 5.311\times10^{-33} & -5.315\times10^{-41} & 0 \\ 5.311\times10^{-33} & 5.276\times10^{-23} & 0 & -5.311\times10^{-33} \\ -5.315\times10^{-41} & 0 & 3.068\times10^{-40} & -5.315\times10^{-41} \\ 0 & -5.311\times10^{-33} & -5.315\times10^{-41} & 1.534\times10^{-40} \end{bmatrix} \tag{7.E7.6.8}$$

On the other hand, summing the reduced matrices given by Eqs. (7.E7.6.5)-(7.E7.6.7) yields the reduced total higher-order mass matrix:

$$\left(\sum[M^1]\right)^* = \left(\sum[M_{I,R,c}]\right)^* + \left(\sum[M_{I,R,nl}]\right)^* + \left(\sum[M_{I,R,HO-nl}]\right)^*$$
$$= \begin{bmatrix} 3.893\times10^{-67} & -3.895\times10^{-59} & 8.409\times10^{-68} & 0 \\ -3.895\times10^{-59} & 5.352\times10^{-50} & 0 & 3.895\times10^{-59} \\ 8.409\times10^{-68} & 0 & 7.786\times10^{-67} & 8.409\times10^{-68} \\ 0 & 3.895\times10^{-59} & 8.409\times10^{-68} & 3.893\times10^{-67} \end{bmatrix} \tag{7.E7.6.9}$$

If the matrices given in Eqs. (7.E7.2.7), (7.E7.6.8). and (7.E7.6.9) are replaced into Eq. (7.71) and the eigenvalue equation is solved, eight positive natural frequencies are reached:

$\omega_{1,\,1} = 121.849$ GHz, $\omega_{1,\,2} = 401.871$ GHz,
$\omega_{1,\,3} = 724.914$ GHz, $\omega_{1,\,4} = 1001.275$ GHz,
$\omega_{2,\,1} = 14517.847$ GHz, $\omega_{2,\,2} = 18484.885$ GHz,
$\omega_{2,\,3} = 26011.260$ GHz, $\omega_{2,\,4} = 36321.965$ GHz.

In a finite element analysis of the Timoshenko beam, $2n$ frequencies are obtained for n freedoms. Among the above values, while the first four values express the nonlocal frequencies of bending mode, the other four represent the nonlocal frequencies of thickness-shear mode. In terms of engineering mechanics, bending mode frequencies need to be known and thickness-shear mode frequencies makes no sense.

Table 7.6 presents the comparison of nondimensional frequencies of first five modes with the NL-FEM approach for the free vibration of Timoshenko simply supported nanobeams. Here the analytical results are obtained from Eq. (4.E4.5.4) and are given in Table 4.2. As can be seen, NL-FEM enables good results for the atomic size-dependent free vibration of Timoshenko nanobeams. In general, the using of many finite elements and therefore computer programming is required for results with high accuracy.

Table 7.6 Comparisons of nondimensional frequencies of simply supported Timoshenko nanobeams for nonlocal transverse free vibration ($e_0a/L = 0.2$)

Mode	Analytical	NL-FEM results				
Number	Eq. (4.E4.5.4)	Manual Solution	Program Code (MATLAB)			
		$n = 2$	$n = 2$	$n = 10$	$n = 15$	$n = 20$
1	8.32404	8.36219	8.35636	8.32409	8.32405	8.32404
2	24.20373	26.70289	26.70038	24.20626	24.20423	24.20389
3	40.23544	48.16788	48.17292	40.25554	40.23950	40.23674
4	55.06336	66.53107	66.53753	55.14368	55.07987	55.06866
5	68.55136	–	–	68.77303	68.59794	68.56644

Example 7.7 Determine the natural frequencies of nonlocal transverse free vibration of fixed supported-transverse mass attached nanobeam in Fig. 4.8. Use two finite elements in the analysis. In addition to this, attachment ratio is $\beta_m = 10$ and defined as $\beta_m = m_a/\rho AL$. Firstly, ignore the environmental effects. Then, repeat the analysis for the thermo-elastic environment. Physical and mechanical parameters are as in Examples 7.1 and 7.3.

Solution. The matrices in Eqs. (7.38)–(7.68) govern the nonlocal free vibration of nanobeam under a thermo-elastic environment. To give the results of these matrices one by one in this application causes unnecessary crowd. Instead, a simplification described as follows is preferred.

If NL-FEM solutions containing two finite elements are examined from previous applications, only four different 6×6 size matrices are observed. The only difference between these matrices is the multipliers that is in front of matrices and contain the different physical/mechanical expressions. Therefore, the following matrices that are reduced by the geometric conditions subject to this example are defined:

$$\sum[A_1] = \begin{bmatrix} 156 & 22l & 54 & -13l & 0 & 0 \\ 22l & 4l^2 & 13l & -3l^2 & 0 & 0 \\ 54 & 13l & 312 & 0 & 54 & -13l \\ -13l & -3l^2 & 0 & 8l^2 & 13l & -3l^2 \\ 0 & 0 & 54 & 13l & 156 & -22l \\ 0 & 0 & -13l & -3l^2 & -22l & 4l^2 \end{bmatrix}, \quad \left(\sum[A_1]\right)^* = \begin{bmatrix} 312 & 0 & 54 & -13l \\ 0 & 8l^2 & 13l & -3l^2 \\ 54 & 13l & 156 & -22l \\ -13l & -3l^2 & -22l & 4l^2 \end{bmatrix} \tag{7.E7.7.1}$$

$$\sum[A_2]=\begin{bmatrix} 36 & 3l & -36 & 3l & 0 & 0 \\ 3l & 4l^2 & -3l & -l^2 & 0 & 0 \\ -36 & -3l & 72 & 0 & -36 & 3l \\ 3l & -l^2 & 0 & 8l^2 & -3l & -l^2 \\ 0 & 0 & -36 & -3l & 36 & -3l \\ 0 & 0 & 3l & -l^2 & -3l & 4l^2 \end{bmatrix}, \quad \left(\sum[A_2]\right)^* = \begin{bmatrix} 72 & 0 & -36 & 3l \\ 0 & 8l^2 & -3l & -l^2 \\ -36 & -3l & 36 & -3l \\ 3l & -l^2 & -3l & 4l^2 \end{bmatrix} \tag{7.E7.7.2}$$

$$\sum[A_3]=\begin{bmatrix} 12 & 6l & -12 & 6l & 0 & 0 \\ 6l & 4l^2 & -6l & 2l^2 & 0 & 0 \\ -12 & -6l & 24 & 0 & -12 & 6l \\ 6l & 2l^2 & 0 & 8l^2 & -6l & 2l^2 \\ 0 & 0 & -12 & -6l & 12 & -6l \\ 0 & 0 & 6l & 2l^2 & -6l & 4l^2 \end{bmatrix}, \quad \left(\sum[A_3]\right)^* = \begin{bmatrix} 24 & 0 & -12 & 6l \\ 0 & 8l^2 & -6l & 2l^2 \\ -12 & -6l & 12 & -6l \\ 6l & 2l^2 & -6l & 4l^2 \end{bmatrix} \tag{7.E7.7.3}$$

$$\sum[A_4]=\begin{bmatrix} 4 & 2l & -4 & 2l & 0 & 0 \\ 2l & l^2 & 2l & l^2 & 0 & 0 \\ -4 & -2l & 8 & 0 & -4 & 2l \\ 2l & l^2 & 0 & 2l^2 & -2l & l^2 \\ 0 & 0 & -4 & -2l & 4 & -2l \\ 0 & 0 & 2l & l^2 & -2l & l^2 \end{bmatrix}, \quad \left(\sum[A_4]\right)^* = \begin{bmatrix} 8 & 0 & -4 & 2l \\ 0 & 2l^2 & -2l & l^2 \\ -4 & -2l & 4 & -2l \\ 2l & l^2 & -2l & l^2 \end{bmatrix} \tag{7.E7.7.4}$$

where various stiffness matrices are expressed as follows:

$$\left(\sum[K_{be}]\right)^* = \frac{EI}{l^3}\left(\sum[A_3]\right)^* \tag{7.E7.7.5}$$

$$\left(\sum[K_{W,c}]\right)^* = \frac{k_W l}{420}\left(\sum[A_1]\right)^* \tag{7.E7.7.6}$$

$$\left(\sum[K_{W,nl}]\right)^{*}=\frac{(e_0a)^2k_W}{30l}\left(\sum[A_2]\right)^{*} \tag{7.E7.7.7}$$

$$\left(\sum[K_{P,c}]\right)^{*}=\frac{k_P}{30l}\left(\sum[A_2]\right)^{*} \tag{7.E7.7.8}$$

$$\left(\sum[K_{P,nl}]\right)^{*}=\frac{(e_0a)^2k_P}{l^3}\left(\sum[A_3]\right)^{*} \tag{7.E7.7.9}$$

$$\left(\sum[K_{W,R,c}]\right)^{*}=\frac{k_WEI}{30kGAl}\left(\sum[A_2]\right)^{*} \tag{7.E7.7.10}$$

$$\left(\sum[K_{W,R,nl}]\right)^{*}=\frac{(e_0a)^2k_WEI}{kGAl^3}\left(\sum[A_3]\right)^{*} \tag{7.E7.7.11}$$

$$\left(\sum[K_{P,R,c}]\right)^{*}=\frac{k_PEI}{kGAl^3}\left(\sum[A_3]\right)^{*} \tag{7.E7.7.12}$$

$$\left(\sum[K_{P,R,nl}]\right)^{*}=\frac{36(e_0a)^2k_PEI}{kGAl^5}\left(\sum[A_4]\right)^{*} \tag{7.E7.7.13}$$

$$\left(\sum[K_{T,c}]\right)^{*}=\frac{N_t}{30l}\left(\sum[A_2]\right)^{*} \tag{7.E7.7.14}$$

$$\left(\sum[K_{T,nl}]\right)^{*}=\frac{(e_0a)^2N_t}{l^3}\left(\sum[A_3]\right)^{*} \tag{7.E7.7.15}$$

$$\left(\sum[K_{T,R,c}]_e\right)^{*}=\frac{N_tEI}{kGAl^3}\left(\sum[A_3]\right)^{*} \tag{7.E7.7.16}$$

$$\left(\sum[K_{T,R,nl}]\right)^{*}=\frac{36(e_0a)^2N_tEI}{kGAl^5}\left(\sum[A_4]\right)^{*} \tag{7.E7.7.17}$$

The components of simple mass matrix can be formulated below:

$$\left(\sum[M_{A,c}]\right)^{*}=\frac{\rho Al}{420}\left(\sum[A_1]\right)^{*} \tag{7.E7.7.18}$$

$$\left(\sum[M_{A,nl}]\right)^{*}=\frac{(e_0a)^2\rho A}{30l}\left(\sum[A_2]\right)^{*} \tag{7.E7.7.19}$$

$$\left(\sum[M_{I,c}]\right)^* = \frac{\rho I}{30l}\left(\sum[A_2]\right)^* \tag{7.E7.7.20}$$

$$\left(\sum[M_{I,nl}]\right)^* = \frac{(e_0a)^2\rho I}{l^3}\left(\sum[A_3]\right)^* \tag{7.E7.7.21}$$

$$\left(\sum[M_{A,R,c}]\right)^* = \frac{\rho EI}{30kGl}\left(\sum[A_2]\right)^* \tag{7.E7.7.22}$$

$$\left(\sum[M_{A,R,nl}]\right)^* = \frac{(e_0a)^2\rho EI}{kGl^3}\left(\sum[A_3]\right)^* \tag{7.E7.7.23}$$

$$\left(\sum[M_{I,W,c}]\right)^* = \frac{k_W\rho Il}{420kGA}\left(\sum[A_1]\right)^* \tag{7.E7.7.24}$$

$$\left(\sum[M_{I,W,nl}]\right)^* = \frac{(e_0a)^2k_W\rho I}{15kGAl}\left(\sum[A_2]\right)^* \tag{7.E7.7.25}$$

$$\left(\sum[M_{I,P,c}]\right)^* = \frac{k_P\rho I}{30kGAl}\left(\sum[A_2]\right)^* \tag{7.E7.7.26}$$

$$\left(\sum[M_{I,P,nl}]\right)^* = \frac{2(e_0a)^2k_P\rho I}{kGAl^3}\left(\sum[A_3]\right)^* \tag{7.E7.7.27}$$

$$\left(\sum[M_{I,T,c}]\right)^* = \frac{N_t\rho I}{30kGAl}\left(\sum[A_2]\right)^* \tag{7.E7.7.28}$$

$$\left(\sum[M_{I,T,nl}]\right)^* = \frac{2(e_0a)^2N_t\rho I}{kGAl^3}\left(\sum[A_3]\right)^* \tag{7.E7.7.29}$$

$$\left(\sum[M_{I,W,HO-nl}]\right)^* = \frac{(e_0a)^4k_W\rho I}{kGAl^3}\left(\sum[A_3]\right)^* \tag{7.E7.7.30}$$

$$\left(\sum[M_{I,P,HO-nl}]\right)^* = \frac{36(e_0a)^4k_P\rho I}{kGAl^5}\left(\sum[A_4]\right)^* \tag{7.E7.7.31}$$

$$\left(\sum[M_{I,T,HO-nl}]\right)^* = \frac{36(e_0a)^4N_t\rho I}{kGAl^5}\left(\sum[A_4]\right)^* \tag{7.E7.7.32}$$

On the other hand, different higher-order mass matrices are expressed as follows:

$$\left(\sum[M_{I,R,c}]\right)^* = \frac{\rho^2 Il}{420kG}\left(\sum[A_1]\right)^* \tag{7.E7.7.33}$$

$$\left(\sum[M_{I,R,nl}]\right)^* = \frac{(e_0a)^2\rho^2 I}{15kGl}\left(\sum[A_2]\right)^* \tag{7.E7.7.34}$$

$$\left(\sum[M_{I,R,HO-nl}]\right)^* = \frac{(e_0a)^4\rho^2 I}{kGl^3}\left(\sum[A_3]\right)^* \tag{7.E7.7.35}$$

Also, the additional mass matrix due to mass attachment can be written as:

$$\sum[M_a] = \begin{bmatrix} 0 & 0 & 0 & 0 & 0 & 0 \\ 0 & 0 & 0 & 0 & 0 & 0 \\ 0 & 0 & 0 & 0 & 0 & 0 \\ 0 & 0 & 0 & 0 & 0 & 0 \\ 0 & 0 & 0 & 0 & m_a & 0 \\ 0 & 0 & 0 & 0 & 0 & 0 \end{bmatrix}, \quad \left(\sum[M_a]\right)^* = \begin{bmatrix} 0 & 0 & 0 & 0 \\ 0 & 0 & 0 & 0 \\ 0 & 0 & m_a & 0 \\ 0 & 0 & 0 & 0 \end{bmatrix} \tag{7.E7.7.36}$$

here the mass of end attachment is calculated as follows:

$$\begin{aligned} m_a = \beta_m \rho AL &= (10) \times \left(2300\,\frac{\text{kg}}{\text{m}^3}\right) \times (2\times10^{-18}\,\text{m}^2) \times (20\times10^{-9}\,\text{m}) \\ &= 9.2\ \times 10^{-22}\,\text{kg} \end{aligned} \tag{7.E7.7.37}$$

First, if the matrices given in Eqs. (7.E7.7.5)–(7.E7.7.17) are subjected to the process defined in Eq. (7.34), the reduced global stiffness matrix is obtained:

$$\left(\sum[K]_e\right)^* = \begin{bmatrix} 6.175 & 0 & -3.035 & 1.352\times10^{-8} \\ 0 & 1.851\times10^{-16} & -1.172\times10^{-8} & 4.075\times10^{-17} \\ -3.035 & -1.352\times10^{-8} & 3.087 & -1.361\times10^{-8} \\ 1.370\times10^{-8} & -4.3\times10^{-17} & -1.361\times10^{-8} & 9.389\times10^{-17} \end{bmatrix} \tag{7.E7.7.38}$$

After Eqs. (7.E7.7.18)–(7.E7.7.32) are substituted into Eq. (7.35), resulting matrix are added to the Eq. (7.E7.7.36). The result of these operations gives the reduced global simple mass matrix:

$$\left(\sum[M^0]\right)^* = \begin{bmatrix} 5.276\times10^{-23} & 0 & -3.381\times10^{-24} & -5.304\times10^{-33} \\ 0 & 3.068\times10^{-40} & 4.535\times10^{-32} & -1.032\times10^{-40} \\ -3.381\times10^{-24} & 5.304\times10^{-33} & 9.464\times10^{-22} & -3.303\times10^{-32} \\ -5.298\times10^{-33} & -5.313\times10^{-41} & -3.303\times10^{-32} & 1.535\times10^{-40} \end{bmatrix} \tag{7.E7.7.39}$$

The higher-order mass matrix is obtained by using the matrices calculated with Eqs. (7.E7.7.33)–(7.E7.7.35) into Eq. (7.36):

$$\left(\sum[M^1]\right)^* = \begin{bmatrix} 5.352\times10^{-50} & 0 & -1.417\times10^{-50} & 3.895\times10^{-59} \\ 0 & 7.786\times10^{-67} & 4.034\times10^{-60} & 3.036\times10^{-68} \\ -1.417\times10^{-50} & -3.895\times10^{-59} & 2.676\times10^{-50} & -5.993\times10^{-59} \\ 3.895\times10^{-59} & 8.409\times10^{-68} & -5.993\times10^{-59} & 3.893\times10^{-67} \end{bmatrix} \tag{7.E7.7.40}$$

Finally, natural frequencies are computed as a result of the utilizing Eqs. (7.E7.7.38)–(7.E7.7.40) into Eq. (7.71):

$\omega_1 = 21.412$ GHz, $\omega_2 = 255.456$ GHz, $\omega_3 = 590.740$ GHz, $\omega_4 = 1126.612$ GHz.

As it is understood, the analytical solution of this nonlocal vibration analysis whose analytical solution is impossible is done with NL-FEM. To obtain results with as high accuracy as possible in such wide-ranging problems depends on the using of many finite elements. For this, the computer programming is recommended.

Problems

7.1. Compute the first three modes nondimensional frequencies of nanobeams given in Fig. 4.9 according to nonlocal EBBT under parameters as follows. Use 2 finite elements for manual solution. Also, if you have a program code or package program based on finite element analysis, try to obtain the approximated values by using 5, 10, and 15 finite elements. Discuss the effects of nonlocal parameter and boundary conditions on differences between results.

Nanobeam length, $L = 20$ nm, height and thickness of rectangular cross-section, $h = 1$ nm and $b = 2$ nm, modulus of elasticity, $E = 1$ TPa, mass of unit volume, $\rho = 2300$ kg/m^3.

Compute according to three different nondimensional nonlocal parameters: $\alpha = 0$, 0.1, 0.2.

Compute the nondimensional frequencies as: $\overline{\omega} = \omega L^2 \sqrt{\rho A / EI}$.

7.2. Examine the Problem 7.1 by considering also thermo-elastic environment under parameter as follows. Discuss the effect of elastic medium on differences between results.

Nondimensional stiffness of Winkler foundation: $K_W = 10$, nondimensional stiffness of Pasternak foundation: $K_P = 1$, temperature change: $\Delta T = 300$ K, and thermal expansion coefficient: $\alpha_t = 1.6 \times 10^{-6}$ K^{-1}.

Compute the nondimensional stiffness of elastic medium as $K_W = k_W L^4/EI$ and $K_P = k_P L^2/EI$.

7.3. Compute the first three modes nondimensional frequencies of nanobeams given in Fig. 4.9 according to nonlocal TBT. Use 2 finite elements for manual solution. Also, if you have a program code or package program based on finite element analysis, try to obtain the approximated values by using 5, 10, and 15 finite elements. Discuss the effects of nonlocal parameter and boundary conditions on differences between results.

7.4. Examine Problem 7.3 according to nonlocal TBT.

7.5. Develop the NL-FEM analysis for nonlocal EBBT nanobeams according to a three-node linear bending element. Then, compute the first five modes nondimensional frequencies of simply supported and simply supported-guided supported by using this NL-FEM procedure. Consider the physical and mechanical parameters given in Problem 7.1. Since the operation volume increases, it is recommended to perform the analysis electronically and examine the convergence of results by using 2, 3, and 4 elements.

7.6. Examine Problem 7.5 according to nonlocal EBBT.

7.7. Compute the first three modes nondimensional frequencies of attached nanobeams mentioned under according to nonlocal EBBT and TBT under parameters as follows. Consider the attachments as only transverse effect (in other words, there is not a rotational effect of attachment). Use two finite elements for manual solution. Also, if you have a program code or package program based on finite element analysis, try to obtain the approximated values by using 5, 10, and 15 finite elements. Discuss the effects of nonlocal parameter and boundary conditions on differences between results.

Nanobeam models: (a) Clamped-spring attached, (b) Clamped-mass attached, (c) Clamped-spring and mass attached, (d) Spring attached-mass attached.

Nanobeam length, $L = 20$ nm, height and thickness of rectangular cross-section, $h = 1$ nm and $b = 2$ nm, modulus of elasticity, $E = 1$ TPa, mass of unit volume, $\rho = 2300$ kg/m^3, shear modulus: $G = 0.42$ TPa, shear correction factor: $k = 5/6$, attachment ratios: $\beta_m = 10$ and $\beta_k = 10$, nondimensional stiffness of Winkler foundation: $K_W = 10$, nondimensional stiffness of Pasternak foundation: $K_P = 1$, temperature change: $\Delta T = 300$ K, and thermal expansion coefficient: $\alpha_t = 1.6 \times 10^{-6}$ K^{-1}.

Compute according to three different nondimensional nonlocal parameters: $\alpha = 0$, 0.1, 0.2.

Compute the nondimensional stiffness of elastic medium as $K_W = k_W L^4/EI$ and $K_P = k_P L^2/EI$.

Compute the mass, stiffness, and nondimensional frequencies of nanobeam as: $\beta_k = k_a L^3/EI$, $\beta_m = m_a/\rho AL$, and $\overline{\omega} = \omega L^2\sqrt{\rho A/EI}$.

References

1. M.A. Eltaher, A.E. Alshorbagy, F.F. Mahmoud, Vibration analysis of Euler–Bernoulli nanobeams by using finite element method. Appl. Math. Model. **37**(7), 4787–4797 (2013)
2. M. Tuna, M. Kirca, Bending, buckling and free vibration analysis of Euler-Bernoulli nanobeams using Eringen's nonlocal integral model via finite element method. Compos. Struct. **179**, 269–284 (2017)
3. M.A. Eltaher, S.A. Emam, F.F. Mahmoud, Free vibration analysis of functionally graded size-dependent nanobeams. Appl. Math. Comput. **218**(14), 7406–7420 (2012)
4. Ç. Demir, Ö. Civalek, A new nonlocal FEM via Hermitian cubic shape functions for thermal vibration of nano beams surrounded by an elastic matrix. Compos. Struct. **168**, 872–884 (2017)
5. H.M. Numanoğlu, Thermal vibration of zinc oxide nanowires by using nonlocal finite element method. Int. J. Eng. Appl. Sci. **12**(3), 99–110 (2020)
6. H.M. Numanoğlu, H. Ersoy, O. Civalek, A.J.M. Ferrerira, Derivation of nonlocal FEM formulation for thermo-elastic Timoshenko beams on elastic matrix. Compos. Struct. **273**, 114292 (2021)
7. S.C. Pradhan, U. Mandal, Finite element analysis of CNTs based on nonlocal elasticity and Timoshenko beam theory including thermal effect. Phys. E **53**, 223–232 (2013)
8. A.I. Aria, M.I. Friswell, A nonlocal finite element model for buckling and vibration of functionally graded nanobeams. Compos. Part B **166**, 233–246 (2019)
9. M. Ganapathi, T. Merzouki, O. Polit, Vibration study of curved nanobeams based on nonlocal higher-order shear deformation theory using finite element approach. Compos. Struct. **184**, 821–838 (2018)
10. Ç. Dinçkal, Free vibration analysis of carbon nanotubes by using finite element method. Iran. J. Sci. Technol. Trans. Mech. Eng. **40**, 43–55 (2016)
11. A.I. Aria, M.I. Friswell, T. Rabczuk, Thermal vibration analysis of cracked nanobeams embedded in an elastic matrix using finite element analysis. Compos. Struct. **212**, 118–128 (2019)
12. M. Fakher, S. Hosseini-Hashemi, Vibration of two-phase local/nonlocal Timoshenko nanobeams with an efficient shear-locking-free finite-element model and exact solution. Eng. Comput. **38**, 231–245 (2022)
13. A. Norouzzadeh, R. Ansari, Finite element analysis of nano-scale Timoshenko beams using the integral model of nonlocal elasticity. Phys. E **88**, 194–200 (2017)
14. H.M. Numanoğlu, *Dynamic Analysis of Nano Scaled Continuous and Discrete Structures Based on Nonlocal Finite Element Formulation (NL-FEM), MSc. Thesis* (Akdeniz University, Antalya, 2019) (In Turkish)

Vibration of Discrete Structures: Nanotrusses and Nanoframes

8

8.1 Introduction

Considering nanostructures only within the framework of one-dimensional and continuous mechanical models may be an insufficient perspective. It cannot be completely denied that such mechanical models should govern the analysis process in many NEMS organizations. However, in some nanostructures, one-dimensional and continuous mechanical models should either be examined indirectly or completely denied since those are not suitable for the problem. Nanostructures which such mechanical models cannot be considered at all, are approached with the nanoplate or nanoshell model. For example, while graphene sheet is a well-known example of nanoplate, some two-dimensional curved nanomaterials emphasize nanoshell model. Mechanical analyses of such nanostructures are further investigations and are excluded from the book.

On the other hand, some examples of nanostructures bring to mind the example of a planar truss or planar frame and these examples are presented in the first chapter. Truss and frame structures are defined as discrete mechanical models in the structural engineering. According to continuum mechanics, a continuous system is a whole, that is, no point is independent of the other. However, the structure can be separated into several members in discrete structure mechanics. In the mechanical analysis of discrete structures, continuous mechanical models have an indirect but still dominant using.

If you pay attention, one-dimensional and continuous nanostructures are investigated so far in the book. These are explained by analytical and NL-FEM approaches. In particular, NL-FEM investigations are the subject of sixth and seventh chapters. When the formulations and examples developed in these sections are considered altogether, we understand that the NL-FEM approach enables a great benefit in nonlocal mechanical

Öm. Civalek et al., *Mechanical Behavior and Vibration of Nano-Scaled Rods, Beams, and Frames*, Synthesis Lectures on Engineering, Science, and Technology,
https://doi.org/10.1007/978-3-032-12023-6_8

analysis. Since, as mentioned above, some NEMS examples suggest discrete structures, this chapter examines the nonlocal free vibration of discrete nanostructures.

In addition to the analytical solution in the mechanical analysis of discrete nanostructures, different numerical solutions are available. It can be said that the finite element formulation is more understandable and applicable for such structures than other approximate analyses. Generally, each member of discrete structures can be modeled with a single finite element or divided into multiple finite elements. The matrix displacement method, which is a special case of finite element method in the discrete structures, governs the process of mechanical analysis of these structures. In order to determine the contributions of discrete members to the global matrices, calculating via the displacements of members is the fundamental basis of formulation.

This chapter presents nonlocal free vibration analyses of nano-scaled planar trusses and frame structures. In the first part of this chapter, a simple rod theory-based nonlocal matrix displacement method in nanotrusses is examined. In the second part, nonlocal free vibrations are given for nanoframes without and with shear deformation. Following the formulations, various numerical examples are analyzed.

It is very important to follow the general flow provided in this chapter from the beginning, because each of the mechanical analysis formulations introduced must be used in the next issue.

8.2 Recent Contributions

When we look at the 20-year period since the early 2000s when the mechanical behavior of nano- and micro-scaled structures began to be investigated through atomic size-dependent continuum mechanics, it can be stated that studies considering the mechanics of discrete structures in this field remain quite limited. The studies carried out on this subject within this period are briefly mentioned below.

In the scientific literature, the first study involving the size-dependent mechanics of discrete nanostructures deals with the free vibration of nanotrusses and nanoframes without shear deformation [1]. Nonlocal free vibration of nanoframes based on Timoshenko beam theory has also been formulated [2]. Additionally, the free vibration of nanoframes has been explained according to the couple stress elasticity theory, which is another continuum theory that considers the atomic size effect, and the Rayleigh-Love and Timoshenko bending approaches [3]. Moreover, a nonlocal stress gradient formulation has been studied for the torsional vibration of a two-member nanoframe [4]. Furthermore, the free vibration of nanoframes with a nonlocal stress-driven approach has been solved by the dynamic stiffness method and Wittrick-Williams algorithm [5].

8.3 Nonlocal Vibration of Nanotrusses

Simple axial rods are structural elements characterized by deformations only in the direction of their own axes and formulated according to continuum mechanics. In addition to this, structures consisting of axial elements (members) but expressing a discrete system rather than a continuous are the type of system frequently encountered in the structural engineering. As it is understood, the NL-FEM formulation of simple axial nanorod should be known for the nonlocal free vibration of nanotrusses. In this section, nonlocal free vibration of nanotrusses based on simple axial rod formulation is presented.

The most fundamental issue in the analysis of discrete systems is the orientation of members. Members can have different orientations depending on the geometry of structure. In order to express the matrix formulation correctly, a transformation should be formulated for the orientation of members. The depiction of axial member on coordinate sets that are the common (global) for all members in the system and specific to (local) any member oriented as θ relative to the horizontal is shown in Fig. 8.1. Here, the end freedom and end forces of axial member are seen. In accordance with general knowledge of finite element, the analysis depends primarily on the transformation of displacements. The equations that transform global displacement components into local displacement components are stated as follows [1]:

$$d_{11} = u_{11}\cos\theta + u_{12}\sin\theta, \quad d_{12} = -u_{11}\sin\theta + u_{12}\cos\theta, \quad d_{21} = u_{21}\cos\theta + u_{22}\sin\theta,$$
$$d_{22} = -u_{21}\sin\theta + u_{22}\cos\theta \tag{8.1}$$

where d_{11} and d_{21} are the local axial motion components, d_{12} and d_{22} are the local transverse motion components. u_{11} and u_{21} are the general horizontal motion components, u_{12} and u_{22} are the general vertical motion components. θ is the orientation angle of member. In this section, the names 1 and 2 are used instead of i and j for the end names in the finite element description. Systems consisting of axial members do not have internal bending effects but can show a linear motion in the shear freedom in the bending members.

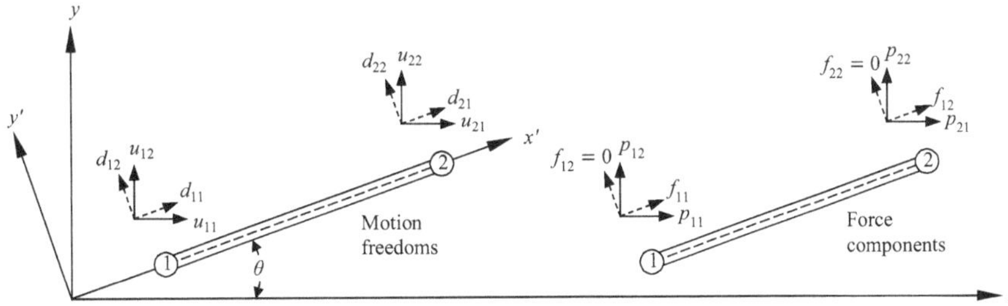

Fig. 8.1 Axial member for NL-FEM analysis of nanotruss (xy: Global and $x'y'$: Local coordinate systems) [1]

The motion of discrete axial rod like the motion of continuous axial rod is defined as follows by considering Eq. (6.2) [1]:

$$\mathbf{u}^{\mathbf{t}} = \phi\mathbf{u} = \begin{bmatrix} \phi_1 & 0 & \phi_2 & 0 \\ 0 & \phi_1 & 0 & \phi_2 \end{bmatrix} \begin{Bmatrix} u_{11} \\ u_{12} \\ u_{21} \\ u_{22} \end{Bmatrix} \tag{8.2}$$

here $\mathbf{u}^{\mathbf{t}}$ is the motion vector, φ is the shape functions matrix, $\mathbf{u}$ is the displacement vector in the global coordinate sets. Eq. (8.1) can be written in matrix form as follows:

$$\mathbf{d} = T\mathbf{u} : \begin{Bmatrix} d_{11} \\ d_{12} \\ d_{21} \\ d_{22} \end{Bmatrix} = \begin{bmatrix} \cos\theta & \sin\theta & 0 & 0 \\ -\sin\theta & \cos\theta & 0 & 0 \\ 0 & 0 & \cos\theta & \sin\theta \\ 0 & 0 & -\sin\theta & \cos\theta \end{bmatrix} \begin{Bmatrix} u_{11} \\ u_{12} \\ u_{21} \\ u_{22} \end{Bmatrix} \tag{8.3}$$

here $\mathbf{d}$ is the displacement vector in the local coordinates. T is known as the transformation matrix. Also, the following expression is given for the end forces:

$$\mathbf{f} = T\mathbf{p} : \begin{Bmatrix} f_{11} \\ f_{12} \\ f_{21} \\ f_{22} \end{Bmatrix} = \begin{bmatrix} \cos\theta & \sin\theta & 0 & 0 \\ -\sin\theta & \cos\theta & 0 & 0 \\ 0 & 0 & \cos\theta & \sin\theta \\ 0 & 0 & -\sin\theta & \cos\theta \end{bmatrix} \begin{Bmatrix} p_{11} \\ p_{12} \\ p_{21} \\ p_{22} \end{Bmatrix} \tag{8.4}$$

where $\mathbf{f}$ and $\mathbf{p}$ mean the nodal force vectors for the local and global coordinates, respectively. At this point, Hooke's Equation are included the NL-FEM formulation:

$$\mathbf{f} = k\mathbf{d} \tag{8.5}$$

where K is the stiffness matrix. Hooke's Equation is introduced for the member based on simple axial rod as follows:

$$\begin{Bmatrix} f_{11} \\ 0 \\ f_{21} \\ 0 \end{Bmatrix} = \frac{EA}{L} \begin{bmatrix} 1 & 0 & -1 & 0 \\ 0 & 0 & 0 & 0 \\ -1 & 0 & 1 & 0 \\ 0 & 0 & 0 & 0 \end{bmatrix} \begin{Bmatrix} d_{11} \\ d_{12} \\ d_{21} \\ d_{22} \end{Bmatrix} \tag{8.6}$$

Since there is no shear force at the ends of the member, the forces in this direction are shown as zero in Fig. 8.1. As a result, the stiffness matrix is defined as follows:

$$k = \frac{EA}{L}\begin{bmatrix} 1 & 0 & -1 & 0 \\ 0 & 0 & 0 & 0 \\ -1 & 0 & 1 & 0 \\ 0 & 0 & 0 & 0 \end{bmatrix} \tag{8.7}$$

On the other hand, the classical and nonlocal mass matrices are written through Eqs. (6.2), (6.13), and (6.14):

$$\begin{aligned} m_c &= \int_0^L \rho A \left(\phi^{\mathrm{T}} \phi\right) \mathrm{d}x = \int_0^L \rho A \begin{bmatrix} \phi_1 & 0 \\ 0 & \phi_1 \\ \phi_2 & 0 \\ 0 & \phi_2 \end{bmatrix} \begin{bmatrix} \phi_1 & 0 & \phi_2 & 0 \\ 0 & \phi_1 & 0 & \phi_2 \end{bmatrix} \mathrm{d}x \\ &= \frac{\rho A L}{6} \begin{bmatrix} 2 & 0 & 1 & 0 \\ 0 & 2 & 0 & 1 \\ 1 & 0 & 2 & 0 \\ 0 & 1 & 0 & 2 \end{bmatrix} \end{aligned} \tag{8.8}$$

$$\begin{aligned} m_{nl} &= \int_0^L (e_0 a)^2 \rho A \left(\mathbf{B}^{\mathrm{T}} \mathbf{B}\right) \mathrm{d}x = \int_0^L (e_0 a)^2 \rho A \begin{bmatrix} \phi_1' & 0 \\ 0 & \phi_1' \\ \phi_2' & 0 \\ 0 & \phi_2' \end{bmatrix} \begin{bmatrix} \phi_1' & 0 & \phi_2' & 0 \\ 0 & \phi_1' & 0 & \phi_2' \end{bmatrix} \mathrm{d}x \\ &= \frac{(e_0 a)^2 \rho A}{L} \begin{bmatrix} 1 & 0 & -1 & 0 \\ 0 & 1 & 0 & -1 \\ -1 & 0 & 1 & 0 \\ 0 & -1 & 0 & 1 \end{bmatrix} \end{aligned} \tag{8.9}$$

Substituting Eqs. (8.3) and (8.4) into Eq. (8.5) yields the following equation:

$$T\mathbf{p} = kT\mathbf{u} \tag{8.10}$$

The following expression can be written by using Eq. (8.10) due to the orthogonality property ($T^{\mathrm{T}} = T^{-1}$) of transformation matrix [1]:

$$\underbrace{T^{-1}T}_{I}\mathbf{p} = \underbrace{T^{\mathrm{T}}kT}_{K}\mathbf{u} \tag{8.11}$$

where I is the unit matrix. Additionally, K is the contribution of axial member to the global stiffness matrix:

$$K = T^{\mathrm{T}} k T = \begin{bmatrix} \cos\theta & -\sin\theta & 0 & 0 \\ \sin\theta & \cos\theta & 0 & 0 \\ 0 & 0 & \cos\theta & -\sin\theta \\ 0 & 0 & \sin\theta & \cos\theta \end{bmatrix} \times \frac{EA}{L} \begin{bmatrix} 1 & 0 & -1 & 0 \\ 0 & 0 & 0 & 0 \\ -1 & 0 & 1 & 0 \\ 0 & 0 & 0 & 0 \end{bmatrix}$$
$$\times \begin{bmatrix} \cos\theta & \sin\theta & 0 & 0 \\ -\sin\theta & \cos\theta & 0 & 0 \\ 0 & 0 & \cos\theta & \sin\theta \\ 0 & 0 & -\sin\theta & \cos\theta \end{bmatrix}$$
$$= \frac{EA}{L} \begin{bmatrix} \cos^2\theta & \cos\theta\sin\theta & -\cos^2\theta & -\cos\theta\sin\theta \\ \cos\theta\sin\theta & \sin^2\theta & -\cos\theta\sin\theta & -\sin^2\theta \\ -\cos^2\theta & -\cos\theta\sin\theta & \cos^2\theta & \cos\theta\sin\theta \\ -\cos\theta\sin\theta & -\sin^2\theta & \cos\theta\sin\theta & \sin^2\theta \end{bmatrix} \tag{8.12}$$

Calculations like the above can also be performed for classical and nonlocal mass matrices. First the classical mass matrix is mentioned:

$$M_c = T^{\mathrm{T}} m_c T = \begin{bmatrix} \cos\theta & -\sin\theta & 0 & 0 \\ \sin\theta & \cos\theta & 0 & 0 \\ 0 & 0 & \cos\theta & -\sin\theta \\ 0 & 0 & \sin\theta & \cos\theta \end{bmatrix} \times \frac{\rho AL}{6} \begin{bmatrix} 2 & 0 & 1 & 0 \\ 0 & 2 & 0 & 1 \\ 1 & 0 & 2 & 0 \\ 0 & 1 & 0 & 2 \end{bmatrix}$$
$$\times \begin{bmatrix} \cos\theta & \sin\theta & 0 & 0 \\ -\sin\theta & \cos\theta & 0 & 0 \\ 0 & 0 & \cos\theta & \sin\theta \\ 0 & 0 & -\sin\theta & \cos\theta \end{bmatrix} = \frac{\rho AL}{6} \begin{bmatrix} 2 & 0 & 1 & 0 \\ 0 & 2 & 0 & 1 \\ 1 & 0 & 2 & 0 \\ 0 & 1 & 0 & 2 \end{bmatrix} \tag{8.13}$$

If we pay attention to the result here, the classical mass matrix is independent of the value of orientation of axial member. Additionally, we can formulate the nonlocal mass matrix for the global coordinates of axial member as follows:

$$M_{nl} = T^{\mathrm{T}} m_{nl} T = \begin{bmatrix} \cos\theta & -\sin\theta & 0 & 0 \\ \sin\theta & \cos\theta & 0 & 0 \\ 0 & 0 & \cos\theta & -\sin\theta \\ 0 & 0 & \sin\theta & \cos\theta \end{bmatrix} \times \frac{e_0 a^2 \rho A}{L} \begin{bmatrix} 1 & 0 & -1 & 0 \\ 0 & 1 & 0 & -1 \\ -1 & 0 & 1 & 0 \\ 0 & -1 & 0 & 1 \end{bmatrix}$$
$$\times \begin{bmatrix} \cos\theta & \sin\theta & 0 & 0 \\ -\sin\theta & \cos\theta & 0 & 0 \\ 0 & 0 & \cos\theta & \sin\theta \\ 0 & 0 & -\sin\theta & \cos\theta \end{bmatrix} = \frac{e_0 a^2 \rho A}{L} \begin{bmatrix} 1 & 0 & -1 & 0 \\ 0 & 1 & 0 & -1 \\ -1 & 0 & 1 & 0 \\ 0 & -1 & 0 & 1 \end{bmatrix} \tag{8.14}$$

where in addition to the classical mass matrix, it is observed that the nonlocal mass matrix is also not affected by the transformation process. Thus, the global mass matrix of an axial member is expressed below:

$$M = M_c + M_{nl} = T^{\mathrm{T}} m_c T + T^{\mathrm{T}} m_{nl} T = \begin{bmatrix} M_{11} & M_{12} & M_{13} & M_{14} \\ & M_{22} & M_{23} & M_{24} \\ & \text{sym.} & M_{33} & M_{34} \\ & & & M_{44} \end{bmatrix} \tag{8.15}$$

where the components of total mass matrix are presented as follows:

$$M_{11} = M_{22} = M_{33} = M_{44} = \rho A \left(\frac{L}{3} + \frac{(e_0 a)^2}{L} \right), \quad M_{12} = M_{14} = M_{23} = M_{34} = 0$$
$$M_{13} = M_{24} = \rho A \left(\frac{L}{6} - \frac{(e_0 a)^2}{L} \right) \tag{8.16}$$

After the global matrices constituted by considering the freedom of system are reduced under the boundary conditions, if they are subjected to the eigenvalue equation known from Chaps. 6 and 7, the nonlocal natural vibration frequencies of nanotruss can be calculated:

$$\det \left[\left(\sum_{e=1}^{n} [K]_e \right)^* - \omega_i^2 \left(\sum_{e=1}^{n} [M]_e \right)^* \right] = 0 \tag{8.17}$$

Example 8.1 Determine the natural frequencies of nonlocal free vibration of nanotruss shown in Fig. 8.2a. Use a single finite element for all member in the analysis (Fig. 8.1b). Length of member **1**: $L = l_{e=\mathbf{1}} = 50$ nm (L is the length parameter of nanotruss), diameter of circular cross-section: $d = 1$ nm, angle between in the members **1** and **3**: $\theta = 45°$, modulus of elasticity: $E = 1$ TPa, and mass of unit volume: $\rho = 2300 \text{kg/m}^3$. Use the nonlocal parameter as $e_0 a / L = 0.3$.

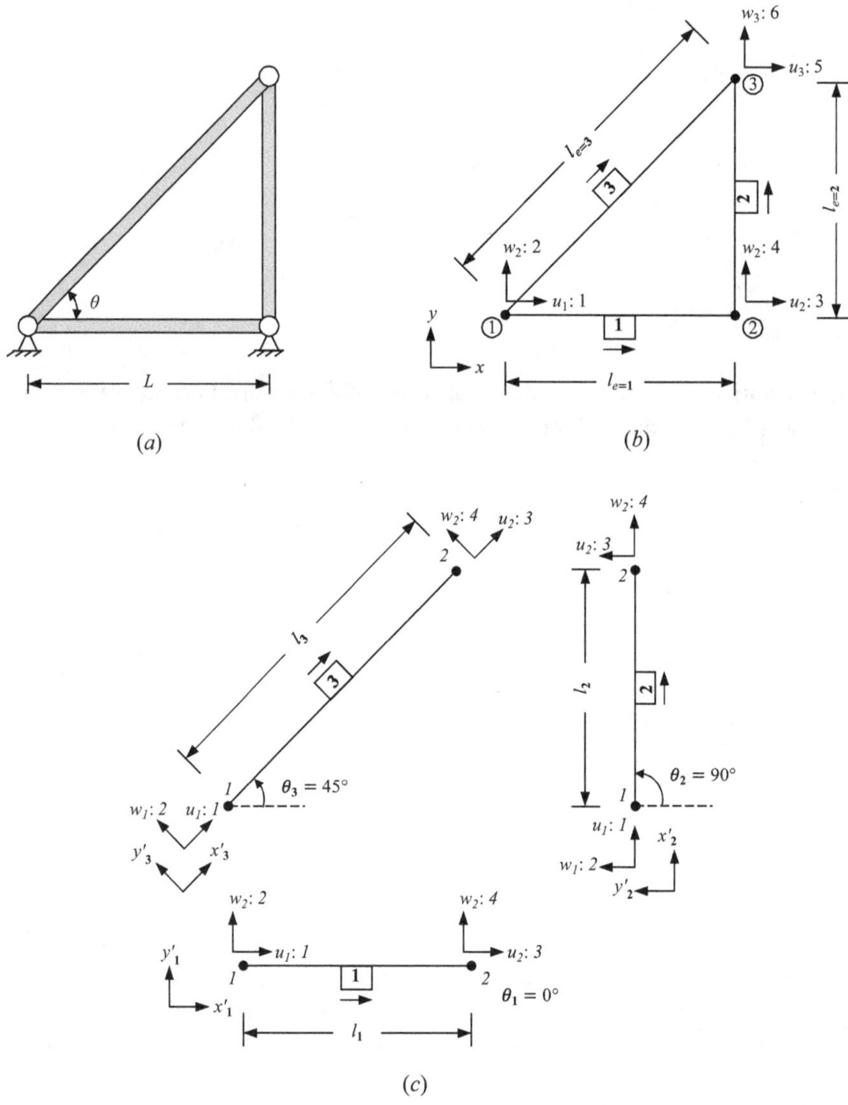

Fig. 8.2 Matrix displacement analysis of a nanotruss with three members. (**a**) Physical system. (**b**) FEM model. (**c**) Axial members in their local coordinates

Solution: The three separate axial members of nanotruss in Fig. 8.1a represent finite element by itself. According to this, the finite element network in the global coordinates is depicted in Fig. 8.2b and these finite elements with their local coordinates are depicted in Fig. 8.2c. Let's write the physical and mechanical parameters of each axial member for analysis:

$$
\begin{aligned}
&\text{Lengths}: l_1 = l_2 = 5\times 10^{-8}\,\text{m},\ \ l_3 = 7.071\times 10^{-8}\,\text{m},\\
&\text{Nonlocal parameters}: (e_0a)_1 = (e_0a)_2 = (e_0a)_3 = 0.3\times L = (0.3)\times(50\ \text{nm})\\
&\quad = 1.5\times 10^{-8}\ \text{m},\\
&\text{Cross}-\text{section areas}: A_1 = A_2 = A_3 = \frac{\pi d^2}{4} = \frac{\pi\times(1^2\,\text{nm}^2)}{4} = 3.142\times 10^{-18}\ \text{m}^2,\\
&\text{Modulus of elasticity}: E_1 = E_2 = E_3 = 1\,\text{TPa} = 1\times 10^{12}\frac{\text{N}}{\text{m}^2},\\
&\text{Unit volume masses}: \rho_1 = \rho_2 = \rho_3 = 2300\frac{\text{kg}}{\text{m}^3}.
\end{aligned}
\tag{8.E8.1.1}
$$

The ends of members should be numbered and this numbering can be observed in Fig. 8.2c due to the kinematics of finite elements expressing axial members. The nodes symbolized by *1* and *2* of each member indicate the orientation of element. Accordingly, the finite element orientation is from end *1* to end *2*. This orientation allows to calculate the positive angle the element makes with the horizontal axis (global x).

On the other hand, it is known that there are six freedoms in this system for matrix displacement analysis. However, since the horizontal and vertical motions of support at nodes (1) and (2) are restricted, the freedoms 1, 2, 3, and 4 in Fig. 8.2b are eliminated. Therefore, there are three freedoms in which the nanotruss can move. These inferences imply that the reduced global matrices of matrix displacement analysis have sizes of 2×2, but this is discussed later. First, let's constitute the global matrices. This process starts with stiffness matrices. As it is known, Eq. (8.12) expresses the contribution of an axial member to the stiffness of nanotruss. θ in this equation is called as the orientation angle of member. The related trigonometric values of orientation angles are calculated as follows:

$$
\begin{aligned}
&\theta_1 = 0^\circ \rightarrow \cos\theta_1 = 1,\ \ \sin\theta_1 = 0,\\
&\theta_2 = \frac{\pi}{2} \rightarrow \cos\theta_2 = 0,\ \ \sin\theta_2 = 1,\\
&\theta_3 = \frac{\pi}{4} \rightarrow \cos\theta_3 = 0.707,\ \ \sin\theta_3 = 0.707.
\end{aligned}
\tag{8.E8.1.2}
$$

Then, the global stiffness matrices of members are obtained below. In order to relate the calculations to the node freedoms, the numbers of local and global freedoms are shown on the stiffness matrix calculations. These representations are not given to the mass matrices to avoid unnecessary crowd, because showing them on the stiffness matrix is sufficient to understand the issue. Meanwhile, the unit of inputs of stiffness matrices is N/m:

$$(K)_1 = \frac{\left(1\times10^{12}\,\frac{\mathrm{N}}{\mathrm{m}^2}\right)\times\left(3.142\times10^{-18}\,\mathrm{m}^2\right)}{5\times10^{-8}\,\mathrm{m}}\times\begin{bmatrix} 1^2 & 1\times0 & -1^2 & -1\times0 \\ 1\times0 & 0^2 & -1\times0 & -0^2 \\ -1^2 & -1\times0 & 1^2 & 1\times0 \\ -1\times0 & -0^2 & 1\times0 & 0^2 \end{bmatrix}$$

$$= \begin{array}{c} \begin{array}{cccc} \mathit{1}\to1 & \mathit{2}\to2 & \mathit{3}\to3 & \mathit{4}\to4 \end{array} \\ \left[\begin{array}{c|c|c|c} 62.840 & 0 & -62.840 & 0 \\ \hline 0 & 0 & 0 & 0 \\ \hline -62.840 & 0 & 62.840 & 0 \\ \hline 0 & 0 & 0 & 0 \end{array}\right] \end{array} \begin{array}{l} \\ 1\leftarrow\mathit{1} \\ 2\leftarrow\mathit{2} \\ 3\leftarrow\mathit{3} \\ 4\leftarrow\mathit{4} \end{array} \qquad (8.E8.1.3)$$

Italic: Local freedoms
Straight: Global freedoms

$$(K)_2 = \frac{\left(1\times10^{12}\,\frac{\mathrm{N}}{\mathrm{m}^2}\right)\times\left(3.142\times10^{-18}\,\mathrm{m}^2\right)}{5\times10^{-8}\,\mathrm{m}}\times\begin{bmatrix} 1^2 & 1\times0 & -1 & -1\times0 \\ 1\times0 & 0^2 & -1\times0 & -0^2 \\ -0^2 & -1\times0 & 1^2 & 1\times0 \\ -1\times0 & -0^2 & 1\times0 & 0^2 \end{bmatrix}$$

$$= \begin{array}{c} \begin{array}{cccc} \mathit{1}\to3 & \mathit{2}\to4 & \mathit{3}\to5 & \mathit{4}\to6 \end{array} \\ \left[\begin{array}{c|c|c|c} 0 & 0 & 0 & 0 \\ \hline 0 & 62.840 & 0 & -62.840 \\ \hline 0 & 0 & 0 & 0 \\ \hline 0 & -62.840 & 0 & 62.840 \end{array}\right] \end{array} \begin{array}{l} \\ 3\leftarrow\mathit{1} \\ 4\leftarrow\mathit{2} \\ 5\leftarrow\mathit{3} \\ 6\leftarrow\mathit{4} \end{array} \qquad (8.E8.1.4)$$

$$(K)_3 = \frac{\left(1\times10^{12}\,\dfrac{\mathrm{N}}{\mathrm{m}^2}\right)\times\left(3.142\times10^{-18}\ \mathrm{m}^2\right)}{7.071\times10^{-8}\,\mathrm{m}}\times$$

$$\begin{bmatrix} 0.707^2 & 0.707\times0.707 & -0.707 & -0.707\times0.707 \\ 0.707\times0.707 & 0.707^2 & -0.707\times0.707 & -0.707^2 \\ -0.707^2 & -0.707\times0.707 & 0.707^2 & 0.707\times0.707 \\ -0.707\times0.707 & -0.707^2 & 0.707\times0.707 & 0.707^2 \end{bmatrix} \tag{8.E8.1.5}$$

$$\begin{array}{c} \\ = \end{array} \begin{array}{cccc|c} \mathit{1}\to 1 & \mathit{2}\to 2 & \mathit{3}\to 5 & \mathit{4}\to 6 & \\ \hline 22.218 & 22.218 & -22.218 & -22.218 & 1\leftarrow \mathit{1} \\ 22.218 & 22.218 & -22.218 & -22.218 & 2\leftarrow \mathit{2} \\ -22.218 & -22.218 & 22.218 & 22.218 & 5\leftarrow \mathit{3} \\ -22.218 & -22.218 & 22.218 & 22.218 & 6\leftarrow \mathit{4} \end{array}$$

Now let's calculate the classical mass matrices:

$$(M_c)_1 = (M_c)_2 = \frac{\left(2300\,\dfrac{\mathrm{kg}}{\mathrm{m}^3}\right)\times(3.142\times10^{-18}\ \mathrm{m}^2)\times(5\times10^{-8}\,\mathrm{m})}{6}\times\begin{bmatrix} 2 & 0 & 1 & 0 \\ 0 & 2 & 0 & 1 \\ 1 & 0 & 2 & 0 \\ 0 & 1 & 0 & 2 \end{bmatrix}$$

$$= \begin{bmatrix} 1.204\times10^{-22} & 0 & 6.022\times10^{-23} & 0 \\ 0 & 1.204\times10^{-22} & 0 & 6.022\times10^{-23} \\ 6.022\times10^{-23} & 0 & 1.204\times10^{-22} & 0 \\ 0 & 6.022\times10^{-23} & 0 & 1.204\times10^{-22} \end{bmatrix} \tag{8.E8.1.6}$$

$$(M_c)_3 = \frac{\left(2300\,\frac{\text{kg}}{\text{m}^3}\right) \times (3.142 \times 10^{-18}\ \text{m}^2) \times (7.071 \times 10^{-8}\,\text{m})}{6} \times \begin{bmatrix} 2 & 0 & 1 & 0 \\ 0 & 2 & 0 & 1 \\ 1 & 0 & 2 & 0 \\ 0 & 1 & 0 & 2 \end{bmatrix}$$

$$= \begin{bmatrix} 1.703 \times 10^{-22} & 0 & 8.517 \times 10^{-23} & 0 \\ 0 & 1.703 \times 10^{-22} & 0 & 8.517 \times 10^{-23} \\ 8.517 \times 10^{-23} & 0 & 1.703 \times 10^{-22} & 0 \\ 0 & 8.517 \times 10^{-23} & 0 & 1.703 \times 10^{-22} \end{bmatrix} \tag{8.E8.1.7}$$

Additionally, the nonlocal mass matrices of axial members in the global coordinates are computed as follows:

$$(M_{nl})_1 = (M_{nl})_2$$

$$= \frac{(1.5^2 \times 10^{-16}\ \text{m}^2) \times \left(2300\frac{\text{kg}}{\text{m}^3}\right) \times (3.142 \times 10^{-18}\ \text{m}^2)}{5 \times 10^{-8}\text{m}} \times \begin{bmatrix} 1 & 0 & -1 & 0 \\ 0 & 1 & 0 & -1 \\ -1 & 0 & 1 & 0 \\ 0 & -1 & 0 & 1 \end{bmatrix}$$

$$= \begin{bmatrix} 3.252 \times 10^{-23} & 0 & -3.252 \times 10^{-23} & 0 \\ 0 & 3.252 \times 10^{-23} & 0 & -3.252 \times 10^{-23} \\ -3.252 \times 10^{-23} & 0 & 3.252 \times 10^{-23} & 0 \\ 0 & -3.252 \times 10^{-23} & 0 & 3.252 \times 10^{-23} \end{bmatrix} \tag{8.E8.1.8}$$

$$(M_{nl})_3 = \frac{\left(2300\frac{\text{kg}}{\text{m}^3}\right) \times (3.142 \times 10^{-18}\ \text{m}^2) \times (7.071 \times 10^{-8}\text{m})}{7.071 \times 10^{-8}\text{m}} \times \begin{bmatrix} 1 & 0 & -1 & 0 \\ 0 & 1 & 0 & -1 \\ -1 & 0 & 1 & 0 \\ 0 & -1 & 0 & 1 \end{bmatrix}$$

$$= \begin{bmatrix} 2.3 \times 10^{-23} & 0 & -2.3 \times 10^{-23} & 0 \\ 0 & 2.3 \times 10^{-23} & 0 & -2.3 \times 10^{-23} \\ -2.3 \times 10^{-23} & 0 & 2.3 \times 10^{-23} & 0 \\ 0 & -2.3 \times 10^{-23} & 0 & 2.3 \times 10^{-23} \end{bmatrix} \tag{8.E8.1.9}$$

It is reminded that inputs of above classical and nonlocal mass matrices are calculated with kg units.

Reduced global stiffness and mass matrices are required to write the eigenvalue equation in Eq. (8.17). The stiffness matrix is determined by the global stiffness matrices of axial members in Eqs. (8.E8.1.3)–(8.E8.1.6). Matrices can be assembled by taking into account the global freedoms of structure. Assembly expresses the global stiffness matrix of nanotruss:

$$\sum[K]=\begin{array}{c} \begin{array}{cccccc} 1 & 2 & 3 & 4 & 5 & 6 \end{array} \\ \left[\begin{array}{cccccc} 62.840+22.218 & 0+22.218 & -62.840 & 0 & -22.218 & -22.218 \\ 0+22.218 & 0+22.218 & 0 & 0 & -22.218 & -22.218 \\ -62.840 & 0 & 62.840+0 & 0+0 & 0 & 0 \\ 0 & 0 & 0+0 & 0+62.840 & 0 & -62.840 \\ -22.218 & -22.218 & 0 & 0 & 0+22.218 & 0+22.218 \\ -22.218 & -22.218 & 0 & -62.840 & 0+22.218 & 62.840+22.218 \end{array}\right] \end{array}\begin{array}{c} \textit{Freedom numbers} \\ 1 \\ 2 \\ 3 \\ 4 \\ 5 \\ 6 \end{array} \quad (8.E8.1.10)$$

where a reduction should be carried out due to the geometric boundary conditions. That is, rows and columns 1–4 in the global stiffness matrix of should be deleted. According to this:

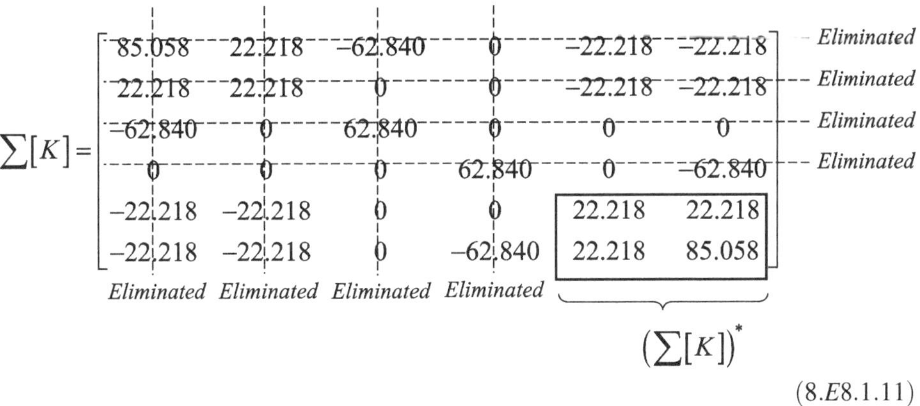

$$\sum[K]=\left[\begin{array}{cccc|cc} 85.058 & 22.218 & -62.840 & 0 & -22.218 & -22.218 \\ 22.218 & 22.218 & 0 & 0 & -22.218 & -22.218 \\ -62.840 & 0 & 62.840 & 0 & 0 & 0 \\ 0 & 0 & 0 & 62.840 & 0 & -62.840 \\ -22.218 & -22.218 & 0 & 0 & 22.218 & 22.218 \\ -22.218 & -22.218 & 0 & -62.840 & 22.218 & 85.058 \end{array}\right]$$

(8.E8.1.11)

Similarly, the reduced global classical and nonlocal mass matrices of nanotruss are obtained as follows:

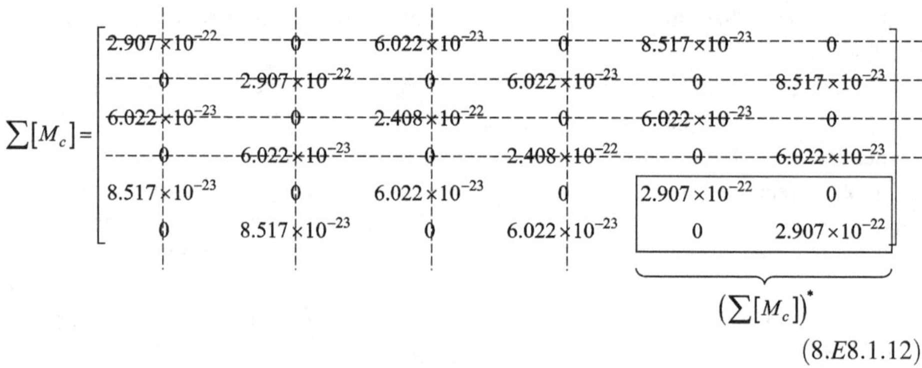

(8.E8.1.12)

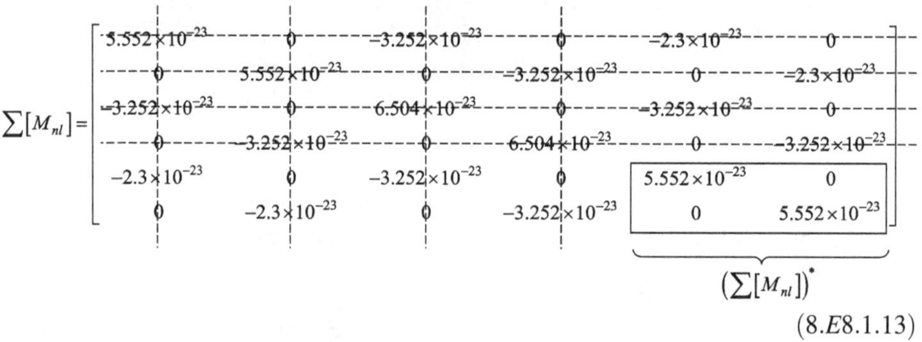

(8.E8.1.13)

As a result of reduction process, the required matrices for Eq. (8.17) are given as follows:

$$\left(\sum[K]\right)^* = \begin{bmatrix} 22.218 & 22.218 \\ 22.218 & 85.058 \end{bmatrix} \tag{8.E8.1.14}$$

$$\left(\sum[M]\right)^* = \left(\sum[M_c]\right)^* + \left(\sum[M_{nl}]\right)^* = \begin{bmatrix} 3.462 \times 10^{-22} & 0 \\ 0 & 3.462 \times 10^{-22} \end{bmatrix} \tag{8.E8.1.15}$$

If Eqs. (8.E8.1.14) and (8.E8.1.15) are replaced into Eq. (8.17), the following equation is obtained:

$$1.199 \times 10^{-46}\omega^4 - 3.714 \times 10^{-23}\omega^2 + 1.396 = 0 \tag{8.E8.1.16}$$

the positive roots of above equation, that is, the natural frequencies of first and second modes of nanotrusses are calculated as follows:

Table 8.1 Nonlocal natural frequencies (GHz) of nanotruss in Fig. 8.2

	Solution manually		Program code (MATLAB)	
	Mode number			
e_0a/L	1	2	1	2
0	–	–	228.302	562.853
0.1			225.919	556.976
0.2			219.192	540.393
0.3	209.233	515.838	209.204	515.767
0.4	–	–	197.267	486.338
0.5			184.552	454.990

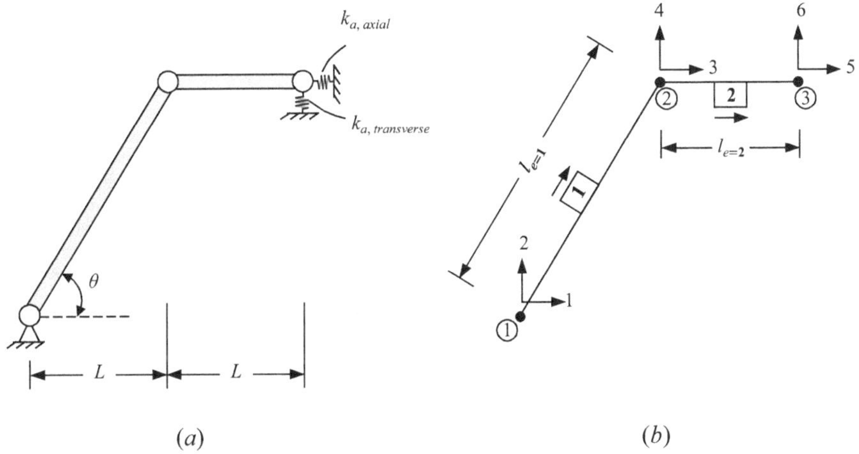

Fig. 8.3 Matrix displacement analysis of a nanotruss with simply and double-deformable supports. (**a**) Physical system. (**b**) FEM model

$$\omega_1 = 209.233\,\text{GHz}, \quad \omega_2 = 515.838\,\text{GHz}.$$

In addition to these results, the natural frequencies of first two modes are calculated through a program code for six different values of nondimensional atomic parameter and listed in Table 8.1. Here it is understood that the nonlocal parameter reduces the frequencies of the nanolattice. Moreover, the nonlocal parameter is more effective in the reducing of frequencies of the second mode. These results reveal that the atomic size-dependency is also important in the nanotrusses, after nanobeams and nanorods.

It should also be stated that member **1** does not affect the mechanical behavior of nanotruss because since both ends are fully clamped, inputs of global matrices addressing this rod are eliminated. To provide diversity in this application, the calculations of this member are also explained.

Example 8.2 Determine the natural frequencies of nonlocal free vibration of nanotruss shown in Figure 8.3a. Nanotruss has simply supported at node (1) and double-deformable supported at node (3). Use a single finite element for all members in the analysis (Fig. 8.3b). Mechanical parameters are as in Example 8.1. Additionally to these, length: $L = l_{e=2} = 20$ nm, diameter of circular cross-section: $d = 1$ nm, stiffnesses of transverse and axial spring attachments: $k_{a,\ \text{axial}} = 10^3$N/m and $k_{a,\ \text{tranverse}} = 10^2$N/m, nonlocal parameter as $e_0a/L = 0.2$, positive-directed angle of member **1**: $\theta = 60°$.

Solution. When the finite element model of physical system is examined, it is understood that the system has six freedoms in total and their number is reduced to four due to the simply support. On the other hand, spring attachments will be added to the reduced global stiffness matrix.

First, let's calculate the stiffness matrices of axial members. Here the length of member **1** is $l_{e=1} = 40$ nm, the cosine and sine of orientation angle are calculated as $\cos\theta_1 = 1/2$ and $\sin\theta_1 = \sqrt{3}/2$. For orientation angles of member **2**, $\cos\theta_2 = 1$ and $\sin\theta_2 = 0$ is obtained. Thus:

$$(K)_1 = \frac{\left(3.142\times10^{-18}\ \text{m}^2\right)\times\left(1\times10^{12}\ \dfrac{\text{N}}{\text{m}^2}\right)}{4\times10^{-8}\ \text{m}}\times \begin{bmatrix} \left(\frac{1}{2}\right)^2 & \left(\frac{1}{2}\right)\times\left(\frac{\sqrt{3}}{2}\right) & -\left(\frac{1}{2}\right)^2 & -\left(\frac{1}{2}\right)\times\left(\frac{\sqrt{3}}{2}\right) \\ \left(\frac{1}{2}\right)\times\left(\frac{\sqrt{3}}{2}\right) & \left(\frac{\sqrt{3}}{2}\right)^2 & -\left(\frac{1}{2}\right)\times\left(\frac{\sqrt{3}}{2}\right) & -\left(\frac{\sqrt{3}}{2}\right)^2 \\ -\left(\frac{1}{2}\right)^2 & -\left(\frac{1}{2}\right)\times\left(\frac{\sqrt{3}}{2}\right) & \left(\frac{1}{2}\right)^2 & \left(\frac{1}{2}\right)\times\left(\frac{\sqrt{3}}{2}\right) \\ -\left(\frac{1}{2}\right)\times\left(\frac{\sqrt{3}}{2}\right) & -\left(\frac{\sqrt{3}}{2}\right)^2 & \left(\frac{1}{2}\right)\times\left(\frac{\sqrt{3}}{2}\right) & \left(\frac{\sqrt{3}}{2}\right)^2 \end{bmatrix}$$

$$= \begin{array}{cccc|c} 1 & 2 & 3 & 4 & \\ \hline 19.638 & 34.013 & -19.638 & -34.013 & 1 \\ 34.013 & 58.913 & -34.013 & -58.913 & 2 \\ -19.638 & -34.013 & 19.638 & 34.013 & 3 \\ -34.013 & -58.913 & 34.013 & 58.913 & 4 \end{array} \tag{8.E8.2.1}$$

$$(K)_2 = \frac{\left(1\times10^{12}\,\frac{\mathrm{N}}{\mathrm{m}^2}\right)\times\left(3.142\times10^{-18}\,\mathrm{m}^2\right)}{2\times10^{-8}\,\mathrm{m}}\times\begin{bmatrix} 1^2 & 1\times0 & -1^2 & -1\times0 \\ 1\times0 & 0^2 & -1\times0 & -0^2 \\ -1^2 & -1\times0 & 1^2 & 1\times0 \\ -1\times0 & -0^2 & 1\times0 & 0^2 \end{bmatrix}$$

$$= \begin{array}{cccc|c} 3 & 4 & 5 & 6 & \\ \hline 157.1 & 0 & -157.1 & 0 & 3 \\ 0 & 0 & 0 & 0 & 4 \\ -157.1 & 0 & 157.1 & 0 & 5 \\ 0 & 0 & 0 & 0 & 6 \end{array} \tag{8.E8.2.2}$$

In this case, the following global matrix is computed by excluding the attachment:

$$\sum[K_{na}] - \begin{bmatrix} 19.638 & 34.013 & -19.638 & -34.013 & 0 & 0 \\ 34.013 & 58.913 & -34.013 & -58.913 & 0 & 0 \\ -19.638 & -34.013 & 176.738 & 34.013 & -157.1 & 0 \\ -34.013 & -58.913 & 34.013 & 58.913 & 0 & 0 \\ 0 & 0 & -157.1 & 0 & 157.1 & 0 \\ 0 & 0 & 0 & 0 & 0 & 0 \end{bmatrix} \tag{8.E8.2.3}$$

On the other hand, the stiffness matrix showing the attachments at node (3) is given as:

$$\sum[K_a] = \begin{bmatrix} 0 & 0 & 0 & 0 & 0 & 0 \\ 0 & 0 & 0 & 0 & 0 & 0 \\ 0 & 0 & 0 & 0 & 0 & 0 \\ 0 & 0 & 0 & 0 & 0 & 0 \\ 0 & 0 & 0 & 0 & 10^2 & 0 \\ 0 & 0 & 0 & 0 & 0 & 10^3 \end{bmatrix} \tag{8.E8.2.4}$$

The global stiffness matrix of system is calculated and reduced as follows:

$$(\sum[K])=(\sum[K_{na}])+(\sum[K_a])=\begin{bmatrix} 19.638 & 34.013 & -19.638 & -34.013 & 0 & 0 \\ 34.013 & 58.913 & -34.013 & -58.913 & 0 & 0 \\ -19.638 & -34.013 & 176.738 & 34.013 & -157.1 & 0 \\ -34.013 & -58.913 & 34.013 & 58.913 & 0 & 0 \\ 0 & 0 & -157.1 & 0 & 257.1 & 0 \\ 0 & 0 & 0 & 0 & 0 & 1000 \end{bmatrix}$$

$$(\sum[K])^* \tag{8.E8.2.5}$$

The sum matrices of axial members should also be calculated. First the nonlocal lengths of members are attained as $(e_0a)_1 = (e_0a)_2 = 0.2 \times L = (0.2) \times (20 \text{ nm}) = 4 \times 10^{-9}$ m. Unlike Example 8.1, calculations of inputs of mass matrix are made according to Eqs. (8.15) and (8.16). According to this, the inputs of mass matrices of members are explained as follows:

$$\begin{aligned} &M_{11,1}=M_{22,1}=M_{33,1}=M_{44,1}=\left(2300\frac{\text{kg}}{\text{m}^3}\right)\times(3.142\times10^{-18}\,\text{m}^2)\times\left(\frac{(4\times10^{-8}\text{m})}{3}+\frac{(4\times10^{-9}\text{m})^2}{(4\times10^{-8}\text{m})}\right)=9.925\times10^{-23}\text{kg},\\ &M_{12,1}=M_{14,1}=M_{23,1}=M_{34,1}=0,\\ &M_{13,1}=M_{24,1}=\left(2300\frac{\text{kg}}{\text{m}^3}\right)\times(3.142\times10^{-18}\,\text{m}^2)\times\left(\frac{(4\times10^{-8}\text{m})}{6}-\frac{(4\times10^{-9}\text{m})^2}{(4\times10^{-8}\text{m})}\right)=4.529\times10^{-23}\text{kg}. \end{aligned} \tag{8.E8.2.6}$$

$$\begin{aligned} &M_{11,2}=M_{22,2}=M_{33,2}=M_{44,2}=\left(2300\frac{\text{kg}}{\text{m}^3}\right)\times(3.142\times10^{-18}\,\text{m}^2)\times\left(\frac{(2\times10^{-8}\text{m})}{3}+\frac{(4\times10^{-9}\text{m})^2}{(2\times10^{-8}\text{m})}\right)=5.396\times10^{-23}\text{kg},\\ &M_{12,2}=M_{14,2}=M_{23,2}=M_{34,2}=0,\\ &M_{13,2}=M_{24,2}=\left(2300\frac{\text{kg}}{\text{m}^3}\right)\times(3.142\times10^{-18}\,\text{m}^2)\times\left(\frac{(2\times10^{-8}\text{m})}{6}-\frac{(4\times10^{-9}\text{m})^2}{(2\times10^{-8}\text{m})}\right)=1.831\times10^{-23}\text{kg}. \end{aligned} \tag{8.E8.2.7}$$

According to the above calculations, the total mass matrices of members can be expressed as follows:

$$(M)_1=\begin{bmatrix} 9.925\times10^{-23} & 0 & 4.529\times10^{-23} & 0 \\ 0 & 9.925\times10^{-22} & 0 & 4.529\times10^{-23} \\ 4.529\times10^{-23} & 0 & 9.925\times10^{-22} & 0 \\ 0 & 4.529\times10^{-23} & 0 & 9.925\times10^{-22} \end{bmatrix} \tag{8.E8.2.8}$$

$$(M)_2 = \begin{bmatrix} 5.396\times10^{-23} & 0 & 1.831\times10^{-23} & 0 \\ 0 & 5.396\times10^{-23} & 0 & 1.831\times10^{-23} \\ 1.831\times10^{-23} & 0 & 5.396\times10^{-23} & 0 \\ 0 & 1.831\times10^{-23} & 0 & 5.396\times10^{-23} \end{bmatrix} \quad (8.E8.2.9)$$

Thus, the reduced spherical mass matrix is given by:

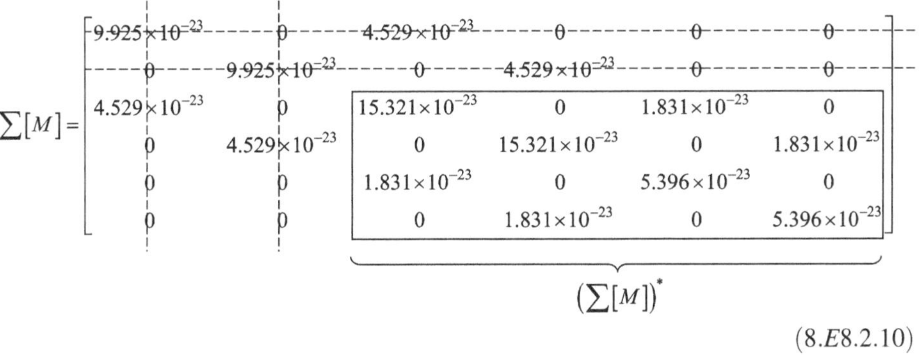

$$\sum[M] = \begin{bmatrix} 9.925\times10^{-23} & 0 & 4.529\times10^{-23} & 0 & 0 & 0 \\ 0 & 9.925\times10^{-23} & 0 & 4.529\times10^{-23} & 0 & 0 \\ 4.529\times10^{-23} & 0 & 15.321\times10^{-23} & 0 & 1.831\times10^{-23} & 0 \\ 0 & 4.529\times10^{-23} & 0 & 15.321\times10^{-23} & 0 & 1.831\times10^{-23} \\ 0 & 0 & 1.831\times10^{-23} & 0 & 5.396\times10^{-23} & 0 \\ 0 & 0 & 0 & 1.831\times10^{-23} & 0 & 5.396\times10^{-23} \end{bmatrix}$$

where the lower-right 4×4 block is $(\sum[M])^*$.

(8.*E*8.2.10)

By using Eqs. (8.E8.2.5) and (8.E8.2.10) into Eq. (8.17), the natural frequencies of nanotruss are reached:

$$\omega_1 = 447.816\,\text{GHz},\quad \omega_2 = 764.472\,\text{GHz},\quad \omega_3 = 2548.087\,\text{GHz},\quad \omega_4 = 4396.846\,\text{GHz}.$$

8.4 Nonlocal Vibration of Nanoframes Without Shear Deformation

Bending members work under the shear force and bending moment in addition to axial forces. Their kinematic relationships may also be affected by shear deformations. The finite element model of a bending member is portrayed in Fig. 8.4. As can be seen, the difference

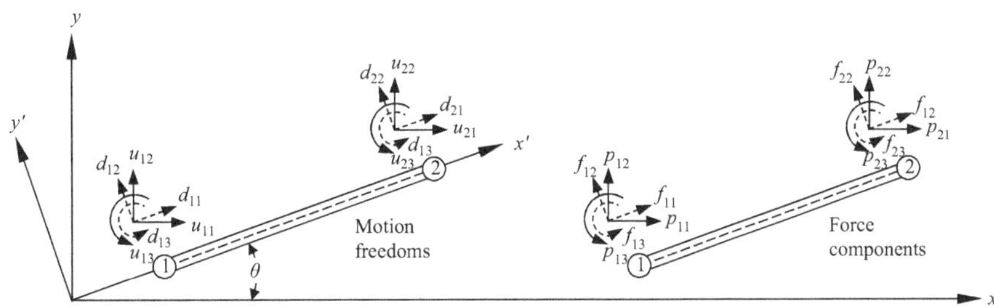

Fig. 8.4 Bending member for NL-FEM analysis of nanoframe (xy: Global and $x'y'$: Local coordinate systems) [1]

of this finite element model according to the axial member is the motion freedom in the direction of rotation and therefore the bending moment is the nodal force.

Starting with this section, the free vibration of nanoframes consisting of bending members is examined within the framework of nonlocal elasticity. First, let's consider the Euler-Bernoulli beam theory-based formulation that neglects shear deformations.

In order to write T expression that performs the transformation operations, in addition to those given in Eq. (8.1), the relations of rotational motions between the global and local coordinates should be constituted. According to this, the following expressions can be written:

$$d_{13} = u_{13}, \quad d_{23} = u_{23} \tag{8.18}$$

where while d_{13} and d_{23} define the rotational freedom in the local coordinates, u_{13} and u_{23} indicate the rotational freedoms in the global coordinates. Thus, the relationship between linear freedoms of motion should be explained for the bending member as follows:

$$\mathbf{d} = T\mathbf{u}: \begin{Bmatrix} d_{11} \\ d_{12} \\ d_{13} \\ d_{21} \\ d_{22} \\ d_{23} \end{Bmatrix} = \begin{bmatrix} \cos\theta & \sin\theta & 0 & 0 & 0 & 0 \\ -\sin\theta & \cos\theta & 0 & 0 & 0 & 0 \\ 0 & 0 & 1 & 0 & 0 & 0 \\ 0 & 0 & 0 & \cos\theta & \sin\theta & 0 \\ 0 & 0 & 0 & -\sin\theta & \cos\theta & 0 \\ 0 & 0 & 0 & 0 & 0 & 1 \end{bmatrix} \begin{Bmatrix} u_{11} \\ u_{12} \\ u_{13} \\ u_{21} \\ u_{22} \\ u_{23} \end{Bmatrix} \tag{8.19}$$

Also, the force components are defined by the following equation:

$$\mathbf{f} = T\mathbf{p}: \begin{Bmatrix} f_{11} \\ f_{12} \\ f_{13} \\ f_{21} \\ f_{22} \\ f_{23} \end{Bmatrix} = \begin{bmatrix} \cos\theta & \sin\theta & 0 & 0 & 0 & 0 \\ -\sin\theta & \cos\theta & 0 & 0 & 0 & 0 \\ 0 & 0 & 1 & 0 & 0 & 0 \\ 0 & 0 & 0 & \cos\theta & \sin\theta & 0 \\ 0 & 0 & 0 & -\sin\theta & \cos\theta & 0 \\ 0 & 0 & 0 & 0 & 0 & 1 \end{bmatrix} \begin{Bmatrix} p_{11} \\ p_{12} \\ p_{13} \\ p_{21} \\ p_{22} \\ p_{23} \end{Bmatrix} \tag{8.20}$$

here T matrix that provides the relationship between displacements is the transformation matrix of a general bending element:

$$T=\begin{bmatrix} \cos\theta & \sin\theta & 0 & 0 & 0 & 0 \\ -\sin\theta & \cos\theta & 0 & 0 & 0 & 0 \\ 0 & 0 & 1 & 0 & 0 & 0 \\ 0 & 0 & 0 & \cos\theta & \sin\theta & 0 \\ 0 & 0 & 0 & -\sin\theta & \cos\theta & 0 \\ 0 & 0 & 0 & 0 & 0 & 1 \end{bmatrix} \tag{8.21}$$

On the other hand, if the freedoms of nanorods are written into the axial freedoms and the own freedoms of nanobeams are written into the inputs of transverse and rotational freedoms of bending freedoms, the stiffness and mass matrices of nanoframe member can be rearranged for the local coordinates:

$$k=\begin{bmatrix} \frac{EA}{L} & 0 & 0 & -\frac{EA}{L} & 0 & 0 \\ 0 & \frac{12EI}{L^3} & \frac{6EI}{L^2} & 0 & -\frac{12EI}{L^3} & \frac{6EI}{L^2} \\ 0 & \frac{6EI}{L^2} & \frac{4EI}{L} & 0 & -\frac{6EI}{L^2} & \frac{2EI}{L} \\ -\frac{EA}{L} & 0 & 0 & \frac{EA}{L} & 0 & 0 \\ 0 & -\frac{12EI}{L^3} & -\frac{6EI}{L^2} & 0 & \frac{12EI}{L^3} & -\frac{6EI}{L^2} \\ 0 & \frac{6EI}{L^2} & \frac{2EI}{L} & 0 & -\frac{6EI}{L^2} & \frac{4EI}{L} \end{bmatrix} \tag{8.22}$$

$$m_c=\frac{\rho AL}{420}\begin{bmatrix} 140 & 0 & 0 & 70 & 0 & 0 \\ 0 & 156 & 22L & 0 & 54 & -13L \\ 0 & 22L & 4L^2 & 0 & 13L & -3L^2 \\ 70 & 0 & 0 & 140 & 0 & 0 \\ 0 & 54 & 13L & 0 & 156 & -22L \\ 0 & -13L & -3L^2 & 0 & -22L & 4L^2 \end{bmatrix} \tag{8.23}$$

$$m_{nl}=(e_0a)^2\rho A\begin{bmatrix} \frac{1}{L} & 0 & 0 & -\frac{1}{L} & 0 & 0 \\ 0 & \frac{6}{5L} & \frac{1}{10} & 0 & -\frac{6}{5L} & \frac{1}{10} \\ 0 & \frac{1}{10} & \frac{2L}{15} & 0 & -\frac{1}{10} & -\frac{L}{30} \\ -\frac{1}{L} & 0 & 0 & \frac{1}{L} & 0 & 0 \\ 0 & -\frac{6}{5L} & -\frac{1}{10} & 0 & \frac{6}{5L} & -\frac{1}{10} \\ 0 & \frac{1}{10} & -\frac{L}{30} & 0 & -\frac{1}{10} & \frac{2L}{15} \end{bmatrix} \tag{8.24}$$

where the calculations in Eqs. (6.10), (6.13), (6.14), (7.10), (7.17), and (7.18) should be investigated. In the matrices given above, L is the length of bending member.

The natural frequencies of nonlocal nanoframe can be calculated through an eigenvalue equation like Eq. (8.17). Of course, the contributions of bending members to the global matrix should be calculated in order to constitute the related global matrices in the eigenvalue problem. For this, the transformation of stiffness and mass matrices of nonlocal bending member is presented with the following equations:

$$K = T^{\mathrm{T}} k T = \begin{bmatrix} k_{11} & k_{12} & k_{13} & k_{14} & k_{15} & k_{16} \\ & k_{22} & k_{23} & k_{24} & k_{25} & k_{26} \\ & & k_{33} & k_{34} & k_{35} & k_{36} \\ & \text{sym.} & & k_{44} & k_{45} & k_{46} \\ & & & & k_{55} & k_{56} \\ & & & & & k_{66} \end{bmatrix} \tag{8.25}$$

$$M = M_c + M_{nl} = T^{\mathrm{T}} m_c T + T^{\mathrm{T}} m_{nl} T = \begin{bmatrix} M_{11} & M_{12} & M_{13} & M_{14} & M_{15} & M_{16} \\ & M_{22} & M_{23} & M_{24} & M_{25} & M_{26} \\ & & M_{33} & M_{34} & M_{35} & M_{36} \\ & \text{sym.} & & M_{44} & M_{45} & M_{46} \\ & & & & M_{55} & M_{56} \\ & & & & & M_{66} \end{bmatrix} \tag{8.26}$$

The inputs of stiffness matrix are calculated as follows [1, 2]:

$$\begin{gathered} k_{11}=k_{44}=\frac{EA}{L}\cos^2\theta+\frac{12EI}{L^3}\sin^2\theta,\ k_{12}=k_{45}=\left(\frac{EA}{L}-\frac{12EI}{L^3}\right)\cos\theta\sin\theta,\ k_{13}=k_{16}=-\frac{6EI}{L^2}\sin\theta, \\ k_{14}=-\frac{EA}{L}\cos^2\theta-\frac{12EI}{L^3}\sin^2\theta,\ k_{15}=k_{24}=\left(-\frac{EA}{L}+\frac{12EI}{L^3}\right)\cos\theta\sin\theta, \\ k_{22}=k_{55}=\frac{EA}{L}\sin^2\theta+\frac{12EI}{L^3}\cos^2\theta,\ k_{23}=k_{26}=\frac{6EI}{L^2}\cos\theta,\ k_{25}=-\frac{EA}{L}\sin^2\theta-\frac{12EI}{L^3}\cos^2\theta, \\ k_{33}=k_{66}=\frac{4EI}{L},\ k_{34}=k_{46}=\frac{6EI}{L^2}\sin\theta,\ k_{35}=k_{56}=-\frac{6EI}{L^2}\cos\theta,\ k_{36}=\frac{2EI}{L}. \end{gathered} \tag{8.27}$$

The entries of mass matrix are also achieved as [1]:

$$M_{11}=M_{44}=\left(\frac{\rho AL}{3}+\frac{(e_0a)^2\rho A}{L}\right)\cos^2\theta+\left(\frac{13\rho AL}{35}+\frac{6(e_0a)^2\rho A}{5L}\right)\sin^2\theta,$$

$$M_{12} = M_{45} = \left(-\frac{4\rho AL}{105} - \frac{(e_0a)^2\rho A}{5L}\right)\cos\theta\sin\theta, \quad M_{13} = \left(-\frac{11\rho AL^2}{210} - \frac{(e_0a)^2\rho A}{10}\right)\sin\theta,$$

$$M_{14} = \left(\frac{\rho AL}{6} - \frac{(e_0a)^2\rho A}{L}\right)\cos^2\theta + \left(\frac{9\rho AL}{70} - \frac{6(e_0a)^2\rho A}{5L}\right)\sin^2\theta,$$

$$M_{15} = M_{24} = \left(\frac{4\rho AL}{105} + \frac{(e_0a)^2\rho A}{5L}\right)\cos\theta\sin\theta, \quad M_{16} = \left(\frac{13\rho AL^2}{420} - \frac{(e_0a)^2\rho A}{10}\right)\sin\theta,$$

$$M_{22} = M_{55} = \left(\frac{\rho AL}{3} + \frac{6(e_0a)^2\rho A}{5L}\right)\cos^2\theta + \left(\frac{13\rho AL}{35} + \frac{(e_0a)^2\rho A}{L}\right)\sin^2\theta,$$

$$M_{23} = \left(\frac{11\rho AL^2}{210} + \frac{(e_0a)^2\rho A}{10}\right)\cos\theta,$$

$$M_{25} = \left(\frac{9\rho AL}{70} - \frac{6(e_0a)^2\rho A}{5L}\right)\cos^2\theta + \left(\frac{\rho AL}{6} - \frac{(e_0a)^2\rho A}{L}\right)\sin^2\theta,$$

$$M_{26} = \left(-\frac{13\rho AL^2}{420} + \frac{(e_0a)^2\rho A}{10}\right)\cos\theta, \quad M_{33} = M_{66} = \frac{\rho AL^3}{105} + \frac{2(e_0a)^2\rho AL}{15},$$

$$M_{34} = \left(-\frac{13\rho AL^2}{420} + \frac{(e_0a)^2\rho A}{10}\right)\sin\theta, \quad M_{35} = \left(\frac{13\rho AL^2}{420} - \frac{(e_0a)^2\rho A}{10}\right)\cos\theta,$$

$$M_{36} = -\frac{\rho AL^3}{140} - \frac{(e_0a)^2\rho AL}{30}, \quad M_{46} = \left(\frac{11\rho AL^2}{210} + \frac{(e_0a)^2\rho A}{10}\right)\sin\theta,$$

$$M_{46} = \left(\frac{11\rho AL^2}{210} + \frac{(e_0a)^2\rho A}{10}\right)\sin\theta, \quad M_{56} = \left(-\frac{11\rho AL^2}{210} - \frac{(e_0a)^2\rho A}{10}\right)\cos\theta \tag{8.28}$$

Finally, the eigenvalue equation that gives the natural frequencies of nanoframe without shear deformation is reminded below:

$$\det\left[\left(\sum_{e=1}^{n}[K]_e\right)^* - \omega_i^2\left(\sum_{e=1}^{n}[M]_e\right)^*\right] = 0 \tag{8.29}$$

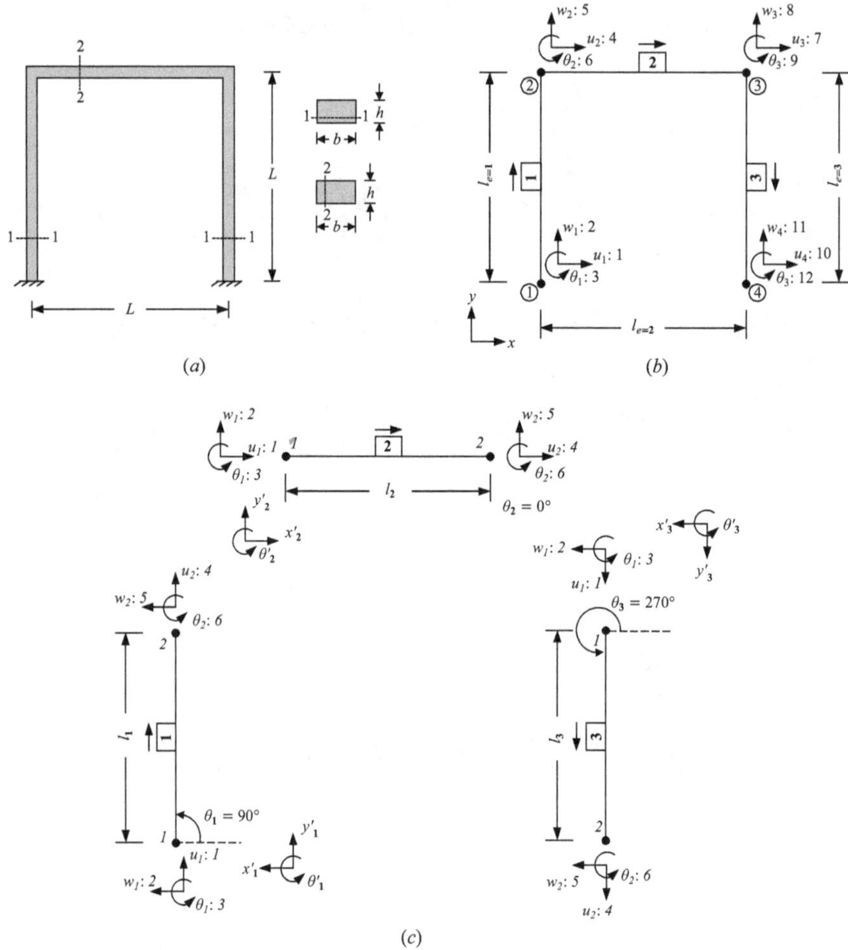

Fig. 8.5 Matrix displacement analysis of a nanoframe with three members. (**a**) Physical system. (**b**) FEM model. (**c**) Bending members in their local coordinates

Example 8.3 Determine the natural frequencies of nonlocal free vibration of nanoframe shown in Fig. 8.5a. Neglect the shear deformation. Use a single finite element for all members in the analysis (Fig. 8.5b). Length parameter of nanoframe: $L = l_{e=1} = l_{e=2} = l_{e=3} = 50$ nm, height and thickness of rectangular cross-section: $h = 1$ nm and $b = 2$ nm, modulus of elasticity: $E = 1$ TPa, and mass of unit volume: $\rho = 2300\text{kg/m}^3$, nonlocal parameter: $e_0a/L = 0.3$.

Solution. The finite element mesh of nanoframe is illustrated in Fig. 8.5b. The freedoms of members of nanoframe according to the local coordinates are as seen in Fig. 8.5c. For matrix displacement analysis, the mechanical and physical parameters of members are written below:

$$\begin{aligned}
&\text{Length}: l_1 = l_2 = l_3 = l = 5\times 10^{-8}\,\text{m},\\
&\text{Nonlocal parameter}: (e_0a)_1 = (e_0a)_2 = (e_0a)_3 = e_0a = 1.5\times 10^{-8}\,\text{m},\\
&\text{Cross-section area}: A_1 = A_2 = A_3 = A = b\times h = (2\,\text{nm})\times(1\,\text{nm}) = 2\times 10^{-18}\,\text{m}^2,\\
&\text{Moment of inertia for in-plane bending}: I_1 = I_2 = I_3 = I = \frac{b\times h^3}{12}\\
&\qquad = \frac{(2\text{nm})\times(1\,\text{nm})^3}{12} = 1.667\times 10^{-37}\,\text{m}^4,\\
&\text{Modulus of elasticity}: E_1 = E_2 = E_3 = E = 1\times 10^{12}\,\frac{\text{N}}{\text{m}^2},\\
&\text{Mass of unit volume}: \rho_1 = \rho_2 = \rho_3 = \rho = 2300\,\frac{\text{kg}}{\text{m}^3}.
\end{aligned} \tag{8.E8.3.1}$$

First, the orientation angles of members and their trigonometric values are presented as follows:

$$\begin{aligned}
&\theta_1 = \frac{\pi}{2} \to \cos\theta_1 = 0, \quad \sin\theta_1 = 1,\\
&\theta_2 = 0^\circ \to \cos\theta_2 = 1, \quad \sin\theta_2 = 0,\\
&\theta_3 = \frac{3\pi}{2} \to \cos\theta_3 = 0, \quad \sin\theta_3 = -1.
\end{aligned} \tag{8.E8.3.2}$$

Within the scope of this application, the global stiffness and mass matrices of members are calculated as introduced in Eqs. (8.25) and (8.26). In these calculations, the local and global coordinates of members are also shown. It is preferred that these representations are not given in the mass matrices. Additionally, the following stiffness parameters are given to prevent unnecessary details in the calculations:

$$\begin{aligned}
&\alpha = \frac{E}{l} = \frac{1\times 10^{12}\text{N/m}^2}{5\times 10^{-8}\,\text{m}} = 2\times 10^{19}\text{N/m}^3,\\
&\alpha A = \left(2\times 10^{19}\text{N/m}^3\right)\times\left(2\times 10^{-18}\,\text{m}^2\right) = 40\text{N/m},\\
&\alpha I = \left(2\times 10^{19}\text{N/m}^3\right)\times\left(1.667\times 10^{-37}\,\text{m}^4\right) = 3.334\times 10^{-18}\,\text{Nm},\\
&\alpha\frac{I}{l} = \left(2\times 10^{19}\text{N/m}^3\right)\times\frac{\left(1.667\times 10^{-37}\,\text{m}^4\right)}{\left(5\times 10^{-8}\,\text{m}\right)} = 6.668\times 10^{-11}\,\text{N},\\
&\alpha\frac{I}{l^2} = \left(2\times 10^{19}\text{N/m}^3\right)\times\frac{\left(1.667\times 10^{-37}\,\text{m}^4\right)}{\left(5\times 10^{-8}\,\text{m}\right)^2} = 1.334\times 10^{-3}\,\text{N/m}
\end{aligned} \tag{8.E8.3.3}$$

$$
(K)_1 = \left(T^{\mathrm{T}}\right)_1 (K)_1 (T)_1 = \begin{bmatrix} 0 & -1 & 0 & 0 & 0 & 0 \\ 1 & 0 & 0 & 0 & 0 & 0 \\ 0 & 0 & 1 & 0 & 0 & 0 \\ 0 & 0 & 0 & 0 & -1 & 0 \\ 0 & 0 & 0 & 1 & 0 & 0 \\ 0 & 0 & 0 & 0 & 0 & 1 \end{bmatrix} \times \begin{bmatrix} 40 & 0 & 0 & -40 & 0 & 0 \\ 0 & 1.6\times10^{-2} & 4.001\times10^{-9} & 0 & -1.6\times10^{-2} & 4.001\times10^{-10} \\ 0 & 4.001\times10^{-10} & 1.334\times10^{-17} & 0 & -4.001\times10^{-10} & 6.668\times10^{-18} \\ -40 & 0 & 0 & 40 & 0 & 0 \\ 0 & -1.6\times10^{-2} & -4.001\times10^{-10} & 0 & 1.6\times10^{-2} & -4.001\times10^{-10} \\ 0 & 4.001\times10^{-10} & 6.668\times10^{-18} & 0 & -4.001\times10^{-9} & 1.334\times10^{-17} \end{bmatrix} \times \begin{bmatrix} 0 & 1 & 0 & 0 & 0 & 0 \\ -1 & 0 & 0 & 0 & 0 & 0 \\ 0 & 0 & 1 & 0 & 0 & 0 \\ 0 & 0 & 0 & 0 & 1 & 0 \\ 0 & 0 & 0 & -1 & 0 & 0 \\ 0 & 0 & 0 & 0 & 0 & 1 \end{bmatrix}
$$

$$
= \begin{array}{cccccc|c} \mathit{1} & \mathit{2} & \mathit{3} & \mathit{4} & \mathit{5} & \mathit{6} & \\ \downarrow & \downarrow & \downarrow & \downarrow & \downarrow & \downarrow & \\ 1 & 2 & 3 & 4 & 5 & 6 & \\ \hline 1.6\times10^{-2} & 0 & -4.001\times10^{-10} & -1.6\times10^{-2} & 0 & -4.001\times10^{-9} & 1 \leftarrow \mathit{1} \\ 0 & 40 & 0 & 0 & -40 & 0 & 2 \leftarrow \mathit{2} \\ -4.001\times10^{-10} & 0 & 1.334\times10^{-17} & 4.001\times10^{-10} & 0 & 6.668\times10^{-18} & 3 \leftarrow \mathit{3} \\ -1.6\times10^{-2} & 0 & 4.001\times10^{-10} & 1.6\times10^{-2} & 0 & 4.001\times10^{-10} & 4 \leftarrow \mathit{4} \\ 0 & -40 & 0 & 0 & 40 & 0 & 5 \leftarrow \mathit{5} \\ -4.001\times10^{-10} & 0 & 6.668\times10^{-17} & 4.001\times10^{-10} & 0 & 1.334\times10^{-17} & 6 \leftarrow \mathit{6} \end{array} \tag{8.E8.3.4}
$$

$$
(K)_2 = \left(T^{\mathrm{T}}\right)_2 (K)_2 (T)_2 = \begin{bmatrix} 1 & 0 & 0 & 0 & 0 & 0 \\ 0 & 1 & 0 & 0 & 0 & 0 \\ 0 & 0 & 1 & 0 & 0 & 0 \\ 0 & 0 & 0 & 1 & 0 & 0 \\ 0 & 0 & 0 & 0 & 1 & 0 \\ 0 & 0 & 0 & 0 & 0 & 1 \end{bmatrix} \times \begin{bmatrix} 40 & 0 & 0 & -40 & 0 & 0 \\ 0 & 1.6\times10^{-2} & 4.001\times10^{-9} & 0 & -1.6\times10^{-2} & 4.001\times10^{-10} \\ 0 & 4.001\times10^{-10} & 1.334\times10^{-17} & 0 & -4.001\times10^{-10} & 6.668\times10^{-18} \\ -40 & 0 & 0 & 40 & 0 & 0 \\ 0 & -1.6\times10^{-2} & -4.001\times10^{-10} & 0 & 1.6\times10^{-2} & -4.001\times10^{-10} \\ 0 & 4.001\times10^{-10} & 6.668\times10^{-18} & 0 & -4.001\times10^{-9} & 1.334\times10^{-17} \end{bmatrix} \times \begin{bmatrix} 1 & 0 & 0 & 0 & 0 & 0 \\ 0 & 1 & 0 & 0 & 0 & 0 \\ 0 & 0 & 1 & 0 & 0 & 0 \\ 0 & 0 & 0 & 1 & 0 & 0 \\ 0 & 0 & 0 & 0 & 1 & 0 \\ 0 & 0 & 0 & 0 & 0 & 1 \end{bmatrix}
$$

$$
= \begin{array}{c} \begin{array}{cccccc} 1 \to 4 & 2 \to 5 & 3 \to 6 & 4 \to 7 & 5 \to 8 & 6 \to 9 \end{array} \\ \begin{bmatrix} 40 & 0 & 0 & -40 & 0 & 0 \\ 0 & 1.6\times10^{-2} & 4.001\times10^{-10} & 0 & -1.6\times10^{-2} & 4.001\times10^{-10} \\ 0 & 4.001\times10^{-10} & 1.334\times10^{-17} & 0 & -4.001\times10^{-10} & 6.668\times10^{-18} \\ -40 & 0 & 0 & 40 & 0 & 0 \\ 0 & -1.6\times10^{-2} & -4.001\times10^{-10} & 0 & 1.6\times10^{-2} & -4.001\times10^{-10} \\ 0 & 4.001\times10^{-10} & 6.668\times10^{-18} & 0 & -4.001\times10^{-10} & 1.334\times10^{-17} \end{bmatrix} \end{array} \begin{array}{l} \\ 4 \leftarrow 1 \\ 5 \leftarrow 2 \\ 6 \leftarrow 3 \\ 7 \leftarrow 4 \\ 8 \leftarrow 5 \\ 9 \leftarrow 6 \end{array} \tag{8.E8.3.5}
$$

$$(K)_3=\left(T^{\mathrm{T}}\right)_3(K)_3(T)_3=\begin{bmatrix}0&1&0&0&0&0\\-1&0&0&0&0&0\\0&0&1&0&0&0\\0&0&0&0&1&0\\0&0&0&-1&0&0\\0&0&0&0&0&1\end{bmatrix}\times\begin{bmatrix}40&0&0&-40&0&0\\0&1.6\times10^{-2}&4.001\times10^{-9}&0&-1.6\times10^{-2}&4.001\times10^{-10}\\0&4.001\times10^{-10}&1.334\times10^{-17}&0&-4.001\times10^{-10}&6.668\times10^{-18}\\-40&0&0&40&0&0\\0&-1.6\times10^{-2}&-4.001\times10^{-10}&0&1.6\times10^{-2}&-4.001\times10^{-10}\\0&4.001\times10^{-10}&6.668\times10^{-18}&0&-4.001\times10^{-9}&1.334\times10^{-17}\end{bmatrix}\times\begin{bmatrix}0&-1&0&0&0&0\\1&0&0&0&0&0\\0&0&1&0&0&0\\0&0&0&0&-1&0\\0&0&0&1&0&0\\0&0&0&0&0&1\end{bmatrix}$$

$$=\begin{array}{c}\begin{array}{cccccc}1&2&3&4&5&6\\\downarrow&\downarrow&\downarrow&\downarrow&\downarrow&\downarrow\\7&8&9&10&11&12\end{array}\\\left[\begin{array}{cccccc}1.6\times10^{-2}&0&4.001\times10^{-10}&-1.6\times10^{-2}&0&4.001\times10^{-9}\\0&40&0&0&-40&0\\4.001\times10^{-10}&0&1.334\times10^{-17}&-4.001\times10^{-10}&0&6.668\times10^{-18}\\-1.6\times10^{-2}&0&-4.001\times10^{-10}&1.6\times10^{-2}&0&-4.001\times10^{-10}\\0&-40&0&0&40&0\\4.001\times10^{-10}&0&6.668\times10^{-17}&-4.001\times10^{-10}&0&1.334\times10^{-17}\end{array}\right]\end{array}\begin{array}{c}\\\\\\7\leftarrow 1\\8\leftarrow 2\\9\leftarrow 3\\10\leftarrow 4\\11\leftarrow 5\\12\leftarrow 6\end{array}\tag{8.E8.3.6}$$

Classical mass matrices of nanoframe members are calculated. For this, the following mass parameters should be calculated:

$$\beta=\frac{\rho Al}{420}=\frac{(2300\,\mathrm{kg/m^3})\times\left(2\times10^{-18}\,\mathrm{m^2}\right)\times\left(5\times10^{-8}\,\mathrm{m}\right)}{420}=5.476\times10^{-25}\,\mathrm{kg},$$

$$140\beta=140\times\left(5.476\times10^{-25}\,\mathrm{kg}\right)=7.667\times10^{-23}\,\mathrm{kg},$$

$$70\beta=70\times\left(5.476\times10^{-25}\,\mathrm{kg}\right)=3.833\times10^{-23}\,\mathrm{kg},$$

$$156\beta=156\times\left(5.476\times10^{-25}\,\mathrm{kg}\right)=8.543\times10^{-23}\,\mathrm{kg},$$

$$22l\beta=22\times\left(5\times10^{-8}\,\mathrm{m}\right)\times\left(5.476\times10^{-25}\,\mathrm{kg}\right)=6.024\times10^{-31}\,\mathrm{kgm},$$

$$54\beta=54\times\left(5.476\times10^{-25}\,\mathrm{kg}\right)=2.957\times10^{-23}\,\mathrm{kg},$$

$$13l\beta = 13 \times \left(5 \times 10^{-8}\,\text{m}\right) \times \left(5.476 \times 10^{-25}\,\text{kg}\right) = 3.56 \times 10^{-31}\,\text{kgm},$$

$$4l^2\beta = 4 \times \left(5 \times 10^{-8}\,\text{m}\right)^2 \times \left(5.476 \times 10^{-25}\,\text{kg}\right) = 5.476 \times 10^{-39}\,\text{kgm}^2,$$

$$3l^2\beta = 3 \times \left(5 \times 10^{-8}\,\text{m}\right)^2 \times \left(5.476 \times 10^{-25}\,\text{kg}\right) = 4.107 \times 10^{-39}\,\text{kgm}^2. \qquad (8.\text{E}8.3.7)$$

The classical mass matrices of members on the global coordinates are reached as follows:

$$(M_c)_1 = \left(T^{\text{T}}\right)_1 (M_c)_1 (T)_1 =$$

$$\begin{bmatrix} 0 & -1 & 0 & 0 & 0 & 0 \\ 1 & 0 & 0 & 0 & 0 & 0 \\ 0 & 0 & 1 & 0 & 0 & 0 \\ 0 & 0 & 0 & 0 & -1 & 0 \\ 0 & 0 & 0 & 1 & 0 & 0 \\ 0 & 0 & 0 & 0 & 0 & 1 \end{bmatrix} \times \begin{bmatrix} 7.667\times10^{-23} & 0 & 0 & 3.833\times10^{-23} & 0 & 0 \\ 0 & 8.543\times10^{-23} & 6.024\times10^{-31} & 0 & 2.957\times10^{-23} & -3.56\times10^{-31} \\ 0 & 6.024\times10^{-31} & 5.476\times10^{-39} & 0 & 3.56\times10^{-31} & -4.107\times10^{-39} \\ 3.833\times10^{-23} & 0 & 0 & 7.667\times10^{-23} & 0 & 0 \\ 0 & 2.957\times10^{-23} & 3.56\times10^{-31} & 0 & 8.543\times10^{-23} & -6.024\times10^{-31} \\ 0 & -3.56\times10^{-31} & -4.107\times10^{-39} & 0 & -6.024\times10^{-31} & 5.476\times10^{-39} \end{bmatrix} \times \begin{bmatrix} 0 & 1 & 0 & 0 & 0 & 0 \\ -1 & 0 & 0 & 0 & 0 & 0 \\ 0 & 0 & 1 & 0 & 0 & 0 \\ 0 & 0 & 0 & 0 & 1 & 0 \\ 0 & 0 & 0 & -1 & 0 & 0 \\ 0 & 0 & 0 & 0 & 0 & 1 \end{bmatrix}$$

$$= \begin{bmatrix} 8.543\times10^{-23} & 0 & -6.024\times10^{-31} & 2.957\times10^{-23} & 0 & 3.56\times10^{-31} \\ 0 & 7.667\times10^{-23} & 0 & 0 & 3.833\times10^{-23} & 0 \\ -6.024\times10^{-31} & 0 & 5.476\times10^{-39} & -3.56\times10^{-31} & 0 & -4.107\times10^{-39} \\ 2.957\times10^{-23} & 0 & -3.56\times10^{-31} & 8.543\times10^{-23} & 0 & 6.024\times10^{-31} \\ 0 & 3.833\times10^{-23} & 0 & 0 & 7.667\times10^{-23} & 0 \\ 3.56\times10^{-31} & 0 & -4.107\times10^{-39} & 6.024\times10^{-31} & 0 & 5.476\times10^{-39} \end{bmatrix} \qquad (8.\text{E}8.3.8)$$

$$
\begin{aligned}
&(M_c)_2=(T^{\mathrm{T}})_2(M_c)_2(T)_2= \\
&\begin{bmatrix} 1\,0\,0\,0\,0\,0 \\ 0\,1\,0\,0\,0\,0 \\ 0\,0\,1\,0\,0\,0 \\ 0\,0\,0\,1\,0\,0 \\ 0\,0\,0\,0\,1\,0 \\ 0\,0\,0\,0\,0\,1 \end{bmatrix} \times \begin{bmatrix} 7.667\times10^{-23} & 0 & 0 & 3.833\times10^{-23} & 0 & 0 \\ 0 & 8.543\times10^{-23} & 6.024\times10^{-31} & 0 & 2.957\times10^{-23} & -3.56\times10^{-31} \\ 0 & 6.024\times10^{-31} & 5.476\times10^{-39} & 0 & 3.56\times10^{-31} & -4.107\times10^{-39} \\ 3.833\times10^{-23} & 0 & 0 & 7.667\times10^{-23} & 0 & 0 \\ 0 & 2.957\times10^{-23} & 3.56\times10^{-31} & 0 & 8.543\times10^{-23} & -6.024\times10^{-31} \\ 0 & -3.56\times10^{-31} & -4.107\times10^{-39} & 0 & -6.024\times10^{-31} & 5.476\times10^{-39} \end{bmatrix} \times \begin{bmatrix} 1\,0\,0\,0\,0\,0 \\ 0\,1\,0\,0\,0\,0 \\ 0\,0\,1\,0\,0\,0 \\ 0\,0\,0\,1\,0\,0 \\ 0\,0\,0\,0\,1\,0 \\ 0\,0\,0\,0\,0\,1 \end{bmatrix} \\
&= \begin{bmatrix} 7.667\times10^{-23} & 0 & 0 & 3.833\times10^{-23} & 0 & 0 \\ 0 & 8.543\times10^{-23} & 6.024\times10^{-31} & 0 & 2.957\times10^{-23} & -3.56\times10^{-31} \\ 0 & 6.024\times10^{-31} & 5.476\times10^{-39} & 0 & 3.56\times10^{-31} & -4.107\times10^{-39} \\ 3.833\times10^{-23} & 0 & 0 & 7.667\times10^{-23} & 0 & 0 \\ 0 & 2.957\times10^{-23} & 3.56\times10^{-31} & 0 & 8.543\times10^{-23} & -6.024\times10^{-31} \\ 0 & -3.56\times10^{-31} & -4.107\times10^{-39} & 0 & -6.024\times10^{-31} & 5.476\times10^{-39} \end{bmatrix}
\end{aligned}
\tag{8.E8.3.9}
$$

$$(M_c)_3 = (T^{\mathrm{T}})_3 (M_c)_3 (T)_3 =$$

$$\begin{bmatrix} 0 & 1 & 0 & 0 & 0 & 0 \\ -1 & 0 & 0 & 0 & 0 & 0 \\ 0 & 0 & 1 & 0 & 0 & 0 \\ 0 & 0 & 0 & 0 & 1 & 0 \\ 0 & 0 & 0 & -1 & 0 & 0 \\ 0 & 0 & 0 & 0 & 0 & 1 \end{bmatrix} \times \begin{bmatrix} 7.667\times10^{-23} & 0 & 0 & 3.833\times10^{-23} & 0 & 0 \\ 0 & 8.543\times10^{-23} & 6.024\times10^{-31} & 0 & 2.957\times10^{-23} & -3.56\times10^{-31} \\ 0 & 6.024\times10^{-31} & 5.476\times10^{-39} & 0 & 3.56\times10^{-31} & -4.107\times10^{-39} \\ 3.833\times10^{-23} & 0 & 0 & 7.667\times10^{-23} & 0 & 0 \\ 0 & 2.957\times10^{-23} & 3.56\times10^{-31} & 0 & 8.543\times10^{-23} & -6.024\times10^{-31} \\ 0 & -3.56\times10^{-31} & -4.107\times10^{-39} & 0 & -6.024\times10^{-31} & 5.476\times10^{-39} \end{bmatrix} \times \begin{bmatrix} 0 & -1 & 0 & 0 & 0 & 0 \\ 1 & 0 & 0 & 0 & 0 & 0 \\ 0 & 0 & 1 & 0 & 0 & 0 \\ 0 & 0 & 0 & 0 & -1 & 0 \\ 0 & 0 & 0 & 1 & 0 & 0 \\ 0 & 0 & 0 & 0 & 0 & 1 \end{bmatrix}$$

$$= \begin{bmatrix} 8.543\times10^{-23} & 0 & 6.024\times10^{-31} & 2.957\times10^{-23} & 0 & -3.56\times10^{-31} \\ 0 & 7.667\times10^{-23} & 0 & 0 & 3.833\times10^{-23} & 0 \\ 6.024\times10^{-31} & 0 & 5.476\times10^{-39} & 3.56\times10^{-31} & 0 & -4.107\times10^{-39} \\ 2.957\times10^{-23} & 0 & 3.56\times10^{-31} & 8.543\times10^{-23} & 0 & -6.024\times10^{-31} \\ 0 & 3.833\times10^{-23} & 0 & 0 & 7.667\times10^{-23} & 0 \\ -3.56\times10^{-31} & 0 & -4.107\times10^{-39} & 6.024\times10^{-31} & 0 & 5.476\times10^{-39} \end{bmatrix} \quad (8.E8.3.10)$$

On the other hand, to obtain the nonlocal mass matrices of members, the following mass parameters should be known:

$$\gamma = (e_0 a)^2 \rho A = \left(1.5\times10^{-8}\,\mathrm{m}\right)^2 \times \left(2300\,\mathrm{kg/m^3}\right) \times \left(2\times10^{-18}\,\mathrm{m^2}\right) = 1.035\times10^{-30}\,\mathrm{kgm},$$

$$\frac{\gamma}{l} = \frac{1.035\times10^{-30}\,\mathrm{kgm}}{5\times10^{-8}\,\mathrm{m}} = 2.07\times10^{-23}\,\mathrm{kg},$$

$$\frac{6\gamma}{5l} = \frac{6\times\left(1.035\times10^{-30}\,\mathrm{kgm}\right)}{5\times\left(5\times10^{-8}\,\mathrm{m}\right)} = 2.484\times10^{-23}\,\mathrm{kg},$$

$$\frac{\gamma}{10}=\frac{1.035\times10^{-30}\,\mathrm{kgm}}{10}=1.035\times10^{-31}\,\mathrm{kgm},$$

$$\frac{2\gamma l}{15}=\frac{2\times\left(1.035\times10^{-30}\,\mathrm{kgm}\right)\times\left(5\times10^{-8}\,\mathrm{m}\right)}{15}=6.9\times10^{-39}\,\mathrm{kgm}^2,$$

$$\frac{\gamma l}{30}=\frac{\left(1.035\times10^{-30}\,\mathrm{kgm}\right)\times\left(5\times10^{-8}\,\mathrm{m}\right)}{30}=1.725\times10^{-39}\,\mathrm{kgm}^2. \tag{8.E8.3.11}$$

Thus, the nonlocal mass matrices for the global coordinates are computed as follows:

$$(M_{nl})_1=\left(T^{\mathrm{T}}\right)_1(M_{nl})_1(T)_1=$$

$$\begin{bmatrix} 0 & -1 & 0 & 0 & 0 & 0\\ 1 & 0 & 0 & 0 & 0 & 0\\ 0 & 0 & 1 & 0 & 0 & 0\\ 0 & 0 & 0 & 0 & -1 & 0\\ 0 & 0 & 0 & 1 & 0 & 0\\ 0 & 0 & 0 & 0 & 0 & 1 \end{bmatrix}\times\begin{bmatrix} 2.07\times10^{-23} & 0 & 0 & -2.07\times10^{-23} & 0 & 0\\ 0 & 2.484\times10^{-23} & 1.035\times10^{-31} & 0 & -2.484\times10^{-23} & 1.035\times10^{-31}\\ 0 & 1.035\times10^{-31} & 6.9\times10^{-39} & 0 & -1.035\times10^{-31} & -1.725\times10^{-39}\\ -2.07\times10^{-23} & 0 & 0 & 2.07\times10^{-23} & 0 & 0\\ 0 & -2.484\times10^{-23} & -1.035\times10^{-31} & 0 & 2.484\times10^{-23} & -1.035\times10^{-31}\\ 0 & 1.035\times10^{-31} & -1.725\times10^{-39} & 0 & -1.035\times10^{-31} & 6.9\times10^{-39} \end{bmatrix}\times\begin{bmatrix} 0 & 1 & 0 & 0 & 0 & 0\\ -1 & 0 & 0 & 0 & 0 & 0\\ 0 & 0 & 1 & 0 & 0 & 0\\ 0 & 0 & 0 & 0 & 1 & 0\\ 0 & 0 & 0 & -1 & 0 & 0\\ 0 & 0 & 0 & 0 & 0 & 1 \end{bmatrix}$$

$$=\begin{bmatrix} 2.484\times10^{-23} & 0 & -1.035\times10^{-31} & -2.484\times10^{-23} & 0 & -1.035\times10^{-31}\\ 0 & 2.07\times10^{-23} & 0 & 0 & -2.07\times10^{-23} & 0\\ -1.035\times10^{-31} & 0 & 6.9\times10^{-39} & 1.035\times10^{-31} & 0 & -1.725\times10^{-39}\\ -2.484\times10^{-23} & 0 & 1.035\times10^{-31} & 2.484\times10^{-23} & 0 & 1.035\times10^{-31}\\ 0 & -2.07\times10^{-23} & 0 & 0 & 2.07\times10^{-23} & 0\\ -1.035\times10^{-31} & 0 & -1.725\times10^{-39} & 1.035\times10^{-31} & 0 & 6.9\times10^{-39} \end{bmatrix} \tag{8.E8.3.12}$$

$$(M_{nl})_2=\left(T^{\mathrm{T}}\right)_2(M_{nl})_2(T)_2=$$

$$\begin{bmatrix} 1&0&0&0&0&0\\ 0&1&0&0&0&0\\ 0&0&1&0&0&0\\ 0&0&0&1&0&0\\ 0&0&0&0&1&0\\ 0&0&0&0&0&1 \end{bmatrix} \times \begin{bmatrix} 2.07\times10^{-23} & 0 & 0 & -2.07\times10^{-23} & 0 & 0\\ 0 & 2.484\times10^{-23} & 1.035\times10^{-31} & 0 & -2.484\times10^{-23} & 1.035\times10^{-31}\\ 0 & 1.035\times10^{-31} & 6.9\times10^{-39} & 0 & -1.035\times10^{-31} & -1.725\times10^{-39}\\ -2.07\times10^{-23} & 0 & 0 & 2.07\times10^{-23} & 0 & 0\\ 0 & -2.484\times10^{-23} & -1.035\times10^{-31} & 0 & 2.484\times10^{-23} & -1.035\times10^{-31}\\ 0 & 1.035\times10^{-31} & -1.725\times10^{-39} & 0 & -1.035\times10^{-31} & 6.9\times10^{-39} \end{bmatrix} \times \begin{bmatrix} 1&0&0&0&0&0\\ 0&1&0&0&0&0\\ 0&0&1&0&0&0\\ 0&0&0&1&0&0\\ 0&0&0&0&1&0\\ 0&0&0&0&0&1 \end{bmatrix}$$

$$= \begin{bmatrix} 2.07\times10^{-23} & 0 & 0 & -2.07\times10^{-23} & 0 & 0\\ 0 & 2.484\times10^{-23} & 1.035\times10^{-31} & 0 & -2.484\times10^{-23} & 1.035\times10^{-31}\\ 0 & 1.035\times10^{-31} & 6.9\times10^{-39} & 0 & -1.035\times10^{-31} & -1.725\times10^{-39}\\ -2.07\times10^{-23} & 0 & 0 & 2.07\times10^{-23} & 0 & 0\\ 0 & -2.484\times10^{-23} & -1.035\times10^{-31} & 0 & 2.484\times10^{-23} & -1.035\times10^{-31}\\ 0 & 1.035\times10^{-31} & -1.725\times10^{-39} & 0 & -1.035\times10^{-31} & 6.9\times10^{-39} \end{bmatrix} \tag{8.E8.3.13}$$

$$(M_{nl})_3 = (T^{\mathrm{T}})_3 (M_{nl})_3 (T)_3 =$$

$$\begin{bmatrix} 0 & 1 & 0 & 0 & 0 & 0 \\ -1 & 0 & 0 & 0 & 0 & 0 \\ 0 & 0 & 1 & 0 & 0 & 0 \\ 0 & 0 & 0 & 0 & 1 & 0 \\ 0 & 0 & 0 & -1 & 0 & 0 \\ 0 & 0 & 0 & 0 & 0 & 1 \end{bmatrix} \times \begin{bmatrix} 2.07\times10^{-23} & 0 & 0 & -2.07\times10^{-23} & 0 & 0 \\ 0 & 2.484\times10^{-23} & 1.035\times10^{-31} & 0 & -2.484\times10^{-23} & 1.035\times10^{-31} \\ 0 & 1.035\times10^{-31} & 6.9\times10^{-39} & 0 & -1.035\times10^{-31} & -1.725\times10^{-39} \\ -2.07\times10^{-23} & 0 & 0 & 2.07\times10^{-23} & 0 & 0 \\ 0 & -2.484\times10^{-23} & -1.035\times10^{-31} & 0 & 2.484\times10^{-23} & -1.035\times10^{-31} \\ 0 & 1.035\times10^{-31} & -1.725\times10^{-39} & 0 & -1.035\times10^{-31} & 6.9\times10^{-39} \end{bmatrix} \times \begin{bmatrix} 0 & -1 & 0 & 0 & 0 & 0 \\ 1 & 0 & 0 & 0 & 0 & 0 \\ 0 & 0 & 1 & 0 & 0 & 0 \\ 0 & 0 & 0 & 0 & -1 & 0 \\ 0 & 0 & 0 & 1 & 0 & 0 \\ 0 & 0 & 0 & 0 & 0 & 1 \end{bmatrix}$$

$$= \begin{bmatrix} 2.484\times10^{-23} & 0 & 1.035\times10^{-31} & -2.484\times10^{-23} & 0 & 1.035\times10^{-31} \\ 0 & 2.07\times10^{-23} & 0 & 0 & -2.07\times10^{-23} & 0 \\ 1.035\times10^{-31} & 0 & 6.9\times10^{-39} & -1.035\times10^{-31} & 0 & -1.725\times10^{-39} \\ -2.484\times10^{-23} & 0 & -1.035\times10^{-31} & 2.484\times10^{-23} & 0 & -1.035\times10^{-31} \\ 0 & -2.07\times10^{-23} & 0 & 0 & 2.07\times10^{-23} & 0 \\ 1.035\times10^{-31} & 0 & -1.725\times10^{-39} & -1.035\times10^{-31} & 0 & 6.9\times10^{-39} \end{bmatrix} \tag{8.E8.3.14}$$

According to Eq. (8.26), the total mass matrices on the global coordinates should be determined as:

$$(M)_1 = (M_c)_1 + (M_{nl})_1$$

$$= \begin{bmatrix} 1.103\times10^{-22} & 0 & -7.059\times10^{-31} & 4.731\times10^{-24} & 0 & 2.525\times10^{-31} \\ 0 & 9.737\times10^{-23} & 0 & 0 & 1.763\times10^{-23} & 0 \\ -7.059\times10^{-31} & 0 & 1.238\times10^{-38} & -2.525\times10^{-31} & 0 & -5.832\times10^{-39} \\ 4.731\times10^{-24} & 0 & -2.525\times10^{-31} & 1.103\times10^{-22} & 0 & 7.059\times10^{-31} \\ 0 & 1.763\times10^{-23} & 0 & 0 & 9.737\times10^{-23} & 0 \\ 2.525\times10^{-32} & 0 & -5.832\times10^{-39} & 7.059\times10^{-31} & 0 & 1.238\times10^{-38} \end{bmatrix} \tag{8.E8.3.15}$$

$$(M)_2=(M_c)_2+(M_{nl})_2$$
$$=\begin{bmatrix} 9.737\times10^{-23} & 0 & 0 & 1.763\times10^{-23} & 0 & 0 \\ 0 & 1.103\times10^{-22} & 7.059\times10^{-31} & 0 & 4.731\times10^{-24} & -2.525\times10^{-31} \\ 0 & 7.059\times10^{-31} & 1.238\times10^{-38} & 0 & 2.525\times10^{-31} & -5.832\times10^{-39} \\ 1.763\times10^{-23} & 0 & 0 & 9.737\times10^{-23} & 0 & 0 \\ 0 & 4.731\times10^{-24} & 2.525\times10^{-31} & 0 & 1.103\times10^{-22} & -7.059\times10^{-31} \\ 0 & -2.525\times10^{-31} & -5.832\times10^{-39} & 0 & -7.059\times10^{-31} & 1.238\times10^{-38} \end{bmatrix} \tag{8.E8.3.16}$$

$$(M)_3=(M_c)_3+(M_{nl})_3$$
$$=\begin{bmatrix} 1.103\times10^{-22} & 0 & 7.059\times10^{-31} & 4.731\times10^{-24} & 0 & -2.525\times10^{-31} \\ 0 & 9.737\times10^{-23} & 0 & 0 & 1.763\times10^{-23} & 0 \\ 7.059\times10^{-31} & 0 & 1.238\times10^{-38} & 2.525\times10^{-31} & 0 & -5.832\times10^{-39} \\ 4.731\times10^{-24} & 0 & 2.525\times10^{-31} & 1.103\times10^{-22} & 0 & -7.059\times10^{-31} \\ 0 & 1.763\times10^{-23} & 0 & 0 & 9.737\times10^{-23} & 0 \\ -2.525\times10^{-32} & 0 & -5.832\times10^{-39} & -7.059\times10^{-31} & 0 & 1.238\times10^{-38} \end{bmatrix} \tag{8.E8.3.17}$$

To prepare the related matrices for the eigenvalue analysis, the stiffness matrices in Eqs. (8.E8.3.8)–(8.E8.3.10) should be assembled in the direction of global freedoms in Fig. 8.5b and then these are reduced according to the geometric boundary conditions:

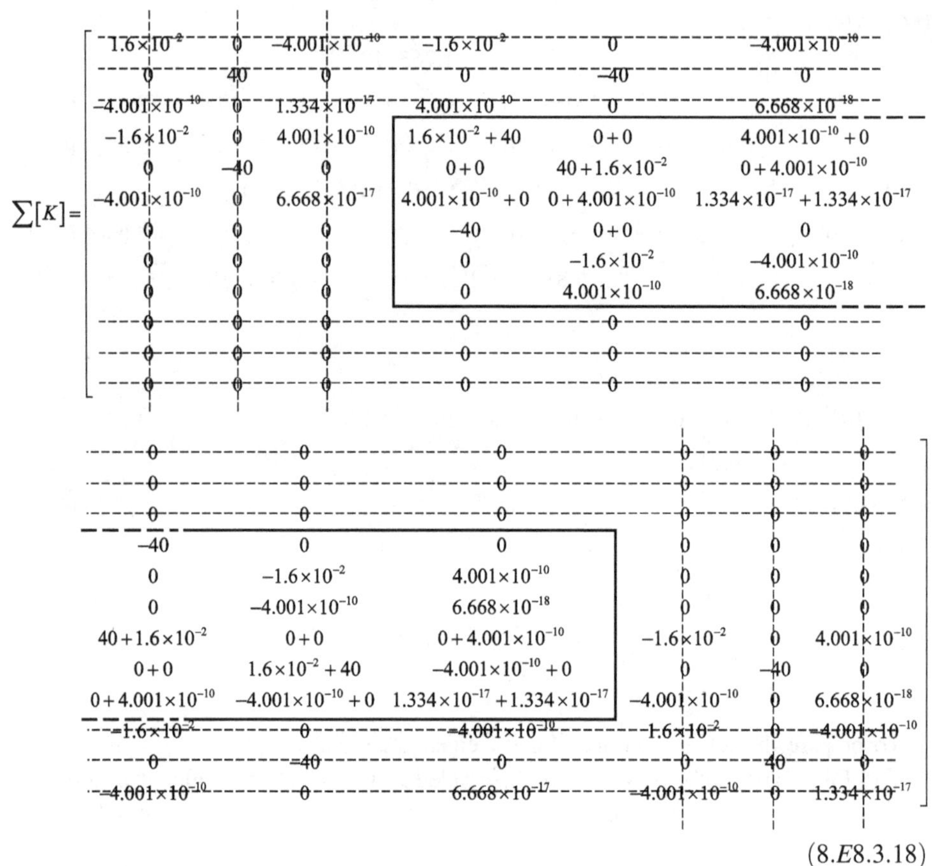

(8.*E*8.3.18)

Similarly, the processes of assembling and reducing the mass matrices in Eqs. (8.E8.3.15)-(8.E8.3.17) yield the following expression:

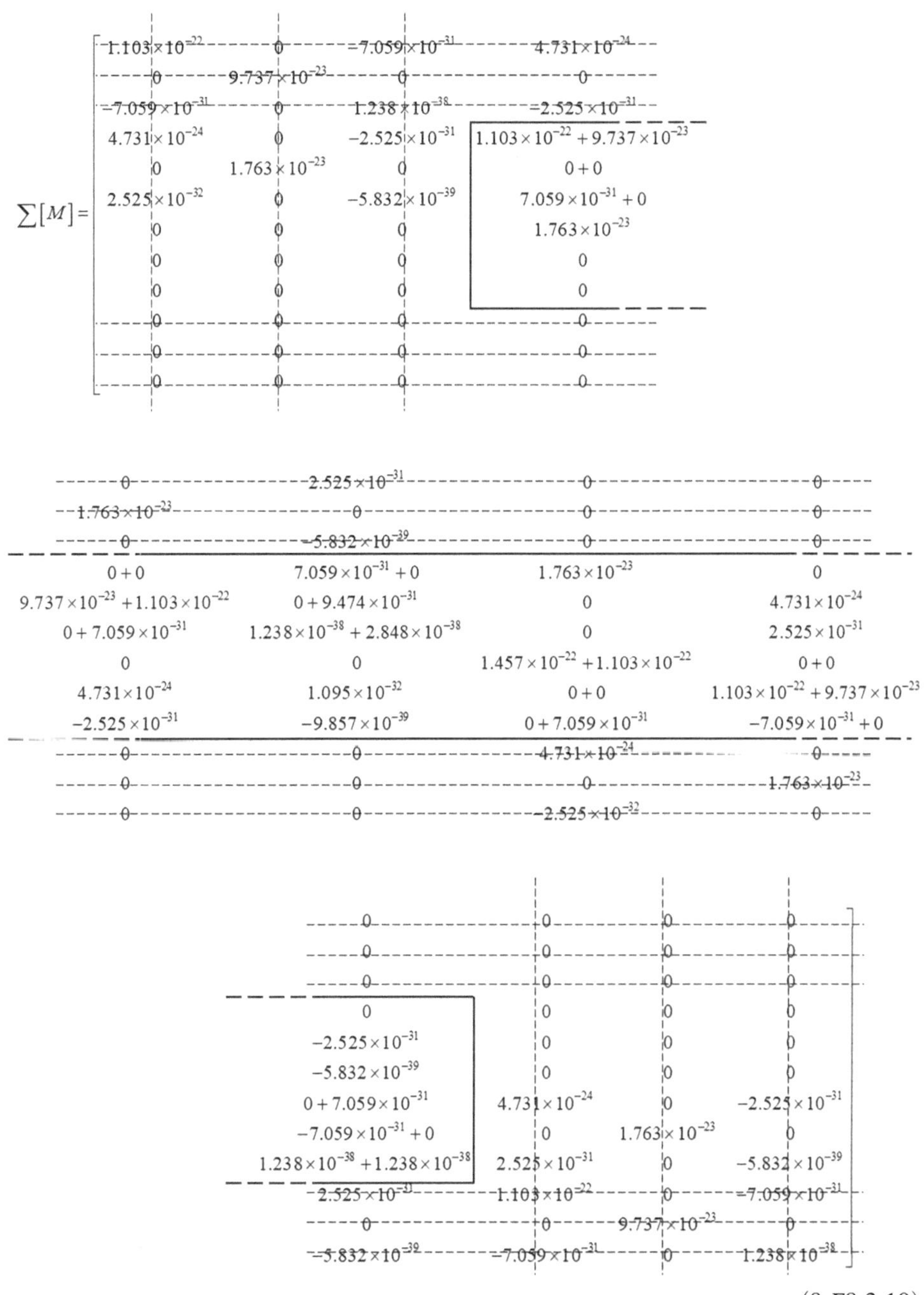

(8.E8.3.19)

Thus, the reduced stiffness and mass matrices are written as follows, respectively:

$$\left(\sum[K]\right)^* = \begin{bmatrix} 40.016 & 0 & 4.001\times10^{-10} & -40 & 0 & 0 \\ 0 & 40.016 & 4.001\times10^{-10} & 0 & -1.6\times10^{-2} & 4.001\times10^{-9} \\ 4.001\times10^{-10} & 4.001\times10^{-10} & 2.667\times10^{-17} & 0 & -4.001\times10^{-10} & 6.668\times10^{-18} \\ -40 & 0 & 0 & 40.016 & 0 & 4.001\times10^{-10} \\ 0 & -1.6\times10^{-2} & -4.001\times10^{-10} & 0 & 40.016 & -4.001\times10^{-10} \\ 0 & 4.001\times10^{-10} & 6.668\times10^{-18} & 4.001\times10^{-10} & -4.001\times10^{-10} & 2.667\times10^{-17} \end{bmatrix} \tag{8.E8.3.20}$$

$$\left(\sum[M]\right)^* = \begin{bmatrix} 2.076\times10^{-22} & 0 & 7.059\times10^{-31} & 1.763\times10^{-23} & 0 & 0 \\ 0 & 2.076\times10^{-22} & 7.059\times10^{-31} & 0 & 4.731\times10^{-24} & -2.525\times10^{-31} \\ 7.059\times10^{-31} & 7.059\times10^{-31} & 2.475\times10^{-38} & 0 & 2.525\times10^{-31} & -5.832\times10^{-39} \\ 1.763\times10^{-23} & 0 & 0 & 2.076\times10^{-22} & 0 & 7.059\times10^{-31} \\ 0 & 4.731\times10^{-24} & 2.525\times10^{-31} & 0 & 2.076\times10^{-22} & -7.059\times10^{-31} \\ 0 & -2.525\times10^{-32} & -5.832\times10^{-39} & 7.059\times10^{-31} & -7.059\times10^{-31} & 2.475\times10^{-38} \end{bmatrix} \tag{8.E8.3.21}$$

The natural frequencies of nanotruss are reached by using the reduced global matrices written with Eqs. (8.E8.3.20) and (8.E8.3.21) into Eq. (8.29):

$$\omega_1 = 8.476\,\text{GHz}, \quad \omega_2 = 25.568\,\text{GHz}, \quad \omega_2 = 43.292\,\text{GHz},$$
$$\omega_4 = 458.241\,\text{GHz}, \quad \omega_5 = 465.383\,\text{GHz}, \quad \omega_6 = 688.732\,\text{GHz}.$$

For a detailed comparison study of the nonlocal natural frequencies of first six modes of nanoframe depicted in Fig. 8.5a, it is recommended to examine Example 8.5.

Example 8.4 Determine the natural frequencies of nonlocal free vibration of nanoframe shown in Fig. 8.6a. Neglect the shear deformation. Use a single finite element for all members in the analysis (Fig. 8.6b). Material parameters are as in Example 8.3. In addition to these, length parameter of nanoframe: $L = l_1 = l_2 = 20$ nm, height and thickness of rectangular cross-section: $h = 1$ nm and $b = 2$ nm, nonlocal parameter: $e_0a/L = 0.2$, positive-directed angle of member **1**: $\theta = 50°$.

Solution. The system has nine freedoms and the number of freedoms is reduced to three by the fixed supports. Free vibration analysis of this nonlocal nanoframe is presented to perform a different application in the arrangement of nodes.

First, let's write the geometric data that are different from the previous example:

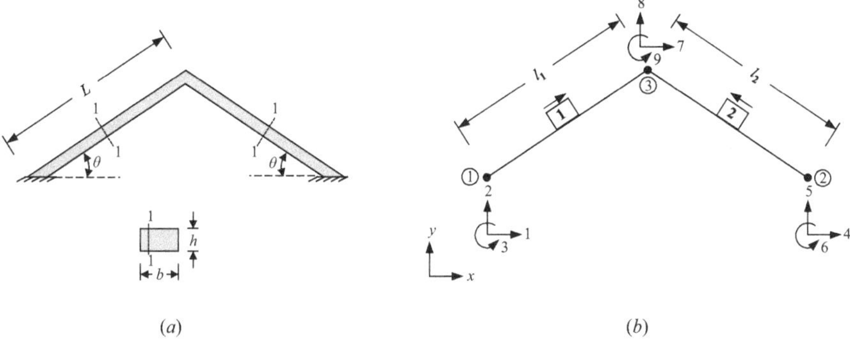

Fig. 8.6 Matrix displacement analysis of a nanoframe with two members. (**a**) Physical system. (**b**) FEM model

$$\begin{aligned}&\text{Length}: l_1 = l_2 = l = 2 \times 10^{-8}\,\text{m},\\&\text{Nonlocal length}: (e_0a)_1 = (e_0a)_2 = e_0a = 4 \times 10^{-9}\,\text{m}.\end{aligned} \tag{8.E8.4.1}$$

First, the orientation angles of nanostructure members and their trigonometric values are written as follows:

$$\begin{aligned}\theta_1 &= \frac{5\pi}{18} \to \cos\theta_1 = 0.643, \quad \sin\theta_1 = 0.766,\\ \theta_2 &= \frac{13\pi}{18} \to \cos\theta_2 = -0.643, \quad \sin\theta_2 = 0.766.\end{aligned} \tag{8.E8.4.2}$$

In this application, it is considered appropriate to calculate the mass and stiffness matrices of members in the global coordinates by using Eqs. (8.27) and (8.28). According to this, we need the following expressions for stiffness matrix:

$$\alpha = \frac{E}{l} = \frac{1 \times 10^{12}\text{N/m}^2}{2 \times 10^{-8}\,\text{m}} = 5 \times 10^{19}\text{N/m}^3,$$

$$\alpha A = \left(5 \times 10^{19}\text{N/m}^3\right) \times \left(2 \times 10^{-18}\,\text{m}^2\right) = 100\text{N/m},$$

$$\alpha I = \left(5 \times 10^{19}\text{N/m}^3\right) \times \left(1.667 \times 10^{-37}\,\text{m}^4\right) = 8.335 \times 10^{-18}\,\text{Nm},$$

$$\alpha\frac{I}{l} = \left(5 \times 10^{19}\text{N/m}^3\right) \times \frac{\left(1.667 \times 10^{-37}\,\text{m}^4\right)}{\left(2 \times 10^{-8}\,\text{m}\right)} = 4.168 \times 10^{-10}\,\text{N},$$

$$\alpha \frac{I}{l^2} = (5 \times 10^{19} \mathrm{N/m^3}) \times \frac{(1.667 \times 10^{-37} \mathrm{m^4})}{(2 \times 10^{-8} \mathrm{m})^2} = 2.084 \times 10^{-2} \mathrm{N/m}. \quad (8.E8.4.3)$$

Thus, the components of stiffness matrix of members are calculated as follows:

$$K_{11,\ 1} = K_{44,\ 1} = (100\mathrm{N/m}) \times (0.643)^2 + 12 \times (2.084 \times 10^{-2} \mathrm{N/m}) \times (0.766)^2 \\ = 41.492\ \mathrm{N/m},$$

$$K_{12,\ 1} = K_{45,\ 1} = [(100\mathrm{N/m}) - 12 \times (2.084 \times 10^{-2} \mathrm{N/m})] \times (0.643) \times (0.766) \\ = 49.131\ \mathrm{N/m},$$

$$K_{13,\ 1} = K_{16,\ 1} = -6 \times (4.168 \times 10^{-10} \mathrm{N}) \times (0.766) = -1.916 \times 10^{-9} \mathrm{N},$$

$$K_{14,\ 1} = (-100\mathrm{N/m}) \times (0.643)^2 - 12 \times (2.084 \times 10^{-2} \mathrm{N/m}) \times (0.766)^2 \\ = -41.492\ \mathrm{N/m},$$

$$K_{15,\ 1} = K_{24,\ 1} = [(-100\mathrm{N/m}) + 12 \times (2.084 \times 10^{-2} \mathrm{N/m})] \times (0.643) \times (0.766) \\ = -49.131\ \mathrm{N/m},$$

$$K_{22,\ 1} = K_{55,\ 1} = (100\mathrm{N/m}) \times (0.766)^2 + 12 \times (2.084 \times 10^{-2} \mathrm{N/m}) \times (0.643)^2 \\ = 58.779\ \mathrm{N/m},$$

$$K_{23,\ 1} = K_{26,\ 1} = 6 \times (4.168 \times 10^{-10} \mathrm{N}) \times (0.643) = 1.608 \times 10^{-9} \mathrm{N},$$

$$K_{25,\ 1} = (-100\mathrm{N/m}) \times (0.766)^2 - 12 \times (2.084 \times 10^{-2} \mathrm{N/m}) \times (0.643)^2 \\ = -58.779\ \mathrm{N/m},$$

$$K_{33,\ 1} = K_{66,\ 1} = 4 \times (8.335 \times 10^{-18} \mathrm{Nm}) = 3.334 \times 10^{-17} \mathrm{Nm},$$

$$K_{34,\ 1} = K_{46,\ 1} = 6 \times (4.168 \times 10^{-10} \mathrm{N}) \times (0.766) = 1.916 \times 10^{-9} \mathrm{N},$$

$$K_{35,\ 1} = K_{56,\ 1} = -6 \times (4.168 \times 10^{-10} \mathrm{N}) \times (0.643) = -1.608 \times 10^{-9} \mathrm{N},$$

$$K_{36,\ 1} = 2 \times (8.335 \times 10^{-18} \mathrm{Nm}) = 1.667 \times 10^{-17} \mathrm{Nm}. \quad (8.E8.4.4)$$

$$K_{11,\ 2} = K_{44,\ 2} = (100\mathrm{N/m}) \times (-0.643)^2 + 12 \times (2.084 \times 10^{-2} \mathrm{N/m}) \times (0.766)^2 \\ = 41.492\ \mathrm{N/m},$$

$$K_{12,\ 2}=K_{45,\ 2}=\left[(100\text{N/m})-12\times\left(2.084\times10^{-2}\text{N/m}\right)\right]\times(-0.643)\times(0.766)$$
$$=-49.131\ \text{N/m},$$

$$K_{13,\ 2}=K_{16,\ 2}=-6\times\left(4.168\times10^{-10}\text{N}\right)\times(0.766)=-1.916\times10^{-9}\text{N},$$

$$K_{14,\ 2}=(-100\text{N/m})\times(-0.643)^2-12\times\left(2.084\times10^{-2}\text{N/m}\right)\times(0.766)^2$$
$$=-41.492\ \text{N/m},$$

$$K_{15,\ 2}=K_{24,\ 2}=\left[(-100\text{N/m})+12\times\left(2.084\times10^{-2}\text{N/m}\right)\right]\times(-0.643)\times(0.766)$$
$$=49.131\ \text{N/m},$$

$$K_{22,\ 2}=K_{55,\ 2}=(100\text{N/m})\times(0.766)^2+12\times\left(2.084\times10^{-2}\text{N/m}\right)\times(-0.643)^2$$
$$=58.779\ \text{N/m},$$

$$K_{23,\ 2}=K_{26,\ 2}=6\times\left(4.168\times10^{-10}\text{N}\right)\times(-0.643)=-1.608\times10^{-10}\text{N},$$

$$K_{25,\ 2}=(-100\text{N/m})\times(0.766)^2-12\times\left(2.084\times10^{-2}\text{N/m}\right)\times(-0.643)^2$$
$$=-58.779\ \text{N/m},$$

$$K_{33,\ 2}=K_{66,\ 2}=4\times\left(8.335\times10^{-18}\text{Nm}\right)=3.334\times10^{-17}\text{Nm},$$

$$K_{34,\ 2}=K_{46,\ 2}=6\times\left(4.168\times10^{-10}\text{N}\right)\times(0.766)=1.916\times10^{-9}\text{N},$$

$$K_{35,\ 2}=K_{56,\ 2}=-6\times\left(4.168\times10^{-10}\text{N}\right)\times(-0.643)=1.608\times10^{-9}\text{N},$$

$$K_{36,\ 2}=2\times\left(8.335\times10^{-18}\text{Nm}\right)=1.667\times10^{-17}\text{Nm}.\tag{8.E8.4.5}$$

According to the inputs calculated above, the related matrices are written as follows:

$$\begin{array}{c|c|c|c|c|c|c}
\mathit{1}\to 1 & \mathit{2}\to 2 & \mathit{3}\to 3 & \mathit{4}\to 7 & \mathit{5}\to 8 & \mathit{6}\to 9 & \\
\hline
41.492 & 49.131 & -1.916\times10^{-9} & -41.492 & -49.131 & -1.916\times10^{-9} & 1\leftarrow \mathit{1}\\
49.131 & 58.779 & 1.608\times10^{-9} & -49.131 & -58.779 & 1.608\times10^{-9} & 2\leftarrow \mathit{2}\\
-1.916\times10^{-9} & 1.608\times10^{-9} & 3.334\times10^{-17} & 1.916\times10^{-9} & -1.608\times10^{-9} & 1.667\times10^{-17} & 3\leftarrow \mathit{3}\\
-41.492 & -49.131 & 1.916\times10^{-9} & 41.492 & 49.131 & 1.916\times10^{-9} & 7\leftarrow \mathit{4}\\
-49.131 & -58.779 & -1.608\times10^{-9} & 49.131 & 58.779 & -1.608\times10^{-9} & 8\leftarrow \mathit{5}\\
-1.916\times10^{-9} & 1.608\times10^{-9} & 1.667\times10^{-17} & 1.916\times10^{-9} & -1.608\times10^{-9} & 3.334\times10^{-17} & 9\leftarrow \mathit{6}
\end{array}$$

$$(K)_1=\begin{bmatrix}
41.492 & 49.131 & -1.916\times10^{-9} & -41.492 & -49.131 & -1.916\times10^{-9}\\
49.131 & 58.779 & 1.608\times10^{-9} & -49.131 & -58.779 & 1.608\times10^{-9}\\
-1.916\times10^{-9} & 1.608\times10^{-9} & 3.334\times10^{-17} & 1.916\times10^{-9} & -1.608\times10^{-9} & 1.667\times10^{-17}\\
-41.492 & -49.131 & 1.916\times10^{-9} & 41.492 & 49.131 & 1.916\times10^{-9}\\
-49.131 & -58.779 & -1.608\times10^{-9} & 49.131 & 58.779 & -1.608\times10^{-9}\\
-1.916\times10^{-9} & 1.608\times10^{-9} & 1.667\times10^{-17} & 1.916\times10^{-9} & -1.608\times10^{-9} & 3.334\times10^{-17}
\end{bmatrix}\tag{8.E8.4.6}$$

$$(K)_2 = \begin{array}{c} \begin{array}{cccccc} 1 & 2 & 3 & 4 & 5 & 6 \\ \downarrow & \downarrow & \downarrow & \downarrow & \downarrow & \downarrow \\ 4 & 5 & 6 & 7 & 8 & 9 \end{array} \\ \left[\begin{array}{cccccc} 41.492 & -49.131 & -1.916\times10^{-9} & -41.492 & 49.131 & -1.916\times10^{-9} \\ -49.131 & 58.779 & -1.608\times10^{-10} & 49.131 & -58.779 & -1.608\times10^{-10} \\ -1.916\times10^{-9} & -1.608\times10^{-10} & 3.334\times10^{-17} & 1.916\times10^{-9} & 1.608\times10^{-9} & 1.667\times10^{-17} \\ -41.492 & 49.131 & 1.916\times10^{-9} & 41.492 & -49.131 & 1.916\times10^{-9} \\ 49.131 & -58.779 & 1.608\times10^{-9} & -49.131 & 58.779 & 1.608\times10^{-9} \\ -1.916\times10^{-9} & -1.608\times10^{-10} & 1.667\times10^{-17} & 1.916\times10^{-9} & 1.608\times10^{-9} & 3.334\times10^{-17} \end{array}\right] \end{array} \begin{array}{c} \\ \\ \\ 4\leftarrow 1 \\ 5\leftarrow 2 \\ 6\leftarrow 3 \\ 7\leftarrow 4 \\ 8\leftarrow 5 \\ 9\leftarrow 6 \end{array} \tag{8.E8.4.7}$$

On the other hand, the following parameters about the total mass matrix should be given:

$$\begin{aligned} \beta &= \rho A l = (2300\,\mathrm{kg/m^3}) \times (2\times10^{-18}\,\mathrm{m^2}) \times (2\times10^{-8}\,\mathrm{m}) = 9.2\times10^{-23}\,\mathrm{kg}, \\ \beta l &= (9.2\times10^{-23}\,\mathrm{kg})(2\times10^{-8}\,\mathrm{m}) = 1.84\times10^{-30}\,\mathrm{kgm}, \\ \beta l^2 &= (9.2\times10^{-23}\,\mathrm{kg}) \times (2\times10^{-8}\,\mathrm{m})^2 = 3.68\times10^{-38}\,\mathrm{kgm^2} \end{aligned} \tag{8.E8.4.8}$$

$$\begin{aligned} \gamma &= (e_0 a)^2 \rho A = (4\times10^{-9}\mathrm{m})^2 \times (2300\,\mathrm{kg/m^3}) \times (2\times10^{-18}\,\mathrm{m^2}) = 7.36\times10^{-32}\,\mathrm{kgm}, \\ \frac{\gamma}{l} &= \frac{7.36\times10^{-32}\,\mathrm{kgm}}{2\times10^{-8}\,\mathrm{m}} = 3.68\times10^{-24}\,\mathrm{kg}, \\ \gamma l &= (7.36\times10^{-32}\,\mathrm{kgm}) \times (2\times10^{-8}\,\mathrm{m}) = 1.472\times10^{-39}\,\mathrm{kgm^2}. \end{aligned} \tag{8.E8.4.9}$$

The inputs of total mass matrices of nanoframe members are listed below:

$$\begin{aligned} M_{11,\,1} = M_{44,\,1} &= \left[\frac{1}{3} \times (9.2\times10^{-23}\,\mathrm{kg}) + (3.68\times10^{-24}\,\mathrm{kg})\right] \times (0.643)^2 \\ &+ \left[\frac{13}{35} \times (9.2\times10^{-23}\,\mathrm{kg}) + \frac{6}{5} \times (3.68\times10^{-24}\,\mathrm{kg})\right] \times (0.766)^2 \\ &= 3.684\times10^{-23}\,\mathrm{kg}, \end{aligned}$$

$$\begin{aligned} M_{12,\,1} = M_{45,\,1} &= \left[-\frac{4}{105} \times (9.2\times10^{-23}\,\mathrm{kg}) - \frac{1}{5} \times (3.68\times10^{-24}\,\mathrm{kg})\right] \times (0.643) \times (0.766) \\ &= -2.089\times10^{-24}\,\mathrm{kg}, \end{aligned}$$

$$\begin{aligned} M_{13,\,1} &= \left[-\frac{11}{210} \times (1.84\times10^{-30}\,\mathrm{kgm}) - \frac{1}{10} \times (7.36\times10^{-32}\,\mathrm{kgm})\right] \times (0.766) \\ &= -7.947\times10^{-32}\,\mathrm{kgm}, \end{aligned}$$

$$M_{14,\ 1}=\left[\frac{1}{6}\times\left(9.2\times10^{-23}\,\text{kg}\right)-\left(3.68\times10^{-24}\,\text{kg}\right)\right]\times(0.643)^2+\left[\frac{9}{70}\times\left(9.2\times10^{-23}\,\text{kg}\right)-\frac{6}{5}\times\left(3.68\times10^{-24}\,\text{kg}\right)\right]\times(0.766)^2=9.167\times10^{-24}\,\text{kg},$$

$$M_{15,\ 1}=M_{24,\ 1}=\left[\frac{4}{105}\times\left(9.2\times10^{-23}\,\text{kg}\right)+\frac{1}{5}\times\left(3.68\times10^{-24}\,\text{kg}\right)\right]\times(0.643)\times(0.766)=2.089\times10^{-24}\,\text{kg},$$

$$M_{16,\ 1}=\left[\frac{13}{420}\times\left(1.84\times10^{-30}\,\text{kgm}\right)-\frac{1}{10}\times\left(7.36\times10^{-32}\,\text{kgm}\right)\right]\times(0.766)=3.799\times10^{-32}\,\text{kgm},$$

$$M_{22,\ 1}=M_{55,\ 1}=\left[\frac{1}{3}\times\left(9.2\times10^{-23}\,\text{kg}\right)+\frac{6}{5}\times\left(3.68\times10^{-24}\,\text{kg}\right)\right]\times(0.643)^2+\left[\frac{13}{35}\times\left(9.2\times10^{-23}\,\text{kg}\right)+\left(3.68\times10^{-24}\,\text{kg}\right)\right]\times(0.766)^2=3.671\times10^{-23}\,\text{kg},$$

$$M_{23,\ 1}=\left[\frac{11}{210}\times\left(1.84\times10^{-30}\,\text{kgm}\right)+\frac{1}{10}\times\left(7.36\times10^{-32}\,\text{kgm}\right)\right]\times(0.643)=6.671\times10^{-32}\,\text{kgm},$$

$$M_{25,\ 1}=\left[\frac{9}{70}\times\left(9.2\times10^{-23}\,\text{kg}\right)-\frac{6}{5}\times\left(3.68\times10^{-24}\,\text{kg}\right)\right]\times(0.643)^2+\left[\frac{1}{6}\times\left(9.2\times10^{-23}\,\text{kg}\right)-\left(3.68\times10^{-24}\,\text{kg}\right)\right]\times(0.766)^2=9.902\times10^{-24}\,\text{kg},$$

$$M_{26,\ 1}=\left[-\frac{13}{420}\times\left(1.84\times10^{-30}\,\text{kgm}\right)+\frac{1}{10}\times\left(7.36\times10^{-32}\,\text{kgm}\right)\right]\times(0.643)=-3.189\times10^{-32}\,\text{kgm},$$

$$M_{33,\ 1}=M_{66,\ 1}=\frac{1}{105}\times\left(3.68\times10^{-38}\,\text{kgm}^2\right)+\frac{2}{15}\times\left(1.472\times10^{-39}\,\text{kgm}^2\right)=5.467\times10^{-40}\,\text{kgm}^2,$$

$$M_{34,\ 1}=\left[-\frac{13}{420}\times\left(1.84\times10^{-30}\,\text{kgm}\right)+\frac{1}{10}\times\left(7.36\times10^{-32}\,\text{kgm}\right)\right]\times(0.766)=-3.799\times10^{-32}\,\text{kgm},$$

$$M_{35,\ 1} = \left[\frac{13}{420} \times \left(1.84 \times 10^{-30}\,\text{kgm}\right) - \frac{1}{10} \times \left(7.36 \times 10^{-32}\,\text{kgm}\right)\right] \times (0.643)$$
$$= 3.189 \times 10^{-32}\,\text{kgm},$$

$$M_{36,\ 1} = -\frac{1}{140} \times \left(3.68 \times 10^{-38}\,\text{kgm}^2\right) - \frac{1}{30} \times \left(1.472 \times 10^{-39}\,\text{kgm}^2\right)$$
$$= -3.119 \times 10^{-40}\,\text{kgm}^2,$$

$$M_{46,\ 1} = \left[\frac{11}{210} \times \left(1.84 \times 10^{-30}\,\text{kgm}\right) + \frac{1}{10} \times \left(7.36 \times 10^{-32}\,\text{kgm}\right)\right] \times (0.766)$$
$$= 7.947 \times 10^{-32}\,\text{kgm},$$

$$M_{56,\ 1} = \left[-\frac{11}{210} \times \left(1.84 \times 10^{-30}\,\text{kgm}\right) - \frac{1}{10} \times \left(7.36 \times 10^{-32}\,\text{kgm}\right)\right] \times (0.643)$$
$$= -6.671 \times 10^{-32}\,\text{kgm}.$$

(8.E8.4.10)

$$M_{11,\ 2} = M_{44,\ 2} = \left[\frac{1}{3} \times \left(9.2 \times 10^{-23}\,\text{kg}\right) + \left(3.68 \times 10^{-24}\,\text{kg}\right)\right] \times (-0.643)^2$$
$$+ \left[\frac{13}{35} \times \left(9.2 \times 10^{-23}\,\text{kg}\right) + \frac{6}{5} \times \left(3.68 \times 10^{-24}\,\text{kg}\right)\right] \times (0.766)^2$$
$$= 3.684 \times 10^{-23}\,\text{kg},$$

$$M_{12,\ 2} = M_{45,\ 2} = \left[-\frac{4}{105} \times \left(9.2 \times 10^{-23}\,\text{kg}\right) - \frac{1}{5} \times \left(3.68 \times 10^{-24}\,\text{kg}\right)\right] \times (-0.643) \times (0.766)$$
$$= 2.089 \times 10^{-24}\,\text{kg},$$

$$M_{13,\ 2} = \left[-\frac{11}{210} \times \left(1.84 \times 10^{-30}\,\text{kgm}\right) - \frac{1}{10} \times \left(7.36 \times 10^{-32}\,\text{kgm}\right)\right] \times (0.766)$$
$$= -7.947 \times 10^{-32}\,\text{kgm},$$

$$M_{14,\ 2} = \left[\frac{1}{6} \times \left(9.2 \times 10^{-23}\,\text{kg}\right) - \left(3.68 \times 10^{-24}\,\text{kg}\right)\right] \times (-0.643)^2$$
$$+ \left[\frac{9}{70} \times \left(9.2 \times 10^{-23}\,\text{kg}\right) - \frac{6}{5} \times \left(3.68 \times 10^{-24}\,\text{kg}\right)\right] \times (0.766)^2$$
$$= 9.167 \times 10^{-24}\,\text{kg},$$

$$M_{15,\ 2} = M_{24,\ 2} = \left[\frac{4}{105} \times \left(9.2 \times 10^{-23}\,\text{kg}\right) + \frac{1}{5} \times \left(3.68 \times 10^{-24}\,\text{kg}\right)\right] \times (-0.643) \times (0.766)$$
$$= -2.089 \times 10^{-24}\,\text{kg},$$

$$M_{16,\ \mathbf{2}} = \left[\frac{13}{420} \times \left(1.84 \times 10^{-30}\,\text{kgm}\right) - \frac{1}{10} \times \left(7.36 \times 10^{-32}\,\text{kgm}\right)\right] \times (0.766)$$
$$= 3.799 \times 10^{-32}\,\text{kgm},$$

$$M_{22,\ \mathbf{2}} = M_{55,\ \mathbf{2}} = \left[\frac{1}{3} \times \left(9.2 \times 10^{-23}\,\text{kg}\right) + \frac{6}{5} \times \left(3.68 \times 10^{-24}\,\text{kg}\right)\right] \times (-0.643)^2$$
$$+ \left[\frac{13}{35} \times \left(9.2 \times 10^{-23}\,\text{kg}\right) + \left(3.68 \times 10^{-24}\,\text{kg}\right)\right] \times (0.766)^2$$
$$= 3.671 \times 10^{-23}\,\text{kg},$$

$$M_{23,\ \mathbf{2}} = \left[\frac{11}{210} \times \left(1.84 \times 10^{-30}\,\text{kgm}\right) + \frac{1}{10} \times \left(7.36 \times 10^{-32}\,\text{kgm}\right)\right] \times (-0.643)$$
$$= -6.671 \times 10^{-32}\,\text{kgm},$$

$$M_{25,\ \mathbf{2}} = \left[\frac{9}{70} \times \left(9.2 \times 10^{-23}\,\text{kg}\right) - \frac{6}{5} \times \left(3.68 \times 10^{-24}\,\text{kg}\right)\right] \times (-0.643)^2$$
$$+ \left[\frac{1}{6} \times \left(9.2 \times 10^{-23}\,\text{kg}\right) - \left(3.68 \times 10^{-24}\,\text{kg}\right)\right] \times (0.766)^2$$
$$= 9.902 \times 10^{-24}\,\text{kg},$$

$$M_{26,\ \mathbf{2}} = \left[-\frac{13}{420} \times \left(1.84 \times 10^{-30}\,\text{kgm}\right) + \frac{1}{10} \times \left(7.36 \times 10^{-32}\,\text{kgm}\right)\right] \times (-0.643)$$
$$= 3.189 \times 10^{\ 32}\,\text{kgm},$$

$$M_{33,\ \mathbf{2}} = M_{66,\ \mathbf{2}} = \frac{1}{105} \times \left(3.68 \times 10^{-38}\,\text{kgm}^2\right) + \frac{2}{15} \times \left(1.472 \times 10^{-39}\,\text{kgm}^2\right)$$
$$= 5.467 \times 10^{-40}\,\text{kgm}^2,$$

$$M_{34,\ \mathbf{2}} = \left[-\frac{13}{420} \times \left(1.84 \times 10^{-30}\,\text{kgm}\right) + \frac{1}{10} \times \left(7.36 \times 10^{-32}\,\text{kgm}\right)\right] \times (0.766)$$
$$= -3.799 \times 10^{-32}\,\text{kgm},$$

$$M_{35,\ \mathbf{2}} = \left[\frac{13}{420} \times \left(1.84 \times 10^{-30}\,\text{kgm}\right) - \frac{1}{10} \times \left(7.36 \times 10^{-32}\,\text{kgm}\right)\right] \times (-0.643)$$
$$= -3.189 \times 10^{-32}\,\text{kgm},$$

$$M_{36,\ \mathbf{2}} = -\frac{1}{140} \times \left(3.68 \times 10^{-38}\,\text{kgm}^2\right) - \frac{1}{30} \times \left(1.472 \times 10^{-39}\,\text{kgm}^2\right)$$
$$= -3.119 \times 10^{-40}\,\text{kgm}^2,$$

$$M_{46,\ 2} = \left[\frac{11}{210} \times \left(1.84 \times 10^{-30}\,\text{kgm}\right) + \frac{1}{10} \times \left(7.36 \times 10^{-32}\,\text{kgm}\right)\right] \times (0.766)$$
$$= 7.947 \times 10^{-32}\,\text{kgm},$$

$$M_{56,\ 2} = \left[-\frac{11}{210} \times \left(1.84 \times 10^{-30}\,\text{kgm}\right) - \frac{1}{10} \times \left(7.36 \times 10^{-32}\,\text{kgm}\right)\right] \times (-0.643)$$
$$= 6.671 \times 10^{-32}\,\text{kgm}. \tag{8.E8.4.11}$$

According to the inputs calculated above, the total mass matrices of members are stated as follows:

$$(M)_1 = \begin{bmatrix} 3.684\times10^{-23} & -2.089\times10^{-24} & -7.947\times10^{-32} & 9.167\times10^{-24} & 2.089\times10^{-24} & 3.799\times10^{-32} \\ -2.089\times10^{-24} & 3.671\times10^{-23} & 6.671\times10^{-32} & 2.089\times10^{-24} & 9.902\times10^{-24} & -3.189\times10^{-32} \\ -7.947\times10^{-32} & 6.671\times10^{-32} & 5.467\times10^{-40} & -3.799\times10^{-32} & 3.189\times10^{-32} & -3.119\times10^{-40} \\ 9.167\times10^{-24} & 2.089\times10^{-24} & -3.799\times10^{-32} & 3.684\times10^{-23} & -2.089\times10^{-24} & 7.947\times10^{-32} \\ 2.089\times10^{-24} & 9.902\times10^{-24} & 3.189\times10^{-32} & -2.089\times10^{-24} & 3.671\times10^{-23} & -6.671\times10^{-32} \\ 3.799\times10^{-32} & -3.189\times10^{-32} & -3.119\times10^{-40} & 7.947\times10^{-32} & -6.671\times10^{-32} & 5.467\times10^{-40} \end{bmatrix} \tag{8.E8.4.12}$$

$$(M)_2 = \begin{bmatrix} 3.684\times10^{-23} & 2.089\times10^{-24} & -7.947\times10^{-32} & 9.167\times10^{-24} & -2.089\times10^{-24} & 3.799\times10^{-32} \\ 2.089\times10^{-24} & 3.671\times10^{-23} & -6.671\times10^{-32} & -2.089\times10^{-24} & 9.902\times10^{-24} & 3.189\times10^{-32} \\ -7.947\times10^{-32} & -6.671\times10^{-32} & 5.467\times10^{-40} & -3.799\times10^{-32} & -3.189\times10^{-32} & -3.119\times10^{-40} \\ 9.167\times10^{-24} & -2.089\times10^{-24} & -3.799\times10^{-32} & 3.684\times10^{-23} & 2.089\times10^{-24} & 7.947\times10^{-32} \\ -2.089\times10^{-24} & 9.902\times10^{-24} & -3.189\times10^{-32} & 2.089\times10^{-24} & 3.671\times10^{-23} & 6.671\times10^{-32} \\ 3.799\times10^{-32} & 3.189\times10^{-32} & -3.119\times10^{-40} & 7.947\times10^{-32} & 6.671\times10^{-32} & 5.467\times10^{-40} \end{bmatrix} \tag{8.E8.4.13}$$

The global freedoms to which the local freedoms of frame members will contribute are shown, e.g., with Eqs. (8.E8.4.6) and (8.E8.4.7), where first the global stiffness matrix is calculated and then reduced as follows:

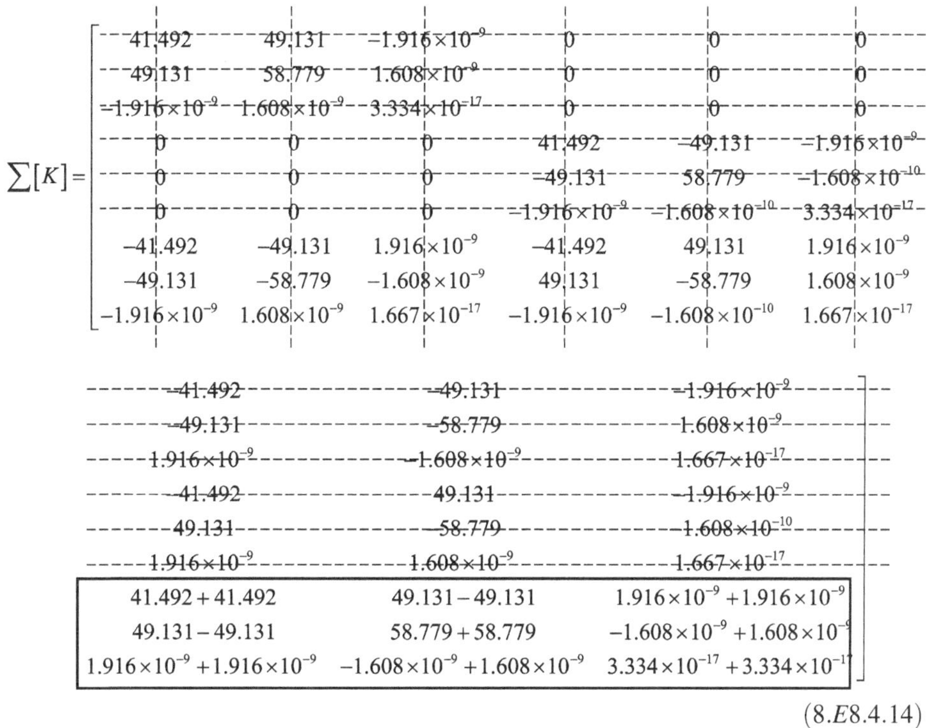

(8.*E*8.4.14)

Additionally, the global mass matrix is written and reduced as follows:

$$\sum[M] = \left[\begin{array}{cccccc|ccc} 3.684\times10^{-23} & -2.089\times10^{-24} & -7.947\times10^{-32} & 0 & 0 & 0 & 9.167\times10^{-24} & 2.089\times10^{-24} & 3.799\times10^{-32} \\ -2.089\times10^{-24} & 3.671\times10^{-23} & 6.671\times10^{-32} & 0 & 0 & 0 & 2.089\times10^{-24} & 9.902\times10^{-24} & -3.189\times10^{-32} \\ -7.947\times10^{-32} & 6.671\times10^{-32} & 5.467\times10^{-40} & 0 & 0 & 0 & -3.799\times10^{-32} & 3.189\times10^{-32} & -3.119\times10^{-40} \\ 0 & 0 & 0 & 3.684\times10^{-23} & 2.089\times10^{-24} & -7.947\times10^{-32} & 9.167\times10^{-24} & -2.089\times10^{-24} & 3.799\times10^{-32} \\ 0 & 0 & 0 & 2.089\times10^{-24} & 3.671\times10^{-23} & -6.671\times10^{-32} & -2.089\times10^{-24} & 9.902\times10^{-24} & 3.189\times10^{-32} \\ 0 & 0 & 0 & -7.947\times10^{-32} & -6.671\times10^{-32} & 5.467\times10^{-40} & -3.799\times10^{-32} & -3.189\times10^{-32} & -3.119\times10^{-40} \\ 9.167\times10^{-24} & 2.089\times10^{-24} & -3.799\times10^{-32} & 9.167\times10^{-24} & -2.089\times10^{-24} & -3.799\times10^{-32} & 3.684\times10^{-23}+3.684\times10^{-23} & -2.089\times10^{-24}+2.089\times10^{-24} & 7.947\times10^{-32}+7.947\times10^{-32} \\ 2.089\times10^{-24} & 9.902\times10^{-24} & 3.189\times10^{-32} & -2.089\times10^{-24} & 9.902\times10^{-24} & -3.189\times10^{-32} & -2.089\times10^{-24}+2.089\times10^{-24} & 3.671\times10^{-23}+3.671\times10^{-23} & -6.671\times10^{-32}+6.671\times10^{-32} \\ 3.799\times10^{-32} & -3.189\times10^{-32} & -3.119\times10^{-40} & 3.799\times10^{-32} & 3.189\times10^{-32} & -3.119\times10^{-40} & 7.947\times10^{-32}+7.947\times10^{-32} & -6.671\times10^{-32}+6.671\times10^{-32} & 5.467\times10^{-40}+5.467\times10^{-40} \end{array}\right] \quad (8.E8.4.15)$$

The following expressions can be expressed according to the reduction operations on the global matrices:

$$\left(\sum[K]\right)^* = \begin{bmatrix} 82.984 & 0 & 3.832\times10^{-9} \\ 0 & 117.558 & 0 \\ 3.832\times10^{-9} & 0 & 6.668\times10^{-17} \end{bmatrix} \quad (8.E8.4.16)$$

$$\left(\sum[M]\right)^* = \begin{bmatrix} 7.368\times10^{-23} & 0 & 1.588\times10^{-31} \\ 0 & 7.342\times10^{-23} & 0 \\ 1.588\times10^{-31} & 0 & 1.093\times10^{-39} \end{bmatrix} \quad (8.E8.4.17)$$

Substituting the above expressions into Eq. (8.29) yields the natural frequencies:

$$\omega_1 = 246.202\,\text{GHz}, \quad \omega_2 = 1265.374\,\text{GHz}, \quad \omega_3 = 1282.940\,\text{GHz}.$$

Example 8.5 Determine the natural frequencies of nonlocal free vibration of nanoframe shown in Fig. 8.5a by using two finite elements for all members (Fig. 8.7). Mechanical and physical parameters are as in Example 8.3.

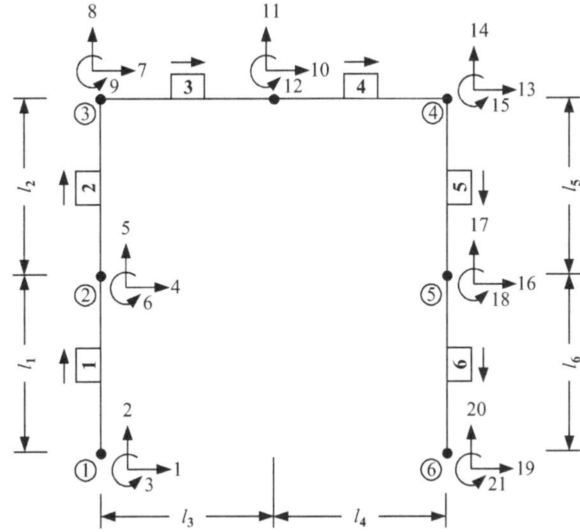

Fig. 8.7 FE mesh with two finite elements for matrix displacement analysis of a nanoframe (Fig. 8.5a)

Solution. This example is discussed to investigate the convergence of NL-FEM used to calculate the frequencies of nonlocal nanoframe. As it is known, high convergence reflects the mechanical behavior of nanostructures more realistically.

Information of all finite elements in the mesh are given as:

$$\begin{aligned} &\text{Length} : l = 2.5\times 10^{-8}\,\text{m},\\ &\text{Nonlocal length} : e_0a = 1.5\times 10^{-8}\,\text{m},\\ &\text{Cross-section area} : A = 2\times 10^{-18}\,\text{m}^2,\\ &\text{Moment of inertia} : I = 6.667\times 10^{-37}\,\text{m}^4,\\ &\text{Modulus of elasticity} : E = 1\times 10^{12}\,\frac{\text{N}}{\text{m}^2},\\ &\text{Mass of unit volume} : \rho = 2300\,\frac{\text{kg}}{\text{m}^3}. \end{aligned} \tag{8.E8.5.1}$$

The orientation angles of finite elements and their trigonometric values can be written as follows

$$\begin{aligned} &\theta_1 = \theta_2 = \frac{\pi}{2} \rightarrow \cos\theta_1 = \cos\theta_2 = 0, \quad \sin\theta_1 = \sin\theta_2 = 1,\\ &\theta_3 = \theta_4 = 0^\circ \rightarrow \cos\theta_3 = \cos\theta_4 = 1, \quad \sin\theta_4 = \sin\theta_4 = 0,\\ &\theta_5 = \theta_6 = \frac{3\pi}{2} \rightarrow \cos\theta_5 = \cos\theta_6 = 0, \quad \sin\theta_5 = \sin\theta_6 = -1. \end{aligned} \tag{8.E8.5.2}$$

In order to calculate the contribution matrices of finite elements to the global stiffness matrix, the following expressions should be obtained:

$$\begin{aligned}
&\alpha = \frac{E}{l} = \frac{1\times 10^{12}\mathrm{N/m^2}}{2.5\times 10^{-8}\,\mathrm{m}} = 4\times 10^{19}\mathrm{N/m^3},\\
&\alpha A = \left(4\times 10^{19}\mathrm{N/m^3}\right)\times\left(2\times 10^{-18}\,\mathrm{m^2}\right) = 80\mathrm{N/m},\\
&\alpha I = \left(2\times 10^{19}\mathrm{N/m^3}\right)\times\left(6.667\times 10^{-37}\,\mathrm{m^4}\right) = 2.667\times 10^{-17}\,\mathrm{Nm},\\
&\alpha\frac{I}{l} = \left(4\times 10^{19}\mathrm{N/m^3}\right)\times\frac{\left(6.667\times 10^{-37}\,\mathrm{m^4}\right)}{\left(2.5\times 10^{-8}\,\mathrm{m}\right)} = 1.067\times 10^{-9}\,\mathrm{N},\\
&\alpha\frac{I}{l^2} = \left(4\times 10^{19}\mathrm{N/m^3}\right)\times\frac{\left(6.667\times 10^{-37}\,\mathrm{m^4}\right)}{\left(2.5\times 10^{-8}\,\mathrm{m}\right)^2} = 4.267\times 10^{-2}\,\mathrm{N/m}.
\end{aligned} \tag{8.E8.5.3}$$

Using Eqs. (8.25), (8.27), (8.E8.5.2), and (8.E8.5.3), the global stiffness matrices of finite elements are presented below:

$$(K)_1 = (K)_2 = \begin{bmatrix}
0.128 & 0 & -1.6\times 10^{-9} & -0.128 & 0 & -1.6\times 10^{-9}\\
0 & 80 & 0 & 0 & -80 & 0\\
-1.6\times 10^{-9} & 0 & 2.667\times 10^{-17} & 1.6\times 10^{-9} & 0 & 1.334\times 10^{-17}\\
-0.128 & 0 & 1.6\times 10^{-9} & 0.128 & 0 & 1.6\times 10^{-9}\\
0 & -80 & 0 & 0 & 80 & 0\\
-1.6\times 10^{-9} & 0 & 1.334\times 10^{-17} & 1.6\times 10^{-9} & 0 & 2.667\times 10^{-17}
\end{bmatrix} \tag{8.E8.5.4}$$

$$(K)_3 = (K)_4 = \begin{bmatrix}
80 & 0 & 0 & -80 & 0 & 0\\
0 & 0.128 & 1.6\times 10^{-9} & 0 & -0.128 & 1.6\times 10^{-9}\\
0 & 1.6\times 10^{-9} & 2.667\times 10^{-17} & 0 & -1.6\times 10^{-9} & 1.334\times 10^{-17}\\
-80 & 0 & 0 & 80 & 0 & 0\\
0 & -0.128 & -1.6\times 10^{-9} & 0 & 0.128 & -1.6\times 10^{-9}\\
0 & 1.6\times 10^{-9} & 1.334\times 10^{-17} & 0 & -1.6\times 10^{-9} & 2.667\times 10^{-17}
\end{bmatrix} \tag{8.E8.5.5}$$

$$(K)_5=(K)_6=\begin{bmatrix} 0.128 & 0 & 1.6\times10^{-9} & -0.128 & 0 & 1.6\times10^{-9} \\ 0 & 80 & 0 & 0 & -80 & 0 \\ 1.6\times10^{-9} & 0 & 2.667\times10^{-17} & -1.6\times10^{-9} & 0 & 1.334\times10^{-17} \\ -0.128 & 0 & -1.6\times10^{-9} & 0.128 & 0 & -1.6\times10^{-9} \\ 0 & -80 & 0 & 0 & 80 & 0 \\ 1.6\times10^{-9} & 0 & 1.334\times10^{-17} & -1.6\times10^{-9} & 0 & 2.667\times10^{-17} \end{bmatrix} \tag{8.E8.5.6}$$

In addition to these, the following parameters should be given for the inputs of total mass matrices:

$$\begin{aligned} \beta&=\rho Al=\left(2300\,\text{kg/m}^3\right)\times\left(2\times10^{-18}\,\text{m}^2\right)\times\left(2.5\times10^{-8}\,\text{m}\right)=1.15\times10^{-22}\,\text{kg}, \\ \beta l&=\left(1.15\times10^{-22}\,\text{kg}\right)\times\left(2.5\times10^{-8}\,\text{m}\right)=2.875\times10^{-30}\,\text{kgm}, \\ \beta l^2&=\left(1.15\times10^{-22}\,\text{kg}\right)\times\left(2.5\times10^{-8}\,\text{m}\right)^2=7.188\times10^{-38}\,\text{kgm}^2. \end{aligned} \tag{8.E8.5.7}$$

$$\begin{aligned} \gamma&=(e_0a)^2\rho A=\left(1.5\times10^{-8}\,\text{m}\right)^2\times\left(2300\,\text{kg/m}^3\right)\times\left(2\times10^{-18}\,\text{m}^2\right)=1.035\times10^{-30}\,\text{kgm}, \\ \frac{\gamma}{l}&=\frac{1.035\times10^{-30}\,\text{kgm}}{2.5\times10^{-8}\,\text{m}}=4.14\times10^{-23}\,\text{kg}, \\ \gamma l&=\left(1.035\times10^{-30}\,\text{kgm}\right)\times\left(2.5\times10^{-8}\,\text{m}\right)=2.588\times10^{-38}\,\text{kgm}^2. \end{aligned} \tag{8.E8.5.8}$$

Employing Eqs. (8.26), (8.28), (8.E8.5.7), and (8.E8.5.8), the contributions of finite elements to the global mass matrix can be reached:

$$(M)_1=(M)_2=$$

$$\begin{bmatrix} 9.239\times10^{-23} & 0 & -2.541\times10^{-31} & -3.489\times10^{-23} & 0 & -1.451\times10^{-32} \\ 0 & 7.973\times10^{-23} & 0 & 0 & -2.223\times10^{-23} & 0 \\ -2.541\times10^{-31} & 0 & 4.135\times10^{-39} & 1.451\times10^{-32} & 0 & -1.376\times10^{-39} \\ -3.489\times10^{-23} & 0 & 1.451\times10^{-32} & 9.239\times10^{-23} & 0 & 2.541\times10^{-31} \\ 0 & -2.223\times10^{-23} & 0 & 0 & 7.973\times10^{-23} & 0 \\ -1.451\times10^{-32} & 0 & -1.376\times10^{-39} & 2.541\times10^{-31} & 0 & 4.135\times10^{-39} \end{bmatrix} \tag{8.E8.5.9}$$

$$(M)_3=(M)_4=$$
$$\begin{bmatrix} 7.973\times10^{-23} & 0 & 0 & -2.223\times10^{-16} & 0 & 0 \\ 0 & 9.239\times10^{-23} & 2.541\times10^{-31} & 0 & -3.489\times10^{-23} & 1.451\times10^{-32} \\ 0 & 2.541\times10^{-31} & 4.135\times10^{-39} & 0 & -1.451\times10^{-32} & -1.376\times10^{-39} \\ -2.223\times10^{-23} & 0 & 0 & 7.973\times10^{-16} & 0 & 0 \\ 0 & -3.489\times10^{-23} & -1.451\times10^{-32} & 0 & 9.239\times10^{-23} & -2.541\times10^{-131} \\ 0 & 1.451\times10^{-32} & -1.376\times10^{-39} & 0 & -2.541\times10^{-31} & 4.135\times10^{-39} \end{bmatrix} \tag{8.E8.5.10}$$

$$(M)_5=(M)_6=$$
$$\begin{bmatrix} 9.239\times10^{-23} & 0 & 2.541\times10^{-31} & -3.489\times10^{-23} & 0 & 1.451\times10^{-32} \\ 0 & 7.973\times10^{-23} & 0 & 0 & -2.223\times10^{-23} & 0 \\ 2.541\times10^{-31} & 0 & 4.135\times10^{-39} & -1.451\times10^{-32} & 0 & -1.376\times10^{-39} \\ -3.489\times10^{-23} & 0 & -1.451\times10^{-32} & 9.239\times10^{-23} & 0 & -2.541\times10^{-31} \\ 0 & -2.223\times10^{-23} & 0 & 0 & 7.973\times10^{-23} & 0 \\ 1.451\times10^{-32} & 0 & -1.376\times10^{-39} & -2.541\times10^{-31} & 0 & 4.135\times10^{-39} \end{bmatrix} \tag{8.E8.5.11}$$

In the finite element mesh with 21 freedoms, the global stiffness and mass matrices have sizes in 21×21. Therefore, it is not easy to write these matrices manually. From here, in order to assemble the matrices of finite elements and the eigenvalue solution, the process is performed on the computer. According to this process, Table 8.2 gives information about the local freedoms of finite elements should contribute to which global freedoms of system where since the freedoms 1, 2, 3, 19, 20, and 21 correspond to the fixed support, the related rows and columns that have these numbers are deleted from global matrices and the natural frequencies of first six modes are obtained by substituting the reduced matrices into Eq. (8.29).

Table 8.2 Local and global freedoms of FE mesh in Fig. 8.7

		Element numbers					
		1	**2**	**3**	**4**	**5**	**6**
Local freedoms of elements		Global freedoms of nanoframe					
1	→	1	4	7	10	13	16
2	→	2	5	8	11	14	17
3	→	3	6	9	12	15	18
4	→	4	7	10	13	16	19
5	→	5	8	11	14	17	20
6	→	6	9	12	15	18	21

Table 8.3 Comparisons of nonlocal natural frequencies (GHz) of nanoframe in Fig. 8.5a

	Mode	Solution manually		Program code (MATLAB)	
e_0a/L	Number	$n=3$	$n=6$	$n=3$	$n=6$
0.3	1	8.476	7.898	7.2777	7.2686
	2	25.568	21.874	25.5656	21.8447
	3	43.292	31.956	43.2716	31.8934
	4	458.241	37.215	458.1755	37.1435
	5	465.383	56.236	465.3081	56.0321
	6	688.732	68.279	688.6390	67.9245
0.5	1	–	–	6.6383	6.6329
	2			18.5388	15.9595
	3			28.9747	22.2489
	4			359.2602	26.5545
	5			412.9411	36.2998
	6			523.7211	43.3313

$$\omega_1=7.898\,\text{GHz},\quad \omega_2=21.874\,\text{GHz},\quad \omega_2=31.956\,\text{GHz},$$
$$\omega_4=37.215\,\text{GHz},\quad \omega_5=56.236\,\text{GHz},\quad \omega_6=68.279\,\text{GHz}.$$

A comparison of above natural frequencies with the results obtained in Example 8.3 and the natural frequencies obtained according to another nondimensional nonlocal parameter value for the same nanoframe are listed in Table 8.3. When the finite element results are examined, although the natural frequencies are not quite different in low modes, this difference increases in high modes. On the other hand, it is understood that the nonlocal parameter has a serious influence on the frequencies of nanoframes. These conclusions clarify that to understand the accurate mechanical behavior of nanoframes, the size effect of discrete members should be exactly considered and the members should be exactly divided into more finite elements.

Example 8.6 Determine the natural frequencies of nonlocal free vibration of nanoframe shown in Fig. 8.8a. Nanoframe has general mass attached at node (1), simply supported at node (2), and horizontally sliding simply supported at node (3). Use a single finite element for all members in the analysis (Fig. 8.8b). Material parameters are as in Example 8.3. In addition to these, length: $L=l_{e\,=\,1}=l_{e\,=\,2}=20$ nm, height and thickness of rectangular cross-section: $h=1$ nm and $b=2$ nm, mass components of mass attachment: $m_{a,\,axial}=10^{-22}$ kg, $m_{a,\,transverse}=10^{-22}$ kg, $m_{a,\,rotation}=10^{-38}$ kgm^2, and nonlocal parameter: $e_0a/L=0.2$.

Solution. This nanoframe example is presented to examine the applications of different geometric boundary condition and end attachment on the nonlocal dynamic analysis. Also, since the previous examples provide that the issue can be understood, computations of these applications are carried out electronically.

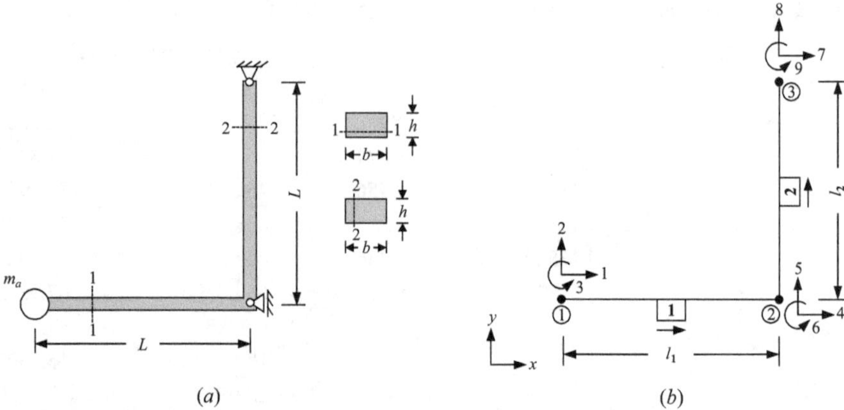

Fig. 8.8 Matrix displacement analysis of a nanoframe with mass attached and simply supported ends. (**a**) Physical system. (**b**) FEM model

Since the geometric information of nanoframe is the same as the previous example, except for member orientations, only the trigonometric values of member orientation angles are calculated as follows by considering Fig. 8.7b:

$$\begin{aligned} &\theta_1 = 0 \rightarrow \cos\theta_1 = 1, \quad \sin\theta_1 = 0, \\ &\theta_2 = \frac{\pi}{2} \rightarrow \cos\theta_2 = 0, \quad \sin\theta_2 = 1. \end{aligned} \tag{8.E8.6.1}$$

If the above values are substituted into Eqs. (8.27) and (8.28) with the values in Eqs. (8.E8.4.3), (8.E8.4.8), and (8.E8.4.9), the global stiffness and mass matrices of members are listed as follows:

$$(K)_1 = \begin{bmatrix} 100 & 0 & 0 & -100 & 0 & 0 \\ 0 & 0.25 & 2.501\times10^{-9} & 0 & -0.25 & 2.501\times10^{-9} \\ 0 & 2.501\times10^{-9} & 3.334\times10^{-17} & 0 & -2.501\times10^{-9} & 1.667\times10^{-17} \\ -100 & 0 & 0 & 100 & 0 & 0 \\ 0 & -0.25 & -2.501\times10^{-9} & 0 & 0.25 & -2.501\times10^{-9} \\ 0 & 2.501\times10^{-9} & 1.667\times10^{-17} & 0 & -2.501\times10^{-9} & 3.334\times10^{-17} \end{bmatrix} \tag{8.E8.6.2}$$

$$(K)_2 = \begin{bmatrix} 0.25 & 0 & -2.501\times10^{-9} & -0.25 & 0 & -2.501\times10^{-9} \\ 0 & 100 & 0 & 0 & -100 & 0 \\ -2.501\times10^{-9} & 0 & 3.334\times10^{-17} & 2.501\times10^{-9} & 0 & 1.667\times10^{-17} \\ -0.25 & 0 & 2.501\times10^{-9} & 0.25 & 0 & 2.501\times10^{-9} \\ 0 & -100 & 0 & 0 & 100 & 0 \\ -2.501\times10^{-9} & 0 & 1.667\times10^{-17} & 2.501\times10^{-9} & 0 & 3.334\times10^{-17} \end{bmatrix} \tag{8.E8.6.3}$$

$$(M)_1 = (M_c)_1 + (M_{nl})_1 =$$
$$\begin{bmatrix} 3.435\times10^{-23} & 0 & 0 & 1.165\times10^{-23} & 0 & 0 \\ 0 & 3.859\times10^{-23} & 1.037\times10^{-31} & 0 & 7.413\times10^{-24} & -4.959\times10^{-32} \\ 0 & 1.037\times10^{-31} & 5.467\times10^{-40} & 0 & 4.959\times10^{-32} & -3.119\times10^{-40} \\ 1.165\times10^{-23} & 0 & 0 & 3.435\times10^{-23} & 0 & 0 \\ 0 & 7.413\times10^{-24} & 4.959\times10^{-32} & 0 & 3.859\times10^{-23} & -1.037\times10^{-31} \\ 0 & -4.959\times10^{-32} & -3.119\times10^{-40} & 0 & -1.037\times10^{-31} & 5.467\times10^{-40} \end{bmatrix} \tag{8.E8.6.4}$$

$$(M)_2 = (M_c)_2 + (M_{nl})_2 =$$
$$\begin{bmatrix} 3.859\times10^{-23} & 0 & -1.037\times10^{-31} & 7.413\times10^{-24} & 0 & 4.959\times10^{-32} \\ 0 & 3.435\times10^{-23} & 0 & 0 & 1.165\times10^{-23} & 0 \\ -1.037\times10^{-31} & 0 & 5.467\times10^{-40} & -4.959\times10^{-32} & 0 & -3.119\times10^{-40} \\ 7.413\times10^{-24} & 0 & -4.959\times10^{-32} & 3.859\times10^{-23} & 0 & 1.037\times10^{-31} \\ 0 & 1.165\times10^{-23} & 0 & 0 & 3.435\times10^{-23} & 0 \\ 4.959\times10^{-32} & 0 & -3.119\times10^{-40} & 1.037\times10^{-31} & 0 & 5.467\times10^{-40} \end{bmatrix} \tag{8.E8.6.5}$$

The matching of local and global freedoms to assemble the global matrices of members is demonstrated in Table 8.4.

The nanoframe in this application has six freedoms and simply supported ends reduce the freedoms to three. Here, while the support at node (2) vanishes the freedom 4, the support at node (3) eliminate the freedoms 7 and 8 where the reduced global stiffness matrix and the mass matrix without attachment are explained respectively as follows:

Table 8.4 Local and global freedoms of FE mesh in Fig. 8.8b

		Member (element) numbers	
		1	**2**
Local freedoms of elements		Global freedoms of nanoframe	
1	→	1	4
2	→	2	5
3	→	3	6
4	→	4	7
5	→	5	8
6	→	6	9

$$\left(\sum[K]\right)^* = \begin{bmatrix} 100 & 0 & 0 & 0 & 0 & 0 \\ 0 & 0.25 & 2.501\times10^{-9} & -0.25 & 2.501\times10^{-9} & 0 \\ 0 & 2.501\times10^{-9} & 3.334\times10^{-17} & -2.501\times10^{-9} & 1.667\times10^{-17} & 0 \\ 0 & -0.25 & -2.501\times10^{-9} & 100.25 & -2.501\times10^{-9} & 0 \\ 0 & 2.501\times10^{-9} & 1.667\times10^{-17} & -2.501\times10^{-9} & 6.668\times10^{-17} & 1.667\times10^{-17} \\ 0 & 0 & 0 & 0 & 1.667\times10^{-17} & 3.334\times10^{-17} \end{bmatrix} \tag{8.E8.6.6}$$

$$\left(\sum[M_{na}]\right)^* = \begin{bmatrix} 3.435\times10^{-23} & 0 & 0 & 0 & 0 & 0 \\ 0 & 3.859\times10^{-23} & 1.037\times10^{-31} & 7.413\times10^{-24} & -4.959\times10^{-32} & 0 \\ 0 & 1.037\times10^{-31} & 5.467\times10^{-40} & 4.959\times10^{-32} & -3.119\times10^{-40} & 0 \\ 0 & 7.413\times10^{-24} & 4.959\times10^{-32} & 7.293\times10^{-23} & -1.037\times10^{-31} & 0 \\ 0 & -4.959\times10^{-32} & -3.119\times10^{-40} & -1.037\times10^{-31} & 1.093\times10^{-39} & -3.119\times10^{-40} \\ 0 & 0 & 0 & 0 & -3.119\times10^{-40} & 5.467\times10^{-40} \end{bmatrix} \tag{8.E8.6.7}$$

Additionally, the reduced global matrix of mass attachment is written as follows:

$$\left(\sum[M_a]\right)^* = \begin{bmatrix} 1\times10^{-22} & 0 & 0 & 0 & 0 & 0 \\ 0 & 1\times10^{-22} & 0 & 0 & 0 & 0 \\ 0 & 0 & 1\times10^{-38} & 0 & 0 & 0 \\ 0 & 0 & 0 & 0 & 0 & 0 \\ 0 & 0 & 0 & 0 & 0 & 0 \\ 0 & 0 & 0 & 0 & 0 & 0 \end{bmatrix} \tag{8.E8.6.8}$$

where $(\sum[M])^* = (\sum[M_{na}])^* + (\sum[M_a])^*$. Natural frequencies are obtained by substituting the reduced global stiffness and mass matrices into Eq. (8.29). Table 8.5 compares the

Table 8.5 Comparisons of nonlocal natural frequencies of nanoframe in Fig. 8.8a (GHz)

m_a	Mode	e_0a/L	
	Number	0	0.2
0	1	34.1836	31.8933
	2	198.8916	157.4057
	3	409.3096	270.3409
	4	914.4808	474.8559
	5	1497.4249	1290.0665
	6	1805.6897	1706.2263
$\begin{pmatrix} \text{Axial}: 10^{-22}\ \text{kg} \\ \text{Transverse}: 10^{-22}\ \text{kg} \\ \text{Rotation}: 10^{-38}\ \text{kgm}^2 \end{pmatrix}$	1	13.6618	13.5000
	2	56.7630	55.8026
	3	207.2010	173.4149
	4	531.1603	377.5871
	5	874.7062	862.9030
	6	1495.8632	1285.4689

natural classical and nonlocal frequencies computed for nanoframe with and without mass attachment.

The values in Table 8.5 show that mass attachment has a significant effect on the nonlocal dynamics of nanoframes.

8.5 Nonlocal Vibration of Shear Deformable Nanoframes

It is emphasized in the fourth chapter that shear deformations are sometimes important in the kinematics of beams. Since the kinematics of beam, that is, the displacement field, affects the formulations of frame structure, the effect of shear deformation on the frames should also be investigated. Because it is understood in the fourth chapter that the shear effect combined with nonlocal elasticity has a greater importance in the atomic scaled structures.

When the displacement field of continuous structural element is studied, it is understood that shear deformation should definitely be considered in the kinematics of beam but can be neglected in the kinematics of axial rod. This inference is expressed based on the displacement field given below:

$$u_x = u - z\varphi = u - z\frac{\partial w_b}{\partial x}, \quad u_y = v = 0, \quad u_z = w = w_b + w_s \tag{8.30}$$

For the definitions here, Sects. 2.3.2 and 4.4.1 should be examined.

If the lateral and axial displacement is dependent on Poisson's ratio, Love-Bishop and Timoshenko theories should be investigated together. In the case of such a combined displacement field, the two-node finite element of axial rod cannot be utilized due to axial vibration. The three-node finite element, which is the lowest-order formulation that should be used, should also be developed for the bending element. Since the part of shear deformable bending element has a large processing intensity in such analysis, the shear deformation of axial rod is neglected for simplicity in this section to introduce the NL-FEM formulation of shear deformable nanoframes.

NL-FEM matrices for nonlocal free vibration analysis of Timoshenko beams are obtained by Eqs. (7.38)–(7.68). After ignoring the environmental effects (elastic foundation and ambient temperature) of these matrices, finite element matrices of nanoframe member with shear deformation can be formulated. By examining Eqs. (7.34)–(7.36), the total stiffness and mass matrices in the local coordinates for only the bending components of bending member are written as follows:

$$\sum k = k \tag{8.31}$$

$$\sum m^0 = m_{A,c} + m_{A,nl} + M_{I,c} + m_{I,nl} + m_{A,R,c} + m_{A,R,nl} \tag{8.32}$$

$$\sum m^1 = m_{I,R,c} + m_{I,R,nl} + m_{I,R,HO-nl} \tag{8.33}$$

Note that, unlike the Euler-Bernoulli beam theory, shear deformation includes new mass matrices into the analysis in response to a single stiffness matrix. Because only the shape function based on bending displacements is used in the bending part of nanoframe with shear deformation.

Moreover, considering the eigenvalue equation presented in Eq. (7.71), the stiffness, simple mass, and higher-order mass matrices in the local coordinates of generalized bending member are represented as follows, respectively [2]:

$$k = \begin{bmatrix} k^a_{11} & 0 & 0 & k^a_{12} & 0 & 0 \\ 0 & k^b_{11} & k^b_{12} & 0 & k^b_{13} & k^b_{14} \\ 0 & k^b_{21} & k^b_{22} & 0 & k^b_{23} & k^b_{24} \\ k^a_{21} & 0 & 0 & k^a_{22} & 0 & 0 \\ 0 & k^b_{31} & k^b_{32} & 0 & k^b_{33} & k^b_{34} \\ 0 & k^b_{41} & k^b_{42} & 0 & k^b_{43} & k^b_{44} \end{bmatrix} \tag{8.34}$$

$$m^0 = \begin{bmatrix} m_{11}^a & 0 & 0 & M_{12}^a & 0 & 0 \\ 0 & m_{11}^{b,0} & m_{12}^{b,0} & 0 & m_{13}^{b,0} & m_{14}^{b,0} \\ 0 & m_{21}^{b,0} & m_{22}^{b,0} & 0 & m_{23}^{b,0} & m_{24}^{b,0} \\ m_{21}^a & 0 & 0 & M_{22}^a & 0 & 0 \\ 0 & m_{31}^{b,0} & m_{32}^{b,0} & 0 & m_{33}^{b,0} & m_{34}^{b,0} \\ 0 & m_{41}^{b,0} & m_{42}^{b,0} & 0 & m_{43}^{b,0} & m_{44}^{b,0} \end{bmatrix} \tag{8.35}$$

$$m^1 = \begin{bmatrix} 0 & 0 & 0 & 0 & 0 & 0 \\ 0 & m_{11}^{b,1} & m_{12}^{b,1} & 0 & m_{13}^{b,1} & m_{14}^{b,1} \\ 0 & m_{21}^{b,1} & m_{22}^{b,1} & 0 & m_{23}^{b,1} & m_{24}^{b,1} \\ 0 & 0 & 0 & 0 & 0 & 0 \\ 0 & m_{31}^{b,1} & m_{32}^{b,1} & 0 & m_{33}^{b,1} & m_{34}^{b,1} \\ 0 & m_{41}^{b,1} & m_{42}^{b,1} & 0 & m_{43}^{b,1} & m_{44}^{b,1} \end{bmatrix} \tag{8.36}$$

where since the simple axial rod does not reveal higher-order mass terms, the mass of axial rod should be represented in the simple mass matrix. On the other hand, operations like Eqs. (8.21) and (8.22) are specified for the NL-FEM analysis of shear deformable frame as follows [2]:

$$K_e = T^{\mathrm{T}} k T \tag{8.37}$$

$$M_e^0 = T^{\mathrm{T}} m^0 T \tag{8.38}$$

$$M_e^1 = T^{\mathrm{T}} m^1 T \tag{8.39}$$

here the transformation matrix is given in Eq. (8.21). According to Eq. (8.31), since shear deformation does not have an effect on the stiffness matrix, the components of matrix that is the stiffness matrix of bending member in the global coordinates and calculated with Eq. (8.32) are as given in Eq. (8.23). Different from the nanoframe without shear deformation, the components of different mass matrices are calculated below. Accordingly, the components of the matrix calculated with Eq. (8.35), which introduces the simple mass, are presented as [2]:

$$M^0_{11}=M^0_{44}=\left(\frac{\rho AL}{3}+\frac{(e_0a)^2\rho A}{L}\right)\cos^2\theta+\left[\frac{13\rho AL}{35}+\frac{6}{5}\left(\frac{(e_0a)^2\rho A}{L}+\frac{\rho I}{L}+\frac{\rho EI}{kGL}\right)+12\left(\frac{(e_0a)^2\rho I}{L^3}+\frac{(e_0a)^2\rho EI}{kGL^3}\right)\right]\sin^2\theta,$$
$$M^0_{12}=M^0_{45}=\left[-\frac{4\rho AL}{105}-\frac{(e_0a)^2\rho A}{5L}-\frac{6}{5}\left(\frac{\rho I}{L}+\frac{\rho EI}{kGL}\right)-12\left(\frac{(e_0a)^2\rho I}{L^3}+\frac{(e_0a)^2\rho EI}{kGL^3}\right)\right]\cos\theta\sin\theta,$$
$$M^0_{13}=\left[-\frac{11\rho AL^2}{210}-\frac{1}{10}\left((e_0a)^2\rho A+\rho I+\frac{\rho EI}{kG}\right)-6\left(\frac{(e_0a)^2\rho I}{L^2}+\frac{(e_0a)^2\rho EI}{kGL^2}\right)\right]\sin\theta,$$
$$M^0_{14}=\left(\frac{\rho AL}{6}-\frac{(e_0a)^2\rho A}{L}\right)\cos^2\theta+\left[\frac{9\rho AL}{35}-\frac{6}{5}\left(\frac{(e_0a)^2\rho A}{L}+\frac{\rho I}{L}+\frac{\rho EI}{kGL}\right)-12\left(\frac{(e_0a)^2\rho I}{L^3}+\frac{(e_0a)^2\rho EI}{kGL^3}\right)\right]\sin^2\theta,$$
$$M^0_{15}=M^0_{24}=\left[-\frac{19\rho AL}{54}+\frac{(e_0a)^2\rho A}{5L}+\frac{6}{5}\left(\frac{\rho I}{L}+\frac{\rho EI}{kGL}\right)+12\left(\frac{(e_0a)^2\rho I}{L^3}+\frac{(e_0a)^2\rho EI}{kGL^3}\right)\right]\cos\theta\sin\theta,$$
$$M^0_{16}=\left[\frac{13\rho AL^2}{420}-\frac{1}{10}\left((e_0a)^2\rho A+\rho I+\frac{\rho EI}{kG}\right)-6\left(\frac{(e_0a)^2\rho I}{L^2}+\frac{(e_0a)^2\rho EI}{kGL^2}\right)\right]\sin\theta,$$
$$M^0_{22}=M^0_{55}=\left(\frac{\rho AL}{3}+\frac{(e_0a)^2\rho A}{L}\right)\sin^2\theta+\left[\frac{13\rho AL}{35}+\frac{6}{5}\left(\frac{(e_0a)^2\rho A}{L}+\frac{\rho I}{L}+\frac{\rho EI}{kGL}\right)+12\left(\frac{(e_0a)^2\rho I}{L^3}+\frac{(e_0a)^2\rho EI}{kGL^3}\right)\right]\cos^2\theta,$$
$$M^0_{23}=\left[\frac{11\rho AL^2}{210}+\frac{1}{10}\left((e_0a)^2\rho A+\rho I+\frac{\rho EI}{kG}\right)+6\left(\frac{(e_0a)^2\rho I}{L^2}+\frac{(e_0a)^2\rho EI}{kGL^2}\right)\right]\cos\theta,$$
$$M^0_{25}=\left(\frac{\rho AL}{6}-\frac{(e_0a)^2\rho A}{L}\right)\sin^2\theta+\left[\frac{9\rho AL}{35}-\frac{6}{5}\left(\frac{(e_0a)^2\rho A}{L}+\frac{\rho I}{L}+\frac{\rho EI}{kGL}\right)-12\left(\frac{(e_0a)^2\rho I}{L^3}+\frac{(e_0a)^2\rho EI}{kGL^3}\right)\right]\cos^2\theta,$$
$$M^0_{26}=\left[-\frac{13\rho AL^2}{210}+\frac{1}{10}\left((e_0a)^2\rho A+\rho I+\frac{\rho EI}{kG}\right)+6\left(\frac{(e_0a)^2\rho I}{L^2}+\frac{(e_0a)^2\rho EI}{kGL^2}\right)\right]\cos\theta,$$
$$M^0_{33}=M^0_{66}=\frac{\rho AL^3}{105}+\frac{2}{15}\left((e_0a)^2\rho AL+\rho IL+\frac{\rho EIL}{kG}\right)+4\left(\frac{(e_0a)^2\rho I}{L}+\frac{(e_0a)^2\rho EI}{kGL}\right),$$
$$M^0_{34}=\left[-\frac{13\rho AL^2}{210}+\frac{1}{10}\left((e_0a)^2\rho A+\rho I+\frac{\rho EI}{kG}\right)+6\left(\frac{(e_0a)^2\rho I}{L^2}+\frac{(e_0a)^2\rho EI}{kGL^2}\right)\right]\sin\theta,$$
$$M^0_{35}=\left[\frac{13\rho AL^2}{420}-\frac{1}{10}\left((e_0a)^2\rho A+\rho I+\frac{\rho EI}{kG}\right)-6\left(\frac{(e_0a)^2\rho I}{L^2}+\frac{(e_0a)^2\rho EI}{kGL^2}\right)\right]\cos\theta,$$
$$M^0_{36}=-\frac{\rho AL^3}{140}-\frac{1}{30}\left((e_0a)^2\rho AL+\rho IL+\frac{\rho EIL}{kG}\right)+2\left(\frac{(e_0a)^2\rho I}{L}+\frac{(e_0a)^2\rho EI}{kGL}\right),$$
$$M^0_{46}=\left[\frac{11\rho AL^2}{210}+\frac{1}{10}\left((e_0a)^2\rho A+\rho I+\frac{\rho EI}{kG}\right)+6\left(\frac{(e_0a)^2\rho I}{L^2}+\frac{(e_0a)^2\rho EI}{kGL^2}\right)\right]\sin\theta,$$
$$M^0_{56}=\left[-\frac{11\rho AL^2}{210}-\frac{1}{10}\left((e_0a)^2\rho A+\rho I+\frac{\rho EI}{kG}\right)-6\left(\frac{(e_0a)^2\rho I}{L^2}+\frac{(e_0a)^2\rho EI}{kGL^2}\right)\right]\cos\theta. \tag{8.40}$$

Also, the inputs of higher-order mass matrix are expressed via Eq. (8.36) as follows [2]:

$$\begin{aligned}
M_{11}^{1} = M_{44}^{1} &= \left(\frac{13\rho^{2}IL}{35kG} + \frac{12(e_{0}a)^{2}\rho^{2}I}{5kGL} + \frac{12(e_{0}a)^{4}\rho^{2}I}{kGL^{3}}\right)\sin^{2}\theta, \\
M_{12}^{1} = M_{45}^{1} &= \left(-\frac{13\rho^{2}IL}{35kG} - \frac{12(e_{0}a)^{2}\rho^{2}I}{5kGL} - \frac{12(e_{0}a)^{4}\rho^{2}I}{kGL^{3}}\right)\cos\theta\sin\theta, \\
M_{13}^{1} &= \left(-\frac{11\rho^{2}IL^{2}}{210kG} - \frac{(e_{0}a)^{2}\rho^{2}I}{5kG} - \frac{6(e_{0}a)^{4}\rho^{2}I}{kGL^{2}}\right)\sin\theta, \\
M_{14}^{1} &= \left(\frac{9\rho^{2}IL}{70kG} - \frac{12(e_{0}a)^{2}\rho^{2}I}{5kGL} - \frac{12(e_{0}a)^{4}\rho^{2}I}{kGL^{3}}\right)\sin^{2}\theta, \\
M_{15}^{1} = M_{24}^{1} &= \left(-\frac{9\rho^{2}IL}{70kG} + \frac{12(e_{0}a)^{2}\rho^{2}I}{5kGL} + \frac{12(e_{0}a)^{4}\rho^{2}I}{kGL^{3}}\right)\cos\theta\sin\theta, \\
M_{16}^{1} &= \left(\frac{13\rho^{2}IL^{2}}{420kG} - \frac{(e_{0}a)^{2}\rho^{2}I}{5kG} - \frac{6(e_{0}a)^{4}\rho^{2}I}{kGL^{2}}\right)\sin\theta, \\
M_{22}^{1} = M_{55}^{1} &= \left(\frac{13\rho^{2}IL}{35kG} + \frac{12(e_{0}a)^{2}\rho^{2}I}{5kGL} + \frac{12(e_{0}a)^{4}\rho^{2}I}{kGL^{3}}\right)\cos^{2}\theta, \\
M_{23}^{1} &= \left(\frac{11\rho^{2}IL^{2}}{210kG} + \frac{(e_{0}a)^{2}\rho^{2}I}{5kG} + \frac{6(e_{0}a)^{4}\rho^{2}I}{kGL^{2}}\right)\cos\theta, \\
M_{25}^{1} &= \left(\frac{9\rho^{2}IL}{70kG} - \frac{12(e_{0}a)^{2}\rho^{2}I}{5kGL} - \frac{12(e_{0}a)^{4}\rho^{2}I}{kGL^{3}}\right)\cos^{2}\theta, \\
M_{26}^{1} &= \left(-\frac{13\rho^{2}IL^{2}}{420kG} + \frac{(e_{0}a)^{2}\rho^{2}I}{5kG} + \frac{6(e_{0}a)^{4}\rho^{2}I}{kGL^{2}}\right)\cos\theta, \\
M_{33}^{1} = M_{66}^{1} &= \frac{\rho^{2}IL^{3}}{105kG} + \frac{4(e_{0}a)^{2}\rho^{2}IL}{5kG} + \frac{4(e_{0}a)^{4}\rho^{2}I}{kGL}, \\
M_{34}^{1} &= \left(-\frac{13\rho^{2}IL^{2}}{420kG} + \frac{(e_{0}a)^{2}\rho^{2}I}{5kG} + \frac{6(e_{0}a)^{4}\rho^{2}I}{kGL^{2}}\right)\sin\theta, \\
M_{35}^{1} &= \left(\frac{13\rho^{2}IL^{2}}{420kG} - \frac{(e_{0}a)^{2}\rho^{2}I}{5kG} - \frac{6(e_{0}a)^{4}\rho^{2}I}{kGL^{2}}\right)\cos\theta, \\
M_{36}^{1} &= -\frac{\rho^{2}IL^{3}}{140kG} - \frac{(e_{0}a)^{2}\rho^{2}IL}{15kG} + \frac{2(e_{0}a)^{4}\rho^{2}I}{kGL}, \\
M_{46}^{1} &= \left(\frac{11\rho^{2}IL^{2}}{210kG} + \frac{(e_{0}a)^{2}\rho^{2}I}{5kG} + \frac{6(e_{0}a)^{4}\rho^{2}I}{kGL^{2}}\right)\sin\theta, \\
M_{56}^{1} &= \left(-\frac{11\rho^{2}IL^{2}}{210kG} - \frac{(e_{0}a)^{2}\rho^{2}I}{5kG} - \frac{6(e_{0}a)^{4}\rho^{2}I}{kGL^{2}}\right)\cos\theta.
\end{aligned} \tag{8.41}$$

Finally, the eigenvalue equation for the free vibration analysis of shear deformable nanoframe is described as follows:

$$\det\left[\left(\sum_{e=1}^{n}[K]_e\right)^{*} - \omega_i^2\left(\sum_{e=1}^{n}[M_0]_e\right)^{*} + \omega_i^4\left(\sum_{e=1}^{n}[M_1]_e\right)^{*}\right] = 0 \tag{8.42}$$

Example 8.7 Determine the natural frequencies of nonlocal free vibration of nanoframe shown in Fig. 8.5a. Consider the shear effect for bending motion but neglect the Poisson's effect for axial motion. Use a single finite element for all members in the analysis. Mechanical and physical parameters are as in Example 8.3. Additionally, Poisson's ratio: $\upsilon = 0.19$, shear modulus: $G = E/2(1 + \upsilon) = 0.42$ TPa, and shear correction factor: $k = 5/6$.

Solution. It is stated in the introduction of issue that if only the bending components of the nonlocal nanoframe members are affected by shear deformation, the stiffness matrix of member does not change according to the global coordinates for free vibration analysis. That is, the related stiffness matrices of members are calculated with Eqs. (8.25) and (8.27). Therefore, in line with the expressions in Eqs. (8.E8.3.1)–(8.E8.3.3), the stiffness matrices of members (Eqs. (8.E8.3.4)–(8.E8.3.6)) can be used and as a result, Eq. (8.E8.3.18) expressing the reduced global stiffness matrix is valid.

On the other hand, the following parameters should be known to calculate the simple mass matrices formulated according to Eqs. (8.38) and (8.40):

$$\begin{aligned}
&\beta_{EB} = \rho A l = \left(2300\,\text{kg/m}^3\right) \times \left(2 \times 10^{-18}\,\text{m}^2\right) \times \left(5 \times 10^{-8}\,\text{m}\right) = 2.3 \times 10^{-22}\,\text{kg},\\
&\beta_{EB} l = \left(2.3 \times 10^{-22}\,\text{kg}\right) \times \left(5 \times 10^{-8}\,\text{m}\right) = 1.15 \times 10^{-29}\,\text{kgm},\\
&\beta_{EB} l^2 = \left(2.3 \times 10^{-22}\,\text{kg}\right) \times \left(5 \times 10^{-8}\,\text{m}\right)^2 = 5.75 \times 10^{-37}\,\text{kgm}^2,\\
&\beta_T = \rho I = \left(2300\,\text{kg/m}^3\right) \times \left(1.667 \times 10^{-37}\,\text{m}^4\right) = 3.834 \times 10^{-34}\,\text{kgm},\\
&\frac{\beta_T}{l} = \frac{3.834 \times 10^{-34}\,\text{kgm}}{5 \times 10^{-8}\,\text{m}} = 7.668 \times 10^{-27}\,\text{kg},\\
&\beta_T l = \left(3.834 \times 10^{-34}\,\text{kgm}\right) \times \left(5 \times 10^{-8}\,\text{m}\right) = 1.917 \times 10^{-41}\,\text{kgm}^2,\\
&\eta = \frac{\rho EI}{kG} = \frac{\left(2300\,\text{kg/m}^3\right) \times \left(1 \times 10^{12}\,\text{kg/ms}^2\right) \times \left(1.667 \times 10^{-37}\,\text{m}^4\right)}{(5/6) \times \left(0.42 \times 10^{12}\,\text{kg/ms}^2\right)} = 1.095 \times 10^{-33}\,\text{kgm},\\
&\frac{\eta}{l} = \frac{1.095 \times 10^{-33}\,\text{kgm}}{5 \times 10^{-8}\,\text{m}} = 2.191 \times 10^{-26}\,\text{kg},\\
&\eta l = \left(1.095 \times 10^{-33}\,\text{kgm}\right) \times \left(5 \times 10^{-8}\,\text{m}\right) = 5.477 \times 10^{-41}\,\text{kgm}^2.
\end{aligned} \tag{8.E8.7.1}$$

$$\begin{aligned}
&\gamma_{EB}=(e_0a)^2\rho A=\left(1.5\times10^{-8}\text{m}\right)^2\times\left(2300\text{kg/m}^3\right)\times\left(2\times10^{-18}\text{m}^2\right)=1.035\times10^{-30}\text{kgm},\\
&\frac{\gamma_{EB}}{l}=\frac{1.035\times10^{-30}\text{kgm}}{5\times10^{-8}\text{m}}=2.07\times10^{-23}\text{kg},\\
&\gamma_{EB}l=\left(1.035\times10^{-30}\text{kgm}\right)\times\left(5\times10^{-8}\text{m}\right)=5.175\times10^{-38}\text{kgm}^2,\\
&\gamma_T=\frac{(e_0a)^2\rho I}{l}=\frac{\left(1.5\times10^{-8}\text{m}\right)^2\times\left(2300\text{kg/m}^3\right)\times\left(1.667\times10^{-37}\text{m}^4\right)}{5\times10^{-8}\text{m}}=1.725\times10^{-42}\text{kgm}^2,\\
&\frac{\gamma_T}{l}=\frac{1.725\times10^{-42}\text{kgm}^2}{5\times10^{-8}\text{m}}=3.451\times10^{-25}\text{kgm},\\
&\frac{\gamma_T}{l^2}=\frac{1.725\times10^{-42}\text{kgm}^2}{\left(5\times10^{-8}\text{m}\right)^2}=6.901\times10^{-28}\text{kg},\\
&\mu=\frac{(e_0a)^2\rho EI}{kGl}=\frac{\left(1.5\times10^{-8}\text{m}\right)^2\times\left(2300\text{kg/m}^3\right)\times\left(1\times10^{12}\text{kg/ms}^2\right)\times\left(1.667\times10^{-37}\text{m}^4\right)}{(5/6)\times\left(0.42\times10^{12}\text{kg/ms}^2\right)\times\left(5\times10^{-8}\text{m}\right)}=4.930\times10^{-42}\text{kgm}^2,\\
&\frac{\mu}{l}=\frac{4.930\times10^{-42}\text{kgm}^2}{5\times10^{-8}\text{m}}=9.859\times10^{-35}\text{kgm},\\
&\frac{\mu}{l^2}=\frac{4.930\times10^{-42}\text{kgm}^2}{\left(5\times10^{-8}\text{m}\right)^2}=1.972\times10^{-27}\text{kg}.
\end{aligned}\tag{8.E8.7.2}$$

Additionally, according to Eqs. (8.39) and (8.41), higher-order mass matrices of shear deformable frames are required the following calculations:

$$\begin{aligned}
&\lambda_1=\frac{\rho^2Il}{kG}=\frac{\left(2300\text{kg/m}^3\right)^2\times\left(1.667\times10^{-37}\text{m}^4\right)\times\left(5\times10^{-8}\text{m}\right)}{(5/6)\times\left(0.42\times10^{12}\text{kg/ms}^2\right)}=1.26\times10^{-49}\text{kgs}^2,\\
&\lambda_1l=\left(1.26\times10^{-49}\text{kgs}^2\right)\times\left(5\times10^{-8}\text{m}\right)=6.299\times10^{-57}\text{kgms}^2,\\
&\lambda_1l^2=\left(1.26\times10^{-49}\text{kgs}^2\right)\times\left(5\times10^{-8}\text{m}\right)^2=3.149\times10^{-64}\text{kgm}^2\text{s}^2,\\
&\lambda_2=\frac{(e_0a)^2\rho^2I}{kG}=\frac{\left(1.5\times10^{-8}\text{m}\right)^2\times\left(2300\text{kg/m}^3\right)^2\times\left(1.667\times10^{-37}\text{m}^4\right)}{(5/6)\times\left(0.42\times10^{12}\text{kg/ms}^2\right)}=5.669\times10^{-58}\text{kgms}^2,\\
&\frac{\lambda_2}{l}=\frac{5.669\times10^{-58}\text{kgms}^2}{5\times10^{-8}\text{m}}=1.134\times10^{-50}\text{kgs}^2,\\
&\lambda_2l=\left(5.669\times10^{-58}\text{kgms}^2\right)\times\left(5\times10^{-8}\text{m}\right)=2.834\times10^{-65}\text{kgm}^2\text{s}^2,\\
&\lambda_3=\frac{(e_0a)^4\rho^2I}{kGl}=\frac{\left(1.5\times10^{-8}\text{m}\right)^4\times\left(2300\text{kg/m}^3\right)^2\times\left(1.667\times10^{-37}\text{m}^4\right)}{(5/6)\times\left(0.42\times10^{12}\text{kg/ms}^2\right)\times\left(5\times10^{-8}\text{m}\right)}=2.551\times10^{-66}\text{kgm}^2\text{s}^2,\\
&\frac{\lambda_3}{l}=\frac{2.551\times10^{-66}\text{kgm}^2\text{s}^2}{5\times10^{-8}\text{m}}=5.102\times10^{-59}\text{kgms}^2,\\
&\frac{\lambda_3}{l^2}=\frac{2.551\times10^{-66}\text{kgm}^2\text{s}^2}{\left(5\times10^{-8}\text{m}\right)^2}=1.134\times10^{-50}\text{kgs}^2.
\end{aligned}\tag{8.E8.7.3}$$

Equations (8.E8.7.1) and (8.E8.7.2) are replaced into Eq. (8.40) and the simple mass matrices represented by Eq. (8.38) are denoted for all members as follows:

$$(M^0)_1=\begin{bmatrix} 1.103\times10^{-22} & 0 & -7.068\times10^{-31} & 4.664\times10^{-24} & 0 & 2.515\times10^{-31} \\ 0 & 9.737\times10^{-23} & 0 & 0 & 1.763\times10^{-23} & 0 \\ -7.068\times10^{-31} & 0 & 1.241\times10^{-38} & -2.515\times10^{-31} & 0 & -5.821\times10^{-39} \\ 4.664\times10^{-24} & 0 & -2.515\times10^{-31} & 1.103\times10^{-22} & 0 & 7.068\times10^{-31} \\ 0 & 1.763\times10^{-23} & 0 & 0 & 9.737\times10^{-23} & 0 \\ 2.515\times10^{-31} & 0 & -5.821\times10^{-39} & 7.068\times10^{-31} & 0 & 1.241\times10^{-38} \end{bmatrix} \tag{8.E8.7.4}$$

$$(M^0)_2=\begin{bmatrix} 9.737\times10^{-23} & 0 & 0 & 1.763\times10^{-23} & 0 & 0 \\ 0 & 1.103\times10^{-22} & 7.068\times10^{-31} & 0 & 4.664\times10^{-24} & -2.515\times10^{-31} \\ 0 & 7.068\times10^{-31} & 1.241\times10^{-38} & 0 & 2.515\times10^{-31} & -5.821\times10^{-39} \\ 1.763\times10^{-23} & 0 & 0 & 9.737\times10^{-23} & 0 & 0 \\ 0 & 4.664\times10^{-24} & 2.515\times10^{-31} & 0 & 1.103\times10^{-22} & -7.068\times10^{-31} \\ 0 & -2.515\times10^{-31} & -5.821\times10^{-39} & 0 & -7.068\times10^{-31} & 1.241\times10^{-38} \end{bmatrix} \tag{8.E8.7.5}$$

$$(M^0)_3=\begin{bmatrix} 1.103\times10^{-22} & 0 & 7.068\times10^{-31} & 4.664\times10^{-24} & 0 & -2.515\times10^{-31} \\ 0 & 9.737\times10^{-23} & 0 & 0 & 1.763\times10^{-23} & 0 \\ 7.068\times10^{-31} & 0 & 1.241\times10^{-38} & 2.515\times10^{-31} & 0 & -5.821\times10^{-39} \\ 4.664\times10^{-24} & 0 & 2.515\times10^{-31} & 1.103\times10^{-22} & 0 & -7.068\times10^{-31} \\ 0 & 1.763\times10^{-23} & 0 & 0 & 9.737\times10^{-23} & 0 \\ -2.515\times10^{-31} & 0 & -5.821\times10^{-39} & -7.068\times10^{-31} & 0 & 1.241\times10^{-38} \end{bmatrix} \tag{8.E8.7.6}$$

Additionally, substituting the parameters in Eq. (8.E8.7.3) into Eq. (8.41) gives the higher-order stiffness matrices of members:

$$
\left(M^1\right)_1 = \begin{bmatrix}
8.62\times10^{-50} & 0 & -7.49\times10^{-58} & -2.325\times10^{-50} & 0 & -2.244\times10^{-58} \\
0 & 0 & 0 & 0 & 0 & 0 \\
-7.49\times10^{-58} & 0 & 3.588\times10^{-65} & 2.244\times10^{-58} & 0 & 9.628\times10^{-67} \\
-2.325\times10^{-50} & 0 & 2.244\times10^{-58} & 8.62\times10^{-50} & 0 & 7.49\times10^{-58} \\
0 & 0 & 0 & 0 & 0 & 0 \\
-2.244\times10^{-58} & 0 & 9.628\times10^{-67} & 7.49\times10^{-58} & 0 & 3.588\times10^{-65}
\end{bmatrix} \tag{8.E8.7.7}
$$

$$
\left(M^1\right)_2 = \begin{bmatrix}
0 & 0 & 0 & 0 & 0 & 0 \\
0 & 8.62\times10^{-50} & 7.49\times10^{-58} & 0 & -2.325\times10^{-50} & 2.244\times10^{-58} \\
0 & 7.49\times10^{-58} & 3.588\times10^{-65} & 0 & -2.244\times10^{-58} & 9.628\times10^{-67} \\
0 & 0 & 0 & 0 & 0 & 0 \\
0 & -2.325\times10^{-50} & -2.244\times10^{-58} & 0 & 8.62\times10^{-50} & -7.49\times10^{-58} \\
0 & 2.244\times10^{-58} & 9.628\times10^{-67} & 0 & -7.49\times10^{-58} & 3.588\times10^{-65}
\end{bmatrix} \tag{8.E8.7.8}
$$

$$
\left(M^1\right)_3 = \begin{bmatrix}
8.62\times10^{-50} & 0 & 7.49\times10^{-58} & -2.325\times10^{-50} & 0 & 2.244\times10^{-58} \\
0 & 0 & 0 & 0 & 0 & 0 \\
7.49\times10^{-58} & 0 & 3.588\times10^{-65} & -2.244\times10^{-58} & 0 & 9.628\times10^{-67} \\
-2.325\times10^{-50} & 0 & -2.244\times10^{-58} & 8.62\times10^{-50} & 0 & -7.49\times10^{-58} \\
0 & 0 & 0 & 0 & 0 & 0 \\
2.244\times10^{-58} & 0 & 9.628\times10^{-67} & -7.49\times10^{-58} & 0 & 3.588\times10^{-65}
\end{bmatrix} \tag{8.E8.7.9}
$$

Reduced global simple and higher-order mass matrices can be constituted as follows:

$$\left(\sum[M^0]\right)^* = \begin{bmatrix} 2.077\times10^{-22} & 0 & 7.068\times10^{-31} & 1.763\times10^{-23} & 0 & 0 \\ 0 & 2.077\times10^{-22} & 7.068\times10^{-31} & 0 & 4.664\times10^{-24} & -2.515\times10^{-31} \\ 7.068\times10^{-31} & 7.068\times10^{-31} & 2.482\times10^{-38} & 0 & 2.515\times10^{-31} & -5.821\times10^{-39} \\ 1.763\times10^{-23} & 0 & 0 & 2.077\times10^{-22} & 0 & 7.068\times10^{-31} \\ 0 & 4.664\times10^{-24} & 2.515\times10^{-31} & 0 & 2.077\times10^{-22} & -7.068\times10^{-31} \\ 0 & -2.515\times10^{-31} & -5.821\times10^{-39} & 7.068\times10^{-31} & -7.068\times10^{-31} & 2.482\times10^{-38} \end{bmatrix} \tag{8.E8.7.10}$$

$$\left(\sum[M^1]\right)^* = \begin{bmatrix} 8.62\times10^{-50} & 0 & 7.49\times10^{-58} & 0 & 0 & 0 \\ 0 & 8.62\times10^{-50} & 7.49\times10^{-58} & 0 & -2.325\times10^{-50} & 2.244\times10^{-58} \\ 7.49\times10^{-58} & 7.49\times10^{-58} & 7.176\times10^{-65} & 0 & -2.244\times10^{-58} & 9.628\times10^{-67} \\ 0 & 0 & 0 & 8.62\times10^{-50} & 0 & 7.49\times10^{-58} \\ 0 & -2.325\times10^{-50} & -2.244\times10^{-58} & 0 & 8.62\times10^{-50} & -7.49\times10^{-58} \\ 0 & 2.244\times10^{-58} & 9.628\times10^{-67} & 7.49\times10^{-58} & -7.49\times10^{-58} & 7.176\times10^{-65} \end{bmatrix} \tag{8.E8.7.11}$$

The natural frequencies of nanoframe can be calculated by substituting Eqs. (8.E8.3.20), (8.E8.7.10), and (8.E8.7.11) into Eq. (8.42). A comparison study on the natural frequencies of nanoframes formulated according to different bending theories is presented in Table 8.6.

In Table 6.8, it is observed that the difference is quite small for the nonlocal bending theories. In the nanoframe shown in Fig. 8.5a, the ratio of lengths of members to their heights is $L/h = 50$ and this is a very high ratio indicating that the shear will lose effect. As it is known, shear effects more influence the dynamic analysis in cases such as low beam length and high section height. In order to learn in more detail the difference between the two nonlocal bending theories for nanoframes, cases that have lower member lengths and higher section heights should be investigated.

Table 8.6 Comparisons of nonlocal natural frequencies (GHz) of nanoframe in Fig. 8.5a for different bending theories ($e_0a/L = 0.3$)

Mode Number	EBBT (Example 8.4)	TBT (Example 8.7)
1	8.476	8.474
2	25.568	25.543
3	43.292	43.193
4	458.241	458.124
5	465.383	465.291
6	688.732	688.624

Problems

8.1. Compute the natural frequencies of nanotrusses that are identical geometrically in Fig. 8.9 under parameters as follows. Discuss the effects of nonlocal parameter and supports on the differences between results.

Length, $L = 20$ nm, diameter of circular cross-section, $d = 1$ nm, inclination angle, $\theta = 45°$, modulus of elasticity, $E = 1$ TPa, mass of unit volume, $\rho = 2300$ kg/m^3.

Compute according to three different nondimensional nonlocal parameters: $\alpha = e_0a/L = 0, 0.1, 0.2$.

8.2. Compute the natural frequencies of capacious nanotrusses in Fig. 8.10 under parameters as follows. Discuss the effect of nonlocal parameter.

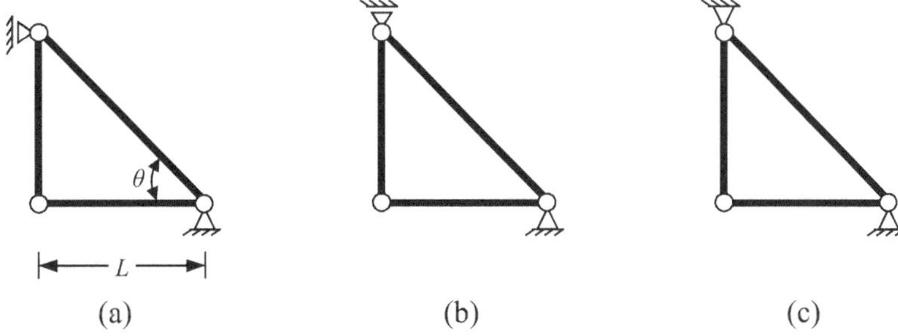

Fig. 8.9 Nanotruss models for Problem 8.1

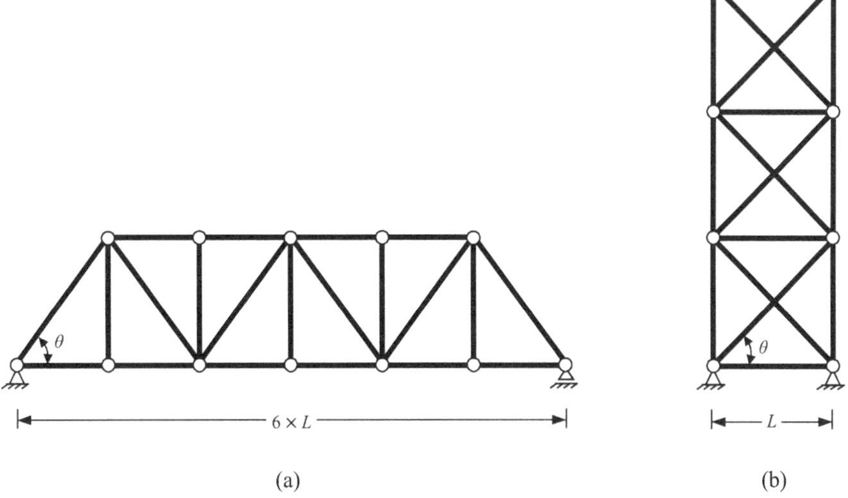

Fig. 8.10 Nanotruss models for Problem 8.2

Lengths, $L = 20$ nm, diameters of circular cross-section, $d = 1$ nm, inclination angles, $\theta = 45°$, modulus of elasticity, $E = 1$ TPa, masses of unit volume, $\rho = 2300$ kg/m^3.

Compute according to three different nondimensional nonlocal parameters: $\alpha = e_0a/L = 0, 0.1, 0.2$.

Note that: Since the operation volume increases (especially global matrices and solution of eigenvalue equation), it is recommended to perform the analysis electronically.

8.3. Compute the natural frequencies of attached nanotrusses in Fig. 8.11 under parameters as follows. Discuss the effects of nonlocal parameter, supports, and attachments on the differences between results.

Lengths, $L_1 = 20$ nm, $L_2 = 15$ nm, $L_3 = L_5 = 10$ nm, $L_4 = 5$ nm, diameters of circular cross-section, $d = 1$ nm, inclination angles, $\theta_1 = 45°$, $\theta_2 = 70°$, $\theta_3 = 60°$, $\theta_4 = 35°$, modulus of elasticity, $E = 1$ TPa, masses of unit volume, $\rho = 2300$ kg/m^3.

Spring attachments, $k_1 = k_3 = 5 \times 10^3$N/m, $k_2 = 1 \times 10^2$N/m, $k_4 = 1 \times 10^4$N/m, mass attachments, $m_1 = 1 \times 10^{-8}$ kg, $m_2 = 1 \times 10^{-14}$ kg.

Compute according to three different nondimensional nonlocal lengths: $e_0a = 0$ nm, 10 nm, 20 nm.

8.4. Compute the natural frequencies of nanoframes in Fig. 8.12 under parameters as follows. Neglect the shear effect. Sections of vertical and horizontal members of all

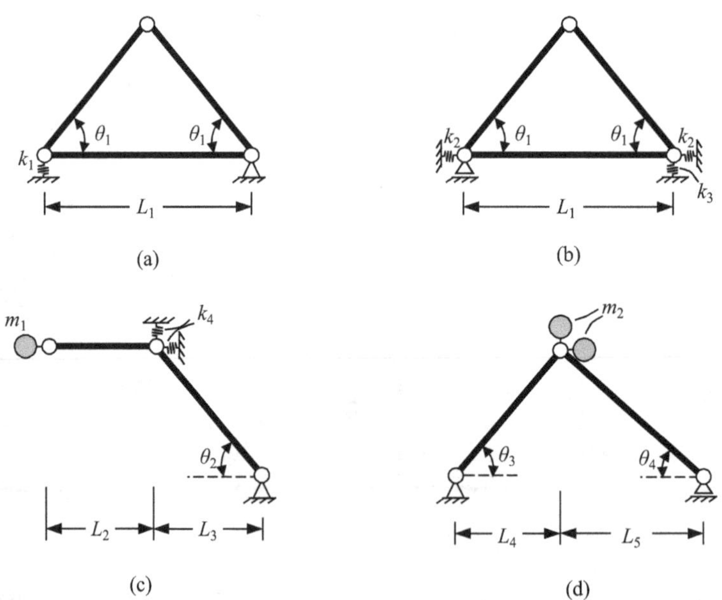

Fig. 8.11 Nanotruss models for Problem 8.3

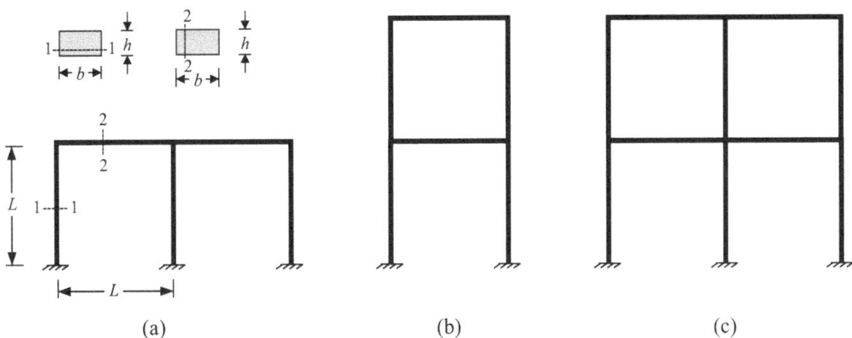

Fig. 8.12 Nanoframe models for Problem 8.4

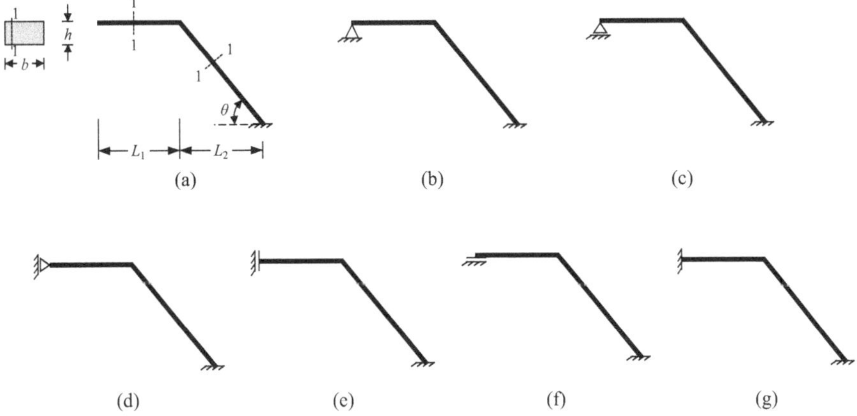

Fig. 8.13 Nanoframe models for Problem 8.5

nanoframes are given as sections 1–1 and 2–2. Discuss the effects of nonlocal parameter and structure hugeness on the differences between results.

Lengths of all members, $L = 20$ nm, heights and widths of rectangular cross-sections, $h = 1$ nm and $b = 2$ nm, modulus of elasticity, $E = 1$ TPa, masses of unit volume, $\rho = 2300$ kg/m^3.

Note that: Since the operation volume increases (especially to compute the global matrices and to solve the eigenvalue equation), it is recommended to perform the analysis electronically.

Compute according to three different nondimensional nonlocal parameters: $\alpha = e_0a/L = 0, 0.1, 0.2$.

8.5. Compute the natural frequencies of nanoframes that are identical geometrically in Fig. 8.13 under parameters as follows. Discuss the effects of nonlocal parameter and supports on the differences between results.

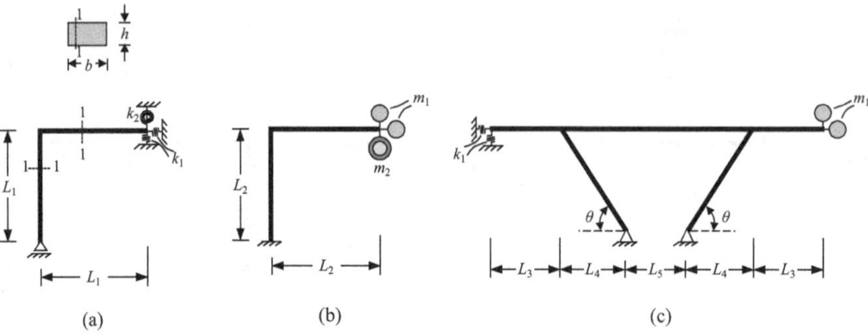

Fig. 8.14 Nanoframe models for Problem 8.6

Lengths, $L_1 = 20$ nm, $L_2 = 20$ nm, heights and widths of rectangular cross-sections, $h = 1$ nm and $b = 2$ nm, inclination angles, $\theta = 60°$, modulus of elasticity, $E = 1$ TPa, masses of unit volume, $\rho = 2300$ kg/m^3.

Compute according to three different nondimensional nonlocal parameters: $\alpha = e_0a/L = 0, 0.1, 0.2$.

Note that: Support types of left end of horizontal members are expressed as: (a) Free (b) Fully simply supported (c) Simply supported at transverse direction only (d) Simply supported at axial direction only (e) Clamped supported at axial direction only (f) Clamped supported at transverse direction only (g) Fully clamped supported. Since the operation volume increases (especially to solve the eigenvalue equation), it is recommended to perform the analysis electronically.

8.6. Compute the natural frequencies of attached nanoframes in Fig. 8.14 under parameters as follows. Discuss the effects of nonlocal parameter and supports on the differences between results. Sections of all members of nanoframes are given as section 1-1 and 2-2.

Lengths, $L_1 = 15$ nm, $L_2 = 15$ nm, $L_3 = L_5 = 10$ nm, $L_4 = 7.5$ nm, heights and widths of rectangular cross-sections, $h = 1$ nm and $b = 2$ nm, inclination angles, $\theta = 60°$, modulus of elasticity, $E = 1$ TPa, masses of unit volume, $\rho = 2300$ kg/m^3.

Spring attachments, $k_1 = 8 \times 10^5$N/m, $k_2 = 4 \times 10^3$ Nm, mass attachments, $m_1 = 1 \times 10^{-15}$ kg, $m_2 = 1 \times 10^{-18}$ kg.

Compute according to three different nonlocal lengths: $e_0a = 0$ nm, 20 nm, 40 nm.

Note that: While the springs with stiffness k_1 are added in the linear direction, the spring with stiffness k_2 is added in the rotational direction. Similarly, masses m_1 are added in the linear direction and mass m_2 are added in the rotational direction. Since the operation volume increases, it is recommended to perform the analysis electronically.

8.7. Recompute the classical and nonlocal natural frequencies by dividing members in Fig. 8.1a, g in Problem 8.5 into two finite elements. Discuss the effects of number of finite elements, nonlocal parameter, and support on the results.

8.8. Examine Problem 8.4 by also considering shear effects. Neglect shear effects in the axial directions of members. Discuss the effects of nonlocal parameter and shear deformation on the results by relating these two factors to each other. In addition to parameters in Problem 8.4, Shear modulus: $G = 0.42$ TPa, shear correction factor: $k = 5/6$.

8.9. Examine Problem 8.5 by also considering shear effects. Neglect shear effects in the axial directions of members. Discuss the effects of nonlocal parameter and shear deformation on the results by relating these two factors to each other. In addition to parameters in Problem 8.5, Shear modulus: $G = 0.42$ TPa, shear correction factor: $k = 5/6$.

8.10. Examine Problem 8.6 by also considering shear effects. Neglect shear effects in the axial directions of members. Discuss the effects of nonlocal parameter and shear deformation on the results by relating these two factors to each other. In addition to parameters in Problem 8.6, Shear modulus: $G = 0.42$ TPa, shear correction factor: $k = 5/6$.

References

1. H.M. Numanoğlu, Ö. Civalek, On the dynamics of small-sized structures. Int. J. Eng. Sci. **145**, 103164 (2019)
2. H.M. Numanoğlu, Ö. Civalek, On shear-dependent vibration of nano frames. Int. J. Eng. Sci. **195**, 103992 (2024)
3. K.B. Mustapha, Free vibration of microscale frameworks using modified couple stress and a combination of Rayleigh-love and Timoshenko theories. J. Vib. Control. **26**(13–14), 1285–1310 (2020)
4. S.M. Hozhabrossadati, N. Challamel, M. Rezaiee-Pajand, A.A. Sani, Free vibration of a nanogrid based on Eringen's stress gradient model. Mech. Based Des. Struct. Mach. **50**(2), 537–555 (2022)
5. A.F. Russillo, G. Failla, G. Alotta, F.M. de Sciarra, R. Barretta, On the dynamics of nano-frames. Int. J. Eng. Sci. **160**, 103433 (2021)

Zeitfracht Medien GmbH
Ferdinand-Jühlke-Straße 7
99095 Erfurt, Deutschland
produktsicherheit@kolibri360.de